HEAT AND MASS TRANSFER
A Biological Context

SECOND EDITION

HEAT AND MASS TRANSFER

TRANSFER

A Biological Context

SECOND EDITION

ASHIM K. DATTA

CRC Press
Taylor & Francis Group
Boca Raton London New York

CRC Press is an imprint of the
Taylor & Francis Group, an **informa** business

CRC Press
Taylor & Francis Group
6000 Broken Sound Parkway NW, Suite 300
Boca Raton, FL 33487-2742

© 2017 by Taylor & Francis Group, LLC
CRC Press is an imprint of Taylor & Francis Group, an Informa business

No claim to original U.S. Government works

Printed on acid-free paper
Version Date: 20161215

International Standard Book Number-13: 978-1-138-03360-3 (Hardback)

Visit the Taylor & Francis Web site at
http://www.taylorandfrancis.com

and the CRC Press Web site at
http://www.crcpress.com

Printed in Canada.

To my parents

Atindra Nath Dutta
and
Bela Rani Dutta

CONTENTS

III APPENDIX   541

PREFACE TO THE SECOND EDITION

This edition was completely driven by student needs and multiple instructor experiences over the past 15 years. The biological engineering student is different. With one foot in biology and the other foot in engineering, they, more than any other engineers, prefer engineering in a context, as opposed to generic, and a more balanced use of math versus application. Engineers are supposed to be problem solvers. However, we do not seem to tell them how to develop this problem solving skill but rely on them to somehow distill this skill through practice. Developing engineering problem solving skills is also an acute need for the biological engineering students. To address these needs, the additions/changes made in this edition can be grouped into the following:

- To keep their interest and show relevance to their interests/lives/careers and to make the point that transport concepts are really broadly applicable, students need to see many more biological applications. To address this, over 160 new problems were developed from scratch going through research papers and practical applications. These problems were not made up simply to have practice in number crunching but they originated from issues important enough to society and the scientific community that papers were published on most of them. Developing these problems that cover important applications, but primarily enhance the engineering fundamentals already in the text, was probably 75% of the effort in this new edition.

- For a third-year student, an engineering problem solving skillset needs to be created. While directly useful and critical for heat and mass transfer, problem solving ability is a life-long learning skill and a must-have for engineers. Using heat and mass transfer as a vehicle, a systematic approach to problem solving was developed—this includes novel templates and algorithms. A university-funded instructional project helped with this creation. Meta-knowledge that shows how the topics of a chapter relate to each other and step-by-step instructions for solving various types of problems should improve the students' problem solving ability significantly and cross over to other subjects.

- While having many more application problems is a good idea, students need to see more solved problems following the problem solving template. This makes the problem solving process less abstract, shows a range of problem types and their nuances and also provides practice in using the problem solving template so that students internalize it. Over 25 solved problems have been developed that use the problem solving template—this covers almost every topic in the book. The template includes an "Evaluating and Interpreting the Solution" step that takes the student deeper into what we learn from the solution.

- Helping the typical (as opposed to the gifted) student excel guided the changes. In addition to assistance with approaches to problem solving, the students also want ways to verify their skills. Answers to all problems are now included at the back—to help the student do problems on his/her own. To help the student develop an eye for problem solving, the contents of each chapter also have been organized in a "Problem Solving" section in terms of the types of problems one can expect to solve in each chapter.

- Problem formulation—making connections between mathematics and the physical problem—is another challenge. Without making this connection, the learner may be doing just the math problem and not the engineering problem. Detailed step-by-step instructions, along with examples, have been developed for problem formulation that show how to take a physical (biological) process and simplify it to a domain with its governing equation and boundary conditions.

Using this book: The solved examples show not just the solution but the logic behind approaching the solution. For every such solved example, the student should pay particular attention to step 2, "Generating and selecting among alternate solutions," and the related problem solving map provided in the chapter. In addition, the section at the end of every chapter dedicated to problem solving discusses all the possible types of problems that can be solved using the contents of the chapter and how to approach each type of problem. The student is strongly encouraged to apply the problem solving logic to try the many problems provided at the end of each chapter. Additional pointers on using the book can be found on page xxiii.

I have been able to significantly enhance teaching and learning in my course using the developments mentioned in the above paragraphs that led me to incorporate them in the text. I sincerely wish the same will be true for others using the book.

Ashim K. Datta, Ithaca, New York

PREFACE

It is very important to give the undergraduate engineer a fundamental education in the context of his/her likely application areas. Transport of energy and mass is fundamental to many biological and environmental processes (see pages xxix to xxxv). Areas from food processing to thermal design of buildings to biomedical devices to pollution control and global warming require knowledge of how energy and mass can be transported through materials. These wide-ranging applications have become part of emerging curricula in biological engineering, and societies such as the Institute of Biological Engineering and the American Society of Agricultural Engineers have recognized the need for a course (and a text) that presents fundamentals while integrating the diverse subject matter.

The basic transport mechanisms of many of these processes are diffusion (or diffusion-like, such as capillary and dispersion) and bulk flow. Additionally, there is radiative heat transfer. It is crucial for the student to see these concepts as comprehensive and unified subject matter (much like fluid mechanics); they are the building blocks for lifelong learning in many of their interest areas. Such a fundamentals-based approach will replace the more empirical and ad hoc teaching that sometimes exists.

Although the concept of teaching transport processes as a unified subject has existed for over forty years in some engineering disciplines, only in recent years have we seen adequate quantitative studies to make such teaching possible in biological and bioenvironmental processes. This book attempts to bring together under one umbrella the unique content, contexts, and parameter regimes of biologically related processes and to emphasize principles and not just mathematical analysis. *Content*, such as bio-heat transfer, thermoregulation, freezing, global warming, capillary flow, and dispersion, are some of the topics not typically included in the undergraduate-level teaching of transport phenomena. *Context*, such as plants, animals, water, soil, and air, is important at this level, because without this information students have an unnecessarily hard time relating to real physical processes. Context also helps students learn about the physical processes themselves in a quantitative way. For example, studying convective transfer of water vapor over a leaf includes a quantitative introduction to transpiration.

(The present text was created by distilling the content of hundreds of research papers and textbooks on similar biological and environmental applications.) The *parameter regimes* of biological processes are also different from those of typical mechanical and chemical processes. For example, biological processes often involve a source term of heat generation or oxygen consumption. The presence of the source or sink term changes the nature of the solution and is emphasized in this text.

How This Book Fits in a Biological Engineering Curriculum

This text is intended for a junior-level engineering science course in curricula that emphasize biology and the environment. The course would build on the prerequisites of partial differential equations and fluid mechanics. Prior knowledge of biological and environmental science, although not required, would be useful. For example, this course can readily build on a course such as Thermodynamics of Biological Systems that has been discussed in the context of a biological engineering curriculum. Mass and heat transfer, much like fluid flow, are just as much building blocks for many of the upper-level courses. Thus, specialized design courses and advanced courses such as bioprocessing, biomedical engineering, food process engineering, environmental processes and their control, and waste management can build on a course that uses this text, greatly reducing the need to teach basic engineering science of mass and heat transfer in these upper-level courses. This text was developed at Cornell University for a junior-level engineering science course.

Approach and Organization of the Book

The overall organization of the book follows the well-tested transport phenomena approach. The chapters and their content on heat and mass transfer are made to follow an almost exact parallel, as shown in the table below. The first two chapters in each part (Part I, "Energy Transfer" and Part II, "Mass Transfer") develop the two building blocks of conservation laws and rate laws, and the next chapter (Chapters 3 and 11) combines them to build the general governing equations and boundary conditions. The next two chapters in each part (as shown in the table below) cover steady-state or transient diffusion, without any flow, while the last one adds the effect of flow. Chapter 7 covers heat transfer with change of phase, and Chapter 8 covers radiative energy transfer. Porous media flow and simple kinetics of zero and first order are included for completeness as they relate to transport. An effort has been made to clarify important processes such as dispersion. Different application areas in biology and environment are included within this framework of chapters, when they are relevant.

| | Chapter number | |
	Energy	Mass
Conservation	1	9
Rate laws	2	10
Governing equation	3	11
Diffusion, steady-state	4	12
Diffusion, transient	5	13
Diffusion (and dispersion) with bulk flow	6	14

How to Use This Book

Students frequently follow individual topics well but have difficulty seeing their relatedness, i.e., the big picture. Thus, a major effort has been made to distill the concepts presented here and to show the connections among chapters. Each chapter begins with a small list of major concepts to be covered, together with important terminologies introduced in that chapter, and each chapter has a map showing how all chapters are interconnected in terms of the topic under consideration. Each chapter has a summary at the end that puts every major concept and equation at the reader's fingertips, providing page numbers for easy access, and a set of descriptive questions checks the reader's understanding of concepts and facts. Summary maps in Appendix A (pages 544–547) show the integration of all the scenarios covered in the text. The first-time reader of the subject is strongly encouraged to use these features.

As curricula in biological and related engineering programs evolve, the core of such curricula will include mass and heat transfer as essential building blocks in students' instruction in a natural and obvious way. The author sincerely hopes that this text will serve the needs of these curricula. He also believes that the text must evolve with the curricula. Additional materials helpful for teaching this subject matter can be seen on the Internet at www.ashimdatta.net. Please do not hesitate to contact the author if you have comments on any aspect of the book.

Ashim K. Datta
akd1@cornell.edu

ACKNOWLEDGMENTS

The author sincerely acknowledges the assistance of many individuals in preparing this manuscript; he hopes that the names included here cover most of them. Jean Hunter, Douglas Haith, Louis Albright, Larry Geohring, Ronald Pitt, Roger Spanswick, Gerald Rehkugler, John Cundiff, Kifle Gebremedhin, Thomas Goldstick, and Arthur Johnson generously brought to bear their particular expertise in discussing transport issues with me. Comments from the instructors who used the prepublication version of this text— Dean Steele, Josse De Baerdemaeker, Guido Wyseure, Sean Kohles, Sudhir Sastry, Bradley Marks, John Nieber, Fu-Hung Hsieh, and Christopher Choi—have been invaluable in the development process. Students of ABEN 350, in the Department of Biological and Environmental Engineering (formerly Agricultural and Biological Engineering) at Cornell University, during the period 1990–1998, as well as students from Katholieke Universiteit Leuven, Belgium, provided feedback that made the text more relevant, manageable, and clear. Development of example problems was facilitated by the Teaching Assistants Kelley Bastian, Haitao Ni, Annie Chi, Steven Lobo, Katherine DeBruin, Jordan Dolande, Indranil Mukerji, and Wenjie Hu. No less important has been the help given to me over the years in typing, drawing, LaTeX, proofreading and related activities from Kevin Hodgson, Sharon Hobbie, Cindy Robinson, Ivan Dobrianov, Andy Ruina, Mark Schroeder, Krishanu Saha, Sue Fredenburg, Jifeng Zhang, Donna Burns, Sandy Bates, and Richard Krizek. The long development process needed encouragement, and the author deeply appreciates everyone who provided it—just a few of whom are Andy Rao, Robert Cooke, Ronald Furry, and Norman Scott. Most important, I want to give my profound thanks to my wife, Anasua, and daughters, Ankurita and Amita, for their patience with my preoccupation for such a long time.

The author also acknowledges financial support from the U.S. Department of Agriculture, subcontracted through Roger Garrett of the University of California, Davis. Also appreciated is support from the Department of Biological and Environmental Engineering at Cornell University.

For the second edition, the author gratefully acknowledges the help of Kimberly Lin in the difficult job of streamlining almost all the problem statements, Hyeongsu Park for painstaking matching the problem statements with their solutions, and Heming Zhao for meticulously checking how the solved examples follow the problem solving approach presented. Discussions with graduate teaching assistants over the years, Jifeng Zhang, Hua Zhang, Vineet Rakesh, Timothy Shelford, Asha Sharma, Amit Halder, Ashish Dhall, Tushar Gulati, Alexander Warning, Mohsen Ranjbaran, and Nayan Mallick, contributed to improvements in the quality of the problems at the end of the chapters. Constant bouncing of ideas from Ankurita Datta and Amita Datta as students, undergraduate teaching assistants, and dinner table companions, helped improve many items. Feedback and comments obtained from Professors Bart Nicolai, Fu-Hung Hsieh, Kunal Mitra, Bradley Marks, and others have helped guide some of the additions. Editorial checking of the newly added solved problems by Margaret Stevens and students Yinhong Liao, Shoshana Das, Karann Putrevu, Gavisha Waidyaratne helped remove many typos and improve clarity. Finally, the author greatly appreciates the assistance of Valorie Adams in making the proofreader's corrections and Margaret Stevens in obtaining the copyright permissions.

INTRODUCTION

Problem Solving in the Transport Processes

This book centers around solving problems in energy and mass transfer in a biological (Biomedical/Plant/Bioprocessing/Bioenvironment) context. Starting from a biological process (Figure 1), after making sufficient assumptions, we formulate it as a mathe-

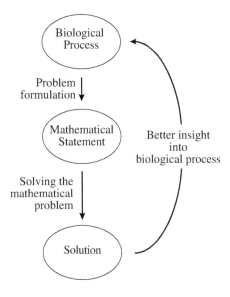

Figure 1: Schematic showing the steps of problem formulation and solution.

matical problem that one can solve using simple analytical solution techniques with which an undergraduate is already familiar. This way, we build a mathematical model of the biological process that provides a more quantitative insight into the process by showing the effects of various process parameters. Having started from fundamentals, the foundation built here carries over to more complex processes that would require numerical solution techniques.

The various application areas in biological engineering can be grouped as Biomedical, Plant, Bioprocessing, and Bioenvironment. Each of these application areas is illustrated a little more in the following pages, which also show a small portion of the examples in this book.

In addition to the specialized texts covering transport processes in various application areas mentioned in the following pages, several well-known books on general transport phenomena can provide the reader with further details:

Bird, R. B., W. E. Stewart, and E. N. Lightfoot. 1960. *Transport Phenomena.* John Wiley & Sons, New York.

Cussler, E. L. 1997. *Diffusion Mass Transfer in Fluid Systems.* Cambridge University Press, Cambridge, UK.

Deen, W. M. 1998. *Analysis of Transport Phenomena.* Oxford University Press, New York.

Eckert, E. R. G. and R. M. Drake. 1987. *Analysis of Heat and Mass Transfer.* Hemisphere Publishing Corporation, New York.

Geankoplis, C. J. 1993. *Transport Processes and Unit Operations.* P T R Prentice Hall, Inc., Englewood Cliffs, NJ.

Geankoplis, C. J. 1972. *Mass Transport Phenomena.* C. J. Geankoplis, Columbus, OH.

Incropera, F. P. and D. P. Dewitt. 1996. *Fundamentals of Heat and Mass Transfer.* John Wiley & Sons, New York.

Middleman, S. 1998. *An Introduction to Mass and Heat Transfer.* John Wiley & Sons, New York.

Slattery, J. C. 1999. *Advanced Transport Phenomena.* Cambridge University Press, Cambridge, UK.

Welty, J. R., C. E. Wicks, R. E. Wilson, and G. Rorrer. 2001. *Fundamentals of Momentum, Heat, and Mass Transfer.* John Wiley & Sons, New York.

Transport in a Biomedical Context

In mammalian systems, transport processes occur at the cellular, tissue, organ, and whole-body levels (Figure 2). At the cellular level, transport across the cell membrane is driven by passive diffusion of solutes and water, hydraulic and osmotic transport of water, carrier-mediated transport, passive ion transport, and active transport. This text includes diffusion, as well as hydraulic and osmotic transport, but not the other important but complex transport mechanisms just mentioned. An extensive treatise on cellular transport is provided in Weiss (1996). At the tissue or organ level, diffusion of oxygen is an important transport process. One of many examples of such transport is the oxygen diffusion from air to the bloodstream in the alveoli of the lungs. At the whole-body level, there is thermoregulation, whereby heat production in the body and heat

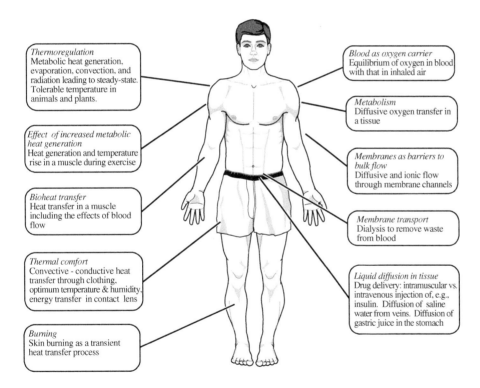

Thermoregulation
Metabolic heat generation, evaporation, convection, and radiation leading to steady-state. Tolerable temperature in animals and plants.

Effect of increased metabolic heat generation
Heat generation and temperature rise in a muscle during exercise

Bioheat transfer
Heat transfer in a muscle including the effects of blood flow

Thermal comfort
Convective - conductive heat transfer through clothing, optimum temperature & humidity, energy transfer in contact lens

Burning
Skin burning as a transient heat transfer process

Blood as oxygen carrier
Equilibrium of oxygen in blood with that in inhaled air

Metabolism
Diffusive oxygen transfer in a tissue

Membranes as barriers to bulk flow
Diffusive and ionic flow through membrane channels

Membrane transport
Dialysis to remove waste from blood

Liquid diffusion in tissue
Drug delivery: intramuscular vs. intravenous injection of, e.g., insulin. Diffusion of saline water from veins. Diffusion of gastric juice in the stomach

Figure 2: Schematic showing examples of transport in mammalian systems covered in this text.

dissipation through behavioral changes (movements of the whole body) or autonomic, reflex-like changes (such as sweating or shivering) are modified. Thermal therapy, either the use of heat (hyperthermia) or the use of freezing temperatures (cryosurgery) to destroy tissue, demonstrates heat transfer in clinical applications. Transport in artificial organs such as the dialyzer is another important group of problems. This area has been exploding since the first edition was written and excellent books such as Truskey et al. (2009) are now available. The reader can consult several specialized books on transport in biomedical systems for further details. Some titles follow.

Berger, S. A., W. Goldsmith, and E. R. Lewis. 1996. *Introduction to Bioengineering.* Oxford University Press, Oxford, UK.

Charny, C. K. 1992. Mathematical models of bioheat transfer. Advances in Heat Transfer 22:19–153.

Cooney, D. O. 1976. *Biomedical Engineering Principles. An Introduction to Fluid, Heat and Mass Transport Processes.* Marcel Dekker, Inc., New York.

Evans, D. H. 1998. *The Physiology of Fishes.* CRC Press, Boca Raton, FL.

Fanger, P. O. 1972. *Thermal Comfort: Analysis and Applications in Environmental Engineering.* McGraw-Hill, New York.

Fournier, R. L. 1998. *Basic Transport Phenomena in Biomedical Engineering.* Taylor & Francis, Philadelphia.

Lightfoot, E. N. 1974. *Transport Phenomena in Living Systems.* John Wiley & Sons, New York.

Lih, M. M. 1975. *Transport Phenomena in Medicine and Biology.* John Wiley & Sons, New York.

Middleman, S. 1972. *Transport Phenomena in the Cardiovascular System.* Wiley-Interscience.

Shitzer, A. and R. C. Eberhart. 1985. *Heat Transfer in Medicine and Biology: Analysis and Applications.* Plenum Press, New York.

Truskey, G. A., F. Yuan and D. F. Katz. 2009. *Transport Phenomena in Biological Systems.* Pearson Prentice Hall, Upper Saddle River, NJ.

Weiss, T. F. 1996. *Cellular Biophysics. Volume 1: Transport.* The MIT Press, Cambridge.

Yang, W. 1989. *Biothermal-Fluid Sciences.* Hemisphere Publishing Corporation, New York.

Transport in a Plant Context

From an engineering standpoint,[1] an annual crop plant may be regarded as a self-replicating structure. The first structures produced, leaves and roots, function primarily in transport and energy acquisition. The leaves intercept solar energy for the fixation of carbon dioxide in photosynthesis, acquiring the carbon dioxide by diffusion through the boundary layer over the leaf and through the stomatal pores, while inevitably losing water by evaporation from the cell surfaces of the wet leaf. The roots extract water from the soil, replacing that lost by the leaves and providing the water that constitutes the bulk of growing plant tissue. For many mineral nutrients essential for plant growth, roots are also the initial site of transport into the plant. The properties of soil-root interface also determine whether pollutants in the soil enter the food chain. Within the plant, transport across the cell membranes involves both diffusion, for water and solutes, and carrier-mediated processes, for ions and organic molecules but not water. Bulk flow of water, ions, and some nitrogen compounds from the roots to the shoot occurs in the specialized vascular tissue called the xylem. Some of the examples of

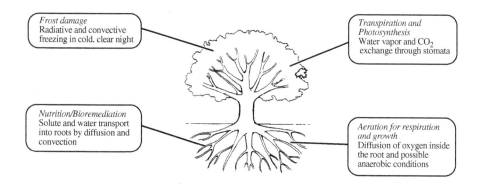

Figure 3: Schematic showing examples of transport plant systems covered in this text.

---

[1] Paragraph based on contribution from Prof. Roger Spanswick, Dept. of Biological and Environmental Engineering, Cornell University.

transport in plant systems covered in this text are shown in Figure 3. (The topic of carrier-mediated transport is outside the scope of this text, and the reader is referred to specialized texts such as Marschner, 1995). Some texts and reference books covering transport in plant systems are noted below.

Baker, D. A. 1978. *Transport Phenomena in Plants*. Chapman and Hall, London.

Buchanan, B., W. Gruissem, and R. L. Jones. 2000. *Biochemistry and Molecular Biology of Plants* (Chapters 3, 15 and 23). American Society of Plant Biologists, Rockville, MD.

Cundiff, J. S. 1999. Simulation of biological systems. Coursenotes for BSE 4144, Virginia Polytechnic and State University, Blacksburg, VA.

Flowers, T. J. and A. R. Yeo. 1992. *Solute Transport in Plants*. Blackie Academic & Professional, London.

Marschner, H. 1995. *Mineral Nutrition of Higher Plants*. Academic Press, San Diego.

Merva, G. E. 1995. *Physical Principles of the Plant Biosystem*. American Society of Agricultural Engineers, St. Joseph, MI.

Nobel, P. S. 2009. *Physiochemical and Environmental Plant Physiology*. Academic Press, San Diego.

Siau, J. F. 1984. *Transport Processes in Wood*. Springer-Verlag, New York.

Transport in a Bioprocessing Context

Transport is at the core of industrial (and domestic) processing of food and biomaterials (see examples in Figure 4). In most food processes heat and moisture transport influences chemical and microbiological changes. Sterilization of food to extend its storage life is done primarily using heat, making heat transfer an extremely important transport process. Freezing of food involves heat transfer with a change of phase. Drying, and related processes such as baking and frying, are also important and involve the phase change of water to vapor and the transport of water and vapor through the food matrix. The diffusion of gases through a packaging material is important in packaging food so that the optimum gas composition (controlled atmosphere) is maintained around the food. Fresh produce respires, and here again heat transfer is important, as heat generation from respiration is important in designing the right storage temperature.

Bioprocessing, such as a fermentation process, involves oxygen transport, for example, from a liquid fermentation medium to the cells. The design of waste treatment facilities also requires knowledge of heat and mass transfer, particularly mass transport. For example, a composting pile must be designed so that the heat generated within the pile is released appropriately and so that temperatures do not become so high as to be lethal to the microbes that are responsible for biological activity. Some specialized texts on engineering aspects of food and bioprocessing are noted here for further reading.

Bailey, O. E. and D. F. Ollis. 1986. *Biochemical Engineering Fundamentals.* McGraw-Hill, New York.

Gekas, V. 1992. *Transport Phenomena of Foods and Biological Materials.* CRC Press, Boca Raton, FL.

Hallström, B., C. Skjöldebrand, and C. Trägårdh. 1987. *Heat Transfer and Food Products.* Elsevier Applied Science Publishers Ltd., Barking, Essex, England.

Heldman, D. R. and D. B. Lund. 2006. *Handbook of Food Engineering.* Marcel Dekker, New York.

Johnson, A. 1999. *Biological Process Engineering.* John Wiley & Sons, New York.

Mujumdar, A. S. 1987. *Handbook of Industrial Drying.* Marcel Dekker, New York.

Singh, R. P. and D. R. Heldman. 2014. *Introduction to Food Engineering.* Elsevier, New York.

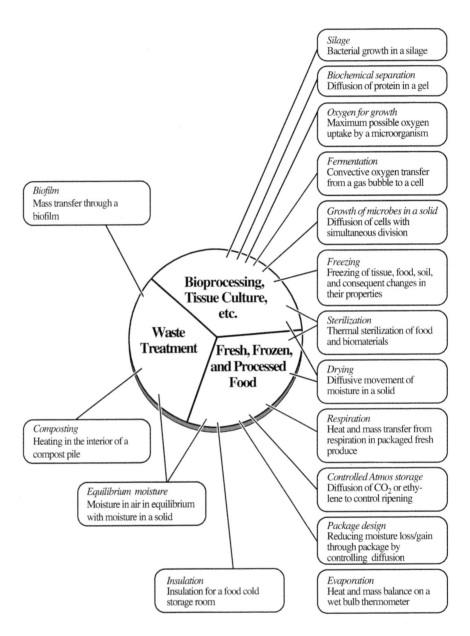

Figure 4: Schematic showing examples of transport in industrial food and biological processing that are covered in this text.

Transport in a Bioenvironmental Context

Transport in the bioenvironmental system (Figure 5) can be seen in three major environmental media—water, air, and soil. Transport in soil can involve bulk flow and dispersion of water and chemicals present in it, eventually reaching groundwater. Such chemicals could be fertilizers or pesticides that are routinely applied. Transport in soil is traditionally covered in textbooks on soil physics, although some specialized transport texts covering transport in soil have appeared in recent years (e.g., Schnoor, 1996). In the air, transport can be of airborne particles such as smoke from a chimney, or of evaporated water, such as from wet soil or a water surface. Specialized texts on air pollution and evaporation also exist, as noted below. The movement of pollutants in surface water occurs in bodies such as streams and lakes and is important in the study of environmental transport. Heat transport in the outdoor environment can involve solar energy— its incidence, as well as its reflection and absorption in the atmosphere, vegetation, and soil. In the indoor environment, heat transport can occur in heat loss through building walls. Heat exchange between humans (or animals) and room air or the interior walls of a building is also important for indoor thermal comfort. Mass transfer indoors can involve transport of odors and indoor pollutants such as smoke, and indoor air quality is a subject of significant interest. Some specialized texts covering transport in the environment are included here for follow-up reading.

Brutsaert, W. 1982. *Evaporation into the Atmosphere: Theory, History and Applications.* Kluwer Academic Publishers, Boston.

Campbell, G. S. 1985. *Soil Physics with Basic Transport Models for Soil-Plant Systems. Developments in Soil Science.* Elsevier Science Publishers, Amsterdam.

Clark, M. M. 1996. *Transport Modeling for Environmental Engineers and Scientists.* John Wiley & Sons, New York.

Ghildyal, B. P. and R. P. Tripathi. 1987. *Soil Physics.* Wiley Eastern Limited, New Delhi, India.

Hillel, D. 1980. *Movement of Solutes and Soil Salinity. Fundamentals of Soil Physics.* Academic Press, New York.

Jumikis, A. R. 1966. *Thermal Soil Mechanics.* Rutgers University Press, New Brunswick, NJ.

Loucks, D. P., J. R. Stedinger, and D. A Haith. 1981. Water quality prediction and simulation. In: *Water Resource Systems Planning and Analysis.* Prentice-Hall, Inc., Englewood Cliffs, NJ.

Monteith, J. L. and M. H. Unsworth. 1990. *Principles of Environmental Physics.* Edward Arnold, New York.

Neprin, S. V. and A. F. Chudnovskii. 1984. *Heat and Mass Transfer in the Plant-Soil-Air System.* Oxonian Press Pvt. Ltd., New Delhi, India.

Novak, V. 2012. *Evapotranspiration in the Soil-Plant-Atmosphere System,* Springer, Dordrecht.

Schnoor, J. L. 1996. *Environmental Modeling: Fate and Transport of Pollutants in Water, Air and Soil.* John Wiley & Sons, New York.

Stern, A. C. 1976. *Air Pollution*, 3rd Ed., Vol. I. Academic Press, New York.

Thibodeaux, L. J. 1979. *Chemodynamics: Environmental Movement of Chemicals in Air, Water, and Soil.* John Wiley & Sons, New York.

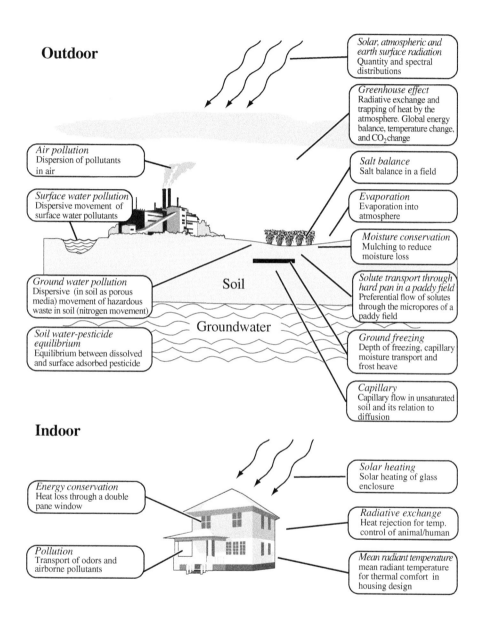

Outdoor

Solar, atmospheric and earth surface radiation
Quantity and spectral distributions

Greenhouse effect
Radiative exchange and trapping of heat by the atmosphere. Global energy balance, temperature change, and CO_2 change

Air pollution
Dispersion of pollutants in air

Surface water pollution
Dispersive movement of surface water pollutants

Salt balance
Salt balance in a field

Evaporation
Evaporation into atmosphere

Moisture conservation
Mulching to reduce moisture loss

Ground water pollution
Dispersive (in soil as porous media) movement of hazardous waste in soil (nitrogen movement)

Soil

Solute transport through hard pan in a paddy field
Preferential flow of solutes through the micropores of a paddy field

Groundwater

Soil water-pesticide equilibrium
Equilibrium between dissolved and surface adsorbed pesticide

Ground freezing
Depth of freezing, capillary moisture transport and frost heave

Capillary
Capillary flow in unsaturated soil and its relation to diffusion

Indoor

Solar heating
Solar heating of glass enclosure

Energy conservation
Heat loss through a double pane window

Radiative exchange
Heat rejection for temp. control of animal/human

Pollution
Transport of odors and airborne pollutants

Mean radiant temperature
mean radiant temperature for thermal comfort in housing design

Figure 5: Schematic showing examples of transport in the bioenvironment covered in this text.

LIST OF SYMBOLS

A	area, m^2
Bi	$= hL/k$, Biot number, dimensionless
Bi_m	$= h_m K^* L/D_{AB}$, mass transfer Biot number, dimensionless
c	total concentration (sometimes abbreviated c_A), kg/m^3
c_A	concentration of component A, kg of A/m^3
$c_{A,s}$	concentration of component A at a surface, kg of A/m^3
$c_{A,\infty}$	concentration of component A in the bulk fluid, kg of A/m^3
$c_{A,i}$	initial concentration, kg/m^3
c_{av}	average concentration, kg/m^3
c_p	specific heat at constant pressure, $kJ/kg\cdot°C$
d	diameter, m
D_{AB}	diffusivity of species A in species B, m^2/s
erf	error function
E	total emissive power, W/m^2
$\dot{E}$	energy of flowing fluid per unit time, kJ/s
E_z	dispersion coefficient in the z direction, m^2/s
E_a	activation energy, $J/mole$
f	frictional coefficient in Stokes equation for diffusivity
F	energy flux, W/m^2
Fo	$= \alpha t/L^2$, Fourier number, dimensionless
g	gravity, m/s^2
Gr	$= g\beta\Delta T\rho^2 L^3/\mu^2$, Grashof number, dimensionless
Gr_{AB}	$= g\rho\Delta\rho_A L^3/\mu^2$, mass transfer Grashof number, dimensionless
h	convective heat transfer coefficient, $W/m^2\cdot°C$
h	Planck's constant, 6.625×10^{-34} $J\cdot s$
h	sum or pressure and matric potential, m
h_m	convective mass transfer coefficient, m/s

H	enthalpy per unit mass, kJ/kg
H	Henry's law constant, atm/mole fraction
$\bar{H}$	humidity of air, kg of vapor/ kg of dry air
$\mathcal{H}$	hydraulic head, m
$j_{A,z}$	diffusive or dispersive mass flux of A in the z direction, kg/m^2·s
j^v	volumetric flux, m^3/m^2·s or m/s
k	thermal conductivity, W/m·°C
k	permeability, m^2
k''	reaction rate constant for first order reaction, 1/s
k_f	thermal conductivity of fluid, W/m·°C
K	hydraulic conductivity, m/s
K^*	distribution coefficient or partition coefficient, units vary
L	half thickness of a slab or characteristic length, m
L_p	membrane permeability, m/Pa·s
m	non-dimensional parameter in Chapters 4 and 12
m	mass, kg
m_s	mass of solids, kg
m	non-dimensional (inverse Biot) number k/hL
M	molecular weight, kg
$\mathcal{M}$	molality or moles of solute per unit mass solvent mol/kg
$\dot{m}$	mass flow rate, kg/s
n	non-dimensional distance
n	order of reaction, dimensionless
$n_{A,z}$	mass flux of species A in the z direction, kg/m^2·s
n^v	volumetric flux, $m^3/m^2 \cdot s$
$N_{A,z}$	mass flow rate of species A in the z direction, kg/s
Nu	Nusselt number, dimensionless
Nu$_x$	Nusselt number at a location x, dimensionless
Nu$_L$	Average Nusselt number based on characteristic length L, dimensionless
P	total pressure, N/m^2 or Pa
p_A	partial pressure of component A, N/m^2 or Pa
p_m	permeability, m/s
P	perimeter, m
Pr	$=\mu c_p/k$, Prandtl number, dimensionless
q_x	heat flow in the x direction, W
q_x''	heat flux in the x direction, W/m^2
Q	volumetric heat generation, W/m^3
r	radial direction
r_A	rate of generation of A per unit volume, kg of A/m^3·s

RH	relative humidity, fraction
R_g	universal gas constant = 8.315 kJ/kmol·K
Ra	= Gr × Pr, Rayleigh number, dimensionless
Ra_m	= Gr_{AB} × Sc, mass transfer Rayleigh number, dimensionless
Re	= $\rho u_\infty L/\mu$, Reynolds number, dimensionless
Sc	= $\mu/\rho D_{AB}$, Schmidt number, dimensionless
Sh	= $h_m L/D_{AB}$, Sherwood number, dimensionless
t	time, s
$t_{1/2}$	half life, s
T	temperature
T_s	surface temperature
T_i	initial or inlet temperature
T_f	freezing point, K
T_∞	fluid or ambient temperature
u	velocity in the x direction, m/s
u_∞	free stream velocity in the x direction, m/s
U	thermal energy per unit volume, J/m^3
U	overall heat transfer coefficient, W/m^2·°C
U_m	overall mass transfer coefficient, m/s
v	velocity in y direction, m/s
V	volume, m^3
w	moisture content, kg of water/kg of dry solids
x	x coordinate, m
x_A	mole fraction of species A in liquid phase
y	y coordinate, m
y_A	mole fraction of species A in vapor phase
z	z coordinate, m

Greek Letters

α	thermal diffusivity, m^2/s; also absorptivity, dimensionless
β	coefficient of thermal expansion
δ	penetration depth of microwaves
δ_{conc}	concentration boundary layer thickness, m
δ_{vel}	velocity boundary layer thickness, m
$\delta_{thermal}$	thermal boundary layer thickness, m
Δ	finite change
ΔH_f	latent heat of fusion, kJ/kmol or kJ/kg
ΔH_{vap}	latent heat of vaporization, kJ/kmol or kJ/kg

$\Delta\beta_i$	volume fraction of pores with radius r_i, dimensionless
ϵ	emissivity, dimensionless
η	non-dimensional quantity in Chapter 5
γ	surface tension, N/m
κ	Boltzmann's constant, 1.380×10^{-23} J/K
λ	wavelength, m
μ	viscosity, kg/m·s
ν	kinematic viscosity, m^2/s
Π	osmotic pressure, Pa
ρ	density or mass concentration, kg/m^3; also reflectivity, dimensionless
σ	Stefan–Boltzmann constant, 5.676×10^{-8} W/m^2·K^4
σ_{AB}	collision diameter, °A
$\Omega_{D,AB}$	dimensionless function defined in Chapter 10
τ	transmissivity, dimensionless
θ	non-dimensional temperature

Part I

Energy Transfer

Chapter 1

EQUILIBRIUM, ENERGY CONSERVATION, AND TEMPERATURE

CHAPTER OBJECTIVES

After you have studied this short introductory chapter, you should be able to

1. Explain thermal equilibrium and how it relates to energy transport.

2. Understand temperature ranges important for biological systems and temperature sensing in mammals.

3. Understand temperature ranges important for the environment.

KEY TERMS

- **first and second laws of thermo-dynamics**

- **thermal equilibrium**

- **energy transfer**

- **tolerable temperatures**

- **deep body temperature**

- **cold and heat receptors**

- **greenhouse effect**

This is the first chapter in the part on energy transfer. It discusses energy transfer as the logical next step after what you have studied in thermodynamics. The relationship of this chapter to other chapters is shown in Figure 1.1.

3

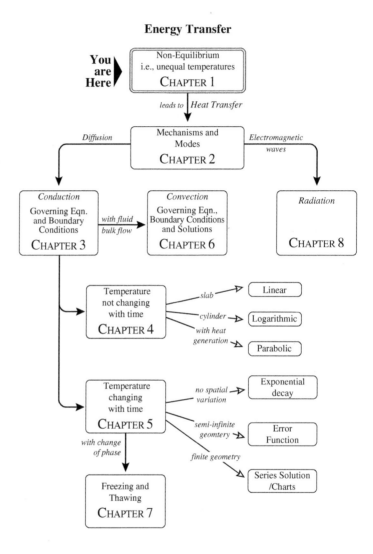

Figure 1.1: Concept map of energy transfer showing how the contents of this chapter relate to other chapters on energy transfer.

1.1 Thermal Equilibrium and the Laws of Thermodynamics

1.1.1 Laws of Thermodynamics

Thermodynamics deals with interchanges between various forms of energy. The first law of thermodynamics states that energy is conserved. The total energy of the system

Figure 1.2: A control volume for energy conservation showing different components.

plus surroundings remains constant. Different forms of energy can inter-convert but their sum remains constant. The second law of thermodynamics states that the total entropy of a system plus surroundings never decreases. This is equivalent to the statement that heat spontaneously flows from a body of higher temperature to one at lower temperature.

1.1.2 Thermal Equilibrium

Two systems are said to be in thermal equilibrium when their temperatures are equal. This is different from *steady state*. In steady state, temperatures do not change with time. In equilibrium, there are no heat flows. This will be explained further in Chapter 4.

1.1.3 Energy Conservation

Conservation of energy is the first law of thermodynamics, as was just mentioned. It is one of the two pillars on which the subject of energy transfer stands. The other pillar is the laws describing the rate of energy transfer, as will be discussed later. To apply the conservation of *thermal* energy to an arbitrary system, as shown in Figure 1.2, we can write a word equation as

$$\begin{matrix} \text{Rate of} \\ \text{Energy In} \end{matrix} - \begin{matrix} \text{Rate of} \\ \text{Energy Out} \end{matrix} + \begin{matrix} \text{Rate of} \\ \text{Energy Generation} \end{matrix} = \begin{matrix} \text{Rate of} \\ \text{Change in} \\ \text{Energy Storage} \end{matrix} \quad (1.1)$$

Note that other forms of energy can convert into thermal energy. When this happens, we think of this source of thermal energy as part of "Energy Generation." An increase in energy storage is manifested as an increase in the temperature or a change in its latent heat. Equation 1.1 will be used in Chapter 3 over a control volume to develop general equations of energy transfer.

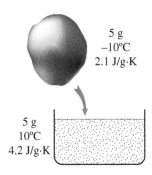

5 g
−10°C
2.1 J/g·K

5 g
10°C
4.2 J/g·K

Figure 1.3: Illustration of energy conservation.

1.1.4 *Example: Simple Energy Conservation*

A solid object of mass 5 g and at −10°C is dropped into a liquid of mass 5 g at 10°C, as shown in Figure 1.3. The specific heats of the solid and the liquid are 2.1 J/kg · K and 4.2 J/kg · K, respectively. Assume zero heat loss to the surroundings. 1) What is the final temperature of the combined solid-liquid system? 2) How would you calculate the final temperature if the solid completely melts, with a latent heat of melting of λ? 3) How would you calculate the final temperature if the solid only partially melts after coming to equilibrium with the liquid?

Solution

Understanding and formulating the problem *1) What is the process?* As the warmer body is brought in contact with the colder body, the warmer body will lose heat while the colder will gain, eventually equilibrating. *2) What are we solving for?* Final temperature when the two bodies are brought in contact. *3) Schematic and given data:* See Figure 1.4 for a schematic with some of the given data superimposed on it. *4) Assumptions:* No heat loss to the surroundings, as provided.

Generating and selecting among alternate solutions Here we have exchange of energy between the two bodies—a process in which energy is conserved. Thus, the principle of conservation of energy can be useful in this case. Also, unlike later chapters, here the only problem solving tool we have covered is energy balance so we can assume that approach would work for this problem.

Implementing the solution The total energy of these two bodies together will stay the same, following the principle of conservation of energy (Eq. 1.1). Another way to see, of course, is that so far in this book all we have dealt with is conservation of energy (Eq. 1.1) so that is all we have to apply.

1) We define the domain or the box (see Figure 1.4). Let m_1, c_{p_1}, T_1 and m_2, c_{p_2}, T_2 be the mass, specific heat, and temperature, respectively, of bodies 1 and 2. Recall that for an incompressible substance (solid or liquid), the internal energy content or enthalpy of a mass, m, and specific heat at constant pressure, c_p, at temperature T is $mc_p(T - T_{ref})$ where T_{ref} is a reference temperature with respect to which the energy content is measured. Using our domain as the dashed line in Figure 1.4, we calculate the individual terms in Eq. 1.1 for this domain as

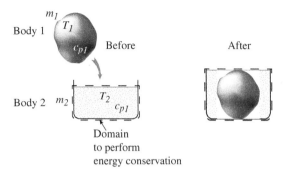

Figure 1.4: The dashed line defines a domain to analyze the problem shown in Figure 1.3.

As Body 1 enters the domain (figure on the right), it brings energy into the domain, based on its temperature. This is written as

$$\text{Energy In}_{\text{(to the domain)}} = m_1 c_{p_1} (T_1 - 0)$$

Body 2 stays in the domain and the problem assumes no heat loss to the surroundings, leading to

$$\text{Energy Out}_{\text{(of the domain)}} = 0$$

No other form of energy is getting converted into heat, e.g., there is no electrical heating or chemical reaction producing or consuming heat, leading to

$$\text{Energy Generation}_{\text{(in the domain)}} = 0$$

The change in the energy stored in the domain is the difference between its energy content at its final state (the "After" figure) that includes both Body 1 and Body 2 in the domain, and the initial state (the "Before" figure) that has only Body 2 in the domain, so

$$\text{Change in Energy Storage}_{\text{(of the domain)}} = \underbrace{(m_1 c_{p_1} + m_2 c_{p_2})(T - 0)}_{\text{final}} - \underbrace{m_2 c_{p_2}(T_2 - 0)}_{\text{initial}}$$

Plugging in Eq. 1.1, we get

$$\underbrace{m_1 c_{p_1}(T_1 - 0)}_{\substack{\text{Energy} \\ \text{in}}} - \underbrace{0}_{\substack{\text{Energy} \\ \text{out}}} + \underbrace{0}_{\substack{\text{Energy} \\ \text{generated}}} = \underbrace{(m_1 c_{p_1} + m_2 c_{p_2})(T - 0) - m_2 c_{p_2}(T_2 - 0)}_{\substack{\text{Change in} \\ \text{energy storage}}}$$

(1.2)

Using the numerical values as shown in Figure 1.3, we get

$$5(2.1)(-10 - 0) = (5(2.1) + 5(4.2))(T - 0) - 5(4.2)(10 - 0)$$

from which we can solve for the final temperature as

$$T = 3.33°C$$

2) When the solid completely melts   Here we need to consider its latent heat, λ, in kJ/kg·°C. Final energy term now needs to be rewritten. If all of the solid melts,

Change in Energy Storage =

$$\underbrace{(m_1 c_{p_1} + m_2 c_{p_2})(T - 0) + m_1 \lambda}_{\text{final}} - \underbrace{\left(m_1 c_{p_1}(T_1 - 0) + m_2 c_{p_2}(T_2 - 0)\right)}_{\text{initial}}$$

So the new balance equation is

$$\underbrace{m_1 c_{p_1}(T_1 - 0)}_{\substack{\text{Energy} \\ \text{in}}} - \underbrace{0}_{\substack{\text{Energy} \\ \text{out}}} + \underbrace{0}_{\substack{\text{Energy} \\ \text{generated}}}$$

$$= \underbrace{(m_1 c_{p_1} + m_2 c_{p_2})(T - 0) + m_1 \lambda - m_2 c_{p_2}(T_2 - 0)}_{\substack{\text{Change in} \\ \text{energy storage}}}$$

3) When the solid partially melts   When only a portion of the solid melts and an equilibrium is reached, it means the system finally reaches its solid-liquid equilibrium temperature. Thus, the final temperature is known. The amount that would melt is the unknown, instead. Thus,

Change in Energy Storage

$$= \underbrace{(m_1 c_{p_1} + m_2 c_{p_2})(T - 0) + m_1^* \lambda}_{\text{final}} - \underbrace{\left(m_1 c_{p_1}(T_1 - 0) + m_2 c_{p_2}(T_2 - 0)\right)}_{\text{initial}}$$

So the new balance equation is

$$\underbrace{m_1 c_{p_1}(T_1 - 0)}_{\substack{\text{Energy} \\ \text{in}}} - \underbrace{0}_{\substack{\text{Energy} \\ \text{out}}} + \underbrace{0}_{\substack{\text{Energy} \\ \text{generated}}}$$

$$= \underbrace{(m_1 c_{p_1} + m_2 c_{p_2})(T - 0) + m_1^* \lambda - m_2 c_{p_2}(T_2 - 0)}_{\substack{\text{Change in} \\ \text{energy storage}}}$$

in which m_1^* is the new unknown that we can solve for.

Evaluating and interpreting the solution *1) Do the computed values (final temperature) make sense?* Since the initial temperatures of the two bodies are $-10°C$ and $10°C$, the final temperature has to be between these two, which it is. *2) What do we learn?* a) We note that energy conservation alone is unable to predict how long it will take to reach the final or equilibrium temperature. b) Our answer above does not and should not depend on our choice of domain. To convince us of this, let us redo the first part above, redefining the domain as in Figure 1.5.

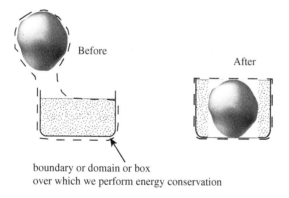

boundary or domain or box
over which we perform energy conservation

Figure 1.5: Redefining the domain.

We now repeat the above process of calculating the individual terms in Eq. 1.1

as

$$
\begin{aligned}
\text{Energy In} &= 0 \\
\text{Energy Out} &= 0 \\
\text{Energy Generation} &= 0 \\
\text{Change in Energy Storage} &= \underbrace{(m_1 c_{p_1} + m_2 c_{p_2})(T - 0)}_{\text{final}} \\
&\quad - \underbrace{\left(m_1 c_{p_1}(T_1 - 0) + m_2 c_{p_2}(T_2 - 0)\right)}_{\text{initial}}
\end{aligned}
$$

Plugging in Eq. 1.1, we get

$$
\underbrace{0}_{\substack{\text{Energy} \\ \text{in}}} - \underbrace{0}_{\substack{\text{Energy} \\ \text{out}}} + \underbrace{0}_{\substack{\text{Energy} \\ \text{generated}}}
$$

$$
= \underbrace{(m_1 c_{p_1} + m_2 c_{p_2})(T - 0) - m_1 c_{p_1}(T_1 - 0) - m_2 c_{p_2}(T_2 - 0)}_{\substack{\text{Change in} \\ \text{energy storage}}}
$$

Using the numerical values as shown in Figure 1.3, we get

$$
0 = (5(2.1) + 5(4.2))(T - 0) - 5(2.1)(-10 - 0) - 5(4.2)(10 - 0)
$$

from which we can solve for the final temperature as

$$
T = 3.33°C
$$

1.2 Non-Equilibrium Thermodynamics and the Transport of Energy

As was just illustrated, the laws of thermodynamics deal with the final or equilibrium state of a process. They provide no information on the nature of the interaction (what type of energy transfer?) or the rate of the process (how long does it take to reach the equilibrium states?). Since energy transfer takes place only when two bodies are not in equilibrium, the subject of energy transfer is sometimes described as non-equilibrium thermodynamics. As thermodynamics does not provide rate information, additional rate laws are defined in non-equilibrium thermodynamics to study the rates of energy transfer. The two rate laws that will be defined in this text are Fourier's law of energy diffusion (Chapter 2) and Fick's law of mass diffusion (Chapter 10).

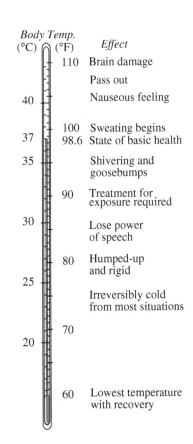

Body Temp.		
(°C)	(°F)	Effect
	110	Brain damage
		Pass out
40		Nauseous feeling
	100	Sweating begins
37	98.6	State of basic health
35		Shivering and goosebumps
	90	Treatment for exposure required
30		Lose power of speech
	80	Humped-up and rigid
25		
		Irreversibly cold from most situations
	70	
20		
	60	Lowest temperature with recovery

Figure 1.6: Effect of various deep body temperatures on humans. Data from Egan (1975).

1.3   Temperature in Living Systems

Most organisms live within a narrow temperature range, with higher temperatures becoming more disastrous than lower temperatures. As shown in Figure 1.7, most biological activity is confined to a rather narrow temperature range of 0–60°C.

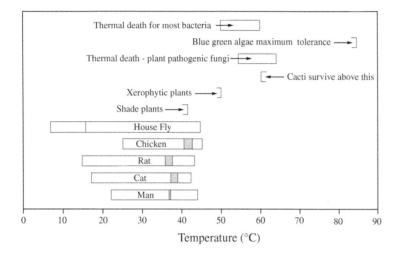

Figure 1.7: Tolerable temperature regimes for many plants and animals. Shaded areas represent the normal temperature for the organism and the unshaded areas represent the extreme tolerances. Adapted from The energy environment in which we live, by D. M. Gates, 1963. Printed by permission of *American Scientist*, Journal of Sigma Xi, The Scientific Research Society.

Although many simple organisms and a few higher forms can remain viable after exposure to very low temperatures, most plants and animals do not carry on biological activity below 0°C. The effects of freezing will be described in detail in Chapter 7. Some of the higher animals maintain a very narrow range in body temperature, as shown in Figure 1.7, through complex physiological controls. Hence, for both plants and animals, the temperature of an organism resulting from environmental influence is of critical importance. The biological systems studied in this text often refer to higher animals and therefore the temperature change in this context is within a narrow range.

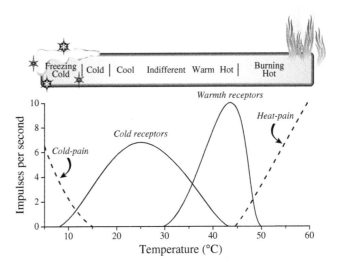

Figure 1.8: Strength of signal from cold, heat, and pain receptors at different temperatures.

1.3.1   Temperature Response of Human Body

As an example of how temperature affects the state of a living system, consider the effect of deep body temperature on humans as shown in Figure 1.6. Man is a constant temperature animal with deep body temperature of about 98.6°F (37°C). It is obvious that temperature needs to be controlled within a very narrow range for effective biological functions.

1.3.2   Temperature Sensation in Humans

The human being can perceive different gradations of cold and heat, as shown in the top portion of Figure 1.8. Thermal gradations are discriminated by two types of temperature receptors—heat receptors and cold receptors. In addition, there are pain receptors that are stimulated only by extreme degrees of heat or cold, as in "burning hot" or "freezing cold" sensations. The brain interprets the input from the different combinations of these receptors as a particular temperature sensation. The heat receptors and cold receptors are free nerve endings located immediately under the skin at discrete but separated points. It is believed that the cold and heat receptors are stimulated by changes in their metabolic rates, these changes resulting from the fact that temperature alters the rates of intracellular chemical reactions more than twofold for each 10°C change.

As shown in Figure 1.8, the heat receptors are most sensitive between about 30 and 45°C. The cold receptors, on the other hand, are most sensitive between about 10 and 40°C. Note that the strength of the signal, measured in terms of nerve impulses per second, is different at different levels of temperature. In the extreme hot or cold region, only pain fibers are stimulated. Extreme degrees of cold or heat can both be painful and both these sensations, when intense enough, may give almost the same quality of sensation— that is, freezing cold and burning hot sensations are almost alike; they are both painful.

Both hot and cold receptors rapidly adapt, so that within about a minute of continuous stimulation, the sensation of heat or cold begins to fade. Thus thermal senses respond markedly to *changes in temperature* in addition to being able to respond to steady states of temperature. Thus when the temperature of the skin is actively falling, a person feels much colder than when the temperature remains cold at that same level. This explains why, on a cold day, it feels much colder when first coming out of a warm house into the outdoors than staying in the same outdoors for longer time. Conversely, when temperature is rising, the person feels much warmer than if that same temperature remained constant. A tub of warm water feels much warmer when first entering it than if staying in the tub for a longer time.

1.3.3 Thermal Comfort of Humans and Animals

Human skin surface should be about 92°F for comfort. The human body maintains a balance with its environment through minor physiological changes (i.e., by increasing or decreasing the flow of blood to the skin). Thermoregulation is the physiological mechanism by which mammals and birds attempt to balance heat gain and loss in order to maintain a constant body temperature. This is described in detail in Chapter 4. When surrounded by air, body heat losses are primarily by convection, evaporation, and radiation, as shown in Figure 1.9.

Body heat losses are affected by air temperature, humidity, velocity, and other factors, as shown in Figure 1.10. These are the thermal comfort factors. Since evaporation (sweating) has a cooling effect, higher evaporation is required at higher temperatures. Lower humidity of air facilitates larger evaporation, which leads to temperature-humidity combination for human comfort, as shown in Figure 1.10.

Information such as in Figure 1.10 is valuable in the design of buildings. Building surface temperatures are important factors for achieving thermal comfort since the body exchanges radiative heat with the walls around it. This is discussed in more detail in Chapter 8. We can also see from Figure 1.10 that, in the comfort zone, human tolerance to humidity is much greater than human tolerance to temperature. Consequently,

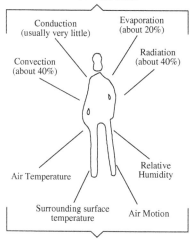

Figure 1.9: Heat loss and thermal comfort factors for humans and animals. Adapted from EGAN, CONCEPTS IN THERMAL COMFORT., 1st Ed., ©1976. Reprinted by permission of Pearson Education, Inc., New York, New York.

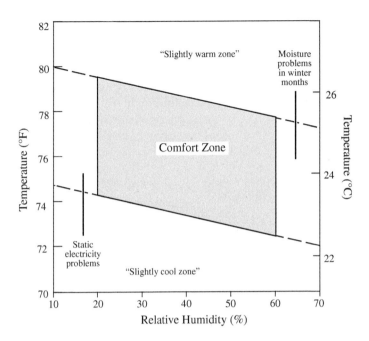

Figure 1.10: Temperature and humidity combinations for human comfort. Adapted from EGAN, CONCEPTS IN THERMAL COMFORT., 1st Ed., ©1976. Reprinted by permission of Pearson Education, Inc., New York, New York.

air temperatures need to be more carefully controlled.

1.4 Temperature in the Environment

1.4.1 The Greenhouse Effect

Gases such as CO_2, methane, and chlorofluorocarbons (CFCs) in our atmosphere let energy from the sun pass through to reach the earth's surface, but stop the energy reflected from the earth's surface from escaping into space. This trapping of energy by the atmosphere, much like the heating up of the insides of cars or greenhouses on a sunny day, is the well-known greenhouse effect. The more these gases are present in our atmosphere, the higher is the earth's temperature. For example, the carbon dioxide concentration and surface temperature have been shown to be closely related, as shown in Figure 1.11. As can be seen from Figure 1.11, during the past 100 years there has

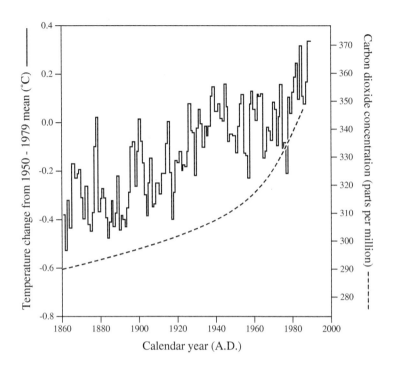

Figure 1.11: History of carbon dioxide concentration and global temperature change for recent times. From *The Changing Climate* by S. H. Schneider. Copyright © 1989 by *Scientific American*. All rights reserved.

been about 0.5°C of **real warming**. The question is how much the temperature will be raised by the greenhouse gases, not whether. Several research centers throughout the world are working on an accurate prediction of the rise in temperature as a function of human activity. **A few degrees of warming** can raise the sea level between 0.2 and 1.5 meters because of widespread melting at the polar icecaps, leading to disastrous consequences.

1.5 Temperature Scales

To refresh our knowledge, some common temperature scales are provided in Table 1.1.

Table 1.1: Various temperature scales and the values of some reference temperatures

	Celsius (°C)	Kelvin (K)	Fahrenheit (°F)	Rankine (°R)
Steam point	100	373.15	212	671.67
Ice point	0	273.15	32	491.67
Absolute zero	−273.15	0	−459.67	0

The relationships between the scales are given as

$$T(°F) = 1.8T(°C) + 32 \tag{1.3}$$
$$T(K) = T(°C) + 273.15 \tag{1.4}$$
$$T(°R) = T(°F) + 459.67 \tag{1.5}$$

Whenever temperature differences are involved, one may use either the absolute scale or the relative scale. For example, a temperature difference of 15°C is equal to 15 K. Thus

$$\Delta T(K) = \Delta T(°C) \tag{1.6}$$
$$\Delta T(°R) = \Delta T(°F) \tag{1.7}$$
$$\Delta T(°F) = 1.8\Delta T(°C) \tag{1.8}$$

Note that wherever a quantity is to be multiplied or divided by temperatures (not temperature differences), absolute scales for temperature must be used except in empirical equations. A feel for various temperature magnitudes is provided in Table 1.2.

1.6   Chapter Summary— Energy Conservation and Temperature

- **Laws of Thermodynamics (page 4)**

  1. The first law of thermodynamics states that energy is conserved.
  2. The second law of thermodynamics is equivalent to the condition that heat spontaneously flows from a body of higher temperature to a body of lower temperature.

Table 1.2: Some temperature values

Lowest achieved temperature	10^{-12} K
Liquid helium	4.2 K
Liquid nitrogen	$-196.15°C$
Liquid CO_2	$-73°C$
Temperature rise in atmosphere due to the greenhouse effect	$0.5°C$
Home refrigerator	
(freezer section)	$-24°C$
(refrigerator section)	$0°C$
Weather	
New York (winter)	$-18°C (\approx 0°F)$
Miami (winter)	$2°C (\approx 35°F)$
New York (summer)	$35°C (\approx 95°F)$
Miami (summer)	$33°C (\approx 91°F)$
Normal human body	$37°C$
Surface gases in the sun	$6000°C$

3. The laws of thermodynamics provide no information on the rate of the energy transfer process.

- ## Temperature and the Living Systems (page 11)

  1. Most living organisms can tolerate only narrow ranges of temperatures.

  2. Temperature sensation in humans is provided by heat and cold receptors located immediately under the skin.

- ## Temperatures in the Environment (page 14)

  1. The greenhouse effect has led to about $0.5°C$ of real warming. A few degrees of warming can be disastrous.

  2. The temperature rise due to the greenhouse effect has been related to an increase in gases such as CO_2, methane, and chlorofluorocarbons in our atmosphere.

1.7   Problem Solving in Energy Conservation

▶**Performing energy conservation to find various quantities (see Example 1.1)**
The steps can be

1. Define the domain over which to perform energy balance.

2. Set up energy balance as

$$\text{In} - \text{Out} + \text{Gen} = \text{Change in Storage}$$

   where the terms can all be either rate or amount over time.

3. Calculate each quantity, **In** and **Out**, making sure of the direction of heat flow, as given by the equations used. For example, the convection equation (introduced in the following chapter), $q = hA(T_1 - T_2)$, is the amount of heat q going from 1 to 2. Whether this heat is **In** or **Out** depends on which body we are referring to as 1 and which one as 2. One does not need to be concerned about which body is hotter and which is colder.

4. If heat is being generated, **Generation** is a positive quantity. If it is being consumed, as in an endothermic reaction, it is negative.

5. Since the energy content of a mass m is $mc_p(T - T_R)$, **Change in storage** is $mc_p\Delta T$, where ΔT is the change in temperature during a chosen period, i.e., $T_{end} - T_{begin}$.

6. Making sure of the directions of heat flow in **In** and **Out** terms, as done in Step 2, and the appropriate sign for **Generation**, plug in all the terms in the above equation. Typically, this will lead to an algebraic equation or a differential equation for temperature.

7. All the terms in the equation need to be in the same units, i.e., they can be in terms of energy, J, or they can be rates such as energy per unit time, J/s or W.

1.8   Concept and Review Questions

1. What additional information is needed beyond the knowledge of thermodynamic equilibrium to decide *how long* it takes for a heating or cooling process?

2. What is the typical temperature range below or above which cell death occurs in human tissue?

3. What has been the approximate temperature rise of the earth's surface in the last 100 years?

4. Is the temperature sensation for a given person always over a fixed temperature range?

5. From a comfort standpoint, which one is a more critical variable—temperature or humidity?

6. Fundamentally speaking, what is the difference between equilibrium thermodynamics that you have studied in previous course(s) and energy transfer that you are studying in this course?

Further Reading

Egan, M. D. 1975. *Concepts in Thermal Comfort*. Prentice Hall, Englewood Cliffs, NJ.

Gates, D. M. 1963. The energy environment in which we live. *American Scientist*. 51:327–348.

Guyton, A. C. and J. E. Hall. 1996. *Textbook of Medical Physiology*. W. B. Saunders Company, Philadelphia.

Nobel, P. S. 1991. *Physiochemical and Environmental Plant Physiology*. Academic Press, San Diego.

Rich, L. G. 1973. *Environmental Systems Engineering*. McGraw-Hill Book Company, New York.

Schneider, S. H. 1989. The changing climate. *Scientific American*. September, 1989.

Shitzer, A. and R. C. Eberhart. 1985. *Heat Transfer in Medicine and Biology*, Volumes I and II. Plenum Press, New York.

1.9 Problems

1.1 Energy Balance including Metabolic Heat under Steady Conditions

By now we are all familiar with how the classroom temperature increases between the time we start and the time we end the lecture. For now, we would like to calculate the possible change in temperature due to the metabolic heat we generate. Consider the walls, including windows, to be perfectly insulating, i.e., no heat flow. Also, for simplicity, consider the entire room to be at one temperature. Colder air comes in and reaches the temperature of the room through mixing, and this air leaves at the same rate through another door of the same size. Consider steady state when the room temperature has reached a constant value. 1) Perform an energy balance for the room in terms of metabolic heat we generate, incoming colder air temperature, air flow rate, air properties, and other dimensions from which you can calculate the room temperature, T. 2) Consider an incoming colder air temperature of 25°C and metabolic heat generation of 60 W per person (a total of 70 persons in the room including the instructor). The door dimensions are 1 m × 2.5 m and the average air velocity through the doors is 0.1 m/s. The properties of air at 25°C are a density of 1.1769 kg/m^3 and specific heat of 1006 J/kgK. Ignore temperature variation in the properties as the incoming air heats up in the room. Solve for room temperature. 3) Comment on how accurate your prediction is and discuss the major assumptions made here that would influence your prediction of temperature. 4) Comment on why the room dimensions, i.e., the volume of air in the room, is not included.

1.2   Energy Balance including Metabolic Heat under Unsteady Conditions

By now we are all familiar with how the classroom temperature increases between the time we start and the time we end the lecture. We would like to have a simple model to describe how temperature changes with time due to the metabolic heat we generate. Consider the walls, including windows, to be perfectly insulating, i.e., no heat flow. Also, for simplicity, consider the entire room to be at one temperature. Colder air comes in and reaches the temperature of the room through mixing, and this air leaves at the same rate through another door of the same size. 1) Perform an energy balance for the room in terms of metabolic heat we generate, incoming colder air temperature, T_i, air flow rate, air properties, and room dimensions from which you can calculate the room temperature, T. 2) For $\Delta t \to 0$, solve for room temperature as a function of time, for an initial temperature of T_i. 3) Consider an incoming colder air temperature of 25°C, metabolic heat generation of 60 W per person (a total of 80 persons in the room including the instructor), and room dimensions of 5m × 3m × 3m. The door dimensions are 1 m × 2.5 m and the average air velocity through the doors is 0.1 m/s. Assume the initial temperature of the room was 25°C. The properties of air at 25°C are a density of 1.1769 kg/m^3 and specific heat of 1006 J/kgK. Ignore temperature variation in the properties as the incoming air heats up in the room. What is the room temperature at the end of a 45 minute lecture? 4) Comment on how accurate your prediction is and discuss the major assumptions made here that would influence your prediction.

1.3   Body Thermoregulation

Improving a player's performance requires an understanding of body heat generation and loss. A squash player is running around in a squash court. We would like to do a simple heat balance on the player by considering convection, radiation, evaporation, and heat production. The convective heat transfer coefficient of the wind on the runner is 20 W/m² · K. The radiative heat loss in watts is given by $\sigma A(T_s^4 - T_w^4)$, where T_s is the body temperature and T_w is the wall temperature in Kelvins. The evaporative heat loss in watts is given by $0.124\sqrt{v}A(p_s - p_\infty)$, where p_s is the surface vapor pressure in Pa and is given by $p_s = 13.33e^{20.386 - 5132/T_s}$. The term p_∞ is the vapor pressure far away and can be considered to be zero. The rate of heat production in watts is given by $4mv$, where m is the mass in kg and v is the speed in m/s of the runner. The player is 180 cm tall with a weight of 80 kg and a surface area of 2 m². His average speed, v, is 1.2 m/s, which is also the air velocity relative to his body (no additional wind). Air temperature, T_∞, and T_w are both 20°C. 1) Assuming the entire body has a uniform temperature, T_s, and specific heat, c_p, that is, considering the problem as a lumped heat transfer one, write the heat balance on the body and develop the differential equation from which we can compute T_s as a function of time. *Do not solve this equation* but make sure T_s is the only unknown. 2) Assuming the body to be at steady state, simplify the equation just developed from which you can calculate body surface temperature, T_s *(do not solve)*. 3) Assume we obtain the solution $T_s = 25.7$°C to the equation in Step 2). If the total heat loss can be written in terms of an effective heat transfer coefficient that includes all modes of heat transfer, i.e., $q_{total} = h_{eff}A(T_s - T_\infty)$, what is the value of h_{eff} for the steady-state situation?

Chapter 2

MODES OF HEAT TRANSFER

CHAPTER OBJECTIVES

After you have studied this chapter, you should be able to

1. Understand the physical processes and the rate laws describing the three modes of heat transfer: conduction, convection, and radiation.

2. Understand the material properties that affect heat conduction in a material.

KEY TERMS

- **conduction**
- **Fourier's law**
- **thermal conductivity**
- **thermal diffusivity**
- **heat flux**
- **heat flow rate**

- **bulk flow**
- **convective heat transfer coefficient**
- **convection**
- **radiation**

In this chapter we will study the fundamental ways energy can be transported. Figure 2.1 shows how the contents of this chapter relate to the overall subject of energy transfer.

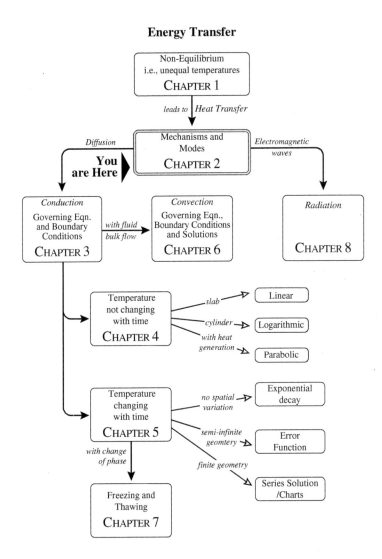

Figure 2.1: Concept map of energy transfer showing how the contents of this chapter relate to other chapters on energy transfer.

2.1 Conductive Heat Transfer

Conductive heat transfer is the movement of thermal energy through a medium from its more energetic particles to the less energetic. The temperature at any location in a material is associated with the energy of the molecules around that location. Higher temperature corresponds to higher molecular energy. In a gas, this energy can be translational, rotational, or vibrational. As the molecules are constantly colliding, they are transferring energy from the more energetic molecules to the less energetic. Thus, energy is transferred from higher to lower temperature. This net transfer of energy due to random molecular motion is termed the *diffusion* of energy. In a solid, translational and rotational motions are restricted. As the temperature of one area of a material increases, the molecules in that area vibrate more and bump into neighboring molecules. This contact between molecules imparts some of the vibrational motion of the first molecule to the second molecule, which then begins to vibrate to a greater extent. This trend continues throughout the material, spreading heat energy by the introduction of increased vibrational motion. In metals there are unbound electrons that translate freely in the material. These electrons allow for a faster and more efficient means of heat conduction in metallic materials than in non-metallic materials.

Conduction heat transfer can be described by Fourier's equation (rate law):

$$\frac{q_x}{A} = -k\frac{dT}{dx} \tag{2.1}$$

where q_x is the rate of heat flow in the x direction, A is the area perpendicular to the x direction through which the heat flows, k is the thermal conductivity of the medium (solid, liquid, or gas), and T is the temperature at a location x. The significance of the negative sign is that heat flows in the direction of decreasing temperature, as shown in Figure 2.2. The quantity q_x/A is called heat flux, which is heat flow per unit time *per unit area*, and will be denoted by the symbol q_x''. The two primes on the symbol q_x'' are simply for convenience and do not carry any special meaning. Thus, in terms of the notation for heat flux, Fourier's law is

$$q_x'' = -k\frac{dT}{dx} \tag{2.2}$$

Note that the heat flux q_x'' or heat flow rate q_x always has a direction associated with it. Thus, if the subscript x is omitted for simplicity, the appropriate direction is given by the direction of the gradient in temperature. The units of thermal conductivity k are given by

$$[k] = \frac{[q_x]}{[A]\left[\frac{dT}{dx}\right]} = \frac{[W]}{[m^2]\left[\frac{°C}{m}\right]} = \frac{W}{m°C} \quad \text{or} \quad \frac{W}{mK}$$

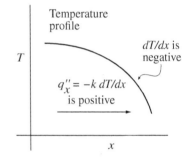

Figure 2.2: Direction of heat flux as related to the temperature gradient.

A note on units— throughout this book, units with a denominator will be written using a slash. For example, the units for thermal conductivity just described will be W/m·K. When multiple units are present in the denominator, as m and K in the case of thermal conductivity, the units in the denominator are separated by a dot (·) and the reader needs to be careful in their use. An alternative is to use a bracket to group the terms in the denominator, such as W/(mK) in the case of thermal conductivity. For a different system of units, thermal conductivity is expressed in

$$[k] = \frac{[\text{Btu/hr}]}{[\text{ft}^2]\left[\frac{\circ F}{\text{ft}}\right]} = \frac{\text{Btu}}{\text{hr ft}\circ F}$$

Only for very simple materials such as simple gases, can thermal conductivities be predicted from molecular knowledge (e.g., kinetic theory of gases). For almost all real materials, available thermal conductivity data is from measurements.

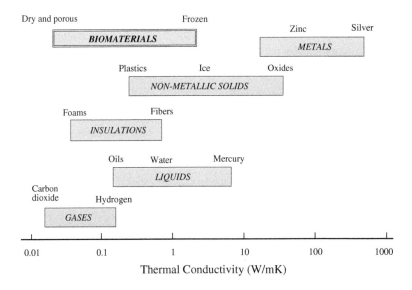

Figure 2.3: Range of thermal conductivity values (at normal temperature and pressure) for various materials compared with those for biomaterials. Adapted from *Fundamentals of Heat and Mass Transfer* by F. P. Incropera and D. P. DeWitt, ©1990 John Wiley & Sons, Inc. Reprinted by permission of John Wiley & Sons, Inc.

Since thermal conductivity is a measure of the efficiency of heat conduction, as explained earlier in this section, it is higher for metallic solids than for non-metallic solids, as shown in Figure 2.3. In a liquid, the intermolecular spacings are generally larger than in solids. Molecules move more randomly. This leads to less effective energy transport and lower values of conductivity, as shown in Figure 2.3. In a gas, the molecules are farther apart. Molecular movements are most random and interactions are less frequent. Thus the lowest thermal conductivity values are found in gases (Figure 2.3).

Consider a simple heat transfer situation of a one-dimensional flow of energy at steady state through a flat surface such as through a wall, as shown in Figure 2.4. To maintain steady state, the same amount of energy has to flow at any location x, making q_x'' a constant. For such a situation, Eq. 2.2 can be integrated over finite distances (between locations 1 and 2) to obtain

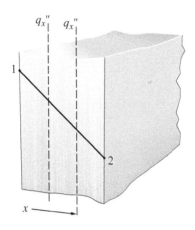

$$\int_1^2 dT = -\frac{q_x''}{k} \int_1^2 dx \qquad (2.3)$$

integrating

$$\Delta T = -\frac{q_x''}{k} \Delta x$$

$$q_x'' = -k\frac{\Delta T}{\Delta x} \qquad (2.4)$$

Figure 2.4: Schematic showing the same value of heat flux at two locations in a slab at steady state.

where $\Delta T = T_1 - T_2$ and $\Delta x = x_1 - x_2$. Equation 2.4 can also be viewed as an approximation of Eq. 2.2 for small distances or when temperature changes linearly with distance as in this steady-state situation.

2.1.1 Example: Fat as a Thermal Insulator

Calculate heat flux through 0.3 cm of skin plus subcutaneous fat tissue and contrast this with that through muscle tissue of the same thickness. The effective thermal conductivities of skin plus subcutaneous fat tissue and muscle tissue are 0.28 W/m · K and 0.56 W/m · K, respectively. The temperatures on the two sides of the tissue are 37°C and 33°C, respectively.

Solution

Understanding and formulating the problem *1) What is the process?* Heat conducts through a tissue layer. *2) What are we solving for?* Heat flux through the tissue layer. *3) Schematic and given data:* A schematic is shown in Figure 2.5 with some of the given data superimposed on it. *4) Assumptions:* Temperature variation is linear between the hot and the cold temperatures (we will learn later

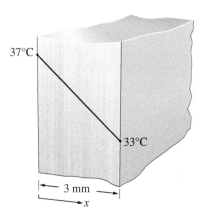

Figure 2.5: Schematic for Example 2.1.1.

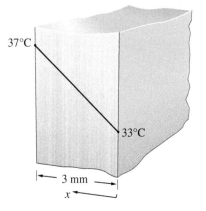

Figure 2.6: Schematic for Example 2.1.1 with the direction of the x axis reversed.

that this can be equivalent to the process being steady). Thermal conductivity does not vary with temperature.

Generating and selecting among alternate solutions We skip this step since it is a simple application of Fourier's law.

Implementing the chosen solution For the skin plus subcutaneous fat tissue layer,

$$q_x'' = -k\frac{\Delta T}{\Delta x} \qquad (2.5)$$

$$= -0.28\left[\frac{W}{m \cdot K}\right]\frac{(33-37)\,[K]}{0.3 \times 10^{-2}\,[m]}$$

$$= 373.3\left[\frac{W}{m^2}\right] \qquad (2.6)$$

For the muscle tissue layer,

$$q_x'' = -k\frac{\Delta T}{\Delta x}$$

$$= -0.56\left[\frac{W}{m \cdot K}\right]\frac{(33-37)\,[K]}{0.3 \times 10^{-2}\,[m]}$$

$$= 746.7\left[\frac{W}{m^2}\right] \qquad (2.7)$$

Evaluating and interpreting the solution *1) Does the change in flux value make sense?* Due to the lower thermal conductivity of the skin plus that of the subcutaneous fat layer, the flux value becomes lower. This is consistent with the everyday understanding that additional fat reduces heat loss. *2) Does the flux direction make sense?* Note that the direction of flux, q_x'', calculated using Eq. 2.5, is always in the positive x direction and does not require knowing which temperature is higher. To check this, we can reverse the direction of the x axis, as shown in Figure 2.6, and perform the same flux calculation:

$$q_x'' = -k\frac{\Delta T}{\Delta x}$$

$$= -0.28\left[\frac{W}{m \cdot K}\right]\frac{(37-33)\,[K]}{0.3 \times 10^{-2}\,[m]}$$

$$= -373.3\left[\frac{W}{m^2}\right]$$

Note we now have a negative flux, which implies that heat is really flowing in the negative x direction. This is the same result we got earlier (Eq. 2.6), both being consistent with thermodynamics (heat flows from hot to cold).

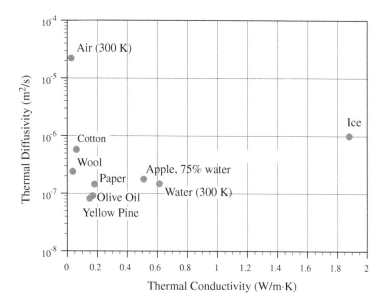

Figure 2.7: Thermal conductivity and thermal diffusivity of some biomaterials.

2.1.2 Thermal Conductivity of Biological and Other Materials

Thermal conductivity of biomaterials as compared to other materials is shown in Figure 2.3. Biomaterials are often mostly water. Therefore, thermal conductivities of unfrozen wet biomaterials are often close to the values for water. It is useful to think of a biomaterial as a composite of its primary ingredients water, ice (for frozen materials), and air (for dry materials), which have large variations in conductivity values, as shown in Figure 2.7. Dry biomaterials have a thermal conductivity much less than water (somewhat closer to air), whereas frozen biomaterials have thermal conductivity values closer to that of ice. See, for example, data for food materials in Appendix C.7 on page 567. Thermal properties for various materials are provided in Sections C.4–C.8 on pages 563–568.

2.1.3 Thermal Diffusivity

Fourier's law (Eq. 2.2) can be rewritten as

$$q_x'' = -k\frac{dT}{dx} = -\frac{k}{\rho c_p}\frac{d(\rho c_p T)}{dx} = -\alpha\frac{dU}{dx} \tag{2.8}$$

Here, ρ is the density, c_p is the specific heat of the material, and

$$U = \rho c_p T \tag{2.9}$$

is the thermal energy per unit volume. The proportionality constant α is called the thermal diffusivity. Thus, the thermal diffusivity is defined as

$$\alpha = \frac{k}{\rho c_p} \tag{2.10}$$

The units of thermal diffusivity α can be calculated as

$$[\alpha] = \frac{[k]}{[\rho c_p]} = \frac{W/m \cdot K}{kg/m^3 \ J/kg \cdot K} = \frac{m^2}{s} \tag{2.11}$$

Since dU/dx is the gradient in energy, α is the proportionality constant between energy flux and energy gradient, i.e.,

$$\begin{array}{c} \text{Flux} \\ \text{of Energy} \end{array} = \alpha \times \begin{array}{c} \text{Gradient} \\ \text{in Energy} \end{array} \tag{2.12}$$

To further interpret thermal diffusivity, note that thermal conductivity provides the rate of heat flow. This alone does not determine temperature change. The amount of energy needed for every degree of temperature change is also a factor in determining temperature rise. Whereas thermal conductivity gives an indication of the ease of heat flow, thermal diffusivity gives an indication of the ease of temperature change in a transient process.

2.1.4 Density and Specific Heat

Density ρ and specific heat c_p are parts of the "thermal mass" of the system. They affect temperature change in the system, as was just discussed under thermal diffusivity. The higher the density and specific heat, the larger the energy it takes to change the temperature. The quantity ρc_p has the units $J/m^3 \cdot K$ and is called the volumetric heat capacity, as can be seen from its units. Examples of data on density and specific heat can be seen in Section C.8.

Two types of densities are used—solid density and bulk density. The distinction between these two densities is important when the material is porous. Solid density (also called mean particle density) is the mass per unit volume of just the solid portion in a porous medium. Dry bulk density is the mass of the dried solid to its total volume (solids and pores together). The porosity of a material is defined as the volume of pores occupied by air and water (if present) divided by the total volume of the solid.

2.2   Convective Heat Transfer

Convective heat transfer is the movement of heat through a medium as a result of the net motion of a material in the medium (e.g., fluid flow over a surface). This is shown in Figure 2.8. Forced convection is due to an external force such as a fan, while free or natural convection is driven by a density difference in the material. Convection over a surface is described by

$$q_{1-2} = hA(T_1 - T_2) \qquad (2.13)$$

where q_{1-2} is the heat flow rate from 1 to 2 (in W or Btu/hr), A is the area normal to the direction of heat flow (m^2 or ft^2), $T_1 - T_2$ is the temperature difference between surface and fluid, and h is the convective heat transfer coefficient, also called the film coefficient. The units of h can be calculated as

$$[h] = \frac{[q]}{[A][\Delta T]} = \frac{\text{W}}{\text{m}^2\,°\text{C}} = \frac{\text{Btu}}{\text{hr ft}^2\,°\text{F}}$$

Equation 2.13 is not a law but a defining equation for h. *It is important to note that the convective heat transfer coefficient h always includes the effect of conduction in the fluid, in addition to its bulk flow (details are in Chapter 6).* The effect of conduction, which is a random molecular effect, is always there, in the presence or absence of bulk flow, and it cannot be stopped. The convective heat transfer coefficient h is a function of system geometry, fluid and flow properties, and magnitude of ΔT. Convection will be studied in more detail in Chapter 6.

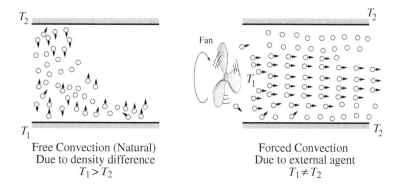

Free Convection (Natural)
Due to density difference
$T_1 > T_2$

Forced Convection
Due to external agent
$T_1 \neq T_2$

Figure 2.8: A schematic comparing free or natural (left) and forced (right) convection.

2.2.1 *Example: Convective Heat Loss from a Human Body*

Calculate the average convective heat flux from a bare body experiencing a gentle breeze leading to a heat transfer coefficient of 50 W/m^2 · K at the skin surface, when the average skin surface temperature is 33°C, and the surrounding air temperature is 20°C. Compare this with the conductive heat flux calculated in Problem 2.2.

Solution

Finding convective heat flux is a straightforward application of Eq. 2.13. Thus, heat flux q'' is given by

$$q''_{\text{body surface}-\text{air}} = h \left(T_{\text{body surface}} - T_{\text{air}} \right)$$
$$= 50 \, \frac{\text{W}}{\text{m}^2\text{K}} \, (33 - 20) \, \text{K}$$
$$= 650 \, \frac{\text{W}}{\text{m}^2}$$

Note that the direction of this flux, from the body surface to air, is enforced by the formula (Eq. 2.13) and does not depend on whether body temperature is higher than the room air. Comparing with the answer in the appendix for conductive heat loss in Problem 2.2, convective heat loss is over six times higher than conductive heat loss. Of course, convective heat loss will be lower for a clothed subject.

2.3   Radiative Heat Transfer

All matter at temperatures above absolute zero emits radiative energy. This radiation is attributed to changes in the electron configuration of the atoms within the matter and is emitted as electromagnetic waves. Unlike conduction and convection, radiation does not require a medium. The maximum flux at which radiation may be emitted by a body at absolute temperature T is given by the Stefan–Boltzmann law:

$$\frac{q}{A} = \sigma T^4 \tag{2.14}$$

where

$$\sigma = 5.670 \times 10^{-8} \, \frac{\text{W}}{\text{m}^2\text{K}^4} = 0.1714 \times 10^{-8} \, \frac{\text{Btu}}{\text{hr ft}^{2\circ}\text{R}^4}$$

Net energy transfer depends on surface and geometric factors. Radiative heat transfer will be studied in detail in Chapter 8.

2.3.1 *Example: Radiative Heat Loss from a Human Body*

Assuming the average skin surface temperature of the human body to be 33°C, find the average energy flux from the human body due to radiation. Is this the *net* radiative heat loss when the person is inside a room? Explain. Compare this with the convective heat loss calculated in Problem 2.2.1.

Solution

Finding radiative heat flux is a straightforward application of Eq. 2.14. Thus, heat flux q'' is given by

$$q''_{\text{body surface}} = \sigma\, T^4_{\text{body surface}}$$

$$= 5.67 \times 10^{-8}\, \frac{\text{W}}{\text{m}^2\text{K}^4}(273.15 + 33)^4\text{K}^4$$

$$= 498.10\, \frac{\text{W}}{\text{m}^2}$$

Note that the direction of this flux is out of the body surface. This is not the net radiative heat loss. This is the energy flux going out of the body. The body also receives energy from the surroundings. The net exchange between the body surface and the *surrounding surfaces* is calculated as

$$q''_{\substack{\text{body} \\ \text{surface}} - \substack{\text{surrounding} \\ \text{surfaces}}} = \sigma\left(T^4_{\substack{\text{body} \\ \text{surface}}} - T^4_{\substack{\text{surrounding} \\ \text{surfaces}}}\right)\text{K}^4 \tag{2.15}$$

$$= 5.676 \times 10^{-8}\, \frac{\text{W}}{\text{m}^2 \cdot \text{K}^4}\left((273.15 + 33)^4 - (273.15 + 20)^4\right)\text{K}^4$$

$$= 79.44\, \frac{\text{W}}{\text{m}^2}$$

Note that the direction of this flux, from the body surface to surrounding surfaces, is enforced by the formula (Eq. 2.15) and does not require knowing whether body temperature is higher than the surfaces. Details of this and other radiation formulas are provided in Chapter 8.

2.4 Chapter Summary—Modes of Heat Transfer

- **Conductive Heat Transfer (page 25)**

    1. Translational, rotational, and vibrational transfer of energy from one molecule to another through physical contact is termed conduction or diffusion.

2. Fourier's law, given by $q''_x = -k\frac{\partial T}{\partial x}$, describes the conduction mode of heat transfer.

3. Thermal conductivity k is a material property that determines the ease of heat conduction. A higher value of k means a higher rate of heat conduction.

4. Thermal diffusivity $\alpha = k/\rho c_p$ determines the ease of temperature change at a location. In analogy to mass diffusivity discussed in Chapter 10, it can be related to the average distance moved by the energy front, $x = \sqrt{2\alpha t}$.

- ## Convective Heat Transfer (page 31)

  1. Convective heat transfer is the transfer of energy when there is net motion, i.e., bulk flow in the medium. The effect of bulk flow is *in addition to* the conductive mode of energy transfer. This mode of energy transport can occur due to the presence of a liquid or a gas.

  2. Convective energy transport is described by $q''_{1-2} = h(T_1 - T_2)$.

- ## Radiative Heat Transfer (page 32)

  1. Radiative energy transport is due to the spontaneous emission of electromagnetic waves by all matter. Such energy transport does not require a medium.

  2. It is described by the equation $q'' = \sigma T^4$.

2.5   Problem Solving in Modes of Heat Transfer

▶ **Calculate fluxes due to various modes.**   This is done using the flux formulas in Table 2.1. Note that every flux has an implied direction, i.e., it is a vector quantity. These directions are also shown in the table.

▶ **Perform energy balance using heat fluxes.**   The heat fluxes above can be combined as part of an energy balance, using what is learned in Chapter 1.

$$\text{In} - \text{Out} + \text{Gen} = \text{Change in Storage}$$

Table 2.1: Various fluxes, q'', and how they are defined

Flux	Expression	Direction	Comments
Conductive	$-k\frac{\partial T}{\partial x}$	Positive x	Conduction alone
Convective, at a surface	$h(T_{\text{surface}} - T_\infty)$	From *surface* to ∞	Conduction in a fluid from a surface, together with flow effect over the surface; h includes the effect of a specific flow situation
Radiative	σT^4	Outward from the surface, all directions	Electromagnetic radiation alone
Due to flow	$u\rho c_p(T - T_R)$	Along that of velocity u	Simply due to being carried with a velocity u
Total, due to conduction and flow	$-k\frac{\partial T}{\partial x}+u\rho c_p(T - T_R)$	Combining terms assumes u is also along the positive x direction	General equation for flux. Not to be confused with the surface convective heat transfer equation (second item above) for a specific flow situation

2.6 Concept and Review Questions

1. Newton's law of viscosity is given by

$$\tau = -\mu\frac{\partial u}{\partial x} \qquad (2.16)$$

where τ is the shear stress, μ is the viscosity, and u is the velocity. Compare the variables in Fourier's law (Eq. 2.2) with this equation and discuss the analogies in the physical quantities and their units. Hint: Multiply and divide u in the above equation by a constant ρ (density) and compare the resulting equation with Eq. 2.8.

2. What is the order of magnitude of the thermal conductivity for gases, liquids, and solids at room temperature and 1 atm pressure?

3. Would you expect wood to have the same thermal conductivity in all three directions? (Hint: see Figure 10.11.)

4. How is convective heat transfer related to conductive heat transfer?

5. What is the most fundamental difference between the conductive and the radiative modes of heat transfer?

6. Describe a heat transfer situation in everyday life and discuss what aspects of it you want to know in more detail.

7. Explain why you would expect gases to have lower thermal conductivity than liquids and liquids lower than that of solids.

8. The fundamental mechanisms (as opposed to modes) of heat transfer are 1) conduction and convection; 2) conduction and radiation; 3) conduction, convection, and radiation.

9. Rank water vapor, ice, and water in terms of their thermal conductivities and provide reasoning.

Further Reading

More references on thermal properties can be found on page 582 in Section D.9.

Adiutori, E. F. 1989. *The New Heat Transfer*. Ventuno Press, West Chester, OH.

Cengel, Y. A. 1998. *Heat Transfer: A Practical Approach*. McGraw-Hill, New York.

Chato, J. C. 1985. Selected thermophysical properties of biological materials. In: *Heat Transfer in Medicine and Biology: Analysis and Applications, Volume 2*. Plenum Press, New York.

Incropera, F. P. and D. P. Dewitt. 1996. *Fundamentals of Heat and Mass Transfer*. Wiley, New York.

Kittel, C. and H. Kroemer. 1980. *Thermal Physics*. W. H. Freeman and Company, San Francisco.

Rao, M. A. and S. S. H. Rizvi. 1986. *Engineering Properties of Foods*. Marcel Dekker, Inc., New York.

Ward, S. and P. J. B. Slater. 2005. Heat transfer and the energetic cost of singing by canaries *Serinus canaria*. *J. Comp. Physiol A*, 191:953–964.

2.7 Problems

2.1 Graphical Implementation of Fourier's Law

Figure 2.9 shows the temperature profile in a cooking pot at some point in time during cooking. Consider the food as a solid (no bulk movement or convection). 1) For a thermal conductivity of 0.45 W/m·K, calculate and plot a profile of heat flux as a function of position on the same figure. 2) Show the direction of positive heat flux along the position axis in the plot. 3) Since the heat flux is lower toward the top, less energy is flowing in the upper regions than lower. Explain whether thermal energy is conserved in this example.

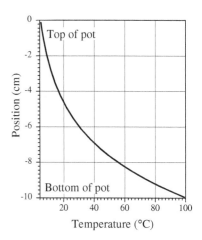

Figure 2.9: Temperature profile in a cooking pot.

2.2 Conductive Heat Flux from a Human Body

Suppose you are sitting on a wooden chair 2.54 cm thick and you are able to maintain the temperature of its wooden surface at the average skin temperature of 33°C. The other side of the chair remains at the surrounding air temperature of 20°C. The thermal conductivity of wood is 0.208 W/m·K. What is the conductive heat flux in the direction toward the surrounding air at steady state?

2.3 Convective Heat Flux from a Human Body

Moved to solved problem

2.4 Energy Flux at the Sun's Surface

Assume the average surface temperature of the sun to be 5800 K. What is the energy flux at the surface of the sun? Compare this to the average solar flux just outside the earth's atmosphere, 1353 W/m^2, and explain the difference.

2.5 Radiative Energy Flux from a Human Body

Moved to solved problem.

2.6   Energy Balance in an Animal at Steady State

We want to calculate the total metabolic heat generated by a singing canary, taking into account heat transfer by radiation, convection, and exhaling air. Assume the canary's body to be a cylinder with a diameter of 7 cm and length of 9 cm, and heat exchange is from the side as well as the top and bottom of the cylinder. The internal and surface temperature of the canary's body is 33°C, and its surface convective heat transfer coefficient is 25.2 W/m² · K. The air temperature is 20°C, the temperature difference between the inhaled and exhaled air is 4.3°C, the ventilation rate is 0.74 cm³ of air per second, the specific heat of air is 1.0066 kJ/kg · K, and the density of air is 1.16 kg/m³. Calculate 1) the net rate of heat lost by radiation, assuming heat gained by the bird through radiation from the surroundings is 11.5 W; 2) the rate of heat transferred by convection to the surrounding air; 3) the rate of heat transferred in the exhaling air without considering any internal evaporation; 4) total metabolic power at steady state.

Chapter 3

GOVERNING EQUATION AND BOUNDARY CONDITIONS OF HEAT TRANSFER

CHAPTER OBJECTIVES

After you have studied this chapter, you should be able to

1. Identify the terms describing storage, convection, diffusion, and generation of energy in the general governing equation for heat transfer.

2. Specify the three common types of heat transfer boundary conditions.

3. Describe heat transfer in mammalian tissue with blood vessels using the bioheat transfer equation.

KEY TERMS

- **governing equation**
- **conduction, convection, generation, storage**
- **boundary temperature specified**

- **boundary heat flux specified**
- **convective boundary condition**
- **bioheat transfer**

In this chapter we will develop a general equation that describes temperature in a material in any kind of heating or cooling situation. Later, in Chapters 4 through 8, we

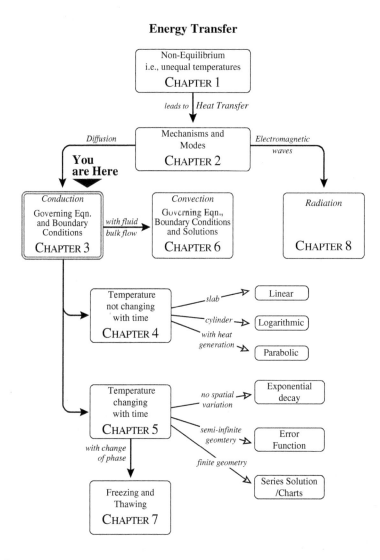

Figure 3.1: Concept map of energy transfer showing how the contents of this chapter relate to other chapters on energy transfer.

will solve this general equation for specific situations. The relationship of this chapter to other chapters on heat transfer is shown in Figure 3.1.

3.1 Governing Equation for Heat Transfer Derived from Energy Conservation and Fourier's Law

We want a general equation that describes temperature distribution in all kinds of situations, not just simple steady-state heat transfer (Eq. 2.4 in Chapter 2). Consider an elemental control volume, as shown in Figure 3.2, where q'' is heat flux (rate of heat flow per unit area) in a particular direction. The heat flux is comprised of conductive and convective heat transport into and out of the control volume. Convective or bulk heat transport arises from the bulk movement of a fluid, if present. As a fluid mass flows into a system, it carries heat (thermal energy) with it at the rate

$$\begin{aligned} E &= \dot{m}c_p\,(T - T_R) \\ &= uA\rho c_p\,(T - T_R) \end{aligned} \tag{3.1}$$

where u is the fluid velocity and T_R is some reference temperature. Thus, the heat flux due to convection is

$$\frac{E}{A} = u\rho c_p(T - T_R) \tag{3.2}$$

Within the control volume, heat is generated at the rate of Q per unit volume. An increase in the stored heat would manifest in the increase in temperature of the control volume. Conversely, if the stored heat in the control volume decreases, its temperature drops. Generation of heat is not to be confused with storage of heat. Generation is the transformation of energy from one form into heat. For example, during exercise, mechanical energy is transformed into heat in the muscle. When electrical current is passed through a material, electrical energy is converted into heat. Other examples of generation are described later in this section. Storage of energy (and rise in temperature) is the effect, whereas generation and conduction of energy are the causes.

Referring to Figure 3.2, we apply the first law of thermodynamics or conservation of energy (Eq. 1.1) over the control volume (Eq. 1.1 is rewritten below for a certain time period):

$$\begin{matrix} \text{Energy} & - & \text{Energy} & + & \text{Energy} & = & \text{Energy} \\ \text{In} & & \text{Out} & & \text{Generated} & & \text{Stored} \end{matrix} \tag{3.3}$$

Note that, in using the term "energy," we are implicitly assuming only the thermal form of energy or heat. For simplicity in deriving the heat transfer equation, let us

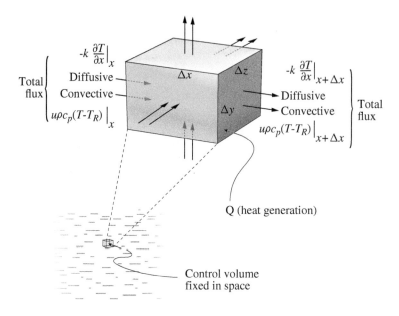

Figure 3.2: Control volume showing energy inflow and outflow by conduction (diffusion) and convection.

consider heat flow only in the x direction. The various quantities in Eq. 3.3 can be written as

$$\text{Energy in during time } \Delta t = \left(q_x'' \Delta y \Delta z + [u\Delta y \Delta z \rho c_p(T - T_R)]_x\right)\Delta t$$

$$\text{Energy out during time } \Delta t = \left(q_{x+\Delta x}'' \Delta y \Delta z + [u\Delta y \Delta z \rho c_p(T - T_R)]_{x+\Delta x}\right)\Delta t$$

$$\text{Energy generated during time } \Delta t = Q\Delta x \Delta y \Delta z \Delta t$$

$$\text{Energy stored during time } \Delta t = \Delta x \Delta y \Delta z \rho c_p \Delta T$$

Substituting in Eq. 3.3, we get

$$
\Delta t \left(q_x'' \, \Delta y \, \Delta z - q_{x+\Delta x}'' \, \Delta y \, \Delta z \right.
$$
$$
+ \rho c_p \, \Delta y \, \Delta z \, [u(T - T_R)_x - u(T - T_R)_{x+\Delta x}]
$$
$$
\left. + Q \, \Delta x \, \Delta y \, \Delta z \right) = \rho c_p \, \Delta x \, \Delta y \, \Delta z \, \Delta T
$$

Dividing throughout by $\Delta x \, \Delta y \, \Delta z \, \Delta t$ and rearranging:

$$
- \frac{q_{x+\Delta x}'' - q_x''}{\Delta x} - \rho c_p \frac{(u(T - T_R)_{x+\Delta x} - u(T - T_R)_x)}{\Delta x} + Q = \rho c_p \frac{\Delta T}{\Delta t}
$$

Making Δx and Δt go to zero and using the definition of a derivative:

$$
- \frac{\partial q_x''}{\partial x} - \rho c_p \frac{\partial}{\partial x} (uT) + Q = \rho c_p \frac{\partial T}{\partial t}
$$

Note that T_R has been dropped since it is a constant. Using Fourier's law for heat conduction (Eq. 2.1) to substitute for the heat flux, q_x'':

$$
- \frac{\partial}{\partial x} \left(-k \frac{\partial T}{\partial x} \right) - \rho c_p \frac{\partial}{\partial x} (uT) + Q = \rho c_p \frac{\partial T}{\partial t}
$$

If k can be assumed constant, this is simplified to

$$
\underbrace{\frac{\partial T}{\partial t}}_{\text{storage}} + \underbrace{\frac{\partial (uT)}{\partial x}}_{\substack{\text{flow or} \\ \text{convection}}} = \underbrace{\frac{k}{\rho c_p} \frac{\partial^2 T}{\partial x^2}}_{\text{conduction}} + \underbrace{\frac{Q}{\rho c_p}}_{\text{generation}} \tag{3.4}
$$

Equation 3.4 is the general governing equation for energy transfer in a one-dimensional cartesian coordinate system with constant thermal properties. This equation is also known as the energy equation or the heat equation. Often, the equation of continuity or mass conservation (see Eq. 11.17 on page 396)

$$
\frac{\partial u}{\partial x} = 0
$$

is used to write an alternate form of Eq. 3.4 as

$$
\underbrace{\frac{\partial T}{\partial t}}_{\text{storage}} + \underbrace{u \frac{\partial T}{\partial x}}_{\substack{\text{flow or} \\ \text{convection}}} = \underbrace{\frac{k}{\rho c_p} \frac{\partial^2 T}{\partial x^2}}_{\text{conduction}} + \underbrace{\frac{Q}{\rho c_p}}_{\text{generation}} \tag{3.5}
$$

Note that the convective term has been simplified.

3.1.1 Meaning of Each Term in the Governing Equation

Although Eq. 3.4 looks complex with all the different terms, the good news is that we will never try (in this text) to solve it keeping all the terms. Depending on the particular situation, we only keep a few of the terms. However, this means we need to be fully aware of what the terms represent, so that we can ignore the ones that are not relevant for a particular situation. For this purpose, we rearrange Eq. 3.4 as

$$\underbrace{\rho c_p \frac{\partial T}{\partial t}}_{\text{storage}} + \underbrace{\rho c_p \frac{\partial}{\partial x}(uT)}_{\text{convection}} = \underbrace{k \left(\frac{\partial^2 T}{\partial x^2} \right)}_{\text{conduction}} + \underbrace{Q}_{\text{generation}} \tag{3.6}$$

Note that, in this rearranged format, each term has an unit of W/m^3 and thus represents energy per unit volume per unit time. The meaning of each term is summarized in Table 3.1.

Table 3.1: Various terms in the governing equation and their interpretations

Term	What it represents	When you can ignore it
Storage	Rate of change of stored energy	Steady state (no variation of temperature with time)
Convection	Rate of net energy transport due to bulk flow	Typically in a solid, with no bulk flow through it
Conduction	Rate of net energy transport due to conduction	Slow thermal conduction in relation to generation or convection. For example, in short periods of microwave heating
Generation	Rate of generation of energy	No internal heat generation due to biochemical reactions, etc.

3.1.2 Examples of the Thermal Source (Generation) Term in Biological Systems

Any living thing is a heat producer. A working muscle such as in the heart or limbs produces heat. The metabolic process itself results in a release of energy in the form of heat (see the table in Section C.1 on page 560). Different activity levels generate body heat to different extents, as shown in Table C.2 on page 561. Fermentation, composting, and other biochemical reactions generate heat. Microwaves are absorbed by a water-containing material to produce heat. An example of this is heating of food in a microwave oven.

3.1.3 Utility of the Energy Equation

It is very general.

1. It is useful for any material.

2. It is useful for any size or shape. Similar equations can be derived for other coordinate systems.

3. It is easier to derive the more general equation and simplify.

4. It is safer—as you drop terms, you are aware of the reasons.

Can we make it more general?

1. To use with compressible fluids.

2. To use when all properties vary with temperature. We need numerical solutions (explained later in the book) to solve such problems.

3. To include mass transfer. For example, the equation cannot predict the temperature inside a steak during cooking in an oven, since the equation does not include water loss from the steak.

3.1.4 *Example: Solution to Specific Situations: Need for Boundary Conditions*

Consider a system such as a slab with temperature variation only in one dimension. Also, consider the system to be at steady state and have no heat generation. The following terms drop out from the general governing equation:

$$\underbrace{\frac{\partial T}{\partial t}}_{\text{steady state}}^{0} = \frac{k}{\rho c_p} \frac{\partial^2 T}{\partial x^2} + \underbrace{\frac{Q}{\rho c_p}}_{\text{no generation}}^{0}$$

leading to the specific governing equation for this problem as

$$\frac{d^2T}{dx^2} = 0 \qquad (3.7)$$

A general solution to this second-order ordinary differential equation can be written as

$$T = C_1 x + C_2 \qquad (3.8)$$

As expected, such general solutions have constants (C_1 and C_2) yet to be determined. More information is needed – need for boundary conditions to evaluate these constants.

Consider the physical situation where the two surfaces (boundaries) of the slab are at two temperatures T_1 and T_2, so that these *boundary conditions* can be written mathematically as

$$T = T_1 \text{ at } x = 0 \qquad (3.9)$$
$$T = T_2 \text{ at } x = L \qquad (3.10)$$

The conditions 3.9 and 3.10 are substituted in Eq. 3.8 to obtain

$$T_1 = C_2$$
$$T_2 = C_1 L + C_2$$

and solve for C_1 and C_2. These constants are then replaced in Eq. 3.8 to obtain the particular solution

$$T = \frac{T_2 - T_1}{L} x + T_1 \qquad (3.11)$$

From the mathematical reasoning provided in this section, it follows that the number of boundary conditions needed equal the highest order derivative in the governing ordinary differential equation. Thus, Eq. 3.7 being a second-order ordinary differential equation, we needed two boundary conditions. Instead of the steady-state heat transfer discussed here, if heat transfer is transient, the governing equation would be a partial differential equation, involving also the time derivative. Thus, we would also need one condition for time (the time derivative is first order). This necessary condition for time is generally provided at the initial time of $t = 0$, hence it is referred to as the initial condition. From a physical standpoint, the boundary conditions are needed since they influence the interior solution. Likewise, the initial condition is needed for a transient problem since the temperatures at a later time are influenced by the initial temperature. Note that initial conditions are not needed for a steady-state problem.

3.2 General Boundary Conditions

As mentioned in the previous section, conditions at the boundary are necessary for obtaining the full details of the solution. The description of a heat transfer problem in a system is not complete without the information on the thermal conditions on the bounding surfaces of the system. The following are the three most common types of thermal conditions that can occur on the boundary of a system in a heat transfer situation:

1. Surface temperature is specified One of the simple thermal conditions that can occur on a surface is a specified temperature. For a one-dimensional heat transfer, as shown in Figure 3.3, this boundary condition is expressed as

$$T\Big|_{x=0} = T_s \tag{3.12}$$

The temperature at the surface, T_s, can be specified as a constant or a function of time. As an example, consider steam condensation on a surface, keeping it at 100°C. This is expressed as

$$T\Big|_{x=0} = 100$$

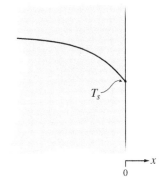

Figure 3.3: A surface temperature specified boundary condition.

2. Surface heat flux is specified Sometimes it is possible to know and specify the rate of heat transfer or *heat flux* on a surface. For a one-dimensional heat transfer, as shown in Figure 3.4, this boundary condition is expressed as

$$-k\,\frac{dT}{dx}\bigg|_{x=0} = q_s'' \tag{3.13}$$

Here the surface heat flux, q_s'', can be specified as a constant or a function of time. For example, consider shining a handheld infrared heat lamp on your back where the skin is receiving energy at the rate of 4000 W/m². This is expressed as

$$-k\,\frac{dT}{dx}\bigg|_{x=0} = 4000$$

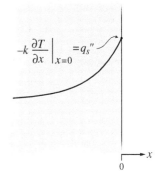

Figure 3.4: A heat flux specified boundary condition.

Note that surface heat flux, q_s'', although a source of energy, is different from the internal heat generation that is included in the governing equation. There are two important special cases of specified surface heat flux. These are now discussed.

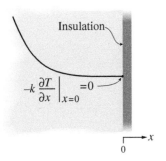

Figure 3.5: An insulated (zero heat flux specified) boundary condition.

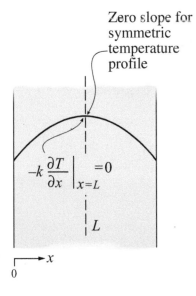

Figure 3.6: A symmetry (zero heat flux specified) boundary condition at the centerline.

2a) Special case: Insulated condition

Surfaces are insulated to reduce heat flux. When a surface is highly insulated, the heat flux through the surface is very small and can be approximated as zero, as illustrated in Figure 3.5. This boundary condition is expressed as

$$-k \left. \frac{dT}{dx} \right|_{x=0} = 0 \tag{3.14}$$

2b) Special case: Symmetry condition

Another common situation arises in a heating or cooling process when the geometry and the boundary conditions are symmetric, as shown in Figure 3.6 for a slab of uniform thickness (therefore symmetric about the centerline) as well as it is cooled symmetrically (same boundary condition on both faces). The resulting temperature profile in the slab will also be symmetric about the centerline, having a zero slope at the centerline. This is expressed as

$$-k \left. \frac{dT}{dx} \right|_{x=L} = 0 \tag{3.15}$$

Note that this symmetry condition resembles the insulated condition mentioned above. To maintain symmetry, heat flux has to be zero at the line of symmetry.

3. Convection at the surface

Perhaps the most common type of thermal condition that can occur on a boundary is convection of a fluid over it. Thus, heat conducted out of the boundary is convected away by the fluid. This condition is written as simply a heat balance at the boundary.

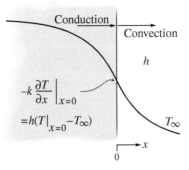

Figure 3.7: A convection boundary condition.

Thus, referring to Figure 3.7, the heat balance at the boundary is

$$\underbrace{-k \left.\frac{dT}{dx}\right|_{x=0}}_{\text{heat conduction}} = \underbrace{h(\left.T\right|_{x=0} - T_\infty)}_{\text{heat convection}} \qquad (3.16)$$

Here h is the heat transfer coefficient mentioned on page 31 and further described in Chapter 6. For example, suppose air at 10°C is blowing over the surface in Figure 3.7, leading to a convective heat transfer coefficient, h, of 50 W/m². The convective boundary condition for this surface can now be written as

$$-k \left.\frac{dT}{dx}\right|_{x=0} = 50(\left.T\right|_{x=0} - 10)$$

Note that the surface temperature $T(x = 0)$ is not known and will come out of the solution. This is in contrast to Eqs. 3.12, where the surface temperature was known.

As a special case of Eq. 3.16, when $h \to \infty$, we get back the temperature specified (first type of) boundary condition mentioned above. To see this, first we rewrite Eq. 3.16 as

$$\left.T\right|_{x=0} - T_\infty = \frac{-k \left.\frac{dT}{dx}\right|_{x=0}}{h} \qquad (3.17)$$

As $h \to \infty$, the right hand side approaches zero, leading to

$$\left.T\right|_{x=0} = T_\infty$$

which is the temperature specified boundary condition.

3.3 Bioheat Transfer Equation for Mammalian Tissue

Although the general governing equation (Eq. 3.4) and its more general versions are valid for most physical systems, it would be hard to apply the equations for a mammalian tissue that has blood vessels of varying sizes and varying amounts of blood flow. In such a system, thermal conductivity, density, and specific heat (the thermal properties) would vary significantly over small distances, among other complications. Researchers have attempted to develop a bioheat transfer equation that is simpler than applying the equations in Section 3.5 directly but still captures the essence of the heat transfer process in a mammalian tissue with blood vessels.

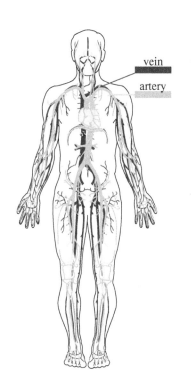

Before we discuss the bioheat transfer equation, let us have a simple look at the mammalian tissue system. The circulatory system in our body comprises two sets of blood vessels— arteries and veins (Figure 3.8) which carry blood from the heart and back. By the pumping action of the heart, blood flows through larger arteries to progressively smaller arteries to arterioles and capillaries to small veins to larger veins and eventually back to the heart. Figure 3.9 shows a schematic of the arterial system with different size vessels. Figure 3.10 shows a schematic of temperature equilibration between the blood and the solid tissue, as the blood traverses different size vessels in the systemic circulation. As blood leaves the heart and travels in the large arteries, its temperature remains essentially constant. This is the arterial blood temperature, T_a. Most of the temperature equilibration occurs as the blood passes through vessels whose diameter is between that of the arterial branch and that of the arteriole. As the blood reaches the latter, blood temperature becomes essentially that of the solid tissue (warmer or colder, as shown). Beyond this point, blood temperature follows the solid tissue temperature through its spatial and time variations until blood reaches the terminal veins. At this point the blood temperature ceases to equilibrate with the

Figure 3.8: Arteries and veins of the circulatory system.

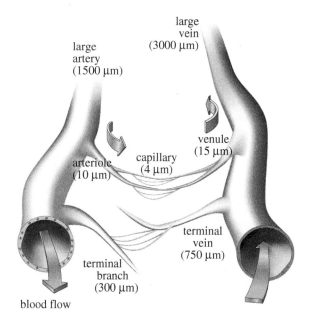

Figure 3.9: Variation of blood vessel sizes.

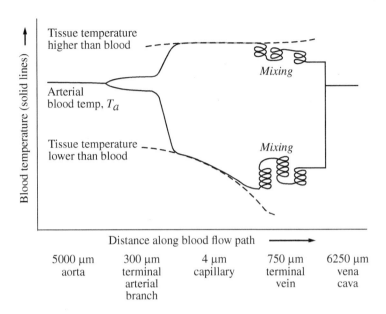

Figure 3.10: Variation of blood temperature in the blood vessels, showing the blood temperature stays primarily unchanged until it reaches the smaller vessels. For additional details, see Chen (1985).

tissue, and remains virtually constant, except as it mixes with other blood of different temperatures at venous confluences. Finally, the cooler blood from peripheral regions and the warmer blood from internal organs mix within the vena cavae and the right atrium and ventricle. Following thermal exchange in the pulmonary circulation and remixing in the left heart, the blood attains the same temperature it had at the start of the circuit.

The bioheat transfer equation can be derived for an idealized tissue system with blood vessels through it, as shown in Figure 3.11, in the same way as Equation 3.4 was derived. Many assumptions that have to be made include 1) homogeneous material with isotropic (same in all directions) thermal properties; 2) large blood vessels are ignored; 3) blood capillaries are isotropic; 4) blood is at arterial temperature but quickly reaches the tissue temperature by the time it reaches the end of the artery system. Using these assumptions, the governing *bioheat equation* for mammalian tissue

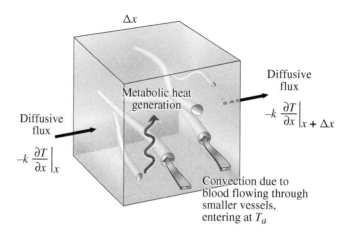

Figure 3.11: Idealized heat transfer in tissue showing metabolic heat generation Q and convective heat transfer due to the passage of blood.

that includes the heat carried by the blood vessels is

$$\underbrace{\rho c \frac{\partial T}{\partial t}}_{\substack{\text{change} \\ \text{in storage}}} = \underbrace{k \nabla^2 T}_{\text{conduction}} + \underbrace{\rho_b c_b \dot{V}_b^v (T_a - T)}_{\substack{\text{convection} \\ \text{due to blood flow}}} + \underbrace{Q}_{\substack{\text{metabolic} \\ \text{heat generation}}} \tag{3.18}$$

where $\dot{V}_b^v$ is the flow rate of blood in m³ of blood / m³ of tissue per second, k, ρ, c are thermal properties of the tissue, ρ_b and c_b are thermal properties of blood, and T_a is arterial blood temperature. Note that the energy carried in the blood is like an additional heat source term of magnitude $\rho_b c_b \dot{V}_b^v (T_a - T)$. This way, Eq. 3.18 can be derived from Eq. 3.4 by rewriting the source term with the energy carried by the blood fluid. Bioheat transfer is an active research area, and alternative equations for bioheat transfer are being formulated. Equation 3.18 is the most widely used bioheat transfer equation today.

The bioheat equation is used generally to solve for tissue temperature distributions. While such considerations as tissue geometry, thermal property values, inhomogeneities, and boundary conditions are very important, this equation can be used to predict the effects of frost bite, determine the depth of damage in a burn victim, and find the amount of heat lost through various parts of the body.

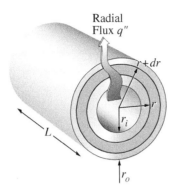

Figure 3.12: Radial heat flow in a long cylinder.

3.4 Governing Equation Derived in Cylindrical Coordinates

Depending on the geometry of the body whose energy transfer we are interested in, a different coordinate system may provide a more convenient way to obtain a solution. A geometrical configuration that is of great importance in engineering is the long cylinder (solid or hollow). The cylindrical coordinate is the natural one to use in this case. We will derive the governing equation in the cylindrical coordinate system by performing energy balance on a control volume, in a procedure completely analogous to that used to derive Eq. 3.4 for the cartesian coordinate system. For convenience, we will drop the convection term.

Using energy conservation (first law of thermodynamics)

$$\begin{array}{ccccccc} \text{Energy} & - & \text{Energy} & + & \text{Energy} & = & \text{Energy} \\ \text{In} & & \text{Out} & & \text{Generated} & & \text{Stored} \end{array} \tag{3.19}$$

these terms can be substituted as (Figure 3.12)

$$\left[2\pi r L q_r'' - 2\pi (r + \Delta r) L q_{r+\Delta r}'' + 2\pi r \Delta r L Q \right] \Delta t = 2\pi r \Delta r L \rho c_p \Delta T$$

Simplifying and rearranging results in

$$- \frac{\left((r + \Delta r)\, q_{r+\Delta r}'' - r q_r'' \right)}{r \Delta r} + Q = \rho c_p \frac{\Delta T}{\Delta t}$$

Making Δr and Δt go to zero and using the definition of a derivative,

$$-\frac{1}{r}\frac{\partial}{\partial r}(rq_r'') + Q = \rho c_p \frac{\partial T}{\partial t}$$

Using Fourier's law,

$$k\frac{1}{r}\frac{\partial}{\partial r}\left(r\frac{\partial T}{\partial r}\right) + Q = \rho c_p \frac{\partial T}{\partial t}$$

Rearranging,

$$\underbrace{\frac{\partial T}{\partial t}}_{\text{storage}} = \underbrace{\frac{k}{\rho c_p}\frac{1}{r}\frac{\partial}{\partial r}\left(r\frac{\partial T}{\partial r}\right)}_{\text{conduction}} + \underbrace{\frac{Q}{\rho c_p}}_{\text{generation}} \qquad (3.20)$$

3.5   Governing Equations for Heat Conduction in Various Coordinate Systems

The governing energy equation in three dimensions in the absence of flow is shown here in some familiar coordinate systems. Control volumes for cartesian, cylindrical, and spherical coordinate systems are shown in Figures 3.13, 3.14, and 3.15, respectively.

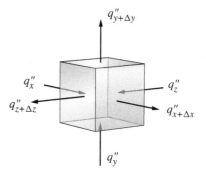

Figure 3.13: Energy balance over a control volume in a cartesian coordinate system.

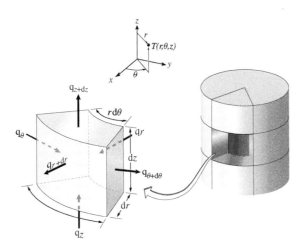

Figure 3.14: Energy balance over a control volume in a cylindrical coordinate system.

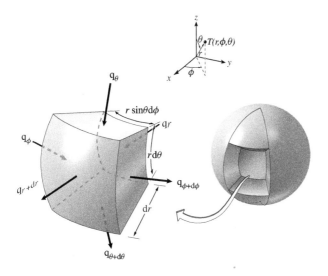

Figure 3.15: Energy balance over a control volume in a spherical coordinate system.

Cartesian

$$\frac{k}{\rho c_p}\left[\frac{\partial^2 T}{\partial x^2} + \frac{\partial^2 T}{\partial y^2} + \frac{\partial^2 T}{\partial z^2}\right] + \frac{Q}{\rho c_p} = \frac{\partial T}{\partial t} \tag{3.21}$$

Cylindrical

$$\frac{k}{\rho c_p}\left[\frac{1}{r}\frac{\partial}{\partial r}\left(r\frac{\partial T}{\partial r}\right) + \frac{1}{r^2}\left(\frac{\partial^2 T}{\partial \theta^2}\right) + \frac{\partial^2 T}{\partial z^2}\right] + \frac{Q}{\rho c_p} = \frac{\partial T}{\partial t} \qquad (3.22)$$

Spherical

$$\frac{k}{\rho c_p}\left[\frac{1}{r^2}\frac{\partial}{\partial r}\left(r^2\frac{\partial T}{\partial r}\right) + \frac{1}{r^2\sin^2\theta}\frac{\partial^2 T}{\partial \phi^2} + \frac{1}{r^2\sin\theta}\frac{\partial}{\partial \theta}\left(\sin\theta\frac{\partial T}{\partial \theta}\right)\right]$$
$$+ \frac{Q}{\rho c_p} = \frac{\partial T}{\partial t} \qquad (3.23)$$

Symbolically (any coordinate system)

$$\frac{k}{\rho c_p}\nabla^2 T + \frac{Q}{\rho c_p} = \frac{\partial T}{\partial t} \qquad (3.24)$$

3.6   Solution Techniques: Analytical vs. Numerical

The governing equations above can be solved using broadly two types of techniques–analytical (what we will do in this book) and numerical. Analytical solutions in simple situations are powerful learning tools in the early stages of learning the physics of heat and mass transfer. Numerical solutions typically do not require the same level of mathematical proficiency from the user and may look more advantageous than analytical solutions. However, just as one should learn the concept of multiplication and do a few multiplications by hand before using the calculator to multiply, one should work with simple analytical solutions to understand the basic physical concepts through the study of simple systems before going for numerical solutions. Numerical solutions are more prone to errors, and only physical understanding (as is easier with analyticals) can help us get through. For detailed solutions in more realistic situations, we use numerical solutions that are actually easier than analytical solutions but are not as efficient for learning the fundamentals.

However, analytical solutions are for simpler situations and therefore have limitations when applied to real-world problems. For example, the series solution for transient heat transfer in a slab, discussed in Section 5.3, has the following limitations at a minimum: 1) constant initial temperature; 2) constant boundary fluid temperature; 3) perfect slab, cylinder, or sphere; 4) far from edges; 5) no heat generation; and 5) constant thermal properties.

In contrast, any geometry, governing equation, boundary condition, properties and other physics can be included in a numerical solution method. Various numerical methods are available—finite difference and finite element are the two most common numerical methods. The COMSOL software, for example, is based on the finite element method. Such numerical method based software is the norm in industrial design and research. The reader can follow up the fundamentals covered in this book with their more realistic applications by simply following the tutorials provided for a user-friendly software such as COMSOL or following a number of texts (e.g., Datta and Rakesh, 2009) that bridge the gap between fundamentals and computer implementations to more complex situations.

3.7 Problem Formulation: An Algorithm to Solve Transport Problems

3.7.1 What Is Problem Formulation?

Problem formulation is creating an equivalent mathematical formulation of a physical problem, i.e., coming up with equations that describe the physical process that constitutes the problem (thus replacing the physical process with its virtual counterpart). It is the first step in modeling. Problem formulation is not unique, and for the same process it is possible to come up with a more complex (typically more realistic) or a less complex (typically less realistic) set of governing equations and boundary conditions. This mathematical problem is subsequently solved. The governing equations and boundary conditions we have learned in this chapter are part of this problem formulation. The steps in this solution process can be summarized as shown in Figure 3.16. Details of each step are discussed in the following sections.

Since the emphasis in this book is on understanding the physics, as opposed to more complex applications, many of the steps in problem formulation have already been completed in the problems at the ends of the chapters. It is still useful to see the process of problem formulation to make the connection between the simplified textbook problems and real physical situations. This should help the reader apply the principles to model biological processes not covered in this book.

Simplification is the key. A very important phrase in this context is "as simple as possible but no simpler." We simplify to achieve the least complex problem formulation possible. For this book, we put the added restriction of being able to solve problems following simple analytical solution techniques. When using numerical solutions, the decision process in problem formulation is less restrictive (see Datta and Rakesh, 2009).

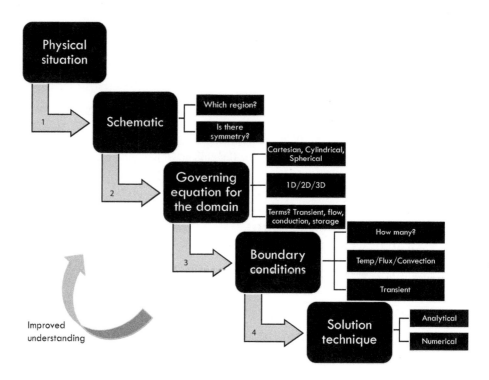

Figure 3.16: A step-by-step procedure to solve heat and mass transfer problems, shown for a heat transfer problem.

3.7.2   Deciding on the Geometry or Domain: Making a Schematic

This is the first step shown in Figure 3.16. The domain is the chosen region over which computations will be performed. The real situation will be replaced by this simplified geometry or the schematic. Not all regions of the physical problem are included in this simplified geometry. For the problems in this book that are chosen to be simple, this step is generally already completed, i.e., a simplified geometry or description is given to you. For example, in Figure 3.17, when there is fluid flowing over a solid, which domain are we interested in—solid (skin) or the water? Although it is obviously the solid region in this problem, it is not always so obvious. The next decision is whether to do a 1D/2D/3D problem. In this book, this simplified geometry is almost always 1D (whereas the real geometry is always 3D). Another decision that often makes the solution easier is to consider whether symmetry is present. Note that symmetry would require symmetry in physics, geometry, and boundary and initial conditions. If symmetry is present, it should be used.

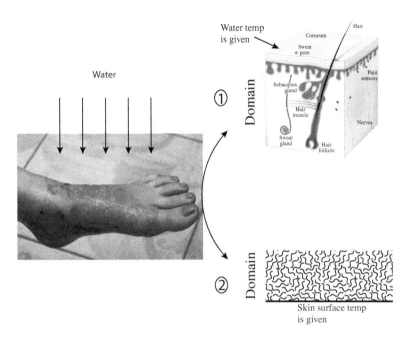

Figure 3.17: Which domain are we interested in? Solid or fluid? Depending on our interests, two of the ways to formulate a problem of hot or cold fluid flow over skin are shown: 1) Skin is the domain and we solve for the skin temperature for a given condition of the fluid. 2) Fluid is the domain and we want to solve for fluid temperature, given a fixed skin surface temperature (not a particularly useful formulation for a skin burn problem but there may be other situations). (iStock.com)

3.7.3 Deciding on the Governing Equation and Its Individual Terms

This is the second step shown in Figure 3.16. Depending on the chosen geometry (slab, cylinder, sphere), choose a governing equation out of Eqs. 3.21–3.23. Note that the flow terms have already been dropped in these equations for simplicity. Once the equation is chosen, we need to decide which terms to keep. This is shown in Table 3.2. Problem description should tell us directly or indirectly which terms to keep. For example, if we are interested in temperature at a time, it is obviously a transient problem. The generation term, Q, in the governing equation typically comes

Table 3.2: Various possibilities of the governing equation

Situation	What is ignored	Final equation
Steady conduction	Transient (storage), flow, generation	$0 = \underbrace{\frac{k}{\rho c_p} \frac{\partial^2 T}{\partial x^2}}_{\text{conduction}}$
Steady conduction with generation	Transient (storage), flow	$0 = \underbrace{\frac{k}{\rho c_p} \frac{\partial^2 T}{\partial x^2}}_{\text{conduction}} + \underbrace{\frac{Q}{\rho c_p}}_{\text{generation}}$
Transient conduction	Flow, generation	$\underbrace{\frac{\partial T}{\partial t}}_{\text{transient}} = \underbrace{\frac{k}{\rho c_p} \frac{\partial^2 T}{\partial x^2}}_{\text{conduction}}$
Transient conduction with generation	Flow	$\underbrace{\frac{\partial T}{\partial t}}_{\text{transient}} = \underbrace{\frac{k}{\rho c_p} \frac{\partial^2 T}{\partial x^2}}_{\text{conduction}} + \underbrace{\frac{Q}{\rho c_p}}_{\text{generation}}$
Transient convection without generation	Generation	$\underbrace{\frac{\partial T}{\partial t}}_{\text{transient}} + \underbrace{u \frac{\partial T}{\partial x}}_{\text{convection}} = \underbrace{\frac{k}{\rho c_p} \frac{\partial^2 T}{\partial x^2}}_{\text{conduction}}$
Transient convection with generation	None	$\underbrace{\frac{\partial T}{\partial t}}_{\text{transient}} + \underbrace{u \frac{\partial T}{\partial x}}_{\text{convection}} = \underbrace{\frac{k}{\rho c_p} \frac{\partial^2 T}{\partial x^2}}_{\text{conduction}} + \underbrace{\frac{Q}{\rho c_p}}_{\text{generation}}$

when energy from other forms, like microwaves in an oven or metabolic reactions in tissue, are turning into heat. If heat is being used up, as in an endothermic reaction, Q will be negative. These have to be provided in the problem description. If there is no mention of heat generation/sink in the problem, that term needs to be dropped. If the domain is a solid, the flow term would be dropped, which is true for most of the situations in this book.

3.7.4   Deciding on the Boundary Conditions

This is the third step shown in Figure 3.16. Setting up boundary and initial conditions correctly is just as important as setting up the governing equations. Here are some steps and pointers we follow in choosing boundary conditions:

- The three types of boundary conditions we will deal with are in Section 3.2. They are 1) temperature given, 2) heat flux given, and 3) convection at a surface. We need to choose one of these for each of the boundaries of our domain.

- In all coordinate systems, the governing equation is second order in space. This would need two boundary conditions in space for each of the x, y or z direction.

- Likewise, the governing equation is first order in time and therefore requires one condition in time which is usually specified to be the value at $t = 0$ (called the initial condition).

- Often apparently one boundary condition seems to be provided. The other one is likely to come from symmetry, if that is mentioned. Symmetry has to be in geometry, material, and boundary conditions. If one of them is not true, symmetry cannot be used.

- Notice that two insulated boundary conditions along a particular dimension can be equivalent to dropping that dimension from the heat equation.

- At the same location and along a particular direction, we generally cannot apply two different boundary conditions (e.g., temperature specified and flux specified) simultaneously.

3.7.5 Getting Property Values

The governing equation shows that the properties we generally need are k, ρ, and c_p. However, for restricted situations, when terms are dropped (Table 3.2), one may not need all the properties. For example, for steady-state problems, ρ and c_p are not needed (see the first two formulations in Table 3.2). The reason for this is that a steady-state situation is not dependent on heat capacity, as the amount of stored energy is not changing. Properties are generally provided as part of the problem description. If not, Appendices C (page 559) and D (page 573) have tabulated property values that can be used. Note that, to keep things simple (obtain simple analytical solutions), this book almost always uses constant properties. More realistic answers can be obtained by having properties vary with temperature and other variables, but analytical solutions typically will not be possible and, instead, numerical solutions are easily obtained.

3.7.6 Solution Technique

In the final step (Figure 3.16), we solve the governing equation (GE) together with the boundary conditions (BC). Solutions in this book are all based on analytical solutions, whether done directly by solving the GE and BC or by using resistance-type formulation (see Chapter 4). How to find which solution technique to use? This is illustrated in a map dedicated to Problem Solving at the end of each chapter (e.g., see Figure 4.19 on Page 105 for steady-state heat conduction). Details of several analytical solution techniques are in the appendix (starting page 593) and the reader is encouraged to go over these as needed.

Solutions we obtain are typically in terms of temperature as function of position and/or time. Once we obtain the temperature, we can obtain heat flux from the various formulas (these are statements of Fourier's law) provided in Table 3.3.

Table 3.3: Conductive heat flux in terms of temperature in various coordinate systems

Coordinate system	Formulas for conductive heat flux
Cartesian	$q''_x = -k\frac{\partial T}{\partial x}; \quad q''_y = -k\frac{\partial T}{\partial y}; \quad q''_z = -k\frac{\partial T}{\partial z}$
Cylindrical	$q''_r = -k\frac{\partial T}{\partial r}; \quad q''_z = -k\frac{\partial T}{\partial z}; \quad q''_\theta = -k\frac{1}{r}\frac{\partial T}{\partial \theta}$
Spherical	$q''_r = -k\frac{\partial T}{\partial r}; \quad q''_\theta = -k\frac{1}{r}\frac{\partial T}{\partial \theta}; \quad q''_\phi = -k\frac{1}{r\sin\theta}\frac{\partial T}{\partial \phi}$

3.7.7 *Example: Problem Formulation—Cooling Using an Ice Pack*

A common injury among athletes is that in soft tissue due to impact, and the traditional treatment for this is cold therapy using an ice pack. This treatment is effective, inexpensive, and easily accessible. The most common advice found in medical literature is a 20 minute on, 20 minute off icing cycle. We would like to develop quantitative information on the temperature in the muscle of an arm.

Physical situation This is described in the problem description above with the associated picture (Figure 3.18).

Schematic Here we want to drastically simplify the problem, assuming we only have simple analytical solutions available to us in our toolbox (the simplifications do not need to be as drastic if we also have numerical solutions at our disposal). Although the region's muscle and ice affect each other's temperature, based on our understanding that ice requires a large amount of latent heat to melt it before raising its temperature, we make the assumption that ice stays at a constant cold

Figure 3.18: Re-usable gel pack for hot or cold therapy to bruises. (Melodia plus photos—Shutterstock.com)

pack temperature. This way, we can exclude the ice from our computational domain. Thus, the domain we want to compute is the tissue, since we are interested in its temperature. If, instead, our interest were in the melting of the ice, that would be our domain.

Next we decide on the number of dimensions needed. While all physical situations are 3D, for the purposes of having a simple analytical solution, we want to typically simplify to a 1D problem. Although we do not have complete symmetry (e.g., the ice pack may not be put completely uniformly around the arm or the tissue blood flow may vary from the top to the bottom), we would have to assume symmetry here to be able to drop the variation in the θ direction (more discussed below). Next we note the dimension along which temperature would change the most—based on the physical process, it is expected that temperature will change primarily in the r direction (we are trying to cool the muscle over its thickness) and temperature probably will not change as much in the axial or the z direction. Thus, we will ignore the z direction. These decisions lead to part of the schematic, as shown in the Figure 3.19. The region in the z direction is arbitrary, as explained below. Note that these decisions need to be supported by the following steps.

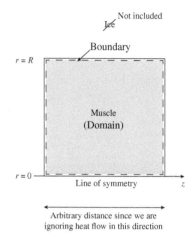

Figure 3.19: Choosing the domain: an intermediate step in making the schematic.

Governing Equation We have three choices of governing equation (see Section 3.5), depending on the coordinate system. Based on the symmetry and the somewhat cylindrical shape, we choose the governing equation in the cylindrical coordinate system:

$$\frac{k}{\rho c_p}\left[\frac{1}{r}\frac{\partial}{\partial r}\left(r\frac{\partial T}{\partial r}\right) + \frac{1}{r^2}\left(\frac{\partial^2 T}{\partial \theta^2}\right) + \frac{\partial^2 T}{\partial z^2}\right] + \frac{Q}{\rho c_p} = \frac{\partial T}{\partial t}$$

We now need to remove the terms not needed (note that the convection term has already been removed). Temperature changes radially so we keep the first term. The symmetry assumption above means T does not vary with θ. Mathematically, this means $\partial T/\partial \theta = 0$ along the circumferential direction that would lead to $\partial^2 T/\partial \theta^2 = 0$, letting us drop the second term. We made the assumption that the hand is long compared to the radius, i.e., temperature does not change as much along the axis. This means $\partial T/\partial z = 0$ along the axis that would lead to $\partial^2 T/\partial z^2 = 0$. This would let us drop the third term on the left-hand side. Note that this assumption may or may not be a good one (there may be significant temperature change along the axis) but to begin, we choose to make this assumption. The schematic includes a region in z but this is arbitrary because, by using $\partial T/\partial z = 0$, we are not allowing any heat flow at the two boundaries in this direction. The term Q would include both metabolic heat generation and that due to blood flow. Finally, since our problem is transient (changing with time), we keep the transient term on the right:

$$\frac{k}{\rho c_p}[\frac{1}{r}\frac{\partial}{\partial r}\left(r\frac{\partial T}{\partial r}\right) + \underbrace{\frac{1}{r^2}\left(\frac{\partial^2 T}{\partial \theta^2}\right)^{\!\!0}}_{\text{symmetry}} + \underbrace{\frac{\partial^2 T}{\partial z^2}^{\!\!0}}_{\text{long cylinder}}] + \frac{Q}{\rho c_p} = \frac{\partial T}{\partial t}$$

which leads to our final governing equation as

$$\frac{k}{\rho c_p}\left[\frac{1}{r}\frac{\partial}{\partial r}\left(r\frac{\partial T}{\partial r}\right)\right] + \frac{Q}{\rho c_p} = \frac{\partial T}{\partial t}$$

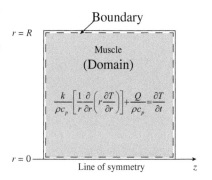

$r = R$

Boundary

Muscle
(Domain)

$\frac{k}{\rho c_p}\left[\frac{1}{r}\frac{\partial}{\partial r}\left(r\frac{\partial T}{\partial r}\right)\right]+\frac{Q}{\rho c_p}=\frac{\partial T}{\partial t}$

$r = 0$

Line of symmetry z

Figure 3.20: Choosing the governing equation or the physics: an intermediate step in making the schematic.

Boundary Conditions Since the above governing equation is second order in r, we need two boundary conditions in r. The two boundaries (ends) in r are $r = 0$ and $r = R$. The symmetry condition at $r = 0$ is written as the first equation below. If the outside of the skin can be assumed to be at the cold pack temperature, T_c, this would lead to the second equation below:

$$\frac{\partial T}{\partial r}\Big|_{r=0} = 0 \ \text{ at } t > 0$$

$$T|_{r=R} = T_c \ \text{ at } t > 0$$

Here we can assume T_c to be a constant since, otherwise, it would get difficult to do an analytical solution. This assumption is not required if we are doing a numerical solution. Note that, at the same boundary, we need *one and only one boundary condition* (temperature, flux, or convection). For example, if the outer boundary has temperature specified, we cannot also simultaneously have a convection condition at the same boundary. This is both a physical and mathematical impossibility. We do not need any more boundary conditions to solve the above governing equation because we are really solving along a line (r direction). In this sense, our geometry should really be a line, but sometimes a 2D representation communicates better visually, as we have done here. However, if we keep our schematic as 2D, we should think of the boundary conditions in the z direction as well. Since the temperature is not changing along the z direction, $\partial T/\partial z = 0$, which leads to heat flux ($= -k\partial T/\partial z$) = 0. This is the implied boundary condition in the z direction when we dropped the z term from the governing equation.

Since the governing equation also has a time derivative (it is transient), we need a condition for time, which is called the initial condition. If we are looking to obtain an analytical solution, this can be simply

$$T = T_i \text{ for all } 0 < r < R$$

where T_i is the constant initial temperature.

Solution Technique In this chapter, we are not discussing solution techniques. It is, however, important to note that a broad assumption of a simple analytical solution approach was implicit in the above formulation. The other approach, a numerical solution that typically would use computational software, would have allowed us to be a lot more flexible in specifying the domain and its properties, governing equation, boundary conditions, and initial condition. For example, we would not have to assume symmetry, constant boundary temperature, or constant initial temperature.

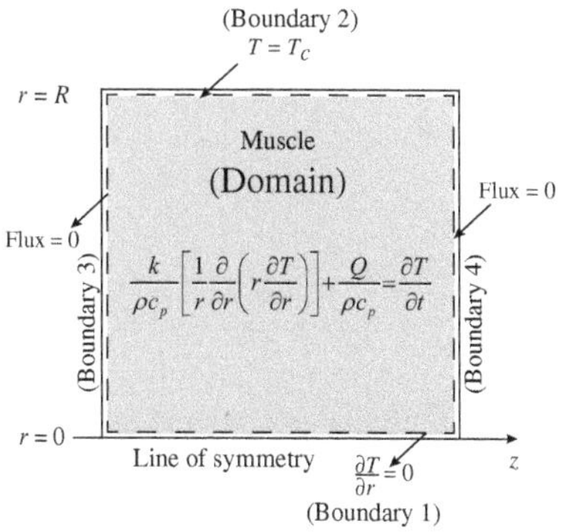

Figure 3.21: Choosing the boundary condition: the final step in making the schematic.

3.8 Problem Solving: A General Approach

Can we learn problem solving? One of the goals of this book is to provide a method to problem solving that can be useful far beyond heat and mass transfer. Figure 3.22 shows the generic, big picture of problem solving, of which the chart in Figure 3.16 is a part. Although we will apply them to heat and mass transfer problems, the concepts in Figure 3.22 come from a generic problem solving approach that cuts across disciplines, heat and mass transfer being just one application. *The reader is encouraged to see this generic aspect.* The generic steps shown in this figure are as follows: comprehend and formulate the problem, generate alternate solutions and select the most feasible solution among the alternates, implement the chosen solution, and, finally, evaluate and interpret the solution. Throughout the book, worked-out examples will follow these generic steps.

3.9 Chapter Summary—Governing Equations and Boundary Conditions

- **Governing Equation (page 41)**

  1. It is a mathematical statement of energy conservation. It is obtained by combining conservation of energy with Fourier's law for heat conduction.

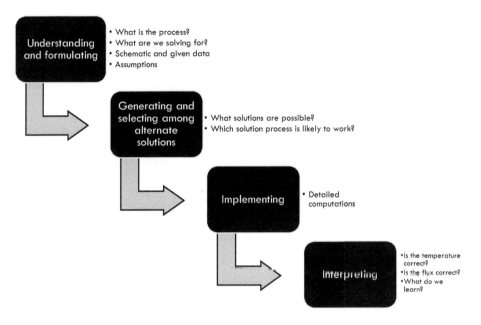

Figure 3.22: Generic steps in problem solving with some questions that adapt to heat and mass transfer. Figure 3.16 belongs to the third step here of *Implementing* when our chosen strategy is developing an analytical solution from scratch (i.e., GE and BC).

 2. Depending on the appropriate geometry of the physical problem, choose a governing equation in a particular coordinate system from the equations in Section 3.5.

 3. Different terms in the governing equation can be identified with conduction, convection, generation, and storage. Depending on the physical situation, some terms may be dropped.

- **Boundary Conditions (page 47)**

 1. Boundary conditions are the conditions at the surfaces of a body.

 2. Initial conditions are the conditions at time $t = 0$.

 3. Boundary and initial conditions are needed to solve the governing equation for a specific physical situation.

 4. One of the following three types of heat transfer boundary conditions typically exists on a surface:

(a) Temperature at the surface is specified (Eq. 3.12)

(b) Heat flux at the surface is specified (Eq. 3.13)

(c) Convective heat transfer condition at the surface (Eq. 3.16)

- **Problem Formulation (page 57)**

    1. It is the development of a mathematical formulation of a physical problem, written in terms of governing equation and boundary conditions.

    2. Follow the steps as shown in Figure 3.16.

3.10   Concept and Review Questions

1. Write Fourier's law of heat conduction using spherical coordinates.

2. What are the assumptions in the bioheat transfer equation?

3. What is meant by the term *one dimensional* when applied to conduction problems?

4. How is the generation of energy different from storage?

5. Provide physical reasoning for why the convective boundary condition approaches the temperature-specified boundary condition for a high heat transfer coefficient.

6. What geometry (spherical, rectangular, etc.) would you choose to analyze heat transfer in the soil near the earth's surface? Why?

7. What are the three types of boundary conditions that we have discussed? You can write in words or equations.

8. Give an example of an energy transfer situation in a solid where, by changing from one type of boundary condition to another, temperatures will vary in a different manner altogether.

9. The bioheat equation is only valid in tissue regions that do not have large blood vessels. Provide the main reasoning for this, referring to the terms in the equation.

10. The human body can be considered analogous to a slab with heat generation inside and heat loss from the surface to the ambient. What is the most important difference between a slab and a human body? (Hint: think of the effect of changes in the ambient temperature.)

11. What two basic physical laws are used to come up with the heat equation?

12. In the bioheat transfer equation

$$\rho c_p \frac{\partial T}{\partial t} = k \frac{\partial^2 T}{\partial x^2} + \underbrace{\rho_b c_b \dot{V}_b^v (T_a - T)}_{\text{term 2}} + Q$$

a) What does the term 2 on the right-hand side represent? b) Does the above equation for the tissue include large arteries? Explain why or why not.

13. When we change boundary conditions but use the same governing equation, we can expect

- Answers that are close
- Completely different answers
- Answers that can look different but can be shown to be equivalent

14. In Figure 3.23, show some of the regions that should not be modeled using the bioheat equation and explain why.

15. How many boundary and initial conditions do you need to find a solution in each of the following cases?

$$\rho c_p \frac{\partial T}{\partial t} = k \frac{\partial^2 T}{\partial x^2} \tag{3.25}$$

$$\frac{\partial T}{\partial t} = -\frac{hA}{mc_p}(T - T_\infty) \tag{3.26}$$

$$0 = k \frac{1}{r} \frac{\partial}{\partial r}\left(r \frac{\partial T}{\partial r}\right) \tag{3.27}$$

$$\rho c_p \frac{\partial T}{\partial t} = k \left(\frac{\partial^2 T}{\partial x^2} + \frac{\partial^2 T}{\partial y^2}\right) \tag{3.28}$$

16. Is a cold-blooded animal with metabolic heat in equilibrium or steady state with its surroundings?

17. What type of a boundary condition would you use to include *just* the effect of the sun shining on a surface?

18. Why are large vessels not included in the bioheat equation (two lines max)?

Figure 3.23: Arteries.

Further Reading

Ben-Hassan, R. M., A. E. Ghaley, and N. Ben-Abdallah. 1993. Heat generation during batch and continuous production of single cell protein from cheese whey. *Biomass and Bioenergy* 4(3):213–225.

Bird, R. B., W. E. Stewart, and E. N. Lightfoot. 1960. *Transport Phenomena*. John Wiley & Sons, New York. (Comprehensive source for governing equations.)

Carslaw, H. S. and J. C. Jaeger. 1959. *Conduction of Heat in Solids*. Oxford University Press, Oxford, UK. (Comprehensive source for analytical solution of the energy equation.)

Charny, C. K. 1992. Mathematical models of bioheat transfer. *Advances in Heat Transfer* 22:19–153.

Chen, M. 1985. The tissue energy balance equation. In: *Heat Transfer in Medicine and Biology*, edited by A. Shitzer and R. C. Eberhart. Plenum Press, New York.

Datta, A. K. and V. Rakesh. 2009. *An Introduction to Modeling of Transport Processes: Applications to Biomedical Systems*. Cambridge University Press, Cambridge, UK.

Diller, K. R. 1992. Modeling of bioheat transfer processes at high and low temperatures. *Advances in Heat Transfer*, 22:157–357. (Comprehensive source for bioheat transfer processes related to biomedical and related applications.)

Eberhart, R. C. 1985. Thermal models of single organs. In: *Heat Transfer in Medicine and Biology*, edited by A. Shitzer and R. C. Eberhart. Plenum Press, New York.

Ozisik, M. N. 1980. *Heat Conduction*. John Wiley & Sons, New York. (Solutions of the energy equation in rectangular, cylindrical, and spherical coordinate system.)

3.11   Problems

3.1   Radial Heat Transfer in a Sphere

Using the control volume shown in Figure 3.15, derive the governing equation for transient heat transfer in a sphere in the radial direction only (Eq. 3.23).

3.2   Equivalence between Boundary Conditions

You have been introduced to the following two boundary conditions at the surface.

$$T|_{\text{surface}} \quad = \quad T_\infty \tag{3.29}$$

$$-k \left. \frac{\partial T}{\partial x} \right|_{\text{surface}} \quad = \quad h(T|_{\text{surface}} - T_\infty) \tag{3.30}$$

1) Show that the two boundary conditions are equivalent for large values of h. 2) Give physical reasoning as to why you can't use Equation 3.29 when h is small.

Problem Formulations

Problem formulation is one of the most important concepts that we should learn in preparing to solve practical problems. Often much of the problem is already formulated in an undergraduate class, as in typical homework. Here we will practice problem formulation. For each of the problems below, you need to provide 1) a schematic, 2) governing equation (do not solve), and 3) boundary and/or initial conditions. There is no right or wrong answer. The final governing equation and boundary conditions should only have terms relevant to this problem, i.e., you should not leave them as the general equations having all the terms. You will not be penalized for the degree to which you simplify. You are free to assume any parameter that you need, and you may need to make many approximations, but you should include a few words of justification for every assumption you make. To do these problems in the most realistic way, you need to know a lot more of the processes—this is not the goal here. You only need to be consistent so that your final set of governing equations and boundary conditions is solvable.

3.3   Problem Formulation for Problem 4.8.18

(See instructions under Problem Formulations above.)

Chapter 4

CONDUCTION HEAT TRANSFER: STEADY STATE

CHAPTER OBJECTIVES

After you have studied this chapter, you should be able to

1. Formulate and solve for a conductive heat transfer process in common geometries where temperature does not change with time (steady state).

2. Extend the steady state heat transfer concept for a slab to a composite slab using the overall heat transfer coefficient.

3. Extend the steady-state heat transfer in a slab to include internal heat generation.

4. Formulate and solve for steady-state conductive heat transfer in a fin.

KEY TERMS

- **steady state**
- **thermal resistance**
- **thermal resistances in series**
- **overall heat transfer coefficient**
- **heat transfer with energy generation**
- **thermoregulation and core body temperature**
- **extended surfaces—fins**

This chapter will deal with steady-state heat transfer. Steady state is defined as the situation when variables such as temperature are not changing with time. Note that this is not the same as equilibrium, when the system is at the same temperature everywhere. In steady state, heat can continue to flow if there is a temperature difference. It's just that the temperature or heat flow is not a function of time. The relationship of this chapter to other chapters in energy transfer is shown in Figure 4.1.

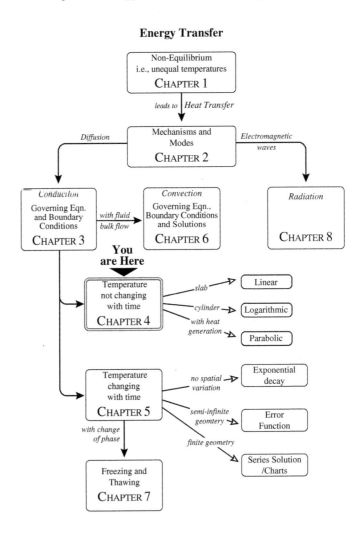

Figure 4.1: Concept map of energy transfer showing how the contents of this chapter relate to other chapters on energy transfer.

Although temperatures can vary along three spatial variables, for simplicity, we will consider only the situations that can be simplified to variations along one dimension.

4.1 Steady-State Heat Conduction in a Slab

Although all geometries are three-dimensional in reality, under certain conditions variations in two of the dimensions can be neglected. In a slab geometry, as shown in Fig. 4.2, if the surface temperatures are uniform, at locations not too close to the edges the temperature variation is only along the thickness of the slab. In other words, the heat transfer is one dimensional. For no heat generation and steady-state heat transfer, the general governing equation

$$\underbrace{\cancel{\frac{\partial T}{\partial t}}^{0}}_{\text{steady state}} + \underbrace{\cancel{u\frac{\partial T}{\partial x}}^{0}}_{\text{no bulk flow}} = \underbrace{\frac{k}{\rho c_p}\frac{\partial^2 T}{\partial x^2}}_{\text{conduction}} + \underbrace{\cancel{\frac{Q}{\rho c_p}}^{0}}_{\text{no generation}}$$

becomes

$$\frac{d^2 T}{dx^2} = 0 \tag{4.1}$$

after eliminating the terms. For the simplest kind of boundary condition where constant temperatures of T_1 and T_2 can be assumed at the two surfaces, we can write the boundary conditions as

$$T(x = 0) = T_1 \tag{4.2}$$
$$T(x = L) = T_2 \tag{4.3}$$

The solution to Eq. 4.1 is given by (see Example 3.1.4)

$$T = \frac{T_2 - T_1}{L}x + T_1 \tag{4.4}$$

which shows a linear change in temperature from T_1 to T_2 at steady state. The heat flow is given by

$$q_x = -kA\frac{dT}{dx}$$

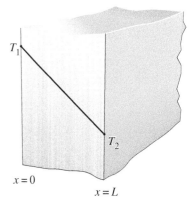

Figure 4.2: A linear temperature profile at steady state in a slab geometry.

since the temperature profile is linear (dT/dx is constant),

$$
\begin{aligned}
q_x &= -kA\frac{T_2 - T_1}{L - 0} \\
&= kA\frac{T_1 - T_2}{L} \\
&= \frac{T_1 - T_2}{L/kA}
\end{aligned}
\tag{4.5}
$$

which is the same at all locations x along the thickness.

4.1.1   Steady-State Heat Conduction in a Slab: Effect of Boundary Conditions

Boundary conditions change the problem and thus affect the temperature profile and heat flux. Thus, the temperature profiles and heat fluxes obtained above will change if the boundary conditions are changed in the same slab. These results are provided in Table 4.1. The reader will find it instructive to derive these solutions, which require only a small amount of effort.

4.1.2   One-Dimensional Conduction through a Composite Slab— Thermal Resistance and Overall Heat Transfer Coefficient

Quite often in practice the slab geometry studied in the previous chapter occurs with two or more materials in series, as shown in Figure 4.3. Examples can be heat transfer through layers of skin, fat, and muscle, all having different properties. Another common example is the wall of a house in a cold climate that can have layers of wood, insulation, sheet rock, etc. Here again, if the surface temperatures are uniform at locations not too close to the edges, the temperature variation is one dimensional. If we ignore heat generation and consider steady-state heat transfer, the same constant heat flow q_x appears at any section. For conduction heat transfer through the solid layers, we can write, using the schematic in Figure 4.3

$$
q_x = \frac{T_1 - T_2}{\frac{L_1}{k_1 A}}
\tag{4.6}
$$

$$
q_x = \frac{T_2 - T_3}{\frac{L_2}{k_2 A}}
\tag{4.7}
$$

$$
q_x = \frac{T_3 - T_4}{\frac{L_3}{k_3 A}}
\tag{4.8}
$$

Table 4.1: Steady-state temperatures profiles and heat fluxes in a slab for the three boundary conditions on its left face. The right face always has the temperature-specified boundary condition. Note heat flux is same at all positions, x.

Boundary Condition	Temperature Profile	Heat flux	Notes
T_1 $x=0$ $x=L$ T_2	$T = (T_2 - T_1)\frac{x}{L} + T_1$	$q_x'' = -k\frac{dT}{dx}$ $= -k\frac{(T_2 - T_1)}{L}$	Flux is set by the two boundary temperatures
T_1 q_s'' $x=0$ $x=L$ T_2	$T = \frac{q_s''}{k}(L - x) + T_2$	$q_x'' = q_s''$	Flux is that given on the left boundary
T_1 T_∞ $x=0$ $x=L$ T_2	$T = (-\frac{h}{k}(T_2 - T_\infty)x + T_2 - \frac{hL}{k}T_\infty) / (1 - hL/k)$	$q_x'' = -\frac{(h/k)(T_2 - T_\infty)}{1 - hL/k}$	Flux depends on fluid temperature, fluid h and right side boundary temperature

For the convective heat transfer at the two surfaces, the same heat flow q_x can be written in terms of the heat transfer coefficients h,

$$q_x = \frac{T_h - T_1}{\frac{1}{h_h A}} \tag{4.9}$$

$$q_x = \frac{T_4 - T_c}{\frac{1}{h_c A}} \tag{4.10}$$

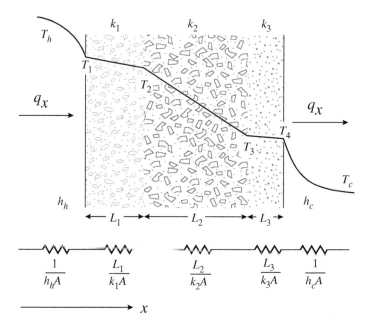

Figure 4.3: Temperature profile and thermal resistances for steady-state conduction through a composite slab with convection at the two surfaces.

Equations 4.6 through 4.10 can be written as

$$T_1 - T_2 = q_x \frac{L_1}{k_1 A}$$

$$T_2 - T_3 = q_x \frac{L_2}{k_2 A}$$

$$T_3 - T_4 = q_x \frac{L_3}{k_3 A}$$

$$T_h - T_1 = q_x \frac{1}{h_h A}$$

$$T_4 - T_c = q_x \frac{1}{h_c A}$$

Adding the left and the right hand side of the equations, we get

$$T_h - T_c = q_x \left(\frac{1}{h_h A} + \frac{L_1}{k_1 A} + \frac{L_2}{k_2 A} + \frac{L_3}{k_3 A} + \frac{1}{h_c A} \right)$$

Rewriting the equation in terms of flow, we get

$$q_x = \frac{T_h - T_c}{\underbrace{\frac{1}{h_h A}}_{\text{convective}} + \underbrace{\frac{L_1}{k_1 A} + \frac{L_2}{k_2 A} + \frac{L_3}{k_3 A}}_{\text{conductive}} + \underbrace{\frac{1}{h_c A}}_{\text{convective}}} \tag{4.11}$$

$$= \frac{\text{temperature difference}}{\sum \text{thermal resistance}} \tag{4.12}$$

In order to understand Eq. 4.12, consider the analogy from electricity where the current flow is given by

$$\text{current flow} = \frac{\text{total potential difference}}{\sum \text{electrical resistance}} \tag{4.13}$$

From the similarity of the two equations, it can be said that, if current flow is analogous to heat flow and temperature is analogous to potential, then the terms in the denominator in Eq. 4.12 denote thermal resistances. The terms $1/h_h A$ and $1/h_c A$ represent thermal resistance due to convection in the fluid. Similarly, the term $L_1/k_1 A$ represents thermal resistance due to the first layer of solid. The heat transfer system depicted in Figure 4.3 is analogous to an electrical system with resistances in series. As a further analogy to electricity, we note that the inverse of resistance is conductance. Applying this to heat transfer, we find that the inverse of the resistance $1/hA$ or hA or simply h (the heat transfer coefficient) is like a conductance. Just as higher conductance increases the current flow, higher h leads to higher heat flow.

R-value as a common measure of thermal resistance In everyday language, the concept of thermal resistance due to a layer of solid is often expressed as the R-value, defined in terms of the thickness of the layer, L, and its thermal conductivity, k, as

$$R\text{-value} = \frac{L}{k} \tag{4.14}$$

which is the thermal resistance per unit area. Although the R-value would be measured in SI units in $\text{m}^2 \cdot \text{K/W}$, in practice (in the U.S.) it is often assumed to have the units of $\text{hr} \cdot \text{ft}^2 \cdot {}^\circ\text{F/Btu}$. For example, a building materials store may tell you that you need an R-value of 30 for the ceiling while you need an R-value of 11 for the outside walls. It makes sense to talk about the *R-value* or thermal resistance (per unit area) rather than thickness of insulation, because, if you are only told about thickness of insulation, you would also need to know its thermal conductivity to find its resistance. *R-value* is the final parameter, thermal resistance, that includes both thickness and thermal conductivity.

Overall heat transfer coefficient   Instead of dealing with the different layers of resistances, as in Eq. 4.12, it is sometimes useful to talk about a total or overall resistance. If we define an overall heat transfer coefficient U by the equation

$$q_x = UA(T_h - T_c)$$
$$= \frac{T_h - T_c}{\frac{1}{UA}} \tag{4.15}$$

then, in order for Eq. 4.15 and Eq. 4.11 to be the same, U has to be given by

$$\frac{1}{UA} = \frac{1}{h_h A} + \frac{L_1}{k_1 A} + \frac{L_2}{k_2 A} + \frac{L_3}{k_3 A} + \frac{1}{h_c A} \tag{4.16}$$

Equation 4.16 shows that the overall heat transfer coefficient U is made up of individual conductive and convective resistances, i.e.,

$$R_{eff} = R_h + R_1 + R_2 + R_3 + R_c \tag{4.17}$$

Oftentimes in practice a U value is given for a heat transfer situation, without mentioning the individual convective and conductive components that make up such a value. In such a situation, only Eq. 4.15 is needed to calculate the total heat transfer q_x.

Resistances in series   The above development, leading to Eq. 4.17, is when the resistances are in series, i.e., heat flows through each of the layers, one after the other. The individual resistances add up to make the total resistance.

Resistances in parallel   The series arrangement of resistances is not the only possible arrangement. Another distinct arrangement is when the resistances are in parallel (Figure 4.4).

The basic difference between the two arrangements are that, whereas the rate of heat flow is the same through all the resistances in series at steady state, for a parallel arrangement, heat flow is divided into the respective parts—together they add up to the total heat flow. Thus, the analysis of this arrangement will also involve setting up individual heat flows, adding them up to get the total heat flow, and writing the total heat flow in terms of an effective resistance that will provide the same flow.

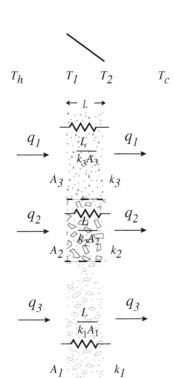

Figure 4.4: Temperature profile and thermal resistances for steady-state conduction through three slabs in parallel with convection at the two surfaces.

Heat flows through the three individual layers in Figure 4.4 are given by

$$q_1 = \frac{T_1 - T_2}{\frac{L}{k_1 A_1}} \tag{4.18}$$

$$q_2 = \frac{T_1 - T_2}{\frac{L}{k_2 A_2}} \tag{4.19}$$

$$q_3 = \frac{T_1 - T_2}{\frac{L}{k_3 A_3}} \tag{4.20}$$

Adding up both sides and noting $q_1 + q_2 + q_3 = q_x$, the total heat flow,

$$q_x = (T_1 - T_2)\left(\frac{1}{L/(k_1 A_1)} + \frac{1}{L/(k_2 A_2)} + \frac{1}{L/(k_3 A_3)}\right) \tag{4.21}$$

$$= (T_1 - T_2)\left(\frac{1}{R_1} + \frac{1}{R_2} + \frac{1}{R_3}\right) \tag{4.22}$$

where R_1, R_2, and R_3, respectively, are the resistances of the individual sections. The same heat flow, q_x, can be written in terms of an effective thermal resistance, R_{eff}, of the three sections combined:

$$q_x = \frac{T_1 - T_2}{R_{eff}} \tag{4.23}$$

Equating the right-hand sides of this and Equation. 4.22,

$$\frac{1}{R_{eff}} = \frac{1}{R_1} + \frac{1}{R_2} + \frac{1}{R_3} \tag{4.24}$$

The form of this equation, how the reciprocals of the resistances are added to get the reciprocal of the effective resistance, provides us with a simple way to calculate the effective resistance of several resistances in parallel. The resistances are in parallel because heat flow is occurring independently through each one, unlike the series arrangement where heat flow through one layer depends on heat flow through the other layers. Contrast Eq. 4.24 with Eq. 4.17, where the resistances were in series.

Instead of thinking in terms of an effective thermal resistance of the three sections, we can also think in terms of their effective thermal conductivity. The same heat flow can be written in terms of an effective thermal conductivity, k_{eff}, and total area, $A = A_1 + A_2 + A_3$, as

$$q_x = \frac{T_1 - T_2}{\frac{L}{k_{eff} A}} \tag{4.25}$$

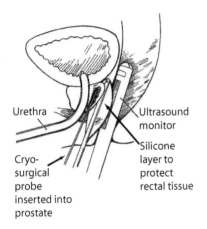

Figure 4.5: Cryosurgery. Adapted from Bischof et al. (1997).

Urethra

Ultrasound monitor

Silicone layer to protect rectal tissue

Cryo-surgical probe inserted into prostate

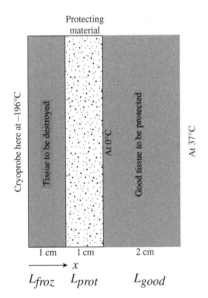

Figure 4.6: Schematic showing the two tissue layers and the protection layer.

Equating the right-hand sides of Eq. 4.21 and Eq. 4.25,

$$\frac{k_{eff} A}{L} = \frac{k_1 A_1}{L} + \frac{k_2 A_2}{L} + \frac{k_3 A_3}{L} \tag{4.26}$$

Simplifying, the effective thermal conductivity, k_{eff}, of the parallel arrangement of the three sections having thermal conductivities of k_1, k_2, and k_3, respectively, is given by

$$k_{eff} = k_1 \frac{A_1}{A} + k_2 \frac{A_2}{A} + k_3 \frac{A_3}{A} \tag{4.27}$$

4.1.3 *Example: Thermal Protection during Cryosurgery*

Consider a cryosurgery problem as shown in Figure 4.5. The cryosurgical probe maintains a temperature of $-196°C$ on the tissue to be frozen and thus destroyed. In order to protect good tissue, a layer of protecting material such as silicone is inserted between the tissue to be destroyed and the good tissue. For the purpose of this problem, consider the layers to be modeled as slabs (see Figure 4.6), and the far end of the good tissue to be maintained at the body temperature of 37 °C. The thermal conductivities of the frozen and unfrozen tissues are 2.0 W/m · K and 0.552 W/m · K, respectively. Assume the heat transfer process reaches a steady state (which may not be true as the surgical procedure may not last that long). 1) What should the thermal conductivity of the protecting material be so that the temperature of the good tissue is 0°C at its coldest point? 2) Some researchers show that complete destruction of tissue can only be guaranteed at temperatures lower than $-50°C$. Where is the $-50°C$ temperature located in this problem (so we know all tissue on the left layer is indeed frozen)?

Solution

Understanding and formulating the problem *1) What is the process?* Heat is being conducted at steady state from the warmer tissue to the colder cryoprobe through three layers that are touching each other—frozen tissue, silicone, and unfrozen tissue. *2) What are we solving for?* The thermal conductivity of the protecting layer and the location where the temperature is $-50°C$. *3) Schematic and given data:* A schematic is shown in Figure 4.6 with some of the given data superimposed on it. *4) Assumptions:* The thicknesses of the slabs are small compared to the other dimensions so they can be considered one dimensional. Thermal properties do not vary with temperature.

Generating and selecting among alternate solutions *1) What solutions are possible?* For steady-state heat transfer, the solutions we have available are shown in the solution chart in Figure 4.19. *2) What approach is likely to work?* The domain of our interest is the combined thickness of the three slabs. As we look at the solution chart in Figure 4.19, we stay to the left side since there is no heat generation mentioned. The problem does not ask us to start from scratch, so a resistance analysis should work. Note that the resistance analysis is a shortcut that is a simple rewrite of the solution from scratch (Eq. 4.5, which can be interpreted as flow equal to temperature difference divided by resistance, is the solution from scratch). Resistance analysis being generally less work than from scratch, it is worth trying. As we will see below, problems involving multiple layers are easier to solve using the resistance formulation.

Implementing the chosen solution 1) Continuing with the resistance analysis in Figure 4.19, we notice that that there are three resistances (frozen tissue, silicone, and unfrozen tissue); they are in slab geometry, and they are in series. A critical piece of information is that, at steady state, the same amount of heat flows through each of the layers. Thus, it seems we can approach this in one of two ways. If q'' is the heat flux, we can write

Strategy 1

$$q'' = \frac{T_{x=4} - T_{x=0}}{\frac{L_{froz}}{k_{froz}} + \frac{L_{prot}}{k_{prot}} + \frac{L_{good}}{k_{good}}} \tag{4.28}$$

We cannot solve for k_{prot} directly from this equation since q'' is also unknown here. But q'' can be written in terms of any of the individual resistances. It makes sense to write q'' in terms of a layer where the temperatures as well as resistance are known. We observe that this is the case with good tissue since $T_{x=2}$ is provided to be 0°C. Writing in terms of the resistance of the good tissue,

$$q'' = \frac{T_{x=4} - T_{x=2}}{\frac{L_{good}}{k_{good}}} \tag{4.29}$$

Equating the two,

$$\frac{T_{x=4} - T_{x=0}}{\frac{L_{froz}}{k_{froz}} + \frac{L_{prot}}{k_{prot}} + \frac{L_{good}}{k_{good}}} = \frac{T_{x=4} - T_{x=2}}{\frac{L_{good}}{k_{good}}} \tag{4.30}$$

where the only unknown is L_{prot}. Substituting numerical values,

$$\frac{37 - (-196) \text{ K}}{\frac{0.01 \text{ m}}{2 \text{ W/m·K}} + \frac{0.01 \text{ m}}{k_{prot}} + \frac{0.02 \text{ m}}{0.552 \text{ W/m·K}}} = \frac{37 - 0 \text{ K}}{\frac{0.02 \text{ m}}{0.552 \text{ W/m·K}}} \tag{4.31}$$

From which we solve for k_{prot} as

$$k_{prot} = 0.0535 \frac{\text{W}}{\text{m} \cdot \text{K}} \qquad (4.32)$$

Strategy 2

A completely equivalent way of doing the above is to write the fluxes for each layer. For the tissue to be frozen,

$$q'' = \frac{T_{x=1} - T_{x=0}}{\frac{L_{froz}}{k_{froz}}} \qquad (4.33)$$

For the protecting layer,

$$q'' - \frac{T_{x=2} - T_{x=1}}{\frac{L_{prot}}{k_{prot}}} \qquad (4.34)$$

and finally for the tissue to be protected,

$$q'' = \frac{T_{x=4} - T_{x=2}}{\frac{L_{good}}{k_{good}}} \qquad (4.35)$$

The three equations above have three unknowns as q'', $T_{x=1}$, and k_{prot}, and thus can be used to solve for k_{prot}.

2) To locate the point where the temperature is $-50°C$, we know that, for steady state in a slab, the temperature profile is linear (same heat flux at any location). To find the temperature profile, we first find the junction temperature, $T_{x=1}$. Again, since the heat flux is the same at every section, we can write heat flux in terms of $T_{x=1}$ (Eq. 4.33 or Eq. 4.34) and equate it to the known heat flux from Eq. 4.35. Thus, using Eq. 4.33 and Eq. 4.35, we write

$$\frac{T_{x=1} - T_{x=0}}{\frac{L_{froz}}{k_{froz}}} = \frac{T_{x=4} - T_{x=2}}{\frac{L_{good}}{k_{good}}}$$

Substituting numerical values,

$$\frac{T_{x=1} - (-196)}{\frac{0.01 \text{ m}}{2 \text{ W}/\text{m} \cdot \text{K}}} = \frac{37 - 0}{\frac{0.02 \text{ m}}{0.552 \text{ W}/\text{m} \cdot \text{K}}} \qquad (4.36)$$

and simplifying,

$$T_{x=1} = -190.9 \qquad (4.37)$$

The $-50°C$ temperature location, being between $0°C$ and $-190.9°C$, lies in the middle (insulation) layer. To locate this, we again use the condition that, at steady state in a slab, flux at any point is the same. We can calculate this flux at the location where the temperature is $-50°C$ and equate it to the known flux from Eq. 4.36:

$$\frac{-50 - (-190.9)}{\dfrac{(x-1)\text{ m}}{0.0535\text{ W/m} \cdot \text{K}}} = \frac{37 - 0}{\dfrac{0.02\text{ m}}{0.552\text{ W/m} \cdot \text{K}}}$$

Solving for x, we get

$$x = 0.01738\text{ m}$$
$$= 1.738\text{ cm}$$

i.e., 0.738 cm into the protecting layer.

Evaluating and interpreting the solution *1) Do the computed values make sense?* The calculated insulation value of $0.552\text{ W/m} \cdot \text{K}$ is within the range for thermal conductivity of solids in Figure 2.3, although that range is huge. We can probably only say that the value is not obviously wrong. *2) What do we learn?* Plugging in k_{prot} and other values in Eq. 4.28, we get

$$q'' = \frac{37 - (-196)\text{ K}}{\dfrac{0.01\text{ m}}{2\text{ W/m·K}} + \dfrac{0.01\text{ m}}{0.0535\text{ W/m·K}} + \dfrac{0.02\text{ m}}{0.552\text{ W/m·K}}}$$

$$= \frac{37 - (-196)\text{ K}}{\underbrace{0.005\text{ m}^2 \cdot \text{K/W}}_{\substack{\text{Resistance of} \\ \text{frozen tissue}}} + \underbrace{0.187\text{ m}^2 \cdot \text{K/W}}_{\substack{\text{Resistance of} \\ \text{protector}}} + \underbrace{0.036\text{ m}^2 \cdot \text{K/W}}_{\substack{\text{Resistance of} \\ \text{good tissue}}}}$$

$$= 1021\text{ W/m}^2$$

As expected, the resistance of the protecting insulation layer is considerably higher than either the frozen or the normal tissue.

4.2 Steady-State Heat Conduction in a Cylinder

Like the slab geometry we just studied, another simple geometry that is of considerable importance is a solid or a hollow cylinder, as shown in Figure 4.7. Pipes carrying hot water or steam inside with colder air outside are perhaps the most common example of heat transfer in such a geometry. If the temperatures are uniform on the cylinder surface, temperature gradients are only in the radial direction and therefore heat transfer is one dimensional. The general governing equation in the cylindrical coordinate (Eq. 3.22) is simplified for steady state, no bulk flow, no heat generation, and heat transfer only in the radial direction as

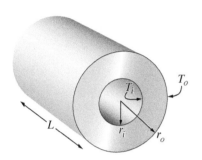

$$\underbrace{\rho c_p \cancel{\frac{\partial T}{\partial t}}^{0}}_{\text{steady state}} = k\left(\frac{1}{r}\frac{\partial}{\partial r}\left(r\frac{\partial T}{\partial r}\right) + \underbrace{\frac{1}{r^2}\cancel{\left(\frac{\partial^2 T}{\partial \phi^2}\right)}^{0}}_{\substack{\text{radial}\\\text{symmetry}}} + \underbrace{\cancel{\frac{\partial^2 T}{\partial z^2}}^{0}}_{\substack{\text{no axial}\\\text{variation}}}\right) + \underbrace{\cancel{\dot{Q}}^{0}}_{\substack{\text{no heat}\\\text{generation}}} \quad (4.38)$$

to obtain the specific governing equation for this problem as

Figure 4.7: Schematic of a hollow cylinder with heat transfer only in the radial direction.

$$\frac{1}{r}\frac{d}{dr}\left(r\frac{dT}{dr}\right) = 0 \qquad (4.39)$$

and the boundary conditions when the inner and outer surfaces are at two different temperatures are

$$T(r = r_0) = T_0 \qquad (4.40)$$
$$T(r = r_i) = T_i \qquad (4.41)$$

To solve Eq. 4.39, it is rewritten as

$$\frac{d}{dr}\left(r\frac{dT}{dr}\right) = 0$$

and integrated once to obtain

$$r\frac{dT}{dr} = c_1$$

Dividing throughout by r, we get

$$\frac{dT}{dr} = \frac{c_1}{r}$$

Integrating for a second time, we get

$$T = c_1 \ln r + c_2 \tag{4.42}$$

where c_1 and c_2 are integration constants. Applying Eqs. 4.40 and 4.41 as the boundary conditions, we get

$$T_0 = c_1 \ln r_0 + c_2$$
$$T_i = c_1 \ln r_i + c_2$$

After solving for c_1 and c_2 from these two equations, they are plugged into Eq. 4.42 to obtain

$$T = T_i - \frac{T_i - T_o}{\ln \frac{r_o}{r_i}} \ln \frac{r}{r_i} \tag{4.43}$$

4.2.1 Thermal Resistance Term for the Hollow Cylinder

Heat flow in Fig. 4.7 is through a cylindrical surface whose surface area increases with radius r. Thus, the effective thermal resistance of the wall is not the same as for a slab of the same thickness where the surface area is constant at any position into the slab. To obtain an expression for the thermal resistance of the hollow cylinder, we write the equation for heat flow as

$$
\begin{aligned}
q_r &= -kA \frac{dT}{dr} \\
&= -2k\pi r L \frac{dT}{dr} \\
&= -2k\pi r L \left(\frac{-(T_i - T_o)}{\ln \frac{r_o}{r_i}} \right) \frac{r_i}{r} \frac{1}{r_i} \\
&= -k \frac{2\pi L (T_o - T_i)}{\ln \frac{r_o}{r_i}} \\
&= \frac{\overbrace{T_i - T_o}^{\text{Driving force}}}{\underbrace{\frac{\ln \frac{r_o}{r_i}}{2\pi k L}}_{\text{Resistance}}}
\end{aligned}
\tag{4.44}
$$

which shows that the thermal resistance term for the hollow cylinder is given by the denominator

$$\frac{\ln \dfrac{r_o}{r_i}}{2\pi k L} \tag{4.45}$$

Note that the thermal resistance term given by Eq. 4.45 is not the *R-value* for the hollow cylinder. Whereas thermal resistance is defined in terms of heat flow, the *R-value* is defined in terms of heat flux or heat flow per unit area (see Eq. 4.14 for a slab). The *R-value* for the hollow cylinder will be defined in Section 4.2.3 below.

4.2.2   Comparison with the Solution for the Slab

The temperature profile for steady-state heat transfer in a hollow cylinder given by Eq. 4.43 is non-linear, unlike the linear temperature profile in a slab given by Eq. 4.4. This is illustrated graphically in Figure 4.8. To see why the temperature profiles have to be different, note that the cross-sectional area $2\pi r L$, through which heat flows increases with radius r. Therefore, the temperature gradient dT/dr has to decrease with radius r to keep the heat flow the same at steady state, which is in fact the case since $dT/dr = (T_0 - T_i)/r \ln \frac{r_0}{r_i}$. In other words,

$$
\begin{aligned}
q_r &= -kA\frac{dT}{dr} \\
&= -k\underbrace{(2\pi r L)}_{\substack{\text{area} \\ \text{increases}}} \underbrace{\frac{T_0 - T_i}{r \ln(r_0/r_i)}}_{\substack{\text{gradient} \\ \text{decreases}}} \\
&= \underbrace{-k\, 2\pi L(T_0 - T_i)/\ln\frac{r_0}{r_i}}_{\text{constant}}
\end{aligned}
$$

For the slab, the cross-sectional area remains constant. This leads to a constant temperature gradient $dT/dx = (T_0 - T_i)/L$ so that the same heat flows at steady state. The temperature profile in the slab is therefore linear.

4.2.3   *BioConnect: Thickness of Fur in Small and Large Animals*

Fur provides a natural thermal insulation for animals. It is interesting to see how the thermal resistance of fur varies with its thickness. Let us consider different size animals that can be treated as if they were approximately of cylindrical shape. Heat

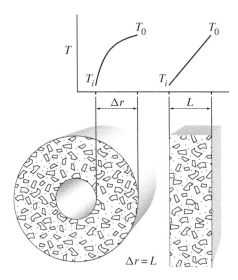

Figure 4.8: Schematic comparing a non-linear temperature profile in a hollow cylinder with a linear temperature profile in a slab at steady state.

loss per unit surface area of the animal is the most relevant because it closely relates to heat loss per unit volume and can be written using Eq. 4.44 as

$$q''_{\substack{body \\ surface}} = \frac{T_i - T_o}{(2\pi r_i L)\left(\dfrac{\ln r_o/r_i}{2\pi k L}\right)} \tag{4.46}$$

where $q''_{\substack{body \\ surface}}$ is the heat flux at the body surface (without the fur), T_i is the temperature at the body surface, and T_o is the temperature at the outer surface of the fur. The body surface area of the animal without the fur is $2\pi r_i L$ (r_i is its radius without the fur). The quantity in the denominator of Eq. 4.46 is the *R-value* for the layer of fur over a cylindrical shaped animal. If the thickness of fur is $\Delta r = r_0 - r_i$, we can write this

R-value for fur as

$$R_{fur} = 2\pi r_i L \frac{\ln\left(1 + \dfrac{\Delta r}{r_i}\right)}{2\pi k L}$$

$$= r_i \frac{\ln\left(1 + \dfrac{\Delta r}{r_i}\right)}{k} \tag{4.47}$$

For small values of $\Delta r / r_i$, the term $\ln(1 + \Delta r/r_i)$ in Eq. 4.47 can be approximated[1] as $\Delta r/r_i$. Using this approximation,

$$R_{fur} = \frac{\Delta r}{k} \tag{4.48}$$

Equation 4.48 shows that R_{fur} increases linearly with the thickness of fur Δr for small values of Δr relative to the radius of the animal, r_i.

Equation 4.47 is plotted in Figure 4.9 for two values of the radius of the animal, r_i. It shows that the initial linear increase in R_{fur} with fur thickness, Δr (as predicted by Eq. 4.48), continues for a larger value of fur thickness when the animal is bigger (large r_i), while for a small animal, R_{fur} quickly levels off as the fur thickness increases. Thus, increased fur thickness beyond a certain value is not as beneficial for a small animal whereas a large animal would continue to benefit significantly (R_{fur} or thermal resistance per unit surface area increases considerably) over a larger thickness of fur. Looking at it differently, the smaller the animal, the more difficult it is to provide insulation. Indeed, small animals may only survive in cold climates by behavioral responses that avoid severe stress and by having very high rates of metabolism (Cena and Clark, 1979).

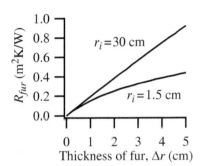

Figure 4.9: Plot of *R*-value of fur, R_{fur}, as related to thickness of fur, Δr, for an assumed fur thermal conductivity of 0.05 W/m·K.

4.2.4   *Example: Estimating Metabolic Heat Generation*

Estimating metabolic heat generation is often difficult. We would like to estimate the metabolic heat generation of a minke whale from a simple measurement of temperature at a certain depth under the skin, assuming steady state. Assume the whale can be approximated as a cylinder, with an outside diameter of 1.5 m, length of 8 m, and a fat layer under the skin with a thickness of 5 cm, with heat transfer only in the radial

Figure 4.10:   A minke whale.   (iStock.com/Moritz Frei)

---

[1]For small values of x, $\ln(1 + x)$ can be approximated as

$$\ln(1 + x) = x - \frac{1}{2}x^2 + \frac{1}{3}x^3 - \cdots$$
$$\approx x$$

direction. The temperature underneath the fat layer is measured to be 37°C, which, for the purposes of this problem, is also the temperature for the entire core of the cylinder under the fat layer. The water temperature is 15°C and convection due to water flow at the surface (skin) maintains a heat transfer coefficient of 200 W/m² · K. The thermal conductivity of the fat layer is 0.25 W/m · K. 1) What is the heat generation per unit volume in the whale body (core) in W/m³? 2) How does the heat generation compare with that in the human heart muscle, typically in the range 30,000–90,000 W/m³?

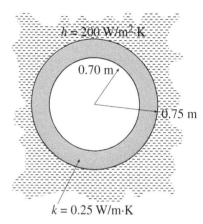

Figure 4.11: Schematic for Example 4.2.4.

Solution

Understanding and formulating the problem *1) What is the process?* Heat is generated in the core of the body and is lost through the fat layer to the surface. The loss equals the generated heat, keeping the process at steady state. If we know the rate of loss, we have found the rate of heat generation. *2) What are we solving for?* Heat generated in the core per unit volume. *3) Schematic and given data:* A schematic is shown in Figure 4.11 with some of the given data superimposed on it. *4) Assumptions:* Cylinder is long so heat transfer is only in the radial direction. Thermal properties do not vary with temperature.

Generating and selecting among alternate solutions *1) What solutions are possible?* For steady-state heat transfer, the solutions we have available are shown in the solution chart in Figure 4.19. *2) What approach is likely to work?* As we traverse the solution chart in Figure 4.19, before we use the chart, we need to make sure of our domain of interest or the domain we are analyzing. While the term heat generation is mentioned in the problem description, our domain of interest is the fat layer (the annulus). *There is no heat generation in the fat layer and therefore our domain has no heat generation.* Another hint in the problem description that steers us away from considering the core as the domain is that the entire core is stated to be at 37°C. This means we are not solving for temperatures in that region.

There are two ways (left side of Figure 4.19) to approach this steady-state heat conduction problem with no heat generation—either use resistance formulation or derive a new solution from scratch. There are two thermal resistances here— that of the fat layer and water. The outside surface of the fat layer is not at the water temperature but slightly warmer than the bulk water temperature of 15°C and there is a thermal resistance of this water layer that is the convective resistance, $1/hA$. These two resistances being in series, they can be combined using formulas for resistances in series. Since the resistance analysis could work and the problem does not ask us to derive from scratch, we choose to do the resistance formulation, which is generally less work.

Implementing the chosen solution Combined heat loss through the two resistances in series (conduction through the fat layer and convection outside) is given by

$$q = \frac{T_c - T_\infty}{\frac{\ln(r_0/r_i)}{2\pi k_{fat} L} + \frac{1}{h 2\pi r_0 L}} \tag{4.49}$$

$$= \frac{(37 - 15)\,\text{K}}{\frac{\ln(0.75/0.70)}{2\pi \times 0.25\ \text{W/m}\cdot\text{K} \times 8\text{m}} + \frac{1}{200\ \text{W/m}^2\cdot\text{K} \times 2\pi \times 0.75\text{m} \times 8\text{m}}}$$

$$= \frac{22}{\underbrace{0.00549}_{\substack{\text{conductive resistance} \\ \text{of fat layer}}} + \underbrace{0.000133}_{\substack{\text{convective resistance} \\ \text{of water}}}}\ \text{W} \tag{4.50}$$

$$= 3912.57\ \text{W} \tag{4.51}$$

This heat loss of 3912.57 W is also the generated heat in the core, making it possible to reach a steady state, as stated in the problem. Assuming this heat generation is uniform over the entire volume of the core, volumetric heat generation in the core, Q, is given by

$$Q = \frac{q}{\pi r_i^2 L}$$

$$= \frac{3912.567\ \text{W}}{\pi \times (0.70\text{m})^2 \times 8\text{m}}$$

$$= 317.7\ \text{W/m}^3$$

Evaluating and interpreting the solution *1) Does the computed value (heat generation) make sense?* From page 560, we see that the approximate basal metabolic rate for a human is $\sim$1000 W/m^3. Thus, the approximate metabolic heat generation estimated from this simple model is in the same order of magnitude so it could be reasonable. *2) What else do we learn?* a) As Eq. 4.50 shows, between the thermal resistance of the fat layer and the convection in the water, the resistance of the fat layer is dominant. b) Even though there is metabolic heat generation implicit in the problem description, we did not include it in our analysis since the metabolic heat is not part of the fat domain. Metabolic heat simply passes through the domain.

4.3   Geometry Effect: Temperature Profile, Heat Flow, and Thermal Resistance Terms for Slab, Cylinder, and Sphere

For comparison and convenience, temperature profile, heat flow, and thermal resistances for a slab, a cylindrical shell, and a spherical shell are provided in Table 4.2. Resistances from slab and cylinder were derived in Section 4.1.2 and 4.2.1, respectively. Resistance for a spherical shell is part of Problem 4.33.

4.4   Steady-State Heat Conduction in a Slab with Internal Heat Generation

All the steady-state heat transfer situations considered in this chapter so far do not have any energy generation. Let us include energy generation in a slab which has temperatures uniform on its two surfaces so that the heat transfer is still one dimensional, as shown in Figure 4.12. An example temperature variation in a compost pile where energy is generated due to biochemical processes. The human body with internal metabolic heat generation is another example (see Section 4.4.2). We are interested in the steady-state heat transfer in a slab with constant internal heat generation. The general governing equation is simplified for steady state

$$\underbrace{\overset{0}{\cancel{\frac{\partial T}{\partial t}}}}_{\text{steady state}} + \underbrace{\overset{0}{u\cancel{\frac{\partial T}{\partial x}}}}_{\text{no bulk flow}} = \underbrace{\frac{k}{\rho c_p}\frac{\partial^2 T}{\partial x^2}}_{\text{conduction}} + \underbrace{\frac{Q}{\rho c_p}}_{\text{heat generation}}$$

to arrive at the specific governing equation

$$\frac{d^2 T}{dx^2} = -\frac{Q}{k} \qquad (4.52)$$

The boundary conditions for a constant temperature of T_1 on both surfaces are given by

$$T(x = L) \quad = \quad T_1 \qquad (4.53)$$

$$\left.\frac{dT}{dx}\right|_{x=0} \quad = \quad 0 \quad \text{(from symmetry)} \qquad (4.54)$$

To solve Eq. 4.52, it is integrated once to obtain

$$\frac{dT}{dx} = -\frac{Q}{k}x + c_1$$

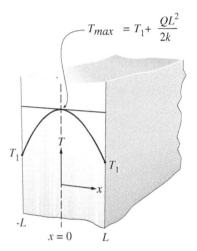

Figure 4.12: Steady-state temperature profile in a slab with constant internal heat generation.

$$T_{max} = T_1 + \frac{QL^2}{2k}$$

Table 4.2: Steady-state temperature, steady-state heat flow, and thermal resistance in various geometries

Geometry	Temperature Profile	Heat flow	Resistance
Slab 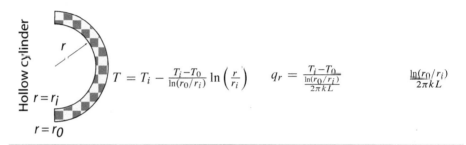	$T = (T_2 - T_1)\frac{x}{L} + T_2$	$q_x = \frac{T_1 - T_2}{L/kA}$	$\frac{L}{kA}$
Hollow cylinder	$T = T_i - \frac{T_i - T_0}{\ln(r_0/r_i)}\ln\left(\frac{r}{r_i}\right)$	$q_r = \frac{T_i - T_0}{\frac{\ln(r_0/r_i)}{2\pi kL}}$	$\frac{\ln(r_0/r_i)}{2\pi kL}$
Hollow sphere 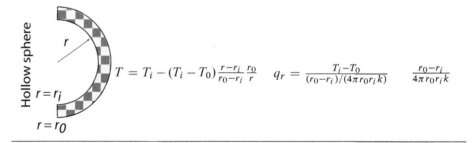	$T = T_i - (T_i - T_0)\frac{r - r_i}{r_0 - r_i}\frac{r_0}{r}$	$q_r = \frac{T_i - T_0}{(r_0 - r_i)/(4\pi r_0 r_i k)}$	$\frac{r_0 - r_i}{4\pi r_0 r_i k}$

Using the second boundary condition (Eq. 4.54),

$$c_1 = 0$$

Integrating again,

$$T = -\frac{Q}{2k}x^2 + c_2 \tag{4.55}$$

Using the first boundary condition,

$$T_1 = -\frac{QL^2}{2k} + c_2 \tag{4.56}$$

Subtracting Eq. 4.56 from Eq. 4.55 to eliminate c_2,

$$
\begin{aligned}
T - T_1 &= -\frac{Q(x^2 - L^2)}{2k} \\
&= \frac{QL^2}{2k}\left(1 - \frac{x^2}{L^2}\right)
\end{aligned}
\tag{4.57}
$$

This solution is plotted in Figure 4.12. The temperatures inside adjust such that the total energy lost through the surface exactly equals the total energy generated inside, leading to a steady state. The heat lost from the surface at $x = L$ is given by the heat flow at that surface:

$$q = -kA \left.\frac{dT}{dx}\right|_{x=L} \tag{4.58}$$

Substituting for temperature from Eq. 4.57, we get

$$q = QLA \tag{4.59}$$

Note that QLA is also the heat generated in half of the slab. The steady-state condition is satisfied since the generated heat is equal to the heat lost through the surface.

As an example, Figure 4.13 shows the near steady-state temperature profile in a compost pile where biochemical activity produces heat inside the pile and it is cooled from the top and the bottom. Note that, although the temperature profile in Figure 4.13 is similar to that in Figure 4.12, it is not symmetric. Why? This is explained in the example problem below.

4.4.1 *Example: Heat Transfer in Composting*

Composting can be used to treat various organic wastes such as grass clippings, leaves, paper, animal manure, and the organic fraction of municipal solid waste. These wastes are converted by microbial degradation to a stabilized compost. The biochemical conversion process generates heat, and consideration of this generated heat is critical in the design of composting systems. Compost managers strive to keep the compost below about 65°C because hotter temperatures cause the beneficial microbes to die off. If the pile gets too hot, turning or aerating will help to dissipate the heat.

Treat a compost pile on the ground as a one-dimensional system of height 2 m in the vertical direction. The top of the pile has wind (at temperature 40°C) blowing on

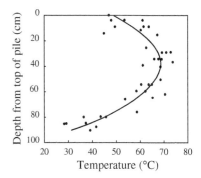

Figure 4.13: Steady-state temperature profile in a compost pile (Lynch and Cherry, 1996).

it, leading to a heat transfer coefficient of 50 W/m²·K. The bottom of the pile can be approximated at the ground temperature of 20°C. Consider only conductive heat transfer inside the pile and neglect any convection inside this porous pile. The volumetric biochemical heat generation is 7 W/m³. The thermal conductivity of the compost material is 0.1 W/m·K. 1) Set up the appropriate governing equation and boundary conditions. 2) Solve for temperature as a function of height from the ground. 3) Calculate the maximum temperature in the pile. 4) Calculate the top surface temperature of the pile.

Solution

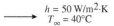

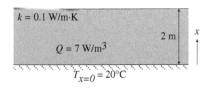

Figure 4.14: Schematic for Example 4.4.1

Understanding and formulating the problem *1) What is the process?* Heat generated in the compost is conducted to the surface where it is convected away by the surrounding colder air, reaching a steady state. *2) What are we solving for?* Temperature profile, maximum temperature, and top surface temperature. *3) Schematic and given data:* A schematic is shown in Figure 4.14 with some of the given data superimposed on it. *4) Assumptions:* Heat generation and thermal conductivity values are uniform in the entire volume. .

Generating and selecting among alternate solutions *1) What solutions are possible?* For steady-state heat transfer, the available solutions are shown in the solution chart in Figure 4.19. *2) What approach is likely to work?* Since we have heat generation, we only look at the right side of the solution chart in Figure 4.19. There are two ways to approach this problem: either use one of the solutions derived in the book or derive a new solution from scratch. To use the solution from the book, not only does the governing equation have to be the same, the boundary conditions also have to be the same. The bottom boundary condition here is temperature specified, and thus different from the insulated boundary condition at the center of the slab done in Section 4.4. The top boundary condition here is a convective boundary condition that is also different from the temperature-specified boundary condition used in Section 4.4. Thus, our approach has to be from scratch.

Implementing the chosen solution Our process will be to write the governing equation and boundary conditions and solve the equation to obtain the temperature. Of the three coordinate systems that we have done, the cartesian coordinate and system is the appropriate one here for the constant thickness of the compost layer:

$$\underbrace{\cancel{\frac{\partial T}{\partial t}}^{0}}_{\text{steady state}} + \underbrace{u\cancel{\frac{\partial T}{\partial x}}^{0}}_{\text{no bulk flow}} = \underbrace{\frac{k}{\rho c_p}\frac{\partial^2 T}{\partial x^2}}_{\text{conduction}} + \underbrace{\frac{Q}{\rho c_p}}_{\text{heat generation}}$$

becomes

$$\frac{d^2 T}{dx^2} = -\frac{Q}{k} \tag{4.60}$$

The boundary conditions at the bottom ($x = 0$) and at the top ($x = L$) are given by

$$T|_{x=0} = T_1 \tag{4.61}$$

$$-k\frac{dT}{dx}\bigg|_{x=L} = h(T_{x=L} - T_\infty) \tag{4.62}$$

Integrating the governing equation, we get

$$\frac{dT}{dx} = -\frac{Qx}{k} + C_1$$

$$T = -\frac{Qx^2}{2k} + C_1 x + C_2$$

Using the first boundary condition at $x = 0$,

$$T_1 = -\frac{Q}{2k}0 + C_1 0 + C_2$$

$$C_2 = T_1$$

$$T = -\frac{Qx^2}{2k} + C_1 x + T_1$$

Using the second boundary condition at $x = L$,

$$-k\frac{dT}{dx}\bigg|_{x=L} = h(T_{x=L} - T_\infty)$$

$$-k\left(\frac{-QL}{k} + C_1\right) = h\left(-\frac{QL^2}{2k} + C_1 L + T_1 - T_\infty\right)$$

$$-C_1(k + hL) = -\frac{QhL^2}{2k} - QL + T_1 h - T_\infty h$$

$$C_1 = \frac{QhL^2}{2k(k + hL)} + \frac{QL}{k + hL} - \frac{T_1 h}{k + hL} + \frac{T_\infty h}{k + hL}$$

Plugging in values,

$$C_1 = \frac{(7 \text{ W/m}^3)(50 \text{ W/m}^2 \cdot \text{K})(2 \text{ m})^2}{2 \times 0.1 \text{ W/m} \cdot \text{K} (0.1 \text{ W/m} \cdot \text{K} + (50 \text{ W/m}^2 \cdot \text{K}) 2 \text{ m})}$$

$$+ \frac{(7 \text{ W/m}^3)(2 \text{ m}) - (20°\text{C})(50 \text{ W/m}^2 \cdot \text{K}) + (40°\text{C})(50 \text{ W/m}^2 \cdot \text{K})}{0.1 \text{ W/m} \cdot \text{K} + (50 \text{ W/m}^2 \cdot \text{K}) (2\text{m})}$$

$$= 80.06 \text{ K/m}$$

Thus, the temperature profile is given by

$$T = -\frac{Qx^2}{2k} + C_1 x + T_1$$

$$= -\frac{7 \text{ W/m}^3 x^2}{2 (0.1 \text{ W/m} \cdot \text{K})} + 80.06x + 20$$

$$= -35x^2 + 80.06x + 20°\text{C}$$

Maximum temperature occurs where the gradient of the temperature is zero, i.e.,

$$\frac{dT}{dx} = -70x + 80.06 = 0 \text{ from which we solve for } x \text{ as}$$

$$x = \frac{80.06}{70} = 1.14 \text{ m}$$

The temperature at this location is given by

$$T_{x=1.14} = -35(1.14)^2 + 80.06 (1.14) + 20 = 65.8°\text{C} \qquad (4.63)$$

The temperature at the top surface is given by

$$T_{x=2\text{m}} = -35(2\text{m})^2 + 80.06 \times 2\text{m} + 20 = 40.1°\text{C} \qquad (4.64)$$

Evaluating and Interpreting the solution *1) Do the computed values make sense?* The temperature profile is not symmetric since the top and bottom boundary conditions are not the same. The maximum temperature inside compost should be higher than the bottom boundary or the top air temperatures to be able to release the generated heat (so steady state can be reached), which is what we have. Since the inside of compost is warmer than air, the top boundary should also be warmer than air, which is the case. *2) What do we learn?* Depending on the height of the pile and the rate of heat generation, the maximum temperature inside can become lethal for the microorganisms needed for composting.

4.4.2 BioConnect: Thermoregulation—Maintaining the Core Body Temperature of Humans and Animals

Thermoregulation is maintaining a constant temperature Thermoregulation is the physiological mechanism by which mammals and birds attempt to balance heat gain and loss in order to maintain a constant body temperature when exposed to variations in the cooling power of the external medium. Such animals are referred to as warm-blooded animals. Some other animals, such as fish, reptiles, and invertebrates, whose body temperature varies with, and is usually higher than, the temperature of the environment do not have thermoregulation and are referred to as *poikilotherms* or cold-blooded animals.

Thermoregulation is an active process Maintenance of a constant body temperature is equivalent to a steady-state heat transfer situation achieved through a balance between heat production, gain, and loss. However, it cannot depend totally on diffusion, which is a passive process. The rate of metabolic heat generation varies with the animal, as can be seen in Table C.1 on page 560. The environmental temperature can also vary. In a passive system as in cold-blooded animals, the steady-state temperature would simply be a result of heat balance and would shift to a different value depending on the sustained variation in environmental temperature or metabolic generation. In contrast, in a regulated system as in warm-blooded animals, a particular steady-state temperature is defended by the activation of appropriate thermoregulatory effectors.

Components of a thermoregulatory system Thermoregulatory processes can be behavioral or autonomic. Behavioral processes of thermoregulation are those which involve the movements of the whole body of the organism relative to the environment, such as moving to a preferred temperature or making changes in posture to change heat gain or loss. Autonomic responses are automatic, reflex-like production or loss of heat in response to heat stress or cold stress. Autonomic heat production can come from shivering or supposedly from the metabolic activity of localized deposits of adipose tissue. Autonomic heat loss can be non-evaporative or evaporative. Non-evaporative loss is effected through changing the thermal resistance of the peripheral tissues by changing the extent of constriction of the blood vessels through them. When the environmental temperature is warmer and heat loss needs to be maintained, the precapillary arterioles in the layer of fatty tissue beneath the skin are fully dilated and the arterial blood passes through the layer of the fatty tissue unimpeded, keeping the skin temperature quite close to the core and allowing a higher heat loss. When an animal is cold-stressed, the precapillary arterioles are fully constricted and the blood flow to the fatty layer is almost zero. This fatty layer becomes inert with high thermal resistance and the skin temperature is also reduced due to the lack of blood flow, both of which reduce the heat loss. Evaporative heat losses are from panting or sweating. Panting, which is open-mouthed rapid shallow breathing, is evaporation, particularly from the tongue, by the inhaled air. Sweating is an aqueous secretion that spreads over the skin in response to heat stress.

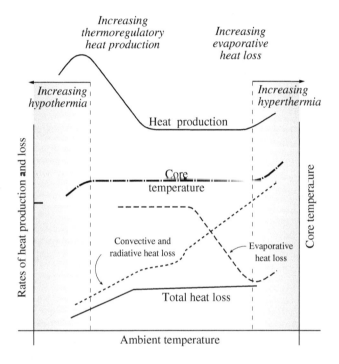

Figure 4.15: Regulation of body core temperature through manipulation of heat production and loss.

Limits of thermoregulation   The thermoregulatory functions described above try to maintain the core body temperature in response to changes in ambient temperature, as shown in Figure 4.15. With a fall in ambient temperature, thermoregulatory heat production increases until the animal reaches its maximum ability to generate heat. Any further fall in ambient temperature leads to a decline in core temperature, eventually at an increasing rate since the maximum level of heat production may not be sustainable at this lower temperature. This leads to hypothermia. With an increase in ambient temperature, the maximum rate of evaporative heat loss is eventually reached. Beyond this point, body temperature can rise and may do so at an increasing rate since the high rate of evaporative loss is not sustainable and also the metabolic rate increases significantly at higher temperatures. This can lead to heat exhaustion with symptoms such as fatigue, thirst, and muscle cramps. In most species, when temperatures reach 42–43°C, there is damage to the central nervous system, with fatal consequences.

4.5 Steady-State Heat Transfer from Extended Surfaces: Fins

From the definition of convective heat transfer coefficient $q = hA(T - T_\infty)$, we know that to increase the heat transfer rate from a surface, one possibility is to increase the convective heat transfer coefficient, h. However, sometimes, situations arise where even the maximum value of h is insufficient or the cost of obtaining pumps, etc., to increase fluid velocity is prohibitive. If we look at this convective heat transfer equation, another choice is to reduce the surrounding temperature, T_∞, which is often impractical. The only other variable that can possibly be manipulated is the area A. In other words, the heat transfer rate can be increased by increasing the surface area over which heat transfer takes place. A *fin* can do precisely this, by extending surfaces from the wall into the surrounding fluid. A schematic of this is shown in Figure 4.16. Everyday examples include radiator fins and fins attached to electronic chips. Fins are also present in the biological world, an example of which is the elephant ear discussed in the next section.

At steady state, the governing equation for a simple fin with constant cross-sectional area (Figure 4.16) is derived as follows:

$$q_x - q_{x+\Delta x} - \Delta q_{conv} = 0$$

where Δq_{conv} is the convective heat loss over the surface area $P\Delta x$ over distance Δx, P being the perimeter. Let A_c be the cross-sectional area of the fin and k the thermal conductivity of the fin material. Substituting the heat flux q_x in terms of temperature gradient and the convective heat loss Δq_{conv} in terms of convective heat transfer coefficient, we get

$$-kA_c \left.\frac{\partial T}{\partial x}\right|_x - \left(-kA_c \left.\frac{\partial T}{\partial x}\right|_{x+\Delta x}\right) - hP\Delta x(T - T_\infty) = 0$$

$$kA_c \left(\left.\frac{\partial T}{\partial x}\right|_{x+\Delta x} - \left.\frac{\partial T}{\partial x}\right|_x\right) - hP\Delta x(T - T_\infty) = 0$$

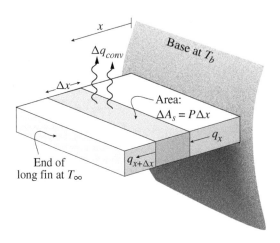

Figure 4.16: Block diagram of a fin showing the base and the extended surface.

Dividing by $kA_c\Delta x$ and taking limits as $\Delta x \to 0$, the first term becomes the second derivative in temperature, and the governing equation for the fin can be written as

$$\frac{\partial^2 T}{\partial x^2} - \frac{hP}{kA_c}(T - T_\infty) = 0 \tag{4.65}$$

Several boundary conditions are possible. For the special case of a long fin, the end of the fin can be assumed to be at the surrounding temperature. For this situation, the boundary conditions can be written as

$$T(x = 0) \;\; = \;\; T_b \tag{4.66}$$
$$T(x \to \infty) \;\; = \;\; T_\infty \tag{4.67}$$

To solve the above governing equation, it is easier to transform it using a new temperature variable $\theta = T - T_\infty$ and a transformed base temperature $\theta_b = T_b - T_\infty$. Using these transformations, the new governing equation is

$$\frac{\partial^2 \theta}{\partial x^2} - \frac{hP}{kA_c}\theta = 0 \tag{4.68}$$

and the new boundary conditions are

$$\theta(x = 0) = \theta_b \tag{4.69}$$
$$\theta(x \to \infty) = 0 \tag{4.70}$$

The solution to the above governing equation is given by (see also Section 12.2 on page 422 for another solution)

$$\theta = c_1 e^{mx} + c_2 e^{-mx} \tag{4.71}$$

where $m^2 = hP/kA_c$ and c_1 and c_2 are constants to be determined using the boundary conditions (Eqs. 4.69 and 4.70). To satisfy the second boundary condition (Eq. 4.70), c_1 has to be zero. Using the first boundary condition (Eq. 4.69), $c_2 = \theta_b$ and the solution is

$$\theta = \theta_b e^{-mx} \tag{4.72}$$

This shows an exponential drop in temperature.

Heat loss with and without a fin To compare heat loss in the presence of a fin to heat loss in its absence, let us first calculate heat loss with a fin. All of the lost energy has to come through the base of the fin, where the rate of heat flow is

$$
\begin{aligned}
q_{\text{fin}} &= -kA_c \left. \frac{\partial \theta}{\partial x} \right|_{x=0} \\
&= -kA_c\, \theta_b(-m)e^{-mx} \big|_{x=0} \\
&= km A_c \theta_b \\
&= \sqrt{hPkA_c}\ \theta_b \tag{4.73}
\end{aligned}
$$

Without a fin, the heat loss is simply what can come out through the surface area equal to the cross-sectional area at the base:

$$q_{\text{no fin}} = hA_c \theta_b \tag{4.74}$$

The effectiveness of the fin can be defined as

$$\frac{q_{\text{fin}}}{q_{\text{no fin}}} = \sqrt{\frac{kP}{hA_c}} \tag{4.75}$$

We can choose higher thermal conductivity k of the fin material to increase the fin effectiveness or to increase the efficiency of heat loss. We can also choose a higher ratio of perimeter P to cross-sectional area A_c or thin geometry to increase the efficiency or heat loss. Note that, under some conditions, the ratio on the right-hand side of Eq. 4.75 can be less than 1. This means we can actually decrease the heat loss. This makes sense because, for example, if the thermal conductivity of the fin material is very low, it will in fact insulate the surface or reduce heat loss instead of increasing it.

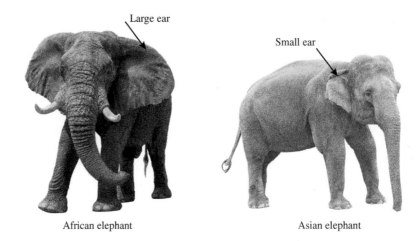

African elephant   Asian elephant

Figure 4.17: The large ears (pinnas) of elephants are needed for thermoregulation to lose large amounts of metabolic heat generated by these large animals since they cannot sweat. The ears of an African elephant are larger in contrast to those of an Asian elephant due to the hotter African climate and the larger amount of metabolic heat that needs to be lost. (iStock.com/ Valerii Kaliuzhnyi)

4.5.1   BioConnect: Fins and Bioheat Transfer

Warm-blooded animals appear to have adapted to new or changing environments by varying the size and shape of their bodies and extremities. Larger animals need to develop means of dealing with the great amounts of heat that they produce (Figure 4.17). The African elephant, the largest land mammal, has accordingly developed the largest thermoregulatory organ known in any animal, the pinna or external ear, which it uses as a radiator-convector. The pinna is considered to be the main external organ responsible for the temperature regulation of the body.

The combined surface area of both sides of both ears of an African elephant is about 20% of its total surface area. The high surface to volume ratio (see discussion following Eq. 4.75), large surface area, and extensive vascular network of subcutaneous vessels in the medial side of the ear make it behave like a fin and play a role in temperature regulation. Temperature distribution patterns measured in a pinna are shown in Figure 4.18. Note how the temperature changes from where the ear attaches to the head (base of the fin) to the outer edges, characteristic of a fin. Movement of the pinna (flapping) also increases the heat loss due to increased air flow.

A 4000 kg elephant needs to maintain a heat loss of 4.65 kW or more while moving and feeding. This large amount of heat cannot be dissipated by surface evaporation

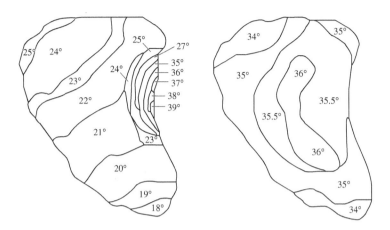

Figure 4.18: Heat transfer from an elephant pinna. Here the pinna acts like a fin for heat transfer purposes. Shown is the right pinna at two different ambient temperatures of 18°C (left) and 32.1°C (right). The change in pattern indicates that a change in blood flow occurs at higher temperatures. Reprinted from COMPARATIVE BIO-CHEMISTRY & PHYSIOLOGY, Phillips & Heath, "Heat Exchange by the Pinna of the African Elephant (Loxodonta Africana)," page 698, (1992), with permission from Elsevier Science.

since elephants do not have sweat glands. Thus, the pinna plays a great role in heat dissipation and by some estimates (Phillips and Heath, 1992), up to 100% of an African elephant's heat loss can be met by movement of its pinna and by vasodilation. The use of pinna for heat loss by convection and radiation is not unique to elephants. New Zealand white rabbits are known to do the same. Also, the Asiatic elephant, with the same size and metabolic rate but with a much smaller pinna, loses about a third as much heat through the ears.

4.6 Chapter Summary—Steady-State Heat Conduction with and without Heat Generation

- **Steady State (page 71)**

    1. A heat transfer process is at steady state if the temperature does not change with time.

2. Steady state can be achieved a number of ways. Using Eq. 1.1, we see the different combinations that would lead to steady state for a positive generation term as 1) In − Out + Gen = 0; 2) In − Out = 0; 3) − Out + Gen = 0. If generation is negative, instead of 3), we will have In + Gen = 0.

- **Steady-State Heat Conduction in a Slab (page 73)**

  1. Steady-state heat conduction in a slab results in a linear temperature profile given by Eq. 4.4.

  2. A composite slab can be treated as series or parallel combinations of the thermal resistances of individual slabs.

- **Thermal Resistance (page 74)**

  1. Thermal resistance is the resistance to heat flow. It is the analog of electrical resistance to current flow. Thermal resistances for conduction and convection are defined by the terms shown in Eq. 4.11.

  2. Conductive thermal resistance decreases with an increase in thermal conductivity and area of heat flow and a decrease in the path length for heat flow.

- **Steady-State Heat Conduction in a Cylinder (page 83)**

  1. The steady-state radial temperature profile for heat transfer through a hollow cylinder is given by Eq. 4.43. The temperature profile is not linear as it has to adjust with the varying cross-sectional area to keep the total heat flow a constant.

- **Steady-State Heat Conduction in a Slab with Internal Heat Generation (page 91)**

  1. The presence of constant internal heat generation with cooling from the surfaces leads to a parabolic temperature profile (Eq. 4.57).

  2. Thermoregulation (page 97) in mammals to maintain a constant internal body temperature can be seen as a steady state achieved through a balance between internal heat generation and heat loss at the surface. However, passive heat conduction is not generally sufficient for achieving this steady state, and active thermoregulatory effectors in the body are necessary.

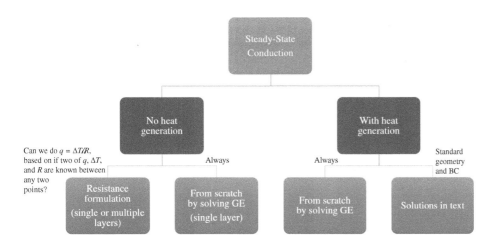

Figure 4.19: Problem solving in steady-state heat conduction. Resistances for various coordinate systems are provided in Table 4.2.

- **Steady-State Heat Conduction from Extended Surfaces (p. 99)**

  1. One way to increase the rate of heat transfer or decrease the thermal resistance is to increase the surface area for heat transfer. This goal is achieved in fins.

  2. The temperature profile in a long fin is given by Eq. 4.72.

4.7 Problem Solving—Steady-State Heat Conduction

▶**Calculate temperature profiles and heat flows.** The solution chart in Figure 4.19 shows an approach to solving steady-state heat conduction problems. Steady-state problems can be divided into those with heat generation and those without. It is important to solve problems corresponding to each situation in the solution chart from those provided at the end of this chapter.

▶▶**Resistance formulation** Calculation steps for resistance formulation can be as follows (Note: problems involving multiple materials are much easier to do using resistance formulation):

  1. How many resistances are there?

2. Resistances in what geometry? Conductive resistances for each geometry are in Table 4.2. Convective resistance is $1/hA$.

3. Are the resistances in series, parallel, or mixed? For series and parallel, use formulas derived in Section 4.1.2.

4. Once heat flow, q, is known, using $q = \Delta T / R$, temperature can be found at any point when the resistance, R, is known.

▶▶ **Solution from scratch**   Solution from scratch (i.e., starting from Governing Equation and Boundary Conditions) can be necessary when a resistance analysis won't work (for example, when thermal conductivity is not a constant) or the problem does not match those in the text or the problem asks for a derivation from scratch.

1. Follow Figure 3.16

2. Use Fourier's law to get heat flux from the temperature profile

3. When there is generation, heat flow at the surface is equal to heat generated inside the domain

4.8 Concept and Review Questions

1. Explain the difference between steady state and equilibrium.

2. What are the components (parameters) in thermal resistance?

3. Can you think of a heat transfer situation where the thermal resistances are in parallel instead of being in a series, as in Eq. 4.11?

4. Does the core temperature in a cold-blooded organism, such as a fish, depend on heat conduction?

5. Why is passive heat conduction not sufficient for thermoregulation in warm-blooded animals?

6. Why is a straight line profile in the case of steady heat transfer through an annulus not physically realistic?

7. What is the purpose of using fins?

8. Can you actually reduce heat loss by adding fins? Explain under what circumstances this can happen.

9. For a slab at steady state, the temperature profile is linear. For a cylindrical annulus at steady state, a linear temperature profile (with radius) is not physically realistic. Provide clear physical reasoning for this, using a sketch.

10. Can we *always* obtain the steady-state temperature of our body from simply a balance between heat generation and heat loss (from conduction, convection, and radiation)?

11. Using the expression for the thermal resistance in cylindrical coordinates, explain why increasing the thickness of fur makes more sense in a large animal than in a small animal. Use $\ln(1 + x) \approx x$.

12. As you know, mammals have to thermoregulate or maintain a steady-state body temperature within a narrow range. Otherwise their body temperature would result from simply a balance between their metabolic heat generation and convective heat loss at the body surface, as for the fish problem above. Explain how the blood vessels under the skin are actively used to thermoregulate when a) the environment is much colder than the body (cold stress); b) the environment is much warmer than the body (heat stress).

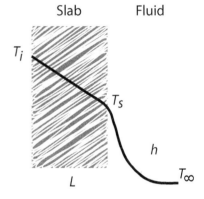

Figure 4.20: Slab with convection at surface.

13. Consider steady-state heat transfer through a slab to a fluid flowing over the slab with temperature T on one side of the slab being fixed, as shown in the schematic in Figure 4.20. The thermal conductivity of the slab is k and the surface heat transfer coefficient is h. a) Find T_s as a function of other temperatures and parameters given; b) How would T_s change if $h \to \infty$? c) Give a physical example of what b) above means.

14. From our common knowledge, adding insulation reduces heat loss. From the homework that you did on insulation over a pipe, you found that there is a range of radius for which adding more insulation actually increases heat loss. Is this result completely counterintuitive? Explain why this can make physical sense.

15. Consider steady-state heat transfer with uniform heat generation of Q in a cylinder, having radius R and length L. What should be the rate of heat loss at the surface?

16. Provide physical reasoning as to why a steady-state temperature profile does not depend on density and specific heat.

17. Write a word equation (no symbols) involving thermal resistance and heat flow.

18. Why is an initial condition not needed in a steady-state calculation (two lines max)?

19. Mention if steady state will be reached in the following cases, giving reasoning: 1) Temperature inside a deceased person with a cold surrounding; 2) A heat lamp shining on your arm, delivering heat; 3) Temperature inside a termite mound with termites generating heat inside; and 4) A perfectly insulated box with fermentation inside that generates heat.

Further Reading

Albright, L. D. 1990. *Environment Control for Animals and Plants.* ASAE, St. Joseph, Michigan.

Bakken, G. S. 1991. Wind speed dependence of the overall thermal conductance of fur and feather insulation. *J. Therm. Biol.* 16(2):121–126.

Basak, T., K. K. Rao, and A. Bejan. 1996. A model for heat transfer in a honey bee swarm. *Chemical Engineering Science* 51(3).387–400.

Bischof, J. C., N. Merry, and J. Hulbert. 1997. Rectal protection during prostate cryosurgery: Design and characterization of an insulating probe. *Cryobiology* 34:80–92.

Bligh, J. 1983. Temperature regulation. In: *Stress Physiology in Livestock: Volume 1, Basic Principles*, edited by M. K. Yousef. CRC Press, Boca Raton, FL.

Bligh, J. 1985. Regulation of body temperature in man and other mammals. In: *Heat Transfer in Medicine and Biology*, edited by A. Shitzer and R. C. Eberhart. Plenum Press, New York.

Bruce, J. M. and J. J. Clark. 1979. Models of heat production and critical temperature for growing pigs. *Animal Production* 28:353–369.

Cena, K. and J. A. Clark. 1979. Transfer of heat through animal coats and clothing. *Environmental Physiology III* 20:1–42.

Dewar, H., J. B. Graham, and R. W. Brill. 1994. Studies of tropical tuna swimming performance in a large water tunnel. II. Thermoregulation. *Journal of Experimental Biology* 192:33–44.

Gebremedhin, K. G. 1987. A model of sensible heat transfer across the boundary layer of animal hair coat. *J. Therm. Biol.* 12(1):5–10.

Gilbert, R. D., H. Schröder, T. Kawamura, P. S. Dale, and G. G. Power. 1985. Heat transfer pathways between fetal lamb and ewe. *Journal of Applied Physiology* 59:634–638.

Hartwig, M. K. and K. Kiessl. 1997. Calculation of heat and moisture transfer in exposed building components. *Int. J. Heat and Mass Transfer* 40(1):159–167.

Haslam, R. A. and K. C. Parsons. 1988. Quantifying the effects of clothing for models of human response to the thermal environment. *Ergonomics* 31(12):1787–1806.

Iberall, A. S. and A. M. Schindler. 1973. On the physical basis of a theory of human thermoregulation. *J. of Dynamic Systems, Measurement and Control, Trans. of ASME* March 1973: 68–75.

Kawashima, Y. 1993. Characteristics of the temperature regulation system in the human body. *J. Therm. Biol.* 18(5/6):307–323.

Kvadsheim, P. H. and J. J. Aarseth. Thermal function of phocid seal fur. *Marine Mammal Science* 18(4):952–962.

Lynch, N. J. and R. S. Cherry. 1996. Winter composting using the passively aerated windrow system. *Compost Science and Utilization* 4(3):44–52.

Oosterhout, G. R. and G. A. Spolek. 1989. Transient heat and mass transfer in layered walls. In *Collected Papers in Heat Transfer* HTD-Vol. 123, ASME, New York.

Pennes, H. H. 1948. Analysis of tissue and arterial blood temperatures in the resting human forearm. *J. of Applied Physiology* 1(2):93–122.

Phillips, P. K. and J. E. Heath. 1992. Heat exchange by the pinna of the African elephant (Loxodonta Africana). *Comparative Biochemistry and Physiology* 101A (4):693–699.

Robertshaw, D. and V. A. Finch. 1984. Heat loss and gain in artificial and natural environments. *Thermal Physiology*, J. R. S. Hales, Ed., Raven Press, New York.

Shitzer A. and R. C. Eberhart. 1985. *Heat Transfer in Medicine and Biology*. Plenum Press, New York.

van Waversveld, J., A. D. F. Addink, G. van den Thillart, and H. Smit. 1989. Heat production of fish: A literature review. *Comparative Biochemistry and Physiology Part A: Physiology* 92(2):159–162.

Zhu, L. and C. Diao. 2001. Theoretical simulation of temperature distribution in the brain during mild hypothermia treatment for brain injury. *Medical & Biological Engineering Computing* 39:681–687.

4.9   Problems

4.1   Reduction of Heat Loss Using a Stagnant Air Film

Compare the heat loss in W from two windows of a room, one having a single sheet of glass 10 mm thick while the other has two sheets of glass, 5 mm each, separated by 5 mm of air. The second window is called a thermopane. The window sizes are 1 m × 1 m. The air temperature in the room is 20°C and the outside air temperature is −5°C. The average value of the thermal conductivity of glass is 0.85 W/m·K and that of air is 0.025 W/m·K over this temperature range of interest. The heat transfer coefficient corresponding to slower moving air inside the room is 20 W/m²·K and that of the faster moving air outside is 200 W/m²·K. Consider the thin layer of air between the glass sheets to be stagnant.

4.2   Clothing and Conduction Heat Transfer

Compare the total heat loss from a person dressed in summer clothes to the heat loss from a person dressed in winter clothing under the same ambient conditions. The insulating effect of the layer or layers of air trapped between the layers of clothing and between the clothing and the body is reflected in the resistance values of the clothing ensembles. The average temperature of skin is 33°C, the ambient temperature is 20°C, the total surface area of the body is 1.7 m², the area of the body covered by summer clothing is 1 m², the area of the body covered by winter clothing is 1.6 m², the heat transfer coefficient of bare skin is 27.3 W/m²·K, the overall heat transfer coefficient of summer clothing is 18.4 W/m²·K, and the overall heat transfer coefficient of winter clothing is 4.3 W/m²·K (Shitzer and Eberhart, 1985).

4.3   Insulation in a Home Freezer

Consider a home freezer compartment with dimensions of height 60 cm, width 40 cm, and depth 40 cm. Assume heat loss to be from the five sides exposed to outside air, i.e., do not consider any heat loss through the side which the freezer sits on. For a styrofoam insulation of thermal conductivity 0.03 W/m·K, what should be the minimum thickness of this insulation so that the total heat loss from the freezer compartment is less than 120 W? Assume the freezer air is at −20°C and the inside surface heat transfer coefficient is 5 W/m²·K, while the outside air temperature is 30°C and the outside surface heat transfer coefficient is 20 W/m²·K.

4.4 Steady-State Heat Transfer in a Slab with Convection on the Surface

Calculate the temperature profile in a slab of thickness ΔL and thermal conductivity k whose one side is maintained at temperature T_2 while the other side exchanges energy by convection with a fluid at a heat transfer coefficient of h. The fluid temperature is T_∞.

4.5 Reduction of Heat Loss in a Pipe

A circular sheet metal duct carries refrigerated air to a cold storage room for apples. The duct itself is 250 mm in outer diameter. The duct wall thickness is 1 mm. To reduce the heat gain from the surrounding air, we need to wrap the duct with insulation. The flowing air maintains the inner surface of the duct at 0°C. The outer surface temperature of insulation would be maintained at 25°C by the room air. The thermal conductivity of the sheet metal is 60 W/m·K and that of the insulation is 0.04 W/m·K. Assuming the heat transfer to be at steady state, what thickness of insulation should be put on the duct to keep the rate of heat gain by the refrigerated air per meter length of the duct to 30 W (Albright, 1990).

4.6 Convective and Conductive Heat Loss in a Radial Geometry

Consider steady-state heat transfer from a hollow metallic cylindrical pipe covered with insulation, as discussed in Section 4.2. Only the thermal resistance of the insulation and the external convective resistance are important. The inside wall of insulation is at T_i while the outside air temperature is T_0. 1) Write the expression for steady-state heat flow in terms of the temperature difference and the total resistance. 2) Show that the total thermal resistance has a minimum value with respect to the thickness of insulation (i.e., below and above this thickness, thermal resistance increases and heat loss decreases) when the pipe diameter (i.e., the outer diameter of the metallic part, without the insulation) is kept fixed. 3) Discuss the physical meaning of such a minimum value. 4) For a heat transfer coefficient of 50 W/m²·K and a thermal conductivity of insulation of 0.2 W/m·K, calculate the radius for minimum thermal resistance and discuss the practical utility of such a result.

4.7 Thickness of Insulation in a Radial Geometry

Refrigerant flows in a copper tube of outer diameter 4.8 cm. The inside surface temperature is −15°C and the room temperature is 20°C. The surface heat transfer coefficient between the tube and the room air is 25 W/m²·K. Ignore the thermal resistance of the copper tube, so that the outside wall temperature of the tube is the same as the inside wall temperature. 1) What is the heat gained by the refrigerant per unit length? 2) To reduce the heat gained from the room air, someone decided to put 4 mm of insulation (thermal conductivity 0.75 W/m·K) around this tube. Show that the heat loss per unit length actually increases after adding this insulation. 3) Explain why you can expect the heat loss to increase.

4.8   Heat Transfer in the Body

At rest, the human body is producing heat at a constant rate as a byproduct of basal metabolism. This heat is dissipated to the surroundings and a steady-state temperature profile is reached in the body. Here we will approximate the body as a slab with uniform heat generation throughout. Assuming that the heat is transferred to the surface primarily by conduction along the smaller dimension (thickness) of the slab, find and plot the temperature distribution in the body. Assume the body to be symmetrical about its vertical axis. The average rate of metabolic heat generation in the body is 1.4 kW/m^3, the average half thickness of the slab approximating the body is 7.5 cm, body surface temperature is 33°C, and the average thermal conductivity of the material is 1.05 W/m·K.

4.9   Temperature Equilibration in a Fish

Fish are poikilothermic, animals whose body temperature depends heavily on the temperature of their environment (some fish can have a somewhat higher level of control of body temperature—see, for example, the tropical tuna study by Dewar et al., 1994). Metabolic heat is produced by the fish, and blood flowing through the tissues picks up this heat and carries it to the gills, where the blood exchanges oxygen and carbon dioxide as well as heat with the surrounding water. Large volumes of water move over the gills, so the blood quickly equilibrates to the temperature of the water. Because of this temperature dependence, it is difficult to measure the metabolic heat production of the fish and determine the heat transfer coefficient across the gills.

Using a lumped parameter analysis, find the heat transfer coefficient across the gills of a fish. Assume the process is at steady state. The mass of the fish is 2 kg, the total surface area of the gills is 300 cm^2, the temperature differential between the gills and the water flowing over the gills is 0.5°C, and the total metabolic heat generation in the fish is given by $0.194m^{0.85}$ W where m is the mass of the fish in kg. Assume heat is lost only at the gills.

4.10 Heat Transfer in Composting

Moved to Solved Problem in the Front

4.11 Heat Transfer Prior to a Surgery

Heart disease is one of the leading causes of death, and cholesterol contributes heavily to this problem. However, this same agent can also cause poor circulation and peripheral artery disease in the legs. In some cases the damage to the leg arteries is so severe that the only option is amputation. In that case the leg is often cooled in a refrigerant prior to surgery to reduce the blood flow even further and numb the nerve endings to make surgery and recovery easier for the patient.

Find and plot the leg's temperature profile in the radial direction after the cooling process has reached steady state. Assume that the skin temperature is equal to the refrigerant temperature, which is unchanged by the cooling process. The leg can be approximated as a cylinder with constant thermal properties in the radial direction. The metabolic heat generation rate is slowed by the cooling, but is still present uniformly throughout the leg, at a value of 1.1 kW/m^3. The thermal conductivity of the leg muscle is 0.6 W/m·K, the radius of leg is 5.0 cm, and the refrigerant temperature is 10°C.

4.12 Steady State and Metabolic Heat Generation

Replaced by Problem 4.16

4.13 Heat Dissipated during Exercise

Calculate the energy dissipated at steady state per unit length at the surface of a working cylindrical muscle. The heat generated in the muscle is 5.8 kW/m^3, the thermal conductivity of the muscle is 0.419 W/m·K, and the radius of the muscle is 1 cm. What is the maximum temperature rise ($T_{max} - T_{surface}$) in the muscle?

Figure 4.21: Exercising generates heat in the muscle. (iStock.com/Milan Stojanovic)

4.14 Fire Walking

Figure 4.22 shows a picture of fire walking. It is thought that when the person's foot touches the fire, a small amount of water evaporates from the skin and this layer of water vapor protects the skin from being burned. We would like to verify this hypothesis. For convenience, a schematic of the approximate problem is provided in Figure 4.23.

Assume that the coal temperature at 10 mm depth and skin temperature at 5 mm depth remain constant at 600°C and 37°C, respectively. The properties of the different layers are in the table below.

Figure 4.22: Fire walking.

1) What will the temperature in the skin be 1 mm from the skin–steam interface? 2) If the steam layer is absent, what will the temperature be at the same location in question 1), i.e., now 1 mm from the skin–coal interface? 3) Did the steam layer make a significant difference?

	Sp. Heat (J/kgK)	Thermal conductivity (W/m · K)	Density (kg/m^3)
Skin	3600	0.43	1102
Steam	2060	0.026	0.5863
Coal	1800	0.16	300

4.15   Heat Generation in MRI

An active MRI implant is a small device incorporated into metal surgical implants to make imaging easier by producing local interference to stop the disruptive eddy currents. However, the MRI implant produces heat and it is important to know whether tissue temperature can rise too high. Here we do a simplified analysis where the implant is considered a sphere of diameter 40 mm that is completely surrounded by tissue.

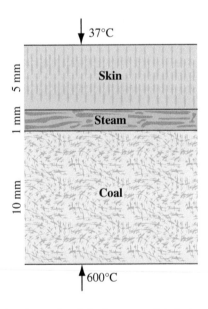

Figure 4.23: A schematic of the heat transfer during fire walking.

The implant has a constant, equally distributed rate of heat generation of 1 W that is conducted from the boundary of the implant into the tissue at steady state. Consider the heat transfer process to be completely symmetrical. The tissue temperature far from the implant is 36°C. The thermal conductivity of the tissue material is 0.5 W/m · K. 1) Make a schematic. 2) Write the governing equation in the appropriate coordinate system and with only the needed terms. *Note that it is the heat transfer in the tissue that we are interested in.* 3) Write the boundary conditions needed to solve the problem, taking particular care for the boundary condition where the tissue touches the implant. 4) Solve the equation to obtain a temperature profile as a function of position. 5) Find the largest temperature rise in the tissue (at steady state) due to the heat generation, in °C. 6) If the heat generation from the implant is doubled, does the largest temperature rise double (yes/no)?

4.16 Steady-State Temperature Distribution in a Tissue

The steady-state radial temperature distribution inside a cylindrical limb being cooled with a refrigerant at 2°C is (obtained from solution of the appropriate energy equation)

$$T = 2 + 0.65\left[1 - \left(\frac{r}{R}\right)^2\right] \qquad (4.76)$$

where T is in °C. Assume constant thermal properties and uniform metabolic heat generation throughout the limb. The radius, R, of the limb is 5.0 cm and its effective thermal conductivity is 0.6 W/mK. 1) Write the appropriate governing equation and boundary conditions assuming only 1D heat transfer that was used to obtain Eq. (4.76). 2) Use this governing equation to calculate the rate of metabolic heat generation in W/m³. *Hint: You do not need to integrate the governing equation.* 3) What is the rate of heat loss from the surface of the limb per unit length?

4.17 Frostbite in Fingers

Frostbite is damage to the skin from freezing due to prolonged exposure to cold temperatures. Model one of your fingers as a long cylinder (radial heat flow only), as illustrated in Figure 4.24, where the inner boundary of the muscle/fat layer remains at 37°C and the heat loss to the cold air reaches a steady state. The outside temperature is −25°C and the heat transfer coefficient is 200 W/m² · K. The bone diameter is 16×10^{-3} m; other dimensions and properties of the tissue and glove layers are given below. 1) Assume frostbite occurs when any part of the tissue in the finger drops below 0°C. Determine how serious the frostbite will be (which tissue layer the 0°C would penetrate to) when a) no glove is present and b) a woolen glove covers the finger. 2) Which of the three layers has the highest thermal resistance?

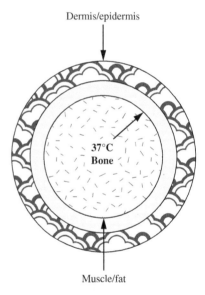

Dermis/epidermis

37°C
Bone

Muscle/fat

Figure 4.24: Schematic showing various layers over which heat transfer is taking place.

	Thickness [m]	k [W/m·K]	c_p [J/kg·K]	ρ [kg/m^3]
muscle/fat	1.25×10^{-3}	0.32	975	310
dermis-epi	4×10^{-4}	0.1	3660	1100
wool glove	4×10^{-4}	0.05	1360	1100

4.18   Thermal Protection during Cryosurgery

Moved to example problems

4.19   Body Temperature of Fish

Unlike temperature regulation in the mammals, fish do not regulate body temperature. Instead, their body temperatures vary with the temperature of the environment (water). Consider a fish that can be approximated as a slab of thickness 5 cm and thermal conductivity 0.5 W/m·K. If the difference between the center (core) and the surface temperature of the fish is 0.5°C, what is the metabolic heat generation in W/m^3?

Figure 4.25: A Tilapia fish. (iStock.com/metalpitt)

4.20   Body Temperature of Fish: Variation

Body temperatures of fish vary with the environment (water). Consider a fish that can be approximated as a slab whose total metabolic heat generation in W is given by $0.194 m^{0.85}$ where m is the mass of the fish in kg. Its movement in water leads to a surface heat transfer coefficient of 50 W/m^2·K. The fish weighs 2 kg, the thickness of the fish (slab) is 5 cm, and its thermal conductivity is 0.5W/m·K. The density of the fish can be assumed to be that of water, 1000 kg/m^3. At steady state, what is the temperature difference between 1) the center (core) and the surface of this fish? 2) The center of the fish and the bulk water?

4.21   Localized heating in hyperthermia using nanoparticles

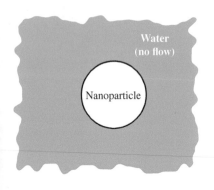

Heat transfer from nanoparticles to a fluid environment has been a research area with growing interest. Results from such studies could be applied to localized heating by absorption of energy from heated nanoparticles, with potential biomedical hyperthermia uses. We want to model steady heat transfer to the surrounding fluid from a solid spherical nanoparticle; however, as an extreme case, no flow is assumed in the fluid. The nanoparticle radius is R [m] and the total heat generated in a nanoparticle is Q [W]. Thermal conductivity of fluid is k [W/m·K]. 1) Provide the governing equation for the problem, keeping only the relevant terms. Be careful about the choice of your domain in this and the following step. 2) Provide the boundary conditions needed to solve the above problem. 3) Solve for temperature profile as a function of radius. 4)

Figure 4.26: Schematic of a nanoparticle with water (non-moving) outside.

If the rate of heat generation, Q, has to be 600 nW or less to preserve the nanoparticle structure, and the maximum temperature can only be 100°C, what is the largest radius of nanoparticle possible for a surrounding water temperature of 37°C? Thermal conductivity of water is 0.6 W/m · K.

4.22   Heat Generation in Composting

In a compost pile, the biochemical conversion process from waste to compost is accompanied by heat generation. Composting systems need to be designed properly so that the temperature rise due to generated heat does not reach above 65 °C, which is when the beneficial microbes start to die off. Consider a compost pile on the ground (assume a pile to be a horizontal layer of a certain thickness) as a steady-state, one-dimensional system with heat transfer along the vertical direction. For simplicity, consider the top and bottom of the pile to be maintained at the same temperature, 20°C, making the layer symmetric. The volumetric heat generation is 7 W/m³ and the thermal conductivity of the compost material is 0.1 W/m·K. 1) Write the *final* governing equation for this problem, keeping only the relevant terms. 2) Solve the equation for the particular boundary condition to obtain the solution $T - T_s = (QL^2/2k)(1 - x^2/L^2)$ where the symbols have the usual meanings. 3) What should the maximum height of the pile be so that the beneficial microbes do not die off at the warmest location?

4.23   Heat Generation in Composting: Variation

Successful composting systems are based on an understanding of the heat (and mass) transfer in the system. Consider 1D heat transfer at steady state in the composting system shown in Figure 4.13. At the top surface, air at 40°C blows over the pile, leading to a heat transfer coefficient of 50 W/m² · K. At the bottom surface, resting on layered material, the boundary can be considered to be insulated. The biochemical conversion generates heat at $Q = 7$ W/m³. The thermal conductivity of the compost material is 0.1 W/m · K

   1) Write the final governing equation for the 1D heat transfer process in the compost, keeping only the relevant terms. 2) Write the boundary conditions needed to solve the equation. 3) Solve the governing equation and obtain the temperature as a function of position from the ground. 4) Plot approximately the temperature variation, as you just solved. Label known axes and intercepts. 5) What is the highest temperature in the pile and where is it located? 6) Does the highest temperature exceed the desired temperature for mesophilic bacteria needed for the composting, 20 to 40 °C, and, if it does, what can be done to lower the temperature (no calculations needed)?

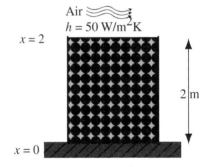

Figure 4.27: Schematic for compost.

4.24 Reducing Brain Damage by Lowering Temperature

Severe asphyxia of the fetus or newborn is known to be an important cause of injury to the development of the brain. Lowering the cerebral temperature over several days after a hypoxic-ischemic insult can markedly protect the brain from damage. Consider a newborn head as a sphere of 8 cm diameter and with average thermal conductivity of 0.5 W/m·K. Assume that its metabolic heat production of 4500 W/m^3 and a positive heat contribution from blood flow of 5500 W/m^3 have radial symmetry. The air surrounding the infant circulates with a heat transfer coefficient of 15 W/m^2·K. The normal temperature of the center of the head without cooling is 37°C. What should the environmental (air) temperature be so that the center of the head is cooled by 2°C, i.e., it maintains a steady-state temperature of 35°C?

4.25 Estimating Metabolic Heat Generation

Moved to Solved Problems

4.26 Thermoregulation in Termite Mounds

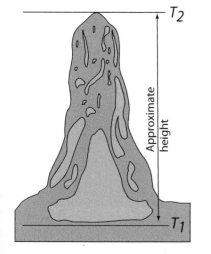

Termites are able to regulate the environment of their nest to an optimal temperature through heat generation. A termite mound (see Figure 4.28, with termites uniformly distributed in it and generating heat uniformly) covering a large area on the ground has a height 1 m, with air blowing over its top surface that maintains the top surface at 27°C, leading to the heat transfer reaching a steady state. The bottom surface of the mound on the ground can be assumed to be at 29°C. Assume the thermal conductivity of the mound is 0.4 W/m·K. 1) Write the governing equation and boundary conditions for the problem. 2) Solve the governing equation to obtain the temperature as a function of vertical position from the ground. Be sure that the terms in your temperature solution have consistent units. 3) Find the location above the ground where the temperature is maximum. Leave your answer in terms of other parameters. 4) Provide reasons for why the location of the maximum temperature is not at mid-height. 5) If the maximum temperature is 30°C (the optimum temperature for the termites), give the two equations from which you can calculate the numerical values for the amount of heat that the termites are producing (in W/m^3 of mound volume) and the location of the maximum temperature (no need to plug in numbers).

Figure 4.28: A termite mound. Assume a large horizontal dimension for this problem.

4.27 Effect of h on Surface Temperature

Consider steady-state conduction heat transfer inside a slab with convective heat transfer over its surface (see Figure 4.29). Temperature T_i on one side of the slab is fixed, as shown in the schematic.

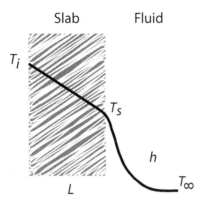

Figure 4.29: Schematic of slab with convection showing qualitative temperature profile.

1) Find the surface temperature, T_s, as a function of other temperatures and parameters given. 2) How would T_s change if $h \to \infty$? 3) Give a physical explanation of what your answer in 2) means.

4.28 Hypothermia to Protect from Ischemia

It has been proposed that mild or moderate hypothermia ($> 30°C$) is a viable treatment in protecting patients from cerebral ischemia resulting from traumatic brain injury. As shown in Figure 4.30, we will develop a very simple hemispherical model of this heat transfer process in the brain tissue. While blood flow and metabolic heat contribute as heat sources in the domain, heat is also lost from the outside spherical surface, leading to a steady state. Because this heat transfer only occurs in the radial direction, there is no heat transfer through the bottom surface. The brain tissue diameter is 53 mm (for an infant) and its uniform thermal conductivity and density are 0.5 W/m · K and 1050 kg/m^3, respectively. The constant blood perfusion rate in the brain tissue is 50 ml/100 g of brain tissue per minute, and the arterial blood temperature is 37°C. The density and specific heat of blood are 1050 kg/m^3 and 3800 J/kg·K, respectively. The temperature of the entire hemispherical outer brain surface is 30°C. The metabolic heat generation rate in the brain tissue is 10437 W/m^3.

1) Assuming the same arterial blood temperature everywhere in the brain tissue, compute the heat source term due to blood flow, in W/m^3. 2) Compare this heat source term due to blood flow with the metabolic heat generation provided, explaining

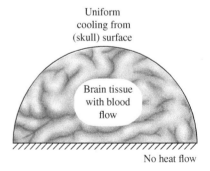

Figure 4.30: Schematic of brain tissue with blood as source of heat.

whether one of them can be ignored in comparison to the other. 3) For the purpose of obtaining an analytical solution, we will assume a constant heat source term due to blood flow equal to half of what you calculated in step 1. What is the *total* heat source term that needs to be used in the governing equation? 4) Write the appropriate governing equation for this problem, keeping only the terms needed. 5) What are the boundary conditions needed to solve the above equation? 6) Solve the equation to obtain the temperature profile. 7) Obtain the average temperature in the brain tissue and compare it with the maximum temperature. *(Hint: Consider a hemispherical shell of thickness dr at a distance r, with volume $2\pi r^2 dr$, and use the definition of volume average.)* 8) If the desired average temperature in the entire brain tissue is 30°C or lower, is this reachable? 9) If the blood flow rate increases, what will happen to this maximum temperature and how would you correct for it (answer qualitatively)?

4.29   Heat Transfer through Clothing

Body heat dissipates through perfused tissue, the subcutaneous fat layer, skin, and clothing into the surrounding air, as illustrated in Figure 4.31. These layers may have very different thermal properties. We will assume heat conduction alone through the intermediate layers of clothing, skin, and subcutaneous fat. The blood flow effect in the perfused tissue will be approximated using an effective conductivity so that heat transfer through this layer is also treated as conduction only. Ambient air flows over the exposed surface (of the clothing) at −5°C, leading to a heat transfer coefficient of $220\ \text{W/m}^2 \cdot \text{K}$. The core temperature is 37°C. The thicknesses of the layers and their thermal conductivities are as shown in Figure 4.31.

1) Calculate the thickness of clothing that will prevent the skin surface (skin-clothing interface) from freezing (stay at 0°C). 2) If clothing with a different material is used, for which the thermal conductivity is three times that of the material shown in the figure, what should the thickness of this material be to keep the skin surface at the same 0°C?

4.30   Heat Loss from a Little Penguin

Figure 4.32 shows the heat transfer modeling from the skin surface of a little penguin that is the smallest penguin species. It is modeled as a long cylinder of length L, diameter D, and plumage thickness t, also shown in the figure. Heat loss from the skin surface occurs first by conduction through the plumage layer and then by both radiation and convection from the surface of the plumage to the surroundings. Note that the radiative heat loss can be approximated as $q = \sigma A(T^4 - T_\infty^4) \approx 4\sigma T_\infty^3 A(T - T_\infty)$ where T_∞ is the environmental temperature and T is the temperature of the plumage.

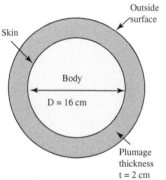

Figure 4.32: A young penguin and the simplified model cylinder. (Ina Raschke–Shutterstock.com)

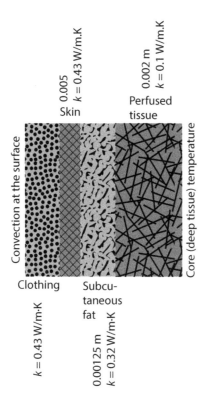

Figure 4.31: Schematic showing various layers near skin.

1) Using the definition of thermal resistance, write the expression for radiative heat transfer resistance using the radiative heat transfer formula provided. 2) As in step 1, write the convective heat transfer resistance between the outside surface and the surrounding air. 3) Write the expression for the total thermal resistance due to convection and radiation between the outside surface and air. *Hint: Calculate the total heat loss by convection and radiation and rewrite this total heat loss in terms of total temperature difference and total resistance.* 4) Calculate the total thermal resistance between the *skin* surface (includes plumage thickness) and the environment. *Leave the answer in terms of symbols and do not plug in numbers.* 5) If the surface of the skin is at 34°C, the environmental temperature is 5°C, the outside convective heat transfer coefficient is 70 W/m$^2 \cdot$ K, the thermal conductivity of plumage is 0.0386 W/m $\cdot$ K, the length of the penguin is 43 cm, and the body and plumage dimensions are as shown in the figure, what is the rate of heat loss in W from the penguin? 6) What

is the total amount of heat generation of the penguin in W, including metabolic and other sources of the penguin, assuming it to be warm-blooded? 7) If the definition of thermal resistance were not available to you, a) What governing equation and boundary conditions will you have started from, to obtain the temperature profile? and b) How would you obtain the heat flow (not flux) from the temperature profile?

4.31   Thermal Therapy Using Nanoparticles

Thermal therapy uses laser-heated gold spherical nanoparticles to cause necrosis of cancer cells through heat-induced lyses and rupture. Consider the schematic shown in Figure 4.33, where, at steady state, a nanoparticle embedded in the tumor tissue reaches a constant temperature that is uniform throughout the sphere. The temperature far from the nanoparticle stays at the body temperature.

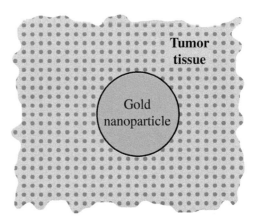

Figure 4.33: Heated gold nanoparticle in a tissue.

1) Write the governing equation and the boundary conditions for the tissue, using your own symbols (be sure to define them), and keeping only the correct terms. 2) Solve the equation to obtain temperature, T, as a function of radius, r. 3) Sketch temperature, T, as a function of radius, r, using your solution from part 2. Be sure to label your axes and intercepts. 4) The thermal conductivity of the tumor tissue is 0.48 W/m · K, the diameter of the nanoparticles is 6 nanometers, and the body temperature is 310 K. Determine the temperature of the nanoparticle shell, so that cells at a distance of 5 nanometers from its surface can reach the lethal temperature of 323 K. 5) What is the energy absorbed per nanoparticle, in W?

4.32 Can Baby Harp Seals Afford To Be in the Water?

In phocid (earless) harp seals (Figure 4.34), blubber serves as the main thermal insulation instead of the fur. Thus, many scientists question the thermal function of the fur. We will investigate this on baby harp seals, which can be modeled as cylindrical layers of blubber, skin, and fur (see Figure 4.35). Assume steady-state heat transfer from a core temperature of 37°C to the surroundings. The length of the seals is 0.9 m.

Figure 4.34: A baby harp seal. (iStock.com)

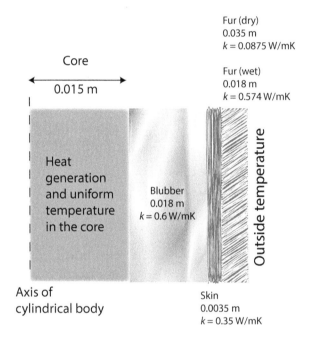

Fur (dry)
0.035 m
$k = 0.0875$ W/mK

Fur (wet)
0.018 m
$k = 0.574$ W/mK

Core

0.015 m

Heat generation and uniform temperature in the core

Blubber
0.018 m
$k = 0.6$ W/mK

Outside temperature

Axis of cylindrical body

Skin
0.0035 m
$k = 0.35$ W/mK

Figure 4.35: Schematic showing dimensions for various layers.

1) What fraction of the total thermal resistance is contributed by the fur when it is dry? 2) Compare this to the fraction of total thermal resistance contributed by the fur when it is wet and discuss why it is not a good idea for a newborn to be in the water. 3) Assume the inside surface of the blubber is at core temperature. Estimate the metabolic heat generation in W/m^3 in the core of a *dry pup* at steady state when the outer surface of the fur is at the air temperature of 4°C. 4) The pup in question 3) is now placed in 4°C water. Assume its fur surface temperature to be that of the water. Theoretically speaking, if this pup is to maintain the same core temperature on the inside surface of the blubber as the pup in question 3), by what factor must its metabolic heat generation be increased?

4.33 Reducing Heat Loss Using Advanced Insulation

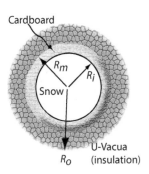

Figure 4.36: Schematic of spherical snow and package.

In January 2011, a Bahraini girl wanted her younger brother to see snow. Panasonic's SPARKS team answered the call by shipping a snowman from Japan (5,314 miles over 40 hours) without refrigeration. They accomplished this by using U-Vacua, an insulation material with an insulation capability that is many times greater than our home insulation.

To understand how this feat was accomplished, the package can be simplified as concentric spheres, as shown in Figure 4.36, with heat transfer at steady state. The thermal conductivities of cardboard and U-Vacua are 0.067 and 0.00070 W/m·K, respectively. The values of R_i, R_m, and R_o, respectively, are 50, 60, and 62 cm. At the radius R_i, the snow is assumed to be at 0°C always. The outside air temperature is 43.3°C and the heat transfer coefficient is 10 W/m^2 · K.

1) Write the governing equation and boundary conditions for steady-state heat conduction in a hollow sphere, as shown in Figure 4.37, with uniform temperatures specified on both surfaces. 2) Solve the governing equation to obtain an algebraic expression for temperature and heat flow. 3) From the expression for heat flow, derive an expression for the conductive thermal resistance of a spherical shell. 4) Calculate the rate of heat gain in W from the ambient air to snow, as shown in Figure 4.37. 5) The latent heat of melting of snow is 333 kJ/kg, and the density of snow is 700 kg/m^3. What is the heat gain over the 40-hour period (in J), *and* what fraction of snow will melt, assuming snow is always at 0°C?

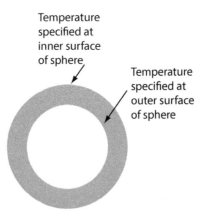

Temperature specified at inner surface of sphere

Temperature specified at outer surface of sphere

Figure 4.37: A hollow sphere.

4.34 Thermoregulation in Arctic Shore Birds

For shore birds that live in the cold regions of the world, chicks can experience thermoregulatory issues because of their small size and poor insulation. We wish to do a heat balance on such a chick. The chick's body can be approximated as a cylinder, as shown in Figure 4.38, with feathers surrounding it. The external resistance between the feather outer surface and the surroundings consists of two resistances in parallel—convective and radiative.

Figure 4.38: Arctic shore bird Ruddy Turnstones. (iStock.com/Brian E Kushner)

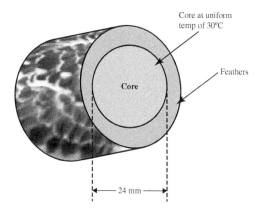

Core at uniform temp of 30°C

Feathers

Core

24 mm

Figure 4.39: Schematic: A crude approximation for heat transfer analysis.

1) If the radiative heat transfer from the outer surface of the feathers can be approximated as $q = 4\sigma T_\infty^3 A(T_s - T_\infty)$, where T_s is the temperature at the outer surface of the feathers and T_∞ is the ambient and air temperature, write an expression for the radiative resistance, making sure the variables used are explained. 2) Develop an expression for total resistance between the inner surface of the feathers and the surroundings at T_∞ that includes both convective and radiative heat loss on the outside and conductive heat loss through the feathers. 3) In addition to heat loss corresponding to step 2), 18% of the total metabolic heat is lost through evaporation. What is the total metabolic heat generation needed by the chick to maintain the core temperature of 30°C? The diameter of the core in Figure 4.39 is 24 mm and the height of the cylinder is 4 cm, the thickness of the feathers is 4 mm, the outside air temperature is 0°C, the outside heat transfer coefficient is 200 W/m² · K, $\sigma = 5.670 \times 10^{-8}$W/m²K⁴, the thermal conductivity of the feathers is 0.02 W/m · K. 4) For the situation in 3), if the thickness of the insulation is halved, what would have to be the new rate of heat generation to keep the core temperature at 30°C? Assume evaporation is to remain constant.

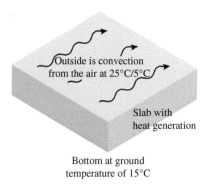

Figure 4.40: Schematic of a slab of waste.

4.35   Heat Generation in Waste as Bioreactor

Bioreactor landfills aim to optimize the conditions within waste to enhance waste stabilization. Consider constant distributed heat generation at steady state in a waste pile of height L whose bottom is kept at the ground temperature of T_g, and air at temperature T_∞ is blowing at the top surface with a heat transfer coefficient h. The uniform rate of heat generation in the waste is Q W/m^3.

1) Write the governing equation and boundary conditions for the problem. 2) Solve to provide temperature as a function of position in the waste. Sketch the solution and mention whether the maximum temperature will be in the middle. 3) If the air temperature changes from 25°C in the summer to 5°C in late fall, how much will the maximum temperature in the pile change? The height of the pile is 5 m and the rate of heat generation is 2 W/m^3. Use $T_g = 15$°C, thermal conductivity of waste as 0.67 W/m·K, and a heat transfer coefficient of 5 W/m$^2 \cdot$K. 4) In the winter, heat generation ceases when the surrounding temperature becomes low enough. If we assume all heat generation has ceased (but all other conditions stay the same), what would be the new steady-state maximum temperature and where would it be located?

4.36   Detecting a Skin Lesion from Thermal Response

In a 2010 research paper, measurement of surface temperature of the skin was suggested for possible early detection of melanoma. In a highly simplified analysis (see Figure 4.41), consider the skin as a 1.5 mm thick slab with one side (deep skin) maintained at the core body temperature of 37°C. The surface of the skin has small air flow over it, leading to a heat transfer coefficient of $h = 10$ W/m$^2 \cdot$K while the air temperature is 20°C. Before making the measurement, the skin is first allowed to reach steady state. The average thermal conductivity of the 1.5 mm layer is 0.4 W/m·K and its rate of metabolic heat generation in healthy individuals is 368 W/m^3, which is uniform throughout the skin.

1) Write the governing equation for this problem, keeping only the needed terms. 2) Write the two boundary conditions needed to solve this problem. 3) Solve the governing equation and provide an expression for surface temperature in terms of all other variables (no need to plug in numbers). 4) By some estimate, the metabolic heat generation in the skin layer increases to 3680 W/m^3 when a lesion is present. At steady state, what is the difference in surface temperature that you would expect between normal skin and a skin lesion?

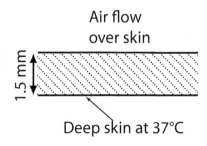

Figure 4.41: Schematic of skin with air flow on surface and fixed temperature at the deep end.

Chapter 5

CONDUCTION HEAT TRANSFER: UNSTEADY STATE

CHAPTER OBJECTIVES

After you have studied this chapter, you should be able to formulate and solve for a time-varying (unsteady) heat conduction process for the following situations:

1. Where temperatures do not change with position.

2. In a simple slab geometry where temperatures vary also with position.

3. Near the surface of a large body (semi-infinite region).

KEY TERMS

- **internal resistance**

- **external resistance**

- **Biot number**

- **lumped parameter analysis**

- **heat conduction in 1D slab, cylinder, and sphere**

- **Heisler charts**

- **heat conduction in a semi-infinite region**

In this chapter, we will consider heat transfer situations where temperature is a function of both position and time. Variation in temperature with time is referred to as unsteady, as opposed to steady state studied in Chapter 4 where temperature did not vary with time. The relationship of this chapter to other chapters on energy transfer is shown in Figure 5.1.

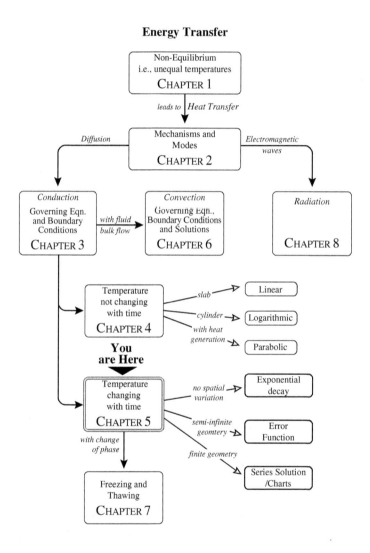

Figure 5.1: Concept map of energy transfer showing how the contents of this chapter relate to other chapters on energy transfer.

The rate of energy transfer in a solid depends on the internal thermal resistance of the solid and the external thermal resistance (usually the convective thermal resistance of the fluid flowing over the surface). The limiting cases are

1. Negligible internal resistance (lumped parameter)

2. Negligible external resistance

3. Neither resistance is negligible

5.1   Transient Heat Transfer When Internal Conductive Resistance Is Negligible: Lumped Parameter Analysis

In some special cases, temperature variation in all three spatial directions can be ignored. When the temperature variations are ignored, the situation is considered lumped. Temperature would then vary only with time. Note that this can only be an approximation and it is physically impossible for this to happen exactly. If temperature does not vary spatially, there can be no heat flow and therefore no change in temperature with respect to time. The transient heating or cooling of the solid is possible only if there are temperature gradients inside it. In reality, there will be temperature variations inside the solid, except these variations will be small compared to the temperature variation outside the solid (in the fluid). In other words, this lumped parameter condition is possible when the internal resistance in the solid is small compared to the external resistance in the fluid.

In Chapter 3, we derived the governing equations for heat transfer and mentioned that they were general enough to consider all kinds of situations, i.e., whether large thermal conductivity or small. It is in fact possible to start from one of these governing equations and derive the governing equation for this special case of lumped parameter or no internal resistance (see page 600 in the appendix). However, since spatial variations are ignored, there is an alternative and much easier way to derive the governing equation for this situation, starting for energy conservation:

$$\boxed{\text{Energy In}} - \boxed{\text{Energy Out}} + \boxed{\text{Energy Generated}} = \boxed{\text{Change in Energy Stored}} \qquad (5.1)$$

Consider the schematic in Figure 5.2. If the temperature change in the solid is ΔT during time Δt, we can write the various quantities in the above word equation for

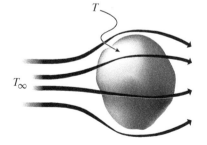

Figure 5.2: A solid with convection over its surface.

energy conservation as

$$\overbrace{0}^{\substack{\text{Energy} \\ \text{in}}} - \overbrace{hA(T - T_\infty)\Delta t}^{\substack{\text{Energy out} \\ \text{by convection}}} + \overbrace{0}^{\substack{\text{Energy} \\ \text{generated}}} = \overbrace{mc_p\Delta T}^{\substack{\text{Change} \\ \text{in storage}}} \tag{5.2}$$

where m is the mass of the solid, c_p is its specific heat, h is the convective heat transfer coefficient at the surface, A is the surface area, and T_∞ is the bulk fluid temperature. Note that the right-hand side of Eq. 5.2 uses T as the surface temperature in the formula for convective heat loss since the entire solid is assumed to be at one temperature, T. Also, it is important to note that, in writing the above equation, we do not need to know whether the solid is hotter or colder than the fluid, i.e., whether T is higher or lower than T_∞. The equation above treats the convection as an energy out and therefore we may have implied T to be greater than T_∞. However, if T is assumed less than T_∞ instead so that heat would be gained, we could have written

$$\overbrace{hA(T_\infty - T)\Delta t}^{\substack{\text{Energy} \\ \text{in}}} - \overbrace{0}^{\substack{\text{Energy out} \\ \text{by convection}}} + \overbrace{0}^{\substack{\text{Energy} \\ \text{generated}}} - \overbrace{mc_p\Delta T}^{\substack{\text{Change} \\ \text{in storage}}} \tag{5.3}$$

Notice that Eqs. 5.3 and 5.2 are the same. Rearranging the terms,

$$-\frac{\Delta T}{\Delta t} = \frac{hA}{mc_p}(T - T_\infty)$$

Taking the limit as $\Delta t \to 0$,

$$\frac{dT}{dt} = -\frac{hA}{mc_p}(T - T_\infty) \tag{5.4}$$

This is the governing equation describing temperature vs. time for a lumped parameter equation. Being a first-order equation, it needs one condition, which in this case is the initial condition

$$T(t = 0) = T_i \tag{5.5}$$

To solve Eqs. 5.4, define transformed temperatures as

$$\begin{aligned} \theta &= T - T_\infty \\ \theta_i &= \theta(t = 0) = T_i - T_\infty \end{aligned}$$

Using the new variable θ, the governing equation 5.4 is transformed as

$$\frac{d\theta}{dt} = -\frac{hA}{mc_p}\theta$$

This equation is integrated over time and the initial condition $t = 0$ is set to $\theta = \theta_i$ to obtain

$$\int_{\theta_i}^{\theta} \frac{d\theta}{\theta} = \int_0^t -\frac{hA}{mc_p} dt \tag{5.6}$$

$$\ln \frac{\theta}{\theta_i} = -\frac{hA}{mc_p} t$$

$$\frac{\theta}{\theta_i} = e^{-\frac{hA}{mc_p} t}$$

Changing the variable from θ to T, we get temperature as a function of time in lumped parameter heat transfer as

$$\frac{T - T_\infty}{T_i - T_\infty} = \exp\left(-\frac{hA}{mc_p} t\right) = \exp\left(-\frac{t}{\frac{mc_p}{hA}}\right) \tag{5.7}$$

Equation 5.7 shows it takes an infinite time to reach the steady state or the final temperature of T_∞. As the temperature of the solid becomes close to the fluid temperature, the rate of heat transfer drops so the solid can never quite reach the fluid temperature T_∞.

Figure 5.3 is a graphical representation of the relationship between time and temperature given by Eq. 5.7. Note that, although much of the total possible change in temperature $T_\infty - T_i$ happens over a short time, the rate of temperature change decreases continuously and the temperature never really reaches T_∞. In practice, our measuring instrument has only a finite resolution so, when the temperature is sufficiently close to T_∞, the measuring instrument will read T_∞, i.e., we say the final temperature has been reached.

5.2 Biot Number: Deciding When to Ignore the Internal Resistance

The lumped parameter solution developed in the previous section is quite simple. It is tempting to use such an approach for many practical situations. We need to decide when such an approach is a valid one, i.e., what is the criterion for ignoring the internal resistance as compared to the external resistance? The parameter that compares the internal and external resistance is

$$\text{Bi (Biot number)} = \frac{hL}{k} = \frac{\frac{L}{kA}}{\frac{1}{hA}} = \frac{\text{conductive resistance}}{\text{convective resistance}} \tag{5.8}$$

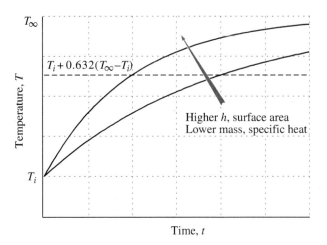

Figure 5.3: Typical temperature change in a lumped parameter analysis.

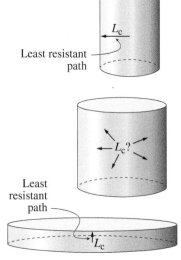

Figure 5.4: Characteristic lengths for heat conduction in various geometries.

Thus, for a large Biot number, the (internal) conductive resistance is higher and therefore the controlling resistance, i.e., we can ignore the (external) convective resistance. Conversely, for a small Biot number, the convective resistance is the controlling one and we can ignore the conductive resistance. For a Biot number Bi < 0.1, i.e.,

$$\frac{h(V/A)}{k} < 0.1 \qquad (5.9)$$

the error in temperature calculation is $< 5\%$ if the lumped parameter approach is used. Qualitatively speaking, lumped parameter is suitable for large surface areas, small volumes, small convective heat transfer coefficients, and large thermal conductivities. Most of the thermal resistance is external, in the fluid surrounding the object.

Characteristic Length   The quantity V/A used in place of L in Eq. 5.9 has the dimension of length and can be called a *characteristic length* of the system. In heat conduction, characteristic length can be interpreted as the path of least thermal resistance for the geometry of concern. Thus, as surface area increases for the same volume, the effective distance through which heat has to diffuse decreases, i.e., the characteristic length decreases, and heating or cooling the body will require less time. As an example, using Table 5.1, for an infinite slab, Eq. 5.9 is written as

$$hL/k < 0.1 \qquad (5.10)$$

Table 5.1: Geometry, characteristic length, and Biot number definition

	Infinite slab	Long cylinder	Sphere
Characteristic length	1/2 thickness	Radius	Radius
Biot number (Bi)	hL/k	hR/k	hR/k

Figure 5.4 illustrates that, while the characteristic length for a tall cylinder is its radius and for a thin cylinder it is its half thickness, at an intermediate length (middle figure), characteristic length is not so obvious.

5.2.1 Example: Keeping Eggs Warm in an Incubator

It is desired to find the transient temperature of chicken eggs when they are placed in an incubator. The incubator air temperature is 38°C. The natural convection heat transfer coefficient for air over the eggs in the incubator is 2 W/m²·K. Calculate the temperature of the egg after 30 minutes if its initial temperature is 20°C. Assume the eggs to be spherical with a volume of 60 cm³. The density, specific heat, and thermal conductivity of the egg are 1035 kg/m³, 3350 J/kg·K, and 0.62 W/m·K, respectively.

Solution

Understanding and formulating the problem *1) What is the process?* It is a transient heat conduction process in a sphere. *2) What are we solving for?* We

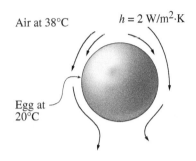

Air at 38°C $h = 2$ W/m^2·K

Egg at
20°C

Figure 5.5: Schematic for Example 5.2.1.

need to find the temperature after 30 minutes, knowing the initial temperature, surrounding fluid temperature, and the surface heat transfer coefficient. *3) Schematic and given data:* A schematic is shown in Figure 5.5 with some of the given data superimposed on it. *4) Assumptions:* The surface heat transfer coefficient provided is an average value and its likely variation over the surface is ignored.

Generating and selecting among alternate solutions *1) What solutions are possible?* For transient heat transfer, the solutions we have available are shown in the solution chart in Figure 5.21. *2) What approach is likely to work?* As we look at the solution chart in Figure 5.21, since our interest is in the center temperature, it is the case of a finite geometry (temperature is changing everywhere). Next we calculate the Biot number to decide whether a lumped parameter solution can be applicable. As shown in the left column of Table 5.2, the Biot number is less than 0.1 and therefore a lumped parameter solution would be appropriate.

Implementing the chosen solution The first column of Table 5.2 shows the computations. The lumped parameter solution (Eq. 5.7) is used to calculate the temperature.

Evaluating and interpreting the solution *1) Does the temperature value make sense?* One obvious check is that the temperature calculated should be between the initial temperature (20°C) and the air temperature (38°C), which it is. The calculated $\sim 2°C$ rise in one hour does seem low—this can be attributed to a very low value of h used in the problem (see Figure 6.19 for a typical range for natural convection in air). Such a low value of h amounts to somewhat of an insulating outside. *2) What do we learn?* The heat transfer resistance in this case is mostly external (in the air) as opposed to internal (within the egg). The temperature variation inside the egg can be ignored compared to the temperature variation outside, in the air.

Table 5.2: How the situations determine which resistance is important and thus which solution is appropriate (Examples 5.2.1 & 5.3.9)

Example 5.2.1	Example 5.3.9
An egg being warmed by slow moving air in an incubator	An egg being heated with fast moving warm water
For a finite geometry, Biot number decides between lumped parameter and series solution	

$$
\begin{aligned}
\text{Bi} &= \frac{hR}{k} \\
&= \frac{(2\ \text{W/m}^2\text{K})\,(0.0243\ \text{m})}{0.62\ \text{W/m} \cdot \text{K}} \\
&= .08 < 0.1
\end{aligned}
$$

$$
\begin{aligned}
\text{Bi} &= \frac{hR}{k} \\
&= \frac{(10^4\ \text{W/m}^2\text{K})\,(0.0243\ \text{m})}{0.62\ \text{W/m} \cdot \text{K}} \\
&= 391.9 \gg 0.1
\end{aligned}
$$

Internal resistance is small compared to external

Internal resistance is large compared to external

Based on the calculated Bi, choose a solution technique

$$
\frac{T - T_\infty}{T_i - T_\infty} = \exp\left(-\frac{t}{mc_p/hA}\right)
$$

A series solution for the spherical geometry, plotted as the Heisler chart (Figure B.3)

Compute the solution

$$
\begin{aligned}
m &= (1035\ \text{kg/m}^3) \\
&\quad \times (4/3)(\pi)(0.0243)^3\ \text{m}^3 \\
&= .0622\ \text{kg} \\
A &= 4\pi(.0243)^2 = 0.00742\ \text{m}^2 \\
mc_p &= (.0622\ \text{kg})(3350\ \text{J/kg} \cdot \text{K}) \\
hA &= (2\ \text{W/m}^2\text{K})(0.00742\ \text{m}^2)
\end{aligned}
$$

$$
\begin{aligned}
n &= r/R = 0 \\
m &= \frac{k}{hR} = \frac{0.62\ \text{W/m} \cdot \text{K}}{(10^4\ \text{W/m}^2 \cdot \text{K})(0.0243\ \text{m})} \\
&= 0.0025 \approx 0 \\
Fo &= \frac{\alpha t}{L^2} = \frac{(1.79 \times 10^{-7}\text{m}^2/\text{s})(1800\ \text{s})}{(0.0243)^2\ \text{m}^2} \\
&= 0.546
\end{aligned}
$$

Plugging in

From chart for sphere

$$
\begin{aligned}
\frac{T - 38}{20 - 38} &= \exp\left(-\frac{1800\ \text{s}}{14041\ \text{s}}\right) = 0.88 \\
T &= 38 + 0.88(20 - 38) \\
&= 22.16°\text{C}
\end{aligned}
$$

$$
\begin{aligned}
\frac{T - T_\infty}{T_i - T_\infty} &= 0.017 \\
T &= 38 + 0.017(20 - 38) \\
&= 37.69°\text{C}
\end{aligned}
$$

5.3 Transient Heat Transfer When Internal Resistance Is Not Negligible: Example of Slab Geometry

In the previous section, it was noted that when the Biot number (which is a function of properties of the solid and fluid and the velocity of the fluid) is not small, internal diffusional resistances are significant. This means temperature variation within the solid is now significant, unlike what we dealt with for the lumped parameter analysis. In this section we will learn how to analyze the spatial variation of temperature in a solid. For simplicity, we will consider a one-dimensional slab where temperature will vary along the thickness, as shown in Figure 5.6. To keep things simple, we will consider the situation where the surface is at a specified temperature, i.e., external fluid resistance is negligible. The governing equation for symmetric heating or cooling of an infinite slab without any heat generation can be simplified from

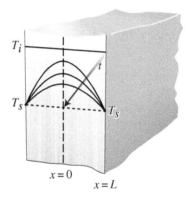

Figure 5.6: Schematic of a slab showing the line of symmetry at $x = 0$ and the two surfaces at $x = L$ and at $x = -L$ maintained at temperature T_s. The material is very large (extends to infinity) in the other two directions.

$$\underbrace{\rho c_p \frac{\partial T}{\partial t}}_{\text{storage}} + \underbrace{u \overbrace{\frac{\partial T}{\partial x}}^{0}}_{\text{no bulk flow}} = \underbrace{k \left(\frac{\partial^2 T}{\partial x^2} \right)}_{\text{conduction}} + \underbrace{\overbrace{Q}^{0}}_{\substack{\text{no heat} \\ \text{generation}}}$$

as

$$\frac{\partial T}{\partial t} = \frac{k}{\rho c_p} \frac{\partial^2 T}{\partial x^2} \tag{5.11}$$

The boundary conditions are

$$\frac{\partial T}{\partial x}\bigg|_{x=0,t} = 0 \quad \text{(from symmetry)} \tag{5.12}$$

$$T(L, t > 0) = T_s \quad \text{(surface temperature is specified)} \tag{5.13}$$

and the initial condition is

$$T(x, t = 0) = T_i \tag{5.14}$$

where T_i is the constant initial temperature and T_s is the constant temperature at the two surfaces of the slab at time $t > 0$. Due to the same temperature T_s on both sides and the same initial temperature everywhere, the temperature profile will always be symmetric, leading to the symmetry condition in Eq. 5.12. Note that we needed two boundary conditions in x since Eq. 5.11 is second order in position x and one (initial) condition in time since it is first order in time. The solution to Eq. 5.11 is based on separation of variables (see page 593 for details) and is given by

$$\frac{T - T_s}{T_i - T_s} = \sum_{n=0}^{\infty} \frac{4(-1)^n}{(2n+1)\pi} \cos\frac{(2n+1)\pi x}{2L} e^{-\alpha\left(\frac{(2n+1)\pi}{2L}\right)^2 t} \tag{5.15}$$

where $\alpha = k/\rho c_p$ is the thermal diffusivity, defined on page 29. Equation 5.15 provides temperature T as a function of position x and time t. Note that the position appears as a non-dimensional quantity x/L and time as another non-dimensional quantity $\alpha t/L^2$. The non-dimensional time is called the Fourier number Fo so that

$$Fo = \frac{\alpha t}{L^2} \tag{5.16}$$

Use of these non-dimensional quantities is discussed below.

5.3.1 How Temperature Changes with Time

We now try to visualize the relationship of temperature vs. position and time provided by Eq. 5.15. The infinite series makes it difficult to visualize this relationship and we will try to simplify it for special situations. As can be seen in Figure G.1 on page 596, we need a large number of terms in the series to satisfy the initial condition. Conversely, if we are interested in times long after $t = 0$, we may not need as many terms. To compare the contribution of the various terms at different times, the relative values of the exponential terms for $n = 1, 2, \ldots$ in Eq. 5.15 are shown in Figure 5.7 for a large and a small value of time. For large value of t (= 600 s), the exponential terms

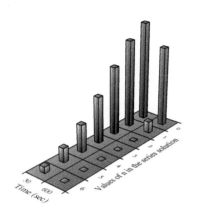

Figure 5.7: The terms in the series (n = 0, 1,... in Eq. 5.15) drop off rapidly for large values of time. Calculations are for Fo = 0.0048 at 30 s and Fo = 0.096 at 600 s for a thickness of L = 0.03 m and a typical α = $1.44 \times 10^{-7} \, \text{m}^2/\text{s}$ for biomaterials.

in Eq. 5.15 decay more rapidly, as shown in this figure. Therefore, the contributions of the subsequent terms to the summation can perhaps be ignored for large values of time and we would have less error as compared to ignoring terms for smaller values of time where the terms decay less rapidly (see Figure 5.7 for t = 30 s). After long times (Figure 5.7 for t = 600 s), often keeping only the first term (n = 0) is a good approximation to the exact solution. It turns out that "long" time is defined as $Fo > 0.2$. Substituting $n = 0$ in Eq. 5.15 to keep only the first term, we get

$$\frac{T - T_s}{T_i - T_s} = \underbrace{\frac{4}{\pi} \cos \frac{\pi x}{2L}}_{\text{spatial}} \underbrace{e^{-\alpha \left(\frac{\pi}{2L}\right)^2 t}}_{\text{time}} \tag{5.17}$$

This expression clearly shows that eventually (for times long after t = 0) the time–temperature relationship is exponential at a given position (x value). It can never reach the final temperature T_s. As the temperature difference between any location and the surrounding (T_s) decreases, the rate of heat flow decreases. If we take the natural logarithm of both sides of Eq. 5.17, we get

$$\ln \frac{T - T_s}{T_i - T_s} = \ln \left(\frac{4}{\pi} \cos \frac{\pi x}{2L} \right) - \alpha \left(\frac{\pi}{2L} \right)^2 t \tag{5.18}$$

This linear relationship between $\ln(T - T_s)/(T_i - T_s)$ and time t can be seen in Figure B.1 on page 555 for various values of position x/L. Note that, at times close to t = 0, they are not linear. This is expected since by dropping terms in the series we fail to satisfy the initial condition.

5.3.2   Temperature Change with Position and Spatial Average

It can be easily seen from Eq. 5.17 that eventually, at a given time, the temperature varies as a cosine function. Note that it stays as a cosine function for all large values of time, except the amplitude of the cosine wave drops exponentially with time. In practice, a spatial average temperature is often needed. A spatial average temperature would be defined by

$$T_{av} = \frac{1}{L} \int_0^L T \, dx \tag{5.19}$$

Applying this definition of average to Eq. 5.18, we obtain an equation for the average temperature as

$$\ln \frac{T_{av} - T_s}{T_i - T_s} = \ln \frac{8}{\pi^2} - \alpha \left(\frac{\pi}{2L} \right)^2 t \tag{5.20}$$

5.3.3 Temperature Change with Size

Consider the equation for average temperature as a function of time for large values of time. It can be rewritten as

$$\frac{\alpha t}{L^2} = -\frac{4}{\pi^2} \ln\left[\frac{\pi^2}{8} \left(\frac{T_{av} - T_s}{T_i - T_s} \right) \right] \tag{5.21}$$

This implies that, for a given change of average temperature (measured in terms of fractional change of the total possible change $T_i - T_s$), the time required increases with the square of the thickness, i.e., $t \propto L^2$. This observation can be generalized for other geometries by saying the time required is proportional to the square of the characteristic dimension.

5.3.4 Charts Developed from the Solutions: Their Uses and Limitations

The solution given by Eq. 5.15 would require much effort in computing every time we need to use it. Fortunately, this series has already been computed and plotted for most practical situations, as shown in Figure B.1 on page 555. We simply need to use these charts. To see how these charts are developed, note that Eq. 5.15 can be rearranged as below:

$$
\begin{aligned}
\frac{T - T_s}{T_i - T_s} &= \sum_{n=0}^{\infty} \frac{4(-1)^n}{(2n+1)\pi} \cos\frac{(2n+1)\pi x}{2L} e^{-\alpha\left(\frac{(2n+1)\pi}{2L}\right)^2 t} \\
&= \sum_{n=0}^{\infty} \frac{4(-1)^n}{(2n+1)\pi} \cos\left[\frac{(2n+1)\pi}{2}\frac{x}{L}\right] e^{-\left(\frac{(2n+1)\pi}{2}\right)^2 \frac{\alpha t}{L^2}} \quad (5.22)
\end{aligned}
$$

Equation 5.22 shows that the non-dimensional temperature $(T - T_s)/(T_i - T_s)$ is a function of non-dimensional position x/L and non-dimensional time $\alpha t/L^2$, and therefore these three non-dimensional variables are sufficient to describe the time–temperature relationship at various locations in the slab. Figure B.1 on page 555 has plots of the relationships between these variables and is known as a Heisler chart. Note, however, Figure B.1 is not a plot of Eq. 5.22, but of Eq. 5.17, which has only the first term in the series and thus the straight lines in this figure are valid at long times. For smaller times, we have to keep more terms in the series. If more terms are kept, the relationship between temperature and time (Eq. 5.22) would not be a straight line near $t = 0$.

In Figure B.1, the set of lines corresponding to $m = 0$ are plots of Eq. 5.17 for the boundary condition of specified surface temperature T_s that is equivalent to $h \to \infty$.

Note that $n(= x/L)$ in the charts has a different meaning than in Eq. 5.22. Other sets of lines corresponding to various values of $m(= k/hL)$ are for convective boundary conditions (finite values of h) and are discussed in Section 5.3.6.

Since the charts were developed from the analytical solution, the assumptions that went into the analytical solution are implicit in the charts. It is important to remind ourselves of these conditions that have to be satisfied in order to be able to use the charts. These are

- Uniform initial temperature

- Constant boundary fluid temperature

- Perfect slab, cylinder, or sphere

- Far from edges

- No heat generation ($Q = 0$)

- Constant thermal properties (k, α, c_p are constants)

- Typically for times long after initial times, given by $\alpha t/L^2 > 0.2$

5.3.5 *Example: Temperatures Reached during Food Sterilization*

Sterilization of food and biological material is a very important application of heat transfer to biological systems. Consider sterilization of a tuna can (Figure 5.8) that is filled with tuna meat and water. Considering the rate of bacterial death, it has been decided to heat this container for 30 minutes to reach sterilization. Calculate the temperature after 30 minutes at the location where the temperature is the lowest, i.e., the location in the container that stays the coldest during heating. The thermal diffusivity of the tuna meat with water is $2 \times 10^{-7} \mathrm{m}^2/\mathrm{s}$. Heating is done in steam that maintains all surfaces of the can at 121°C. For simplicity, ignore any heat transfer from the side, i.e., the radial direction, so that the heat transfer is one dimensional. The initial temperature of the tuna is 40°C. Assume the thermal resistance of the can material to be negligible and ignore any convective movement of water in the can during heating.

Figure 5.8: A cylindrical can containing food to be sterilized.

Understanding and formulating the problem *1) What is the process?* This is an unsteady-state heat transfer process where a can filled with material is heated from the surface. *2) What are we solving for?* The coldest point in the can is the geometric center. Thus, we need to find the temperature at the center of the 25 mm slab after 30 minutes of heating. *3) Schematic and given data:* A schematic is shown in Figure 5.9 with some of the given data superimposed on it. Additionally, time of heating is 1800 s. *4) Assumptions:* Since thickness is half the size of the diameter, we assume heating from the side may be ignored. The thermal diffusivity value does not change with temperature during the heating process (if it did, we are most likely going to use a numerical solution).

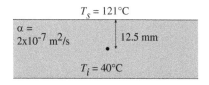

Figure 5.9: Schematic for Example 5.3.5.

Generating and selecting among alternate solutions   As we look at the solution chart in Figure 5.21, we notice our problem is not of a semi-infinite geometry (since the farthest point, the center, is changing in temperature, it is not infinite); therefore, it must be a finite geometry (left side of solution chart). Since the surface temperature is given (which amounts to $h \to \infty$), the Biot number is infinite and not less than 0.01. Therefore, lumped parameter analysis cannot be used. Thus, the series solution is the appropriate one. To decide whether several terms of the series solution would be needed or just one term (Heisler chart), we need to check whether $\alpha t / L^2 > 0.2$ is satisfied, which it is, since $\alpha t / L^2 = 2.3$, letting us use the Heisler chart. Note that we could have proceeded to try to use the Heisler chart anyway since the chart can only be read for $\alpha t / L^2 > 0.2$.

Implementing the chosen solution For transient heat conduction in a slab, the Heisler charts are to be used. The parameters needed to read the chart are the non-dimensional distance n at the center given by

$$n = \frac{x}{x_1} = \frac{0}{.0125} = 0$$

and parameter m signifying the convective condition at the surface given by

$$m = \frac{k}{hx_1} = 0$$

where $h \to \infty$ has been used since the surface temperature is specified (see discussion under the third kind of boundary condition in Section 3.2). The non-dimensional time is calculated as

$$Fo = \frac{\alpha t}{L^2} = \frac{2 \times 10^{-7} \ [\text{m}^2/\text{s}] \ 1800 \ [\text{s}]}{(0.0125)^2 \ [\text{m}^2]} = 2.3$$

With these values of *Fo*, *m*, and *n*, the value of non-dimensional temperature is obtained from the vertical axis of the chart on page 555 as

$$\frac{T - T_\infty}{T_i - T_\infty} = 0.0043$$

So that temperature $T = 120.65°C$ after 30 minutes of heating.

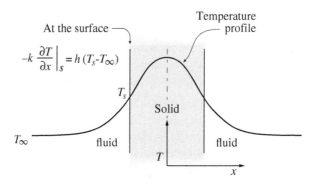

Figure 5.10: In a convective boundary condition, surface temperature is not the same as the bulk fluid temperature, T_∞, signifying additional fluid resistance.

Evaluating and Interpreting the solution *1) Does the temperature value make sense?* One obvious check is that the temperature calculated should be between the initial temperature (40°C) and the air temperature (121°C), which it is. Another check could be to use the one-term series solution (Eq. 5.17) instead of the Heisler chart to do the calculations. Since the Heisler chart is a plot of this equation, this will only check whether the Heisler chart was read correctly. *2) What do we learn?* The actual temperature will be higher (even closer to 121°C) since the can will also heat from the sides. This temperature can be calculated considering heat transfer from both directions using formulas described in Section 5.4.

5.3.6 When Both Internal and External Resistances Are Present: Convective Boundary Condition

So far in Section 5.3 we have considered a negligible external fluid resistance to heat transfer ($h \rightarrow \infty$) represented by the boundary condition of specified surface temperature. A convective boundary condition on the slab surface (denoted by location s in Figure 5.10) is written as

$$-k \left. \frac{\partial T}{\partial x} \right|_s = h(T_s - T_\infty)$$

When h is finite (instead of $h \rightarrow \infty$), external resistance of the fluid ($1/h$) needs to be considered. Here T_∞, the external fluid temperature, is known, but the surface temperature T_s is not. The solution procedure used to obtain Eq. 5.15 can be generalized for any value of h, and, as expected, the right-hand side of Eq. 5.15 would become a function of h as well. In the final solution, T_s is replaced by T_∞, the fluid temperature; i.e., the non-dimensional temperature on the left of Eq. 5.15 is replaced by $(T - T_\infty)/(T_i - T_\infty)$. Thus, additional sets of lines are added in Figures B.1, B.2, and B.3 on pages 555, 556, and 557 for various values of $m = k/hL$ where m is the inverse of the non-dimensional Biot number hL/k.

5.3.7 Effect of Boundary Condition: Analytical Solutions for the Three Different Boundary Conditions

Solutions must change with boundary conditions, since the physical process changes. Mathematically speaking, different conditions at the boundaries will definitely lead to different solutions to the same differential equation. Table 5.3 provides the solutions for a slab for the three common boundary conditions of specified temperature, heat flux, or convection, applied symmetrically.

5.3.8 Effect of Geometry: Analytical Solutions for Slab, Cylinder, and Sphere

Analytical solution in terms of an infinite series, as derived earlier for a slab, can be derived for other geometries in appropriate coordinate systems. Solutions for the three most common shapes of solid slab, cylinder, and sphere, and for a temperature-specified boundary condition, are shown in Table 5.4 for 1D situations. When the boundary condition changes to convection instead of temperature specified, the solutions become more complex than those shown in Table 5.4 (for a slab, see the last row in Table 5.3). However, solutions for both temperature-specified and convection boundary conditions have been included in the respective Heisler charts. To read the Heisler charts, one can use Table 5.5.

Table 5.3: Transient temperature profiles in a slab for various boundary conditions.

Geometry	Temperature Profile[1]

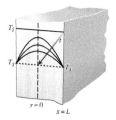

$$\frac{T - T_s}{T_i - T_s} = \sum_{n=0}^{\infty} \frac{4(-1)^n}{(2n+1)\pi} \cos \frac{(2n+1)\pi x}{2L} e^{-\left(\frac{(2n+1)\pi}{2}\right)^2 \frac{\alpha t}{L^2}}$$

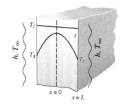

$$T = T_i + 2A\alpha t + Ax^2 - \frac{AL^2}{3}$$
$$- \frac{4AL^2}{\pi^2} \sum_{n=1}^{\infty} \frac{(-1)^n}{n^2} \cos \frac{n\pi x}{L} e^{-(n\pi)^2 \frac{\alpha t}{L^2}}$$

where $A = q_s''/(2kL)$

$$\frac{T - T_\infty}{T_i - T_\infty} = \sum_{n=1}^{\infty} \frac{2(\lambda_n^2 + (h^2/k^2))}{h/k + L(\lambda_n^2 + (h^2/k^2))}$$
$$\times \sin(\lambda_n L) \cos(\lambda_n x) e^{-\lambda_n^2 \alpha t}$$

where λ_n is given by $\lambda_n \tan(\lambda_n L) = h/k$

[1]The constant heat flux solution is taken from Carslaw and Jaeger (1992), page 112.

Table 5.4: Transient temperature profiles in various coordinate systems for surface-temperature-specified boundary conditions

Geometry	Temperature profile
1D slab, x direction	$\dfrac{T-T_s}{T_i-T_s} = \displaystyle\sum_{n=0}^{\infty} \dfrac{4(-1)^n}{(2n+1)\pi} \cos\dfrac{(2n+1)\pi x}{2L} e^{-\left(\frac{(2n+1)\pi}{2}\right)^2 \frac{\alpha t}{L^2}}$
1D cylinder, radial direction	$\dfrac{T-T_s}{T_i-T_s} = \displaystyle\sum_{n=1}^{\infty} \dfrac{2}{\alpha_{n0} J_1(\alpha_{n0})} J_0\left(\alpha_{n0}\dfrac{r}{R}\right) e^{-(\alpha_{n0})^2 \frac{\alpha t}{R^2}}$ α_{n0} are the roots of the equation $J_0(\alpha_{n0}) = 0$
1D sphere, radial direction	$\dfrac{T-T_s}{T_i-T_s} = 2\displaystyle\sum_{n=1}^{\infty} \dfrac{(-1)^{n+1}}{n\pi r} \sin\left(n\pi \dfrac{r}{R}\right) e^{-(n\pi)^2 \frac{\alpha t}{R^2}}$

Table 5.5: Geometry, characteristic length, and solution using Heisler charts

Geometry	Characteristic length	When reading Heisler chart		
		m value	n value	F_o value
Infinite slab	$L = 1/2$ thickness	k/hL	x/L	$\alpha t/L^2$
Long cylinder	$R = $ radius	k/hR	r/R	$\alpha t/R^2$
Sphere	$R = $ radius	k/hR	r/R	$\alpha t/R^2$

5.3.9   Example: Redoing Example 5.2.1 for a Different Heating Situation

It is desired to find the transient temperature of chicken eggs when they are placed in warm water. The warm water temperature is 38°C. The forced convection heat transfer coefficient for water over the eggs is 10^4 W/m$^2 \cdot$ K. Calculate the temperature of the eggs after 30 minutes if the initial temperature is 20°C. Assume the eggs to be spherical with a radius of 2.43 cm. The density, specific heat, and thermal conductivity of the eggs are 1035 kg/m^3, 3350 J/kg $\cdot$ K, and 0.62 W/m $\cdot$ K, respectively.

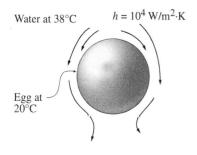

Water at 38°C $h = 10^4$ W/m$^2 \cdot$K

Egg at 20°C

Figure 5.11:   Schematic for Example 5.3.9.

Understanding and formulating the problem *1) What is the process?* It is a transient heat conduction process in a sphere. *2) What are we solving for?* We need to find the temperature after 30 minutes knowing the initial temperature, surrounding fluid temperature, and the surface heat transfer coefficient. *3) Schematic and given data:* A schematic is shown in Figure 5.11 with some of the given data superimposed on it. *4) Assumptions:* The surface heat transfer coefficient provided is an average value and its likely variation over the surface is ignored.

Generating and selecting among alternate solutions *1) What solutions are possible?* For transient heat transfer, the solutions we have available are shown in the solution chart in Figure 5.21. *2) What approach is likely to work?* As we look at the solution chart in Figure 5.21, since our interest is in the center temperature, it is the case of a finite geometry (temperature is changing everywhere). Next we calculate the Biot number to decide whether a lumped parameter solution can be applicable. As shown in the right column of Table 5.2 on page 135, the Biot number is much larger than 0.1 and therefore a series solution is required. To decide whether several terms of the series solution would be needed or just one term (Heisler chart), we need to check whether $\alpha t / L^2 > 0.2$ is satisfied, which it is, since $\alpha t / L^2 = 0.546$, letting us use the Heisler chart. Note that we could have proceeded to try to use the Heisler chart anyway since the chart can only be read for $\alpha t / L^2 > 0.2$.

Implementing the chosen solution The second column of Table 5.2 shows the computations. The Heisler chart for a sphere, Figure B.3, is used.

Evaluating and interpreting the solution *1) Does the temperature value make sense?* We can compare with a similar heating situation, in Problem 5.2.1. We see the temperature calculated is higher than in Problem 5.2.1. This makes sense since the reduced external resistance reduces the total resistance so heat flow rate is increased. *2) What do we learn?* Note that the absolute value of the internal resistance of the egg really did not change from Problem 5.2.1 (it is the same egg), but this internal resistance now is relatively much higher compared to the decreased external resistance in the warm water having a high h value.

5.3.10   Numerical Methods as Alternatives to the Charts

The limitations mentioned in the previous section of the analytical solution can indeed be very serious, particularly when applied to biological materials. Such materials and processes often involve complex natural shapes, variations in properties, initial conditions that vary throughout the material, and boundary conditions that vary with time.

Limitations of the analytical solutions can be overcome using numerical, computer-based solutions. In numerical methods, the governing equation developed here would be discretized and essentially solved at a finite number of locations in the material and at finite time steps, as opposed to solving exactly for all locations in the material and at all times. Such methods are quite versatile and can often accommodate arbitrary shape, size, initial condition, boundary conditions, properties, etc. These methods are rapidly becoming the standard procedure for solving heat and mass transfer problems. A large number of commercial software packages are available for this purpose; see for example, the website http://icemcfd.com/cfd/CFD_codes.html.

5.4   Transient Heat Transfer in a Finite Geometry—Multi-Dimensional Problems

Frequently, situations are encountered where one-dimensional approximation is inadequate and two- and three-dimensional effects need to be considered. Qualitatively speaking, if the location we are interested in is at a comparable distance from surfaces in more than one direction, heat transfer from both of these directions must be considered. To solve for multi-dimensional heat transfer, a finite geometry is considered as the intersection of two or three infinite geometries. A rectangular box, for example, is considered as intersection of three infinite slabs and the temperature $T_{x,y,z,t}$ can be calculated as

$$\frac{T_{xyz,t} - T_s}{T_i - T_s} = \left(\frac{T_{x,t} - T_s}{T_i - T_s}\right)_{\substack{\text{infinite} \\ x \text{ slab}}} \left(\frac{T_{y,t} - T_s}{T_i - T_s}\right)_{\substack{\text{infinite} \\ y \text{ slab}}} \left(\frac{T_{z,t} - T_s}{T_i - T_s}\right)_{\substack{\text{infinite} \\ z \text{ slab}}} \quad (5.23)$$

Similarly, for a finite cylinder (Figure 5.12), the temperature $T_{r,z,t}$ can be calculated as

$$\frac{T_{r,z,t} - T_s}{T_i - T_s} = \left(\frac{T_{r,t} - T_s}{T_i - T_s}\right)_{\substack{\text{infinite} \\ \text{cylinder}}} \left(\frac{T_{z,t} - T_s}{T_i - T_s}\right)_{\substack{\text{infinite} \\ \text{slab}}} \quad (5.24)$$

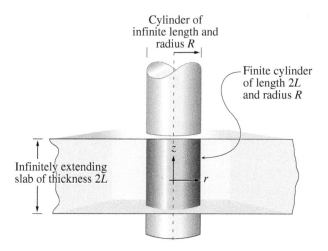

Figure 5.12. A finite cylinder can be considered as an intersection of an infinite cylinder and a slab.

5.5   Transient Heat Transfer in a Semi-Infinite Region

A semi-infinite region extends to infinity in two directions and a single identifiable surface in the other direction. The surface shown in Figure 5.13 extends to infinity in the y and z directions and has an identifiable surface at $x = 0$. The semi-infinite region provides a useful idealization for many practical situations where we are interested in heat transfer for a relatively short time and/or in a relatively thick material. Examples can be heating and cooling near the earth's surface, minor skin burns, etc. The governing equation for conduction heat transfer with constant properties and no heat generation is given by

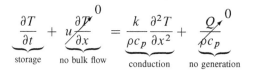

$$\underbrace{\frac{\partial T}{\partial t}}_{\text{storage}} + \underbrace{u\,\cancel{\frac{\partial T}{\partial x}}^{0}}_{\text{no bulk flow}} = \underbrace{\frac{k}{\rho c_p}\frac{\partial^2 T}{\partial x^2}}_{\text{conduction}} + \underbrace{\cancel{\frac{Q}{\rho c_p}}^{0}}_{\text{no generation}}$$

which is simplified to

$$\frac{\partial T}{\partial t} = \alpha \frac{\partial^2 T}{\partial x^2} \tag{5.25}$$

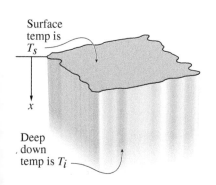

Figure 5.13: Schematic of a semi-infinite region showing only one identifiable surface.

The boundary conditions are given by

$$T(x = 0) = T_s \tag{5.26}$$
$$T(x \to \infty) = T_i \tag{5.27}$$

and the initial condition is given by

$$T(t = 0) = T_i \tag{5.28}$$

The solution to the above equation, although not difficult, is lengthy (see Appendix G.2 on page 598). The temperature T as a function of position x and time t is given by

$$\frac{T - T_i}{T_s - T_i} = 1 - \text{erf}\left[\frac{x}{2\sqrt{\alpha t}}\right] \tag{5.29}$$

where the function $\text{erf}(\eta)$ is called the error function and is given by

$$\text{erf}(\eta) = \frac{2}{\sqrt{\pi}} \int_o^\eta e^{-\eta^2} d\eta$$

To get a sense of what the error function looks like qualitatively, consider Figure 5.14 where the error function erf is compared with an exponential. For more accurate values, the error function is also tabulated in Table B.3 on page 553.

Heat flux at the surface of the semi-infinite region is often an important quantity that is of practical interest. The surface heat flux can be calculated from temperatures given by Eq. 5.29 as

$$\begin{aligned}
q_s'' &= -k \left.\frac{dT}{dx}\right|_{x=0} = -k \left.\frac{dT}{d\eta}\frac{d\eta}{dx}\right|_{x=0} \\
&= -k(T_s - T_i)\left(-\frac{2}{\sqrt{\pi}}e^{-\eta^2}\right)_{\eta=0} \frac{1}{2\sqrt{\alpha t}} \\
&= \frac{k(T_s - T_i)}{\sqrt{\pi \alpha t}} \tag{5.30}
\end{aligned}$$

Equation 5.30 shows that the surface heat flux decreases with time. This is expected since the temperature inside approaches the surface temperature, thus decreasing the temperature gradient at the surface.

When can we use the semi-infinite approximation As mentioned earlier, there is no such thing as a semi-infinite geometry in reality; it is only an idealization. The

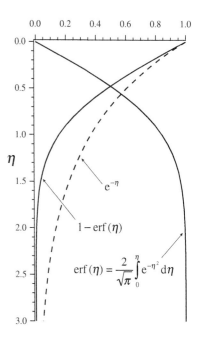

Figure 5.14: Comparison of the complementary error function $(1 - \text{erf}\ \eta)$ with an exponential $e^{-\eta}$.

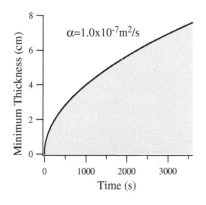

Figure 5.15: Plot of Eq. 5.31, illustrating the minimum thickness of a material for which the error function solution can be used.

question is under what conditions can we approximate a real body as a semi-infinite? If the edges of a real body are far enough from the point of interest on the surface *for the time period of interest*, the propagating energy front does not know that the material is actually finite. The process therefore works as if the body were infinite. This would be true for thin materials over short times or thick materials over longer times. A more quantitative measure of when such approximations are valid can be derived from Figure 5.14. Since for positions and time combinations given by

$$x/2\sqrt{\alpha t} \geq 2$$
$$x \geq 4\sqrt{\alpha t} \tag{5.31}$$

the temperature change is less than 0.5%, any material whose thickness L is larger than $4\sqrt{\alpha t}$ can be approximated as semi-infinite for this purpose. Note that the appropriate thickness L is defined in terms of the time of interest t. For small times, even a small thickness can be considered semi-infinite. As an illustration, for a thermal diffusivity of $1 \times 10^{-7} \mathrm{m^2/s}$, that is, the typical order of magnitude for biomaterials containing a lots of water, the relationship given by Eq. 5.31 is plotted in Figure 5.15.

5.5.1 Effect of Change in Boundary Conditions

The above solution was for a specified surface temperature, i.e., for no external resistance. For a specified surface heat flux boundary condition (see Section 3.2)

$$q''_{surface} = q''_s \tag{5.32}$$

the solution for the temperature is given by

$$T - T_i = \frac{2}{k}q''_s \left(\frac{\alpha t}{\pi}\right)^{1/2} \exp\left(-\frac{x^2}{4\alpha t}\right) - \frac{q''_s x}{k}\left(1 - \mathrm{erf}\left(\frac{x}{2\sqrt{\alpha t}}\right)\right) \tag{5.33}$$

For a convective boundary condition (see Section 3.2),

$$-k\left.\frac{\partial T}{\partial x}\right|_{surface} = h(T_{surface} - T_\infty)$$

the solution is given by

$$\begin{aligned}\frac{T - T_i}{T_\infty - T_i} &= 1 - \mathrm{erf}(\frac{x}{2\sqrt{\alpha t}}) \\ &- \exp(\frac{hx}{k} + \frac{h^2\alpha t}{k^2})(1 - \mathrm{erf}(\frac{x}{2\sqrt{\alpha t}} + \frac{h\sqrt{\alpha t}}{k}))\end{aligned} \tag{5.34}$$

Note that Eq. 5.34 would become Eq. 5.29 when $h \to \infty$, since the specified surface temperature boundary condition is a special case of a convective boundary condition over a surface.

5.5.2 Summary of Solutions for Heat Conduction in a Semi-Infinite Region for the Three Different Boundary Conditions

Temperature and surface heat flux for a semi-infinite region for the three common boundary conditions are summarized in Table 5.6. Note that the surface heat flux approaches zero with time for the temperature-specified boundary condition and convective boundary condition since the interior temperature approaches the surface temperature. For a convective boundary condition, the surface temperature also approaches the surrounding fluid temperature with time.

Table 5.6: Summary of solutions for semi-infinite region for the three different boundary conditions

Boundary condition	Temperature	Flux at surface
Temperature given	Equation 5.29	$\frac{k(T_s - T_i)}{\sqrt{\pi \alpha t}}$; drops with time
Flux given	Equation 5.33	Given value; a constant
Convective condition	Equation 5.34	Calculate $-k \left. \frac{\partial T}{\partial x} \right|_{x=0}$ from Eq. 5.34; drops with time

5.5.3 *BioConnect: Analysis of Skin Burns*

A thermal burn occurs as a result of an elevation in tissue temperature above a threshold value for a finite period of time. The values of both the absolute temperature and the exposure time are crucial in determining the extent of injury. Since heat transfer to inside the skin is limited by thermal conduction, the temperature history in the affected depth is non-uniform, and regions of graded injury develop, with the most acute involvement at the surface. Detailed physiological responses, such as changes in blood flow, coagulation process, etc., are a very significant part of the total burn injury. Figure 5.16 shows a cross-section of normal skin and underlying tissue. Macroscopic characterization of burn injuries (see Krizek et al., 1973) describes a first-degree burn as one involving temporary discomfort, with no permanent scarring or skin discoloration. A superficial second-degree burn involves some but not all of the basal layer. A deep second-degree burn involves complete loss of the basal layer. In a third-degree burn, all epidermal elements are destroyed. A fourth degree burn can extend all the

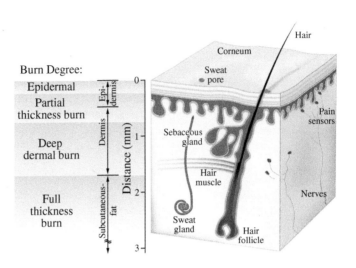

Figure 5.16: Section of skin with degrees of burn superimposed on it. Data from Lawton (1994).

way to the bone. Ignoring the effects of blood flow and other physiological changes, we may analyze the temperature history and thermal injury for up to a third-degree burn, using semi-infinite region approximation. See Problem 5.18 for a numerical calculation of the depth of skin layer affected.

5.5.4   Example: Laser-Heated Destruction of Tissue in Eye Surgery

Figure 5.17 shows a highly simplified schematic for heat transfer modeling during laser surgery. A laser beam of radius 250 μm is incident on the surface of the retina (assumed to be flat relative to the small radius of the laser) and all of it is absorbed on this surface. Consider heat transfer to be one dimensional, for simplicity. For heat-induced tissue damage (by a mechanism called photocoagulation) to occur, the temperature has to reach 60°C or higher. The retinal tissue has a thermal conductivity, density, and specific heat of 0.6 W/m·K, 1000 kg/m³, and 4178 J/kg·K, respectively. The initial temperature of the tissue is 37°C.

1) In order to reach the damage temperature at a depth of 10 μm into the tissue in 20 milliseconds, what laser power (in kW/m²) should be used? 2) Using this laser, what

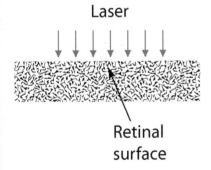

Figure 5.17: Schematic of laser incident on a tissue surface.

will the surface temperature be in 20 milliseconds? 3) How does surface temperature change with time (choose one)—linearly with time, with the square of time, or with the square root of time? 4) Sketch temperature vs. tissue depth for the scenario in step 2. Be sure to label your graph with the values you calculated. 5) If it is desired to keep the top 10 μm of tissue above 60°C for 100 milliseconds during the surgery (note that this is much longer than the time needed to reach 60°C), describe qualitatively (no computation needed) how this can be achieved without substantially overheating the surface. 6) If the temperature solution that you used in steps 1 and 2 was not available, what governing equation (keep only necessary terms), boundary conditions, and initial condition would you start from?

Solution

Understanding and formulating the problem *1) What is the process?* Here one surface is getting heated from the absorption of a constant flux of laser. The temperature rises with time depending on the flux level and heat conducts inside. *2) What are we solving for?* Laser flux needed for the temperature to rise by a certain amount at a given location. *3) Schematic and given data:* A schematic is shown in Figure 5.18 with some of the given data superimposed on it. *4) Assumptions:* Thermal diffusivity of the tissue does not change with temperature.

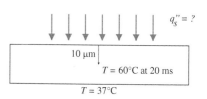

Figure 5.18: Schematic of laser incident on a tissue surface.

Generating and selecting among alternate solutions *1) What solutions are possible?* For transient heat transfer, the solutions we have available are shown in the solution chart in Figure 5.21. *2) What approach is likely to work?* As we look at the solution chart in Figure 5.21, since the thickness of the slab, L, is not provided, we cannot calculate $\alpha t/L^2$, so we cannot really be sure of a finite or semi-infinite geometry solution. However, if we tried to proceed with finite geometry solution, we note that the series solution (for a slab) is not something we can use without L. It is obviously not a lumped parameter problem (the other choice under finite geometry) since that would assume no temperature variation with position while the problem refers to temperatures at various locations. Thus, there is no choice but to treat (at least initially) the problem as one in a semi-infinite geometry that does not require L to solve the problem.

Implementing the chosen solution Note that the boundary condition at the surface is flux given, not temperature given, so the solution for flux given (Eq. 5.33) is the right one. Rewriting in terms of the flux, q_s'', we get

$$q_s'' = \frac{T - T_i}{\frac{2}{k}\left(\frac{\alpha t}{\pi}\right)^{1/2}\exp\left(\frac{-x^2}{4\alpha t}\right) - \frac{x}{k}\left(1 - \operatorname{erf}\left(\frac{x}{2\sqrt{\alpha t}}\right)\right)}$$

To plug in values, we define the following:

$$A = \frac{2}{k}\left(\frac{\alpha t}{\pi}\right)^{1/2} \exp\left(\frac{-x^2}{4\alpha t}\right)$$

$$= \frac{2}{0.6\text{W/mK}}\left(\frac{1.44 \times 10^{-7}[\text{m}^2/\text{s}] \times 0.02\text{s}}{3.14}\right)^{1/2}$$

$$\times \exp\left(\frac{-\left(10^{-5}\right)^2[\text{m}^2]}{4 \times 1.44 \times 10^{-7}[\text{m}^2/\text{s}] \times 0.02[\text{s}]}\right)$$

and

$$B = \frac{x}{k}\left(1 - \text{erf}\left(\frac{x}{2\sqrt{\alpha t}}\right)\right)$$

$$= \frac{10^{-5}[\text{m}]}{0.6\left[\frac{W}{mK}\right]}\left(1 - \text{erf}\left(\frac{10^{-5}[\text{m}]}{2\sqrt{1.44 \times 10^{-7}[\text{m}^2/\text{s}]0.02[\text{s}]}}\right)\right)$$

so that

$$q_s'' = \frac{60 - 37[\text{K}]}{A - B}$$

$$= 265\frac{\text{kW}}{\text{m}^2}$$

2) For part 2, we use Eq. 5.33 at $x = 0$,

$$T_{surface} - T_i = \frac{2}{k}q_s''\left(\frac{\alpha t}{\pi}\right)^{1/2}$$

$$T_{surface} - 37 = \frac{2}{0.6\,[\text{W/m}\cdot\text{K}]}265000\,[\text{W}]\left(\frac{1.44 \times 10^{-7}\,[\text{m}^2/\text{s}]\,0.02\,[\text{s}]}{\pi}\right)^{1/2}$$

$$T_{surface} = 63.7°\text{C}$$

3) Surface temperature changes with the square root of time. 4) See Figure 5.19. 5) Continuous heating would raise the surface temperature considerably higher. One possibility would be to cycle the laser on and off. During the "on" time, the temperature is allowed to reach slightly higher than needed. During the "off" time, conduction will carry heat away, lowering the temperature. When the temperature reaches below a certain threshold, the laser can come back on.

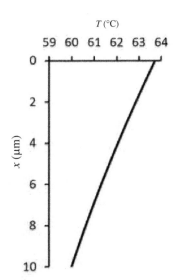

Figure 5.19: Temperature profile at 20 ms for Example 5.5.4.

6) The governing equation and boundary conditions for this problem are given by (these can be solved numerically to obtain the above results instead of the analytical solution (Eq. 5.33)

$$\frac{\partial T}{\partial t} = \frac{\partial^2 T}{\partial x^2}$$

$$-k\frac{\partial T}{\partial x} = q'' \text{ at } x = 0$$

$$T = T_i \text{ at } x \to \infty$$

with the initial condition $T = T_i$.

Evaluating and interpreting the solution *1) Do the computed values (surface temperature) make sense?* Without heat conduction, the energy absorbed would all go into temperature rise. A very quick and highly inaccurate calculation, assuming energy is uniformly absorbed over the 10 μm thickness and the thermal properties of tissue are those of water, would lead to energy balance as

$$\underbrace{265 \times 10^3 \text{ [W/m}^2] \, A \text{ [m}^2] \, 20 \times 10^{-3} \text{ [s]}}_{\text{Energy absorbed from laser}} =$$

$$\underbrace{1000 \text{ [kg/m}^3] \, A \text{ [m}^2] \, 10 \times 10^{-6} \text{ [m] } 4.2 \times 10^3 \text{ [J/kg} \cdot \text{K}] \Delta T}_{mc_p \Delta T}$$

from which we get $\Delta T = 126°C$. Since heat would be conducted away inside the tissue, calculations using the semi-infinite solution to the heat conduction equation should not lead to temperature rise above 126°C, which is the case (it is 26.7°C), so our calculated value is not obviously unreasonable. *2) What do we learn?* Temperature rise is very quick in such heating processes.

5.6 Chapter Summary—Transient Heat Conduction

- ## No Internal Resistance, Lumped Parameter (page 129)

    1. The thermal resistance of the solid can be ignored in comparison with the thermal resistance due to convection at the solid surface if a Biot number (page 131) condition given by $h(V/A)/k < 0.1$ is satisfied.

    2. As thermal resistances are ignored, temperature is not a function of position; it is a function of time only and is given by Eq. 5.7.

- ## Internal Resistance Is Significant

1. When internal resistance is significant (Biot number > 0.1), temperature is a function of both position and time.

2. For an infinite slab (page 136), infinite cylinder, and spherical geometry, the solutions are given by Figures B.1, B.2, and B.3 on pages 555, 556, and 557, respectively.

3. For finite slab and finite cylinder (page 147), the solutions are given by Eq. 5.23 and Eq. 5.24, respectively.

4. Materials with thicknesses $L \geq 4\sqrt{\alpha t}$ are considered effectively semi-infinite (page 148) and the solution is given by Eq. 5.29.

5.7   Problem Solving—Unsteady-State Heat Conduction

▶ **Calculate temperature profiles and heat flows.**   The map in Figure 5.21 shows an approach to solving transient heat conduction problems. Some additional comments for problem solving are

- If L is not provided (or otherwise adequate information is not available), $\alpha t / L^2$ cannot be calculated and a semi-infinite approximation may have to be made to obtain a solution using what we know in this book.

- A semi-infinite solution may work for curved geometries as well, for small times.

- If surface temperature is provided, it amounts to $h \to \infty$ (discussed in Chapter 3), as would be needed in a Heisler chart.

- The non-dimensional quantity $\alpha t / L^2$ implies $t \propto L^2$.

5.8   Concept and Review Questions

1. Contrast the physical meanings of thermal diffusivity and thermal conductivity.

2. What is a semi-infinite solid? Give examples.

3. Explain how the one-dimensional Heisler charts can be used to solve two- and three-dimensional problems.

4. Before you begin any calculation for transient temperatures in a solid with fluid flowing over it, what parameter value should you check, to reduce your work? What is the physical meaning of this parameter?

5. Discuss several advantages and disadvantages of the analytical methods of solutions that you are learning. Why are numerical methods of solution often preferrable for practical problems?

6. A fluid flows over the slab in Figure 5.20 with a heat transfer coefficient h. Consider the thickness, L, of the slab to be small compared to its other two dimensions. Let the surface area be A. 1) Write the formulae for the conductive resistance for the slab and the convective resistance for flow over the slab. 2) Why is it useful to compare the two resistances you just noted above (short answer, please)? 3) What parameter is used to compare the two resistances? 4) Write the expression for this parameter in terms of the thickness and other parameters for a slab noted above.

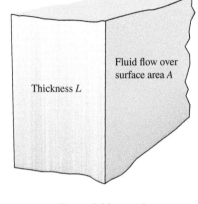

Figure 5.20: A slab.

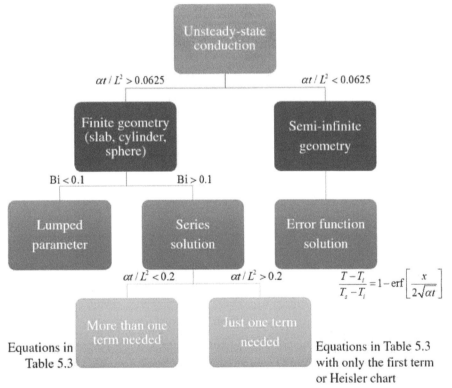

Figure 5.21: Problem solving in unsteady-state heat conduction.

7. We would like to know how heating (or cooling) processes speed up as we reduce size. Consider a slab geometry for which the spatial average temperature is given by

$$\frac{T_{av} - T_s}{T_i - T_s} = \frac{8}{\pi^2} e^{-\alpha \left(\frac{\pi}{2L}\right)^2 t}$$

Consider two slabs, one half the thickness of the other. From this equation, comment on how the required time would change to reach the same average temperature in the two slabs.

8. Soon after a heating or cooling process starts, the temperature inside a slab can be obtained from (select the most appropriate one)

   - Heisler chart
   - Solution with error function in it
   - Complete series solution to heat equation
   - Heisler chart and complete series solution
   - Heisler chart and error function
   - Error function and complete series solution

9. Explain, using the average temperature equation for a slab and the lumped parameter equation given below,

$$\frac{T_{av} - T_s}{T_i - T_s} = \frac{8}{\pi^2} e^{-\alpha \left(\frac{\pi}{2L}\right)^2 t} \qquad (5.35)$$

$$\frac{T - T_\infty}{T_i - T_\infty} = \exp\left(-\frac{t}{\frac{mc_p}{hA}}\right) \qquad (5.36)$$

why the temperature change with time depends on h in one equation but not in the other.

10. A slab of tissue takes 10 minutes to heat up from a certain initial temperature to a certain average temperature. Suppose this tissue lost moisture (got dried) so that its thermal conductivity, density, and specific heat are 50%, 75%, and 75%, respectively, of the non-dried tissue. *All other conditions being the same,* how long will it take to reach the same average temperature?

11. Can you use a Heisler chart for heat flux specified on both surfaces of a slab? Explain how (if yes) or why not (if no).

12. Theoretically speaking, how long would you have to wait in a transient system for steady state to be fully achieved? Why?

13. For the lumped parameter analysis to be valid, for a given material, you can make the mass smaller or you can make the ____ larger.

14. We have studied three different transient solutions to heat conduction—lumped parameter solution, series solution and semi-infinite solution. Given a transient heat transfer situation in a slab, what steps would you take to decide the appropriate solution method? It would be nice if you can provide your answer as a flow chart.

Further Reading

Bairi, A., N. Laraqi, and J. M. Garcia de Maria. 2005. Determination of thermal diffusivity of foods using 1D Fourier cylindrical solution. *Journal of Food Engineering* 78:669–675.

Bird, R. B., W. E. Stewart, and E. N. Lightfoot. 1960. *Transport Phenomena*. John Wiley & Sons, New York.

Carslaw, H. S. and J. C. Jaeger. 1992. *Conduction of Heat in Solids*. Oxford University Press, Oxford.

Cole, G. W. and N. R. Scott. 1977. A mathematical model of the dynamic heat transfer from the respiratory tract of a chicken. *Bulletin of Mathematical Biology* 39:415–433.

Davey, K. R. 1990. Equilibrium temperature in a clump of bacteria heated in a fluid. Applied and Environmental Microbiology, 56(2):566–568.

Diller, K. R. 1985. Analysis of skin burns. In *Heat Transfer in Medicine and Biology* (Vol. 2), edited by A. Shitzer and R. C. Eberhart. Plenum Press, New York.

Geankoplis, C. J. 1983. *Transport Processes and Unit Operations*. Allyn and Bacon, Inc., Boston.

Gordon, N., S. Rispler, S. Sideman, R. Shofty, and R. Beyar. 1998. Thermographic imaging in the beating heart: A method for coronary flow estimation based on a heat transfer model. *Medical Engineering & Physics* 20:443-451.

Incropera, F. P. and D. P. Dewitt. 1990. *Introduction to Heat Transfer*. Wiley, New York.

Krizek, T. J., M. C. Robson, and R. C. Wray, Jr. 1973. Care of the burned patient. In: *Management of Trauma*, edited by W. F. Ballinger, R. B. Rutherford, and G. D. Zuidema. W. B. Saunders Co., Philadelphia.

Lawton, B. 1994. Heat dose to produce skin burns in humans. In *Fundamentals of Biomedical Heat Transfer*, edited by M. A. Ebadian and P. H. Oosthuizen, HTD-vol. 295, ASME, New York.

Lin, M., F. Xu, T. J. Lu, and B. F. Bai. 2010. A review of heat transfer in human tooth. Experimental characterization and mathematical modeling. *Dental Materials* 26:501–513.

Marino, F. E., Z. Mbambo, E. Kortekaas, G. Wilson, M. I. Lambert, T. D. Noakes, and S. C. Dennis. 2000. Advantages of smaller body mass during running in warm, humid environments. *European Journal of Physiology* 441:359–367.

Pages, T., J. F. Fuster, and L. Palacios. 1991. Thermal responses of the fresh water turtle *Mauremys Caspica* to step-function changes in the ambient temperature. *J. Thermal Biol.* 16(6):337–343.

Pan, J. C. and S. R. Bhowmik. 1991. The finite element analysis of transient heat transfer in fresh tomatoes during cooling. *Trans. of ASAE* 34(3):972–976.

Sastry, S. K., G. Q. Shen, and J. L. Blaisdell. 1989. Effect of ultrasonic vibration on fluid-to-particle convective heat transfer coefficients. *Journal of Food Science* 54(1):229–230.

Simpson, W. T., X. Wang, and S. Verrill. 2003. Heat sterilization time of Ponderosa pine and Douglas-Fir boards and square timbers. Research Paper FPL-RP-607. United States Department of Agriculture Forest Service, Forest Products Laboratory, Madison, Wisconsin.

Sien, H. P. and R. K. Jain. 1979. Temperature distributions in normal and neoplastic tissues during hyperthermia: Lumped parameter analysis. *Journal of Thermal Biology* 4:157–164.

Wang, Z.-H. and E. Bou-Zeid. 2012. A novel approach for the estimation of soil ground heat flux. *Agricultural and Forest Meteorology* 154–155:214–221.

5.9 Problems

5.1 Thermocouple and the Lumped Parameter Approach

Figure 5.22 shows a thermocouple for measuring temperature. The junction that produces a voltage due to changes in temperature has a spherical lead "bead" of diameter 1 mm, as shown in the enlarged view in this figure. The thermocouple is exposed to air flow that leads to an h value of 10 W/m²·K for the bead. For lead, the thermal conductivity is 35.3 W/m·K, density is 11340 kg/m³, and the specific heat is 129 J/kg·K. 1) Determine if you can use a lumped parameter approximation for the bead. 2) If the bead is suddenly inserted into the air at 200°C from an initial temperature of 20°C, calculate the time it takes to reach 99.9% of the air temperature. 3) How long does it take for the bead to reach the exact air temperature?

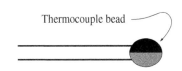

Figure 5.22: Schematic of a thermocouple with a "bead" that measures the temperature.

5.2 Changing Heat Flux with Time

Using the temperature solution for a lumped parameter heat transfer situation, show mathematically that the heat flux at the surface (i.e., energy leaving or entering the body) drops exponentially. *(Hint: think in terms of convective heat flux.)* What is the physical explanation for the reduction in heat flux?

5.3 Heat Transfer in a Leaf

Consider a typical thin leaf on a tree. When the air temperature surrounding the leaf drops suddenly, we would like to calculate how the center temperature of the leaf changes with time. The surface convective coefficient for a 5 mph wind speed is 20 W/m²·K and the thermal conductivity of the leaf is 0.3 W/m·K. 1) What formula would you use and why? 2) List all the additional input parameters needed (you do not need to do any calculations).

5.4 Ultrasound-Aided Food Sterilization

In sterilizing foods that have liquid as well as large solid particles, such as chunky soup, the bottleneck in the rate of heating is the heat transfer from the liquid to the solid. One potential approach to increase the convective heat transfer coefficient, h, between the liquid and the solid particles is the use of ultrasonic agitation, which currently has wide application in industrial cleaning and cell homogenization processes.

To determine this heat transfer coefficient, h, particles made of material with a very high thermal conductivity, such as aluminum, are often used (why?). In heating a mixture of aluminum particles and water, one achieves a heat transfer coefficient of

1323 and 558 $W/m^2 \cdot K$ between the particles and the water, with and without ultrasound, in the absence of any other mechanical agitation. Assume that the temperature of the water bath is constant at 85°C. The initial uniform temperature of each particle is 20°C. The mass of each particle is 18 g and it is spherical. The density, specific heat, and thermal conductivity of aluminum are 2701.1 kg/m^3, 938.3 $J/kg \cdot K$, and 229 $W/m \cdot K$, respectively. Compare the time needed for the temperature of one aluminum particle to equilibrate to within 3°C of the water temperature with ultrasonic agitation, to the time without.

5.5   Sterilization of a Bacterial Clump

To be properly sterilized, all bacteria must be heated for a required duration. Although individual bacterial sizes are small, when present in large clumps, some bacteria located deep inside the clump may require a significant amount of time to reach the medium (air or water) temperature (Davey, 1990). Thus, they cannot be assumed to be at the medium temperature all the time. Consider bacterial clumps of different sizes heated in air. The shape of the clump can be assumed spherical with a radius R, which is related to n, the number of bacteria in a clump, and r, the radius of individual bacteria, by the equation $R = r[n/(1-E)]^{1/3}$ where $E = 0.40$ is porosity. The surface heat transfer coefficient for air flow over the clump is given by $h = 4 \times 10^{-4} k_f / R$ where k_f is the air thermal conductivity and is equal to 0.0317 $W/m \cdot K$. Consider an individual bacterial radius of $1 \mu m$ and the specific heat, density, and thermal conductivity of the bacterial clump of 3.6 $kJ/kg \cdot K$, 1080 kg/m^3, and 0.49 $W/m \cdot K$, respectively. For each of three bacterial clumps given by $n = 10^2, 10^5$, and 10^6, 1) Determine the clump radius and the heat transfer coefficient, h. 2) Show that the lumped parameter approach is valid. 3) Calculate the time required for the temperature of the bacterial clump to increase by 99% of the total possible change in temperature. 4) Comment on whether or not it is always reasonable to assume the bacterial clump to be at air temperature.

5.6   Heating of a Liquid in a Vessel with Agitation

In many industrial applications, liquid is heated or cooled in a closed cylindrical vessel and agitated to speed up the heat transfer. The agitation maintains the inside temperature to be fairly uniform, somewhat analogous to the lumped parameter solution described by Eq. 5.7. In this case, the mass average temperature T_{av} of the liquid will replace the temperature T of the solid in Eq. 5.7. Also, the heat transfer coefficient will now be the inside heat transfer coefficient between the vessel wall and liquid. The container wall temperature is at 121°C and the initial temperature of the fluid is 30°C. The container has a height of 10.67 cm and a radius of 4.19 cm. The density and specific heat of the liquid are 950 kg/m^3 and 4100 J/kgK, respectively. 1) Calculate how the mass average temperature (T_{av}) of the liquid would change with time during agitated heating of a containerized liquid where the inside heat transfer coefficient is 120 W/m^2·K. 2) How long does it take for the average temperature to reach 95% of the maximum possible temperature?

5.7 Rate of Heating in a Transient Process

Consider a vertical cylindrical kettle of diameter 1 m filled with liquid that is constantly stirred, with its vertical surface (excluding top and bottom surfaces) heated with steam. The steam maintains this surface at 110°C. The initial uniform temperature of the fluid is 40°C and the stirring produces a heat transfer coefficient of 200 W/m^2 · K. The density and specific heat of the fluid are 950 kg/m^3 and 4.1 kJ/kg · K, respectively. Calculate the time it will take for the *average* temperature of the fluid to reach 1) 109°C, and 2) 109.9°C. 3) Compare the two answers. *(Hint: Use the lumped parameter approach, replacing the solid temperature by the mass average temperature of the liquid.)*

5.8 Bulk Warming of Soil

The temperature of the topmost layer of soil depends heavily on the relatively small quantity of solar flux that gets absorbed into the soil. Since solar heat is absorbed in the topmost layer, the soil may be assumed to be well insulated beyond that layer. Solar flux absorbed into the soil layer is 40 W/m^2, the depth of the soil layer is 10 cm, the bulk density of the soil is 1200 kg/m^3, and the specific heat of the soil is 2500 J/kg·K. Find the *average* rate of temperature increase in the soil, assuming no convective heat transfer or energy generation in the soil.

5.9 Tissue–Blood Equilibrium

As blood travels through the body, its temperature, T_b, is constantly attempting to equilibrate with the temperature T_t of the surrounding tissue. Assuming that the blood temperature varies only in the direction of blood flow, this variation in T_b with distance can be easily related to the heat exchange between the blood and the tissue by

$$\frac{\pi}{4}d^2\rho_b c_b u \frac{dT_b}{dx} = \pi d h (T_t - T_b) \tag{5.37}$$

where x is the distance along the blood vessel, d is the diameter of the blood vessel, u is the velocity of blood, and ρ_b and c_b are the density and specific heat of blood, respectively. 1) Solve the above equation for blood temperature, T_b, as a function of

x. 2) Consider the following input parameters: the constant tissue temperature is $35°C$, the temperature of blood at the vessel (large artery) inlet is $37°C$, the diameter of the large artery is 1.5 mm, $\rho_b c_b = 4.46 \times 10^6 J/m^3·K$, the velocity of blood is 40 cm/s, the length of the artery is 20 cm, and the heat transfer coefficient between the blood and the artery wall is 200 W/m^2·K. Calculate the blood temperature at the vessel outlet ($x = L$). 3) Assuming equilibrium to be reached when the temperature of the blood is 95% of the total possible increase in temperature, comment on the distance the blood must travel in the artery before reaching equilibrium with the surrounding tissue.

5.10   Experimental Determination of Thermal Property

The solution to the heat equation is also used to solve for unknown thermal properties. Consider a slab 10 cm thick that was initially at a uniform temperature and cooled symmetrically with a surface temperature at $15°C$. A thermocouple fixed at a particular location inside the slab measures the temperature to be $4.23°C$ and $5°C$ after 55 and 60 minutes, respectively. From these measurements, calculate the thermal diffusivity of the slab.

5.11   Average Temperature in a Slab

Starting from the approximate solution for large times for temperature in a symmetrically heated slab as a function of time and position, find the average temperature of the slab.

5.12   Effect of Size on the Rate of Cooling (or Heating)

We would like to know how heating (or cooling) processes speed up as we reduce size. Consider a slab geometry. Starting from the equation for the average temperature of a slab, comment on how the required time to heat (or cool) to the same final average temperature would change when the slab size is halved.

5.13   Changing Heat Flux with Time on a Slab Surface

Consider a slab initially at constant temperature, T_0, with the surface temperature maintained at T_1. 1) Using the expression for temperature in the slab as a function of position and time at long times, calculate the heat flux on the surface of the slab as a function of time. 2) Explain in words how this heat flux varies with time and why.

5.14   Heating Time for Food Sterilization

Consider a product to be sterilized if its coldest location during heating reaches 115°C. Consider sterilization of potatoes (spheres with 4 cm diameter) in metal cans filled with water. The sealed can itself is heated in steam at 121°C. Assume symmetrical heating of potatoes and an *overall heat transfer coefficient between the steam and the potatoes* of 20 W/m^2·K. The thermal diffusivity for potatoes is 1.5×10^{-7} m^2/s and the thermal conductivity is 0.4 W/m·K. The potatoes are initially at 30°C. Calculate the heating time needed for sterilization.

5.15   Cooling of a Donor Heart

Human-to-human heart transplants are becoming more common. Often the heart donor is hundreds or even thousands of miles away from the recipient, so it is very important to quickly preserve the heart to prevent the damage or death of tissue during its cross-country journey. This is done by submerging the heart in a saline solution and packing it in ice inside a commercial picnic cooler.

The heart can be approximated by a solid cylinder with constant, equal thermal properties in all directions. During the time the heart is between bodies and packed in ice, it can be assumed that the tissue activity has slowed enough that the metabolic heat production is negligible. Assume also that the saline solution is agitated so that it maintains the heart surface temperature at the constant value of the saline solution temperature. The thermal diffusivity of the heart tissue is 1.4×10^{-7} m^2/s, the initial temperature is 37°C, the temperature of the saline solution is 0°C, and the radius of the cylinder approximating the heart is 3.75 cm. 1) Assuming an infinite cylinder (negligible heat conduction in the axial direction), find the temperature at the center of the heart after 3 hours. 2) Recalculate the temperature at the center for a finite cylinder with a height of 9 cm and the same radius, in which axial heat conduction cannot be ignored.

5.16   Effect of Heat Transfer Coefficient on Food Sterilization

Assume that a food particle is considered to be sterilized when its center reaches a presumed sterilizing temperature of 115°C. The density, specific heat, and thermal conductivity of food are 850 kg/m^3, 1680 J/kg·K, and 1.42 W/m·K, respectively. For the two heat transfer coefficients given in Problem 5.4, calculate the difference in heating time that would be necessary to sterilize a spherical food particle of 5 mm radius initially at a temperature of 30°C, immersed in water that is at a temperature of 121°C. *Note: Requires interpolation in Heisler's charts.*

5.17   Effect of Heat Transfer Coefficient on Food Sterilization

For simplicity, consider a product is sterilized if its coldest location during heating reaches 115°C. Consider sterilization of spherical potatoes 4 cm in diameter and initially at 30°C, with a thermal conductivity and diffusivity of 0.4 W/m·K and $1.5 \times 10^{-7} \text{m}^2/\text{s}$, respectively. In an old sterilization process for potatoes, metal cans would be filled with water and the potatoes and the can heated in steam at 121°C, giving an overall heat transfer coefficient between the steam and the potatoes of 20 W/m²·K. Assume symmetrical heating of potatoes. 1) Calculate the heating time needed for sterilization. 2) In a newer process, the water and the potatoes are agitated inside a large vessel, to increase the overall heat transfer coefficient to 200 W/m²·K. The vessel is also heated in steam at 121°C. Calculate the heating time needed for sterilization in the new process. Make approximations in reading the charts. 3) Comment on which potatoes you would prefer to eat and why.

5.18   Skin Burn Injury

In modeling a skin burn from an oven, approximate the skin and tissue layer to be infinitely thick compared to the damaged layer. The temperature throughout the entire skin and tissue layer is uniform at 33°C before contact with the oven, and the surface layer of skin increases to the temperature of the oven, 200°C, instantaneously upon contact. Consider that skin becomes damaged when it reaches 62°C. The thermal diffusivity of skin is $2.5 \times 10^{-7} \text{m}^2/\text{s}$. Find the depth of the damaged layer of skin after 2 seconds of exposure to the oven temperature.

5.19   Induction Thermocoagulation in the Brain

Brain tumors are often treated by creating small thermal lesions known as induction thermocoagulation. In this procedure, a small section of the brain is heated to thermally destroy some of the tissue. Assume the tissue to be thick and at an initial temperature of 36°C. The surface of the tissue is maintained at 75°C by a probe tip. The time of exposure is 3 minutes. The thermal conductivity, density, and specific heat of the tissue are 0.586 W/m·K, 1050 kg/m³, and 3763 J/kg·K, respectively. Assuming one-dimensional heat transfer, find the thickness of the thermal lesion if the tissue dies when its temperature reaches 55°C.

5.20   Heat Transfer Near the Surface

Biological materials are often chilled to decrease biological activity and therefore increase their storage life. A slab of a certain biological material 50 mm thick and at an initial temperature of 30°C is to be chilled using an air blast that maintains the surface of both sides of the biomaterial at 0°C. The thermal diffusivity of the biomaterial is $0.625 \times 10^{-4} \text{m}^2/\text{s}$. Calculate the temperature at a depth of 5 mm from either surface at 10 seconds after it is put in the air blast, using 1) the Heisler charts, and 2) the semi-infinite model. 3) Compare the answers to questions 1 and 2 and explain why the answers are within a few degrees.

5.21 Cooling of the Ground

The depth to which freezing temperature is reached inside soil is important for plant growth and construction of buildings. We will postpone the inclusion of latent heat from freezing until the chapter "Heat Transfer with Phase Change." Thus, only cooling is considered in this problem, i.e., no phase change. Approximate the soil temperature to be uniform at 15°C toward the end of fall, when the air temperature suddenly drops to −20°C. The wind over the soil surface produces a convective heat transfer coefficient of 15 $\text{W/m}^2 \cdot \text{K}$. The thermal conductivity and thermal diffusivity of the soil are 0.85 W/m·K and $4.5 \times 10^{-7} \text{m}^2/\text{s}$, respectively. Calculate the following at $t = 3$ hours: 1) the depth at which the soil temperature is 0°C, 2) the surface temperature of the soil, and 3) the heat flux at the surface of the soil.

5.22 Cooling of the Ground with Varying Surface Temperature

The daily periodic variations (diurnal variations) in air temperature and solar radiation lead to periodic variations in soil temperature. Assuming the soil to be semi-infinite in depth, the soil temperature deep inside the soil is T_{av}, while the surface temperature of the soil is $T_{av} + (T_{max} - T_{av}) \cos \omega t$, where T_{max} is the maximum temperature that occurs at the soil surface, and time, t, is measured from the time that T_{max} is reached. The frequency of diurnal variations is given by ω. The solution to the problem is

$$\frac{T - T_{av}}{T_{max} - T_{av}} = \exp\left(-\sqrt{\frac{\omega}{2\alpha}} z\right) \cos\left(\omega t - \sqrt{\frac{\omega}{2\alpha}} z\right)$$

1) Write the governing equation (keeping only necessary terms) and the boundary and initial conditions for this problem. 2) Plot approximately the solution given above (temperature as a function of time) and the surface temperature. 3) Explain the physical meaning of the exponential term and the cosine term in the solution. (If you are interested in the solution procedure, see Bird et al., 1960).

5.23   Heating in a Microwave Oven

In microwave heating of foods and biomaterials, polarized molecules of water oscillate in the electromagnetic field of the microwaves. The rate of heating decays from the surface as you go deeper into the material, since the waves lose their electromagnetic energy (which appears as thermal energy) as they travel. This decay in the rate of heating is often approximated as

$$Q = Q_0 e^{-\frac{x}{\delta}}$$

where Q is the rate of heat generation in W/cm^3 at a distance x from the surface, Q_0 is the rate of heat generation in W/cm^3 at the surface, and δ is the penetration depth, a material property that is a measure of how easily the material absorbs energy. The rate of heating in microwaves is often so fast that diffusion is not significant for the first few seconds of heating. For these first few seconds, 1) find the temperature profile $T(x,t)$, in a thick slab heated with microwaves from both faces. Sketch this approximate temperature profile (T vs. x). Assume any other property or parameter that you need and also that no heat is lost from the surface. 2) Describe qualitatively how you would expect the profile to change with time, and indicate this on your graph.

5.24   Cooling of a Leaf

Consider a typical leaf, as shown in Figure 5.23 from the oak tree outside our building. When the air temperature surrounding the leaf drops suddenly, we would like to calculate how the center temperature of the leaf changes with time. The surface convective heat transfer coefficient for a 5 mph wind speed is 20 W/m^{2}K and the thermal conductivity of the leaf is 0.3 W/m·K. The average thickness of the leaf can be assumed as 1 mm and its density and specific heat can be assumed as 900 kg/m^3 and 3700 J/kg·K, respectively. 1) What final formula would you need to find this temperature, and why? Leave your answer in terms of symbols, and make reasonable assumptions of any parameters not provided. 2) How long does it take for the temperature change $T - T_i$ in the leaf to reach 99% of the possible change in temperature?

Figure 5.23: Leaf from an oak tree.

5.25   Measurement Time for a Thermocouple

Figure 5.24 shows a thermocouple for measuring temperature. The junction that produces the voltage has a "bead" that is a lead sphere of diameter 1 mm, shown enlarged in the figure. The thermocouple is exposed to air flow that leads to an h value of 10 W/m^2 · K for the bead. For lead, the thermal conductivity is 35.3 W/m · K, the density is 11340 kg/m^3, and the specific heat is 129 J/kg · K. 1) If the bead was at an initial temperature of 20°C and is suddenly inserted into air that is at 200°C, calculate

Figure 5.24: A bead on a thermocouple for measuring temperature.

the time it takes to reach 99.9% of the air temperature. 2) Make a sketch of the bead on a graph of temperature vs. time and show on the sketch how long it takes for the bead to reach the exact air temperature. Label axes and intercepts.

5.26 Cryopreservation of Tissue

Using an extremely rapid cooling rate, small amounts of tissue can be successfully preserved in a process called vitrification to avoid ice formation. To do this, chemicals called cryoprotectants are infused into such a tissue before the cooling process to lower the required cooling rate. For example, stem cells harvested from the umbilical cord may be preserved this way. Consider an umbilical cord to be a solid cylindrical tissue of radius 1 cm initially at 21°C that has been injected with propylene glycol as a cryoprotectant. It is dipped in liquid nitrogen at −196°C where vigorous boiling of the liquid nitrogen maintains the surface of the tissue at −196°C. The thermal properties of the infused tissue are given by a specific heat of 1459 J/kg·K, a density of 1097 kg/m^3, and thermal conductivity of 0.45 W/m·K. What is the temperature reached at the center of the tissue after 180 and 240 seconds, respectively?

5.27 Cooling Rate in a Slab

Consider the cooling of a slab of thickness $2L$, starting from initial temperature T_i. Provide an expression of cooling rate (time rate of temperature change) as a function of the distance from the center of the slab, x, for *long times*. Discuss how the cooling rate would vary from the surface of the slab to the inside (increase/decrease/stay constant).

5.28 Effect of h on Temperature Profile

The three situations in Figure 5.25 show temperature profiles at different times in a slab for three different values of heat transfer coefficients on the slab surface. 1) Which slab (of the three) represents the condition where internal resistance is negligible compared to the external resistance? 2)What method would you use to calculate temperature for the condition in 1)? Just write in words. 3) Which slab represents the condition when external resistance is negligible compared to internal resistance? 4) What method would you use to calculate temperature for the condition in 3)? Write in words. 5) Which one represents the condition where neither external nor internal resistance is negligible? 6) What method would you use to calculate temperature for the condition in 5)? Write in words.

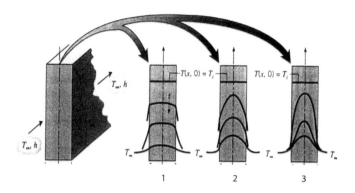

Figure 5.25: Transient temperature profiles in a slab under various situations. Adapted from Fundamentals of Heat and Mass Transfer by F. P. Incropera and D. P. DeWitt, ©1990 John Wiley & Sons, Inc. Reprinted by permission of John Wiley & Sons, Inc.

5.29   Mild Whole Body Hyperthermia

Mild whole body hyperthermia (MWBH) has been shown in several studies to protect against radiation-induced cell death in mice. Therefore, MWBH is of great interest to scientists who study cancer biology. Consider a mouse to be a long cylinder with a radius of 3 cm. The initial uniform body temperature of the mouse is 37°C. The approximate thermal conductivity, specific heat, and density of the mouse body are $0.42 \, \text{W/m} \cdot \text{K}$, $975 \, \text{J/kg} \cdot \text{K}$ and $310 \, \text{kg/m}^3$, respectively. The heat transfer coefficient for the mouse surface is $14 \, \text{W/m}^2 \cdot \text{K}$. 1) As a researcher who works with MWBH, determine the time an anesthetized mouse will need to be placed in a 45°C incubator so that its core body temperature reaches 40°C and it enters MWBH. 2) Sketch the temperature profile (temperature as a function of radius). Label axes and intercepts.

5.30   Whole Body Cryotherapy

Whole body cryotherapy (WBC) is a form of physical therapy in which the entire body is cooled to alleviate symptoms of muscle or joint diseases such as arthritis. Consider a WBC situation[1] where a person is wearing a swimsuit and the entire body is exposed to cold air at −25°C inside a chamber for 10 minutes. To prevent frostbite, the patient's hands, feet, and face are covered. However, his forearm, a cylinder of radius 3 cm that is initially at a temperature of 37°C, stays exposed. The thermal diffusivity and thermal

---

[1]The data for this problem has been changed from reality to be able to solve as an exam problem

conductivity of the tissue are 7.1×10^{-7} m^2/s and 0.592 W/m·K, respectively, and the heat transfer coefficient over the forearm surface for the particular airflow situation is 19.7 W/m^2·K. If skin at a temperature lower than 0°C experiences frostbite, calculate if the forearm surface will experience frostbite.

5.31   Cooking Time for Turkey

A cooking process at high temperature involves a significant amount of water evapo-ration and other modes of water and energy transport in addition to heat conduction. For the purpose of this problem, we are ignoring all modes of energy transport except heat conduction. Imagine you are cooking a turkey with a perfectly spherical geome-try until it reaches a desired final temperature at its coldest location, and you want to compare cooking times. 1) For the same initial temperature of the turkey and the same oven temperature, what is the ratio by which the cooking time should change if the mass of the turkey is doubled, but the density stays the same? 2) In a cookbook, the directions say to cook a 2.5 kg turkey for 3 hrs and a 5 kg turkey for 4.25 hrs for the same scenario as in part 1). How does the ratio you found in part 1) compare with that provided in the cookbook, and what could be the possible explanations for this differ-ence? *(Hint: You can think of this sphere problem as analogous to a slab problem and use the appropriate characteristic length.)*

5.32   Laser Heating of Tissue

To approximate laser heating of tissue with a point source, we can come up with a simple solution for how the heat deposited by the laser spreads into the tissue. The temperature rise in the tissue as a function of position, r, and time, t, can be obtained by solving the heat equation in radial coordinates in an infinite domain to be

$$T = \frac{Q}{8(\pi \alpha t)^{3/2}} e^{-r^2/4\alpha t} \tag{5.38}$$

where Q is the instantaneous amount (as opposed to a persistent amount) of heat de-posited at $r = 0$ at time $t = 0$ and α is the thermal diffusivity of the tissue. 1) From Eq. 5.38, show that, at a given radius, the maximum temperature occurs at a time given by $t = r^2/6\alpha$. 2) Roughly sketch how the radial temperature profile (temperature as a function of radius) evolves with time. 3) From Eq. 5.38, we see that, at a distance $r = \sqrt{4\alpha t}$ from the origin, the temperature decreases to $1/e$ of its peak value. Defin-ing this distance to be the thermal penetration depth for the laser energy that causes thermal tissue damage, calculate the thermal penetration depth for times of 1 μs, 1 ms, and 1 s for an approximate thermal diffusivity of skin of 1.4×10^{-7} m^2/s.

5.33   Sterilization of Wood

Heat sterilization of lumber, timbers, and pallets is used to kill insects to prevent their transfer between countries in international trade. This is analogous to food sterilization by heat. A typical requirement here is that the slowest heating point of any wood configuration be held at 56°C for 30 minutes. Consider hot air heating of wooden boards that maintains their surface temperature at 70°C. The boards are stacked outside and in the winter time they can be considered to be at 0°C when they are brought in for heating. The thermal diffusivity of the wood is $9 \times 10^{-8} \mathrm{m}^2/\mathrm{s}$. 1) Calculate the time from the start of heating for a 2.5 cm thick board to reach a sterilization temperature of 56°C at its slowest heating point. 2) Calculate the heating time when four such boards are stacked together. 3) Calculate the ratio of the two heating times (for single a board versus when they are stacked), and explain the ratio.

5.34   Forced Air Cooling of Strawberries

Strawberries need to be cooled quickly after harvest to ensure high quality. Consider a strawberry (spherical for simplicity, 2.5 cm in diameter) with air at −1°C blowing over it, leading to a large surface heat transfer coefficient. The initial uniform temperature of the strawberry is 23°C. The thermal conductivity, density, and specific heat of strawberries are 0.675 W/m·C, 860 kg/m^3 and 3.96 kJ/kg·C, respectively. In practice, the speed of cooling is determined by the 7/8 cooling time, defined as the time required for the temperature at the center of the strawberry to drop by 7/8 of the difference between the initial strawberry temperature and the air temperature. 1) Calculate the 7/8 cooling time. 2) In a package, the strawberries downstream of cold air, i.e., further away from the source of cold air, experience slightly warmer air temperatures (air has been warmed by the strawberries upstream) and therefore do not cool at the same rate, possibly leading to quality problems. For a strawberry downstream that is experiencing an air temperature of 2°C, what would the time needed to reach the same temperature as in 1) be?

5.35   Lumped Parameter Model of Fingertip Cooling

One study describes a simple lumped parameter model for response of fingertips exposed to cold environments. The fingertip is modeled as a hemisphere that is supplied by a single artery and drained by one vein, as shown in Figure 5.26, such that the entire hemisphere is always at one temperature, T. This hemisphere exchanges heat with surrounding air at T_∞. Heat is supplied by the blood at a temperature T_b and flowing at the rate of $\dot{V}_b$ ml of blood per ml of tissue per second (1 ml = 1 cm^3) and flowing out at the fingertip temperature, T. The initial temperature (before being exposed to the cold environment) of the tissue is the same as the incoming blood temperature, T_b. The density and specific heat of blood are ρ_b and c_{pb}, respectively. The heat transfer coefficient over the hemisphere is h. The hemispherical mass of tissue is m and its specific heat is c_p.

1) Using symbols only, write an energy balance for a time period Δt for the fingertip (the hemisphere) during which its temperature changes by ΔT. 2) By making $\Delta t \to 0$, obtain the governing differential equation. 3) Solve the differential equation to obtain temperature as a function of time. *(Hint: By rearranging the terms, the equation should look similar to the lumped parameter equation done in class.)* 4) Plug the following parameter values into your answer in part 3): $h = 8$ W/m$^2 \cdot$ K, $c_p = 3.9$ kJ/kg $\cdot$ K, $c_{pb} = 3.1$ kJ/kg $\cdot$ K, $\rho_{tissue} = 1057$ kg/m^3, $\rho_b = 1060$ kg/m^3, $\dot{V}_b = 1/3600$ ml of blood per ml of tissue per second, $T_b = 25°$C, $T_\infty = 0°$C, and the radius of the sphere is 7 mm. Write the final solution.

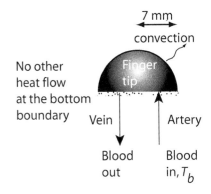

Figure 5.26: Lumped parameter approximation of the finger tip.

5.36 Determining Time of Death from Body Temperature

Many forensic pathologists use measured body temperature during postmortem cooling to estimate the time of death. Consider a thermocouple (measures temperature) inserted in the rectum, as shown in Figure 5.27. This estimation grossly simplifies the process to be heat transfer in a slab of total thickness 20 cm with the temperature measured at the thermocouple tip as shown. The body happens to be left in a position such that the slab can be assumed to cool symmetrically from both sides (front and back). The room ambient temperature is 20°C and the effective heat transfer coefficient between the body surface and the ambient air is 15 W/m$^2 \cdot$ K. Also, assume the live person had a core body temperature (as would be measured by the thermocouple at the same location) of 37°C. The effective thermal conductivity, density, and specific heat of all the materials combined are 0.5 W/m$\cdot$K, 1050 kg/m^3, and 3750 J/kg$\cdot$K, respectively. 1) Estimate the time of death when the measured temperature at the indicated location in the figure is 21.7°C.

5.37 Heat Exchange in a Moth

Body temperature exerts a major influence on the biochemical reaction rates of organisms, which in turn affects important performance capabilities such as locomotion. Quantifying heat exchange in a particular part of the body can be important in determining its importance in thermoregulation. Consider heat transfer in the abdomen of a sphinx moth, as illustrated in Figure 5.28. We would like to treat the abdomen as a sphere (detached from the rest of the body for heat transfer purposes) and use a lumped parameter approach to find the abdomen temperature as a function of time.

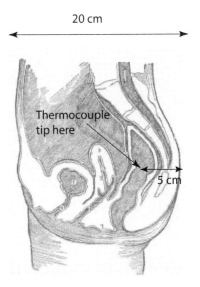

www.gutenberg.org/files/19825/19825-h/19825-h.htm

Figure 5.27: A cross section of a female body showing the placement of the thermocouple in the rectum.

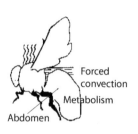

Figure 5.28: Schematic of a sphinx moth.

1) Perform an energy balance on this sphere over time Δt (during which time the temperature changes by ΔT) considering convective heat loss over its surface (with heat transfer coefficient h) and heat generation per unit volume, Q. 2) Let $\Delta t \rightarrow 0$ and develop the differential equation. 3) Solve the differential equation to obtain an expression for temperature as a function of time. 4) The diameter of the abdomen is 3 mm, its volumetric heat generation is 14000 W/m^3, specific heat is 4200 J/kg·K, density is 1000 kg/m^3, and the convective heat transfer coefficient over the abdomen surface is 50 W/m^2·K. Calculate the difference (numerical value) between the steady-state temperature and the ambient temperature and comment on whether it is high or low. 5) Calculate the time constant (numerical value) and comment on why it is high or low.

5.38   Vitrifying a Human Body?

In vitrification, cells or whole tissues are preserved by rapidly cooling to low temperatures, typically to the boiling point of liquid nitrogen. Sometimes you hear of preserving an entire human body using such a procedure. Of the many possible difficulties with human cryopreservation, one is the limitation in cooling rates that can be achieved. Assume a thin body (slab of 20 cm thickness) that is at a uniform initial temperature of 37°C and immersed in liquid nitrogen so that the body surface is at –196°C. The overall thermal diffusivity of the body is 1.27×10^{-7} m^2/s. Assume no freezing, i.e., no latent heat needs to be taken into account. 1) Estimate the cooling rate (in °C/min) at the center of the body between the times of 11 hours and 14 hours, assuming a constant cooling rate during this time interval. 2) If a rate of 100°C/min is required for no ice formation, are we able to achieve close to this rate? 3) Would the cooling rate at the same location be faster or slower in a thinner slab, under the same conditions otherwise? Provide reasoning.

5.39   Measuring Thermal Diffusivity

Thermal properties such as thermal diffusivity need to be measured so that heat transfer can be studied using mathematical models. In an experimental setup to measure thermal diffusivity of some food material (minced red meat), a tube, which can be approximated as a long cylinder, is filled with the material and cooled. The center temperature is measured to be 5°C after 65 s. The initial temperature of the food material was 27°C and its boundary temperature is 0°C. The diameter of the cylinder is 10 mm. 1) What is the thermal diffusivity of the food material? 2) If the time duration in the experiment needs to be reduced by 10%, keeping everything else the same, what must be the new diameter? Show your procedure.

5.40 Infrared Heat Lamp on Skin Surface

An infrared heat lamp is a common device to heal and soothe pain, due to its effects on increasing blood flow. We will ignore the blood flow effects in this problem and also consider all of the thermal radiation to be absorbed on the skin surface (no penetration). For an infrared lamp shining over a large skin area (this provides a constant net radiative heat flux over the surface), as shown in Figure 5.29, a skin surface temperature of 30°C is observed after 25 s. The initial temperature of the surface region of the skin tissue is 25°C. Find the time at which the skin surface temperature would reach 43°C, causing thermal damage to the tissue.

5.41 Calculating Heating Time in an RNA Amplification Procedure

A research group wants to reduce the total assay time of a pathogen detection system that uses an isothermal RNA amplification procedure. The procedure involves heating the sample material to partially denature the target DNA before cooling it (we are only interested in the heating part in this problem). The material to be heated is in a long, cylindrical tube and is viscous enough that any flow inside the tube can be ignored (i.e., the heating is by conduction alone). Also, consider the glass of the tube to be very thin so its thermal resistance can be ignored. The tube diameter is 1 cm and the properties of the material are the same as those of water, given by 0.64 W/m · K, 1000 kg/m³, and 4180 J/kg · K, for thermal conductivity, density, and specific heat, respectively. The water bath temperature is 65°C and the initial temperature of the material is 27°C.

1) When there is no agitation of the water bath, the heat transfer coefficient can be considered to be 128 W/m² · K. Considering the tube to be a long cylinder, how long does it take for the coldest point to heat up to a temperature corresponding to 99% of the total possible temperature increase, i.e., to reach a value of $T_i + 0.99(T_\infty - T_i)$? 2) If the surrounding hot water is agitated using a stirrer, leading to a considerably large surface heat transfer coefficient, how long does it take to reach the same temperature as was used in step 1? 3) In order to reduce the assay time, the diameter of the tube was reduced by 25% to 0.75 cm. Keeping the same final temperature and the same convective boundary condition as in step 2, what is the new heating time? *You are not allowed to use the Heisler chart to answer this question.* 4) If the initial temperature and the water bath temperatures are both different from what is given above but we are still interested in reaching a value that is equal to $T_i + 0.99(T_\infty - T_i)$, would your answers in steps 1 and 2 change? Explain why or why not.

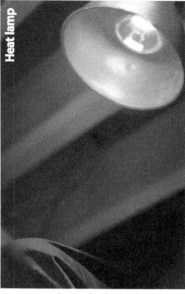

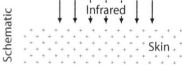

Figure 5.29: Schematic of an infrared heat lamp. (iStock.com/Edward Olive)

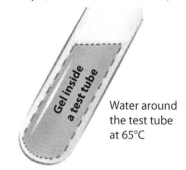

Water around the test tube at 65°C

Figure 5.30: A test tube being heated by the surrounding water.

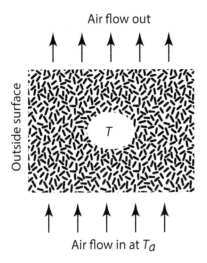

Figure 5.31: A compost pile with air flow through it.

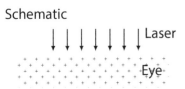

Figure 5.32: Schematic of laser incident on eye. (iStock.com/Arndt Vladimir)

5.42   Simplified Heat Transfer in a Compost Pile

In an aerated compost pile, as shown in Figure 5.31, any mixing of the pile due to air flow leads to a somewhat uniform temperature in the pile that is amenable to a lumped parameter analysis. Let T be this uniform temperature that is averaged over the entire composting matrix of volume, V. A uniform rate of heat generation, Q_m W/m^3 of compost, occurs in the pile due to biological reactions. A constant rate of evaporation of water, $\dot{m}$ kg water/m^3 of compost per second, occurs in the pile, and the latent heat of evaporation is λ kJ/kg of water. I The incoming air temperature is T_a while the outgoing air is at compost temperature, T. The volumetric rate of air flow is $\dot{V}_{air}$ m^3 of air/m^3 of compost per second and the specific heat and density of air are c_{pa} and ρ_a, respectively. The specific heat and density of the solid material in the compost are c_p and ρ, respectively. In addition, convective heat transfer from the surface of the pile (with total surface area A) also takes place, with a convective heat transfer coefficient, h.

1) Starting from an energy balance, considering all sources of heat generation and loss, derive the differential equation for rate of temperature change. *Hint: To include the effect of air flow, think of it in terms of blood flow through tissue.* 2) Solve for temperature, T, as a function of time for an initial temperature of T_i.

5.43   Laser Heated Destruction of Tissue in Eye Surgery

Moved to solved problems

5.44   Laser Heated Destruction of Tissue in Eye Surgery

For laser therapy in ophthalmology, consider a laser beam incident on the eye as shown in Figure 5.32. For small durations of heating, we can ignore the curvature of the eyeball and treat it as a plane surface that is quite thick. Assume the laser energy flux is absorbed entirely on the surface, i.e., no penetration, and it stays constant over time. Thermal necrosis (death of most cells) of the eye tissue occurs at temperatures of 60°C and above. The eye is initially at 37°C. The average thermal conductivity, density, and specific heat of the tissue are 0.54 W/m · K, 1050 kg/m^3, and 3684 J/kg · K, respectively.

1) If we want necrosis *only in the first 250 μm of tissue* after 250 milliseconds, what should the laser flux at the surface be? 2) For the same time and laser flux in question 1, what is the temperature at a depth of 500 μm? 3) You want to determine if there is any benefit to doubling the value of this laser flux at the surface that you found in question 1. In order to do this, calculate the time needed at the higher flux to reach necrosis at a depth of 250 μm. Then, for this time, calculate the temperature at a depth

of 500 μm. Compare this temperature with your result from question 2 to comment on whether or not this is beneficial, and explain your reasoning.

5.45   Bacterial Destruction when Temperature is Changing

Consider a slab of agar gel (Figure 5.33) containing immobilized bacteria *E. coli* O157:H7 being put into a water bath of temperature, T_∞. The slab is initially at temperature, T_i, and has a thickness of $2L$. The heat transfer coefficient between the water and slab surface is very high. The concentration of *E. coli*, N, at any depth in the slab is given as a function of time, t, and the temperature at that depth, T, by first order kinetics as

$$N = N_0 e^{-\int_0^t k_0 e^{-(E/RT)} dt} \qquad (5.39)$$

where other symbols refer to various constants. *Write an expression* for bacterial concentration, N, as a function of time only, for *the coldest location* in the slab. Do not try to integrate. We are only interested in the later times, i.e., we are not interested in the initial times when the temperature is low and the bacterial destruction is negligible.

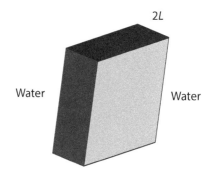

Figure 5.33: Slab of agar gel with the water flowing on both sides.

5.46   Variation of Problem 5.45

Thermal inactivation has been the most popular method of eliminating disease-causing bacteria from food. Consider slabs of agar gel of thickness $2L$, containing immobilized bacteria *E. coli* O157:H7 initially at temperature T_i being put into a water bath of temperature T_∞. The heat transfer coefficient between the water and slab surface is very high. The *E. coli* population is reduced as a first-order reaction given by

$$\frac{dN}{dt} = -k_0 e^{-E/RT} N \qquad (5.40)$$

where k_0 and E/R are known (and constant) values for *E. coli*, and T is absolute temperature (not constant). In our analysis, we are only interested in the later times, i.e., we are not interested in the initial times when the temperature is low and the bacterial destruction is negligible.

1) Write the equation (solution to governing equation) that shows how the temperature of the slab at a depth x changes with time. 2) Using this equation from part 1 with Eq. 5.40, calculate the time needed for the *E. coli* population at the coldest point in the slab to go down by a factor of 10^7, i.e., $N_{\text{final}}/N_{\text{initial}} = 10^{-7}$. Leave all the unknown quantities in terms of their symbols. *Any integration not possible should be left as integrals.*

5.47 Detecting skin lesion from thermal response

Moved to steady-state chapter

5.48 Live Chilling Time of Atlantic Salmon

Figure 5.34: Pink salmon fish (source: Wikipedia).

Live chilling is an experimental technique performed on commercially farmed salmon in order to preserve product quality. Prior to slaughter, the fish are chilled to below 4°C to prevent undesirable chemical changes that occur in muscles after death. We want to build a model that can predict the salmon's core (center) body temperature with time. The fish can be approximated as a slab with dimensions of 100 cm × 25 cm × 10 cm, with symmetrical heat transfer only along the thickness of the shortest dimension. Assume that the fish is at an initial temperature of 16°C (uniform throughout the body) and the chilling water temperature is 2°C. The heat transfer coefficient between fish and water is estimated to be "very large." The thermal conductivity, density, and specific heat of the salmon are 0.53 W/m · K, 400 kg/m^3, and 2.97 kJ/kg·K, respectively.

1) Write the model (governing equation with necessary terms only, and boundary and initial conditions) that can be used to predict the fish's body temperature as a function of position and time. Ignore any metabolic heat generation. 2) Write the solution to the governing equation (provide reasoning for your choice of solution) that predicts the salmon's core temperature, T, at time t. Assume we are only interested in the long time solution. 3) A 2002 research paper uses an empirical predictive model for live chilling given by

$$T = (T_i - T_\infty) Ae^{-Rt} + T_\infty \qquad (5.41)$$

where T and T_i are the core temperature of the fish at time t, and its initial temperature, respectively. Here T_∞ is the chilling water temperature, R is the rate of chilling with a unit of $1/s$, and A is an empirical constant. All temperatures are in °C in Eq. 5.41. Compare Eq. 5.41 to your answer in step 2 and state the expressions that make up A and R in terms of the model parameters from your equation in step 2. 4) The research paper provides the following empirical equation for R:

$$R = 7.73 \times 10^{-4} - 7.06 \times 10^{-5} m^{0.7} \qquad (5.42)$$

where m is the mass in kg, with $A = 1$. When the fish is moved from seawater (where the entire body was at 16°C) into chilling water at 2°C, calculate the predicted chilling time needed to reach 4°C using Eq. 5.41, *and then* using your model in step 2, by calculating the percent error in relation to the model in the research paper. 5) Comment (provide reasons) on whether or not you, as an engineer, would go for the empirical model as opposed to your heat transfer model.

5.49 Can There Be a Temperature Difference across a Protein–Water Interface?

Although our heat equation (the one we always use or its lumped parameter equivalent), *may not be valid* for very small dimensions (<70 nm), a 2010 research paper indeed uses our heat equation to study heat transfer between a protein (assumed to be a sphere of 2 nm radius) and surrounding water. The heat transfer coefficient at the protein–water interface is 10^8 W/m$^2 \cdot$ K. If the entire sphere is at a constant initial temperature that is different from that of the surrounding water, calculate the time it will take for the temperature at the center of the protein to change by 90% of its total possible change. The thermal conductivity of the protein is 0.2 W/m·K and its thermal diffusivity is 4×10^{-8}m^2/s.

5.50 Heat Transfer in a Tooth during Drinking of Warm Fluid

We wish to model how the temperature of a single tooth will change as we drink hot water. This information can be useful for a sensitive tooth, for example. The tooth can be considered a long homogeneous cylinder (Figure 5.35) with a diameter of 7.2 mm, thermal diffusivity of 2.27×10^{-7}m^2/s, specific heat of 1.26×10^3 J/kg·K, and density of 2.20×10^3kg/m^3. The initial temperature of the tooth is 35°C, the tooth is heated symmetrically with the water temperature kept constant at 60°C, and the heat transfer coefficient for the flow of water over a tooth is 3.49×10^2W/m$^2 \cdot$ K.

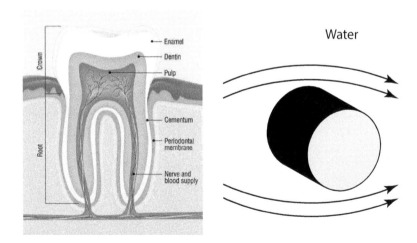

Figure 5.35: Single tooth with water flowing over it.

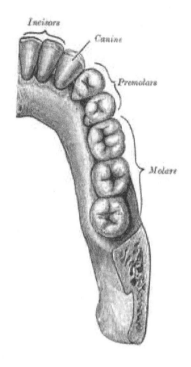

1) Write the governing equation and boundary conditions for this problem. Do not try to solve. 2) Assuming the edge of the nerve that senses temperature is located at 1/3 radial distance from the center to the surface of the tooth (Figure 5.35), what is the temperature at this location after 30 seconds? 3) Due to its tight placement (Figure 5.36), the tooth array can also be treated as a slab (7.2 mm thick), heated from both sides. If the nerve can still be considered at 1/3 distance from the center of the slab to the side, what is the temperature at the nerve location after 30 seconds? 4) Provide a short physical reasoning as to why the temperature in the case of a slab should be lower/higher than that in a cylindrical geometry.

5.51   Heat Transfer over Epicardium

In cardiac surgery, surgeons can obtain important information about coronary blood flow from thermographic measurements of epicardial temperature over time. Cold saline was introduced into an epicardial artery (see Figure 5.37a) that brings down its temperature and the rewarming of the artery due to blood flow is to be modeled. The arterial wall has a surface area A and volume V exposed to blood, a specific heat of c_p, and a density of ρ. The constant temperature of the flowing blood is T_b and the heat transfer coefficient between blood and the arterial wall is h. Ignore any heat loss of the artery to surroundings (on the outside). Also, ignore any metabolic heat generation in the artery.

1) A thin arterial wall allowed a lumped parameter analysis of the artery segment (Gordon et al., 1998). Perform a lumped heat balance on the segment shown in Figure 5.37b, due to heat exchange with blood, to obtain a differential equation for the artery segment temperature with time. 2) Solve this differential equation to obtain the temperature of the artery segment as a function of time. 3) If the saline water had cooled the artery to an initial temperature of 30.5°C and the blood temperature is 32.7°C, sketch approximately the temperature vs. time. 4) If experimentally measured temperature vs. time were available, how could you use it to estimate h using the model that you just developed? Explain in words.

Figure 5.36: Permanent teeth of right half of lower dental arch.   From wikipedia.org/wiki/File:Gray997.png.

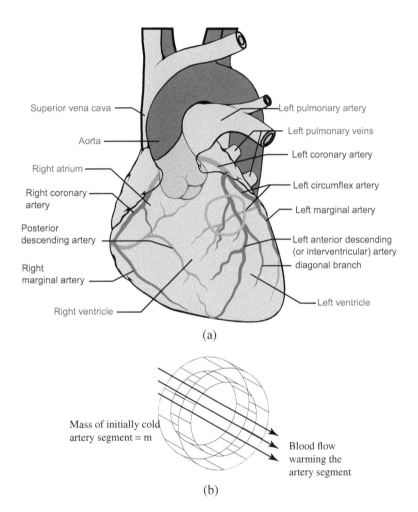

(a)

Mass of initially cold artery segment = m

Blood flow warming the artery segment

(b)

Figure 5.37: a) Epicardial arteries (source: Wikipedia); some on the left side are circled. b) Schematic of a segment of one epicardial artery on which heat balance will be performed.

Chapter 6

CONVECTION HEAT TRANSFER

CHAPTER OBJECTIVES

After you have studied this chapter, you should be able to

1. Explain the process of convection heat transfer in a fluid over a solid surface in terms of the relatively stagnant fluid layer over the surface called the boundary layer.

2. Calculate the convective heat transfer coefficient h knowing the flow situation.

KEY TERMS

- **convective heat transfer coefficient**

- **velocity boundary layer**

- **thermal boundary layer**

- **Reynolds number**

- **Nusselt number**

- **Prandtl number**

- **Grashof number**

- **natural and forced convection**

- **laminar and turbulent flow**

In this chapter we will develop a more general version of the governing equation that we derived in Chapter 3. It is more general since it will now include fluids that can flow and carry energy by virtue of their motion. The relationship of this chapter to other chapters in energy transfer is shown in Figure 6.1.

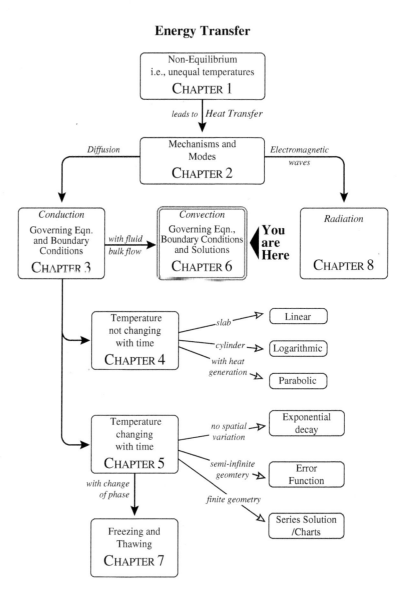

Figure 6.1: Concept map of energy transfer showing how the contents of this chapter relate to other chapters on energy transfer.

6.1 Governing Equation for Convection

As mentioned in Chapter 2, convective heat transfer is the movement of heat through a medium as a result of the net motion of a material in the medium. Convection takes place in most heat transfer applications involving a liquid or gas, *in addition to conduction*. In the previous chapters this convection was considered only as a boundary condition at the surface of a conducting solid. The parameter h, the convective heat transfer coefficient describing the convecting process, was assumed to be known. In this chapter we look at the details of the convection process with one of the goals being to investigate ways to predict the convective heat transfer coefficient h.

The general equation (Eq. 3.5) describing energy transfer is repeated here from page 43 as

$$\underbrace{\frac{\partial T}{\partial t}}_{\text{storage}} + \underbrace{u\frac{\partial T}{\partial x}}_{\text{bulk flow}} = \underbrace{\frac{k}{\rho c_p}\frac{\partial^2 T}{\partial x^2}}_{\text{conduction}} + \underbrace{\frac{Q}{\rho c_p}}_{\text{generation}} \tag{6.1}$$

In this chapter we will be applying the equation to fluids (as opposed to solids in the previous chapters). Thus, for this chapter, all the properties in the above equation are for fluid. We will need to retain the convection term, i.e., we cannot drop it as in the previous chapters. If the convection term is kept, knowledge of fluid velocity u is needed. Thus, the equations governing fluid flow are also needed in addition, to solve the energy equation given by Eq. 6.1. This makes the study of convection heat transfer often more complex than that of simple conduction.

Although boundary layer approximations (described below) are made to simplify the energy and flow equations for many practical situations, the solution to a convective heat transfer problem (Eq. 6.1 and the related flow equations) is often involved. We will not pursue such detailed solutions to the convection equations, but instead, discuss the expected form of the solution and spend our effort in understanding the implications of the solutions.

6.2 Temperature Profiles and Boundary Layers over a Surface

Figure 6.2 is a schematic of flow over a flat plate. When fluid of initial uniform velocity flows over a surface, such as a flat plate, the velocity becomes zero at the surface of the plate. The decrease in velocity, from the stream velocity to zero, takes place in a relatively small layer of fluid known as the boundary layer. The fluid velocity approaches the free stream velocity u_∞ as we move further away from the plate. It

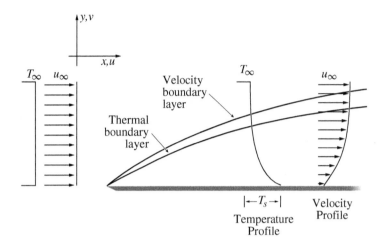

Figure 6.2: Velocity and temperature variation in the boundary layer over a flat plate.

may take a very large distance from the plate for the velocity to be exactly equal to u_∞, but most of this change in velocity takes place over this small distance known as the boundary layer. The thickness of the boundary layer $\delta_{velocity}$ is defined as the distance where the velocity is 99% of the free stream velocity, i.e.,

$$\delta_{velocity} = y|_{u=0.99u_\infty} \tag{6.2}$$

Thus, the fluid can be considered to have two separate regions: a boundary layer, where velocity gradients are large, and outside the boundary layer, where the velocity gradients are small. In other words, the effect of the flat plate on the flow is essentially restricted to the boundary layer. The concept of this boundary layer was first introduced by Prandtl (1904).

Like the velocity boundary layer, a thermal boundary layer develops if the surface temperature for the flat plate is different from the uniform initial fluid temperature T_∞. Since the fluid in contact with the surface will be at rest, it will come to the surface temperature. The temperature will vary from the value T_s at the surface to T_∞ in the free stream. Like the velocity variation, this temperature variation is asymptotic, and a thickness of thermal boundary layer $\delta_{thermal}$ is defined as the distance at which temperature T is given by

$$\frac{T_s - T_{\delta_{thermal}}}{T_s - T_\infty} = 0.99 \tag{6.3}$$

Thus $\delta_{thermal}$ is the distance over which most of the temperature change takes place. Like the velocity boundary layer, the fluid can be considered to have two distinct thermal regions: a thermal boundary layer, where thermal gradients are large, and outside the thermal boundary layer, where the thermal gradients are small and temperatures are uniform. In other words, the thermal effect of the flat plate on the flow is essentially restricted to the thermal boundary layer. While a velocity boundary layer will always exist in a flow situation, a thermal boundary layer exists only if there is a difference in temperature between the surface and the bulk fluid.

The thickness of the velocity boundary layer can be shown to relate to the flow parameter Reynolds number, Re, defined as

$$\mathrm{Re}_x = \frac{u_\infty x \rho}{\mu} \qquad (6.4)$$

where u_∞ is the free stream velocity, as shown in Figure 6.2, x is the distance along the flow from the leading edge of the plate, ρ is the density of the fluid, and μ is the viscosity of the fluid. The thermal boundary layer additionally depends (shown in Section 6.5) on a characteristic number called the Prandtl number, Pr, for the fluid:

$$\mathrm{Pr} = \frac{\mu c_p}{k_{fluid}} = \frac{\mu/\rho}{k_{fluid}/\rho c_p} = \frac{\text{Momentum diffusivity}}{\text{Thermal diffusivity}} \qquad (6.5)$$

where c_p is the specific heat of the fluid and k_{fluid} is the thermal conductivity of the fluid. Thus, Prandtl number depends only on the fluid properties. As an example of boundary layer thicknesses, for laminar flow over a flat plate, the velocity boundary layer thickness is given by

$$\delta_{velocity} = \frac{5x}{\mathrm{Re}_x^{\frac{1}{2}}} \qquad (6.6)$$

Thus, as the velocity (and therefore Re_x) increases, the boundary layer thickness δ decreases. The thermal boundary layer thickness for laminar flow over a flat plate is related to the velocity boundary layer thickness by

$$\frac{\delta_{velocity}}{\delta_{thermal}} = \mathrm{Pr}^{1/3} \qquad (6.7)$$

or

$$\delta_{thermal} = \frac{\delta_{velocity}}{\mathrm{Pr}^{1/3}}$$

which implies that the thermal boundary layer thickness also decreases as the velocity increases. Since $\mathrm{Pr} \approx 5$ for water, the thermal boundary layer will be much thinner than the velocity boundary layer on a flat plate. This is in contrast to gases, for which $\text{Pr} \approx 1$ and the thermal and velocity boundary layers on a flat plate are approximately equal.

6.3   Laminar and Turbulent Flows

In the previous section, we discussed the boundary layer over a flat plate. Although fluid motion on such a flat plate starts out to be orderly (streamlined) at the leading edge, as shown in Figure 6.2, as the fluid moves along the plate, the flow changes to a more chaotic situation at some distance from the leading edge, as illustrated in Figure 6.3. This change from a streamlined to a more chaotic condition changes the nature of the flow quite drastically. In particular, the *mixing process* due to the more chaotic motion or turbulence *enhances the rate of heat transfer.* The region of the plate over which the fluid motion is orderly is called the laminar region, whereas the region with a more chaotic fluid motion is called the turbulent region. The transition from laminar to turbulent can occur over a region, which is called the transition region. In

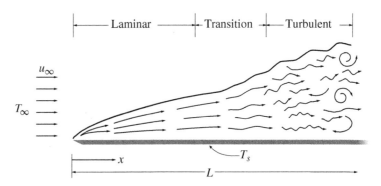

Figure 6.3: Schematic showing laminar, transition, and turbulent flow over a flat plate.

determining the type of boundary layer (laminar or turbulent) that exists in a certain flow situation, it is frequently assumed that the transition from laminar to turbulent occurs at a critical Reynolds number. For example, for flow over a flat plate, the critical Reynolds number based on distance x along the plate ($\text{Re}_x = u_\infty x \rho / \mu$) has been found to vary experimentally between 2×10^5 and 3×10^6, i.e., the following conditions define the laminar, turbulent, and transition regions over a flat plate:

$$
\begin{array}{lll}
\text{laminar region} & \text{Re}_x & < 2 \times 10^5 \\
\text{transition region} & 2 \times 10^5 < \quad \text{Re}_x & < 3 \times 10^6 \\
\text{turbulent region} & 3 \times 10^6 < \quad \text{Re}_x &
\end{array}
$$

As will be shown later, when calculating rates of convective heat transfer, it is critical to consider whether the flow is in the laminar or turbulent region, as characterized by the corresponding Reynolds number.

6.4 Convective Heat Transfer Coefficient Defined

Consider fluid flowing over a surface, as shown in Figure 6.2. Since the layer of fluid in contact with the surface is at rest, heat transfer *in the fluid* at the surface is by conduction only and the flux is given by

$$
\begin{array}{c}
\text{conductive heat} \\
\text{flux in the fluid}
\end{array}
= -k_{fluid} \left. \frac{\partial T}{\partial y} \right|_{y=0,\text{ in fluid}}
$$

where k_{fluid} is the thermal conductivity *of the fluid* and T is the *fluid* temperature at a distance y from the surface. However, by definition of the convective heat transfer coefficient, h, the same heat flux can be written as

$$
\begin{array}{ll}
\text{convective} \\
\text{heat flux}
\end{array}
\begin{array}{ll}
= & h(T_{y=0,\text{ in fluid}} - T_\infty) \\
= & h(T_s - T_\infty)
\end{array}
$$

Since the two expressions represent the same heat flux,

$$
-k_{fluid} \left. \frac{\partial T}{\partial y} \right|_{y=0,\text{ in fluid}} = h\,(T_s - T_\infty)
$$

$$
h = \frac{-k_{fluid} \left. \frac{\partial T}{\partial y} \right|_{y=0,\text{ in fluid}}}{T_s - T_\infty} \tag{6.8}
$$

The temperature profile $T(y)$ can be obtained by solving the energy equation for the fluid (see next section). Knowing $T(y)$, we can calculate the derivative $\partial T/\partial y$ in Eq. 6.8 and therefore the heat transfer coefficient h. Since $T(y)$ includes the effect of conduction and flow, h would also include *conduction and flow effects*.

6.5   Significant Parameters in Convective Heat Transfer

To obtain the convective heat transfer coefficient h for many practical situations, one would like to obtain the solution to the convection equation. In this section we will discuss the expected form of the solution to the convection equation for the boundary layer approximation. We will not solve the equations, as mentioned earlier, for brevity. Such solutions can be found in most heat transfer texts, such as Incropera and Dewitt (1996). As an example, consider steady convection over a flat plate (Figure 6.2); the flow is two dimensional (along and perpendicular to the plate) and the governing energy equation for the fluid without any heat generation and with constant property values is (generalized from Eq. 3.5 in one dimension)

$$\underbrace{u\frac{\partial T}{\partial x} + v\frac{\partial T}{\partial y}}_{\text{bulk flow}} = \underbrace{\alpha\left(\frac{\partial^2 T}{\partial x^2} + \frac{\partial^2 T}{\partial y^2}\right)}_{\text{conduction}} \tag{6.9}$$

Note that the conduction term is included in the above equation, signifying convection heat transfer solutions that will be developed include the conduction effects. The additional governing equation to be solved for fluid flow (refer to your fluid mechanics book) is

$$u\frac{\partial u}{\partial x} + v\frac{\partial u}{\partial y} = -\frac{1}{\rho}\frac{\partial p}{\partial x} + \nu\left(\frac{\partial^2 u}{\partial x^2} + \frac{\partial^2 u}{\partial y^2}\right) \tag{6.10}$$

$$u\frac{\partial v}{\partial x} + v\frac{\partial v}{\partial y} = -\frac{1}{\rho}\frac{\partial p}{\partial y} + \nu\left(\frac{\partial^2 v}{\partial x^2} + \frac{\partial^2 v}{\partial y^2}\right) \tag{6.11}$$

with the continuity equation

$$\frac{\partial u}{\partial x} + \frac{\partial v}{\partial y} = 0 \tag{6.12}$$

where u, v, p, ν are velocity in the x direction, velocity in the y direction, pressure, and kinematic viscosity, respectively. The solution to this set of equations will be considerably simplified by using the boundary layer approximations discussed earlier. The assumptions we will use for the boundary layer are

$$u \gg v \tag{6.13}$$

which signifies the velocity along the plate is much larger than that perpendicular to the plate. The gradient is also much larger in the y direction for u:

$$\frac{\partial u}{\partial y} \gg \frac{\partial u}{\partial x}, \frac{\partial v}{\partial x}, \frac{\partial v}{\partial y} \tag{6.14}$$

For the thermal boundary layer, the thermal gradient in the y direction is much larger than the gradient in the x direction, i.e.,

$$\frac{\partial T}{\partial y} \gg \frac{\partial T}{\partial x} \tag{6.15}$$

With these approximations (Eq. 6.13, 6.14, 6.15), the governing equations for the boundary layer (Eq. 6.9, 6.10) for convection over a flat plate can be approximated as

$$u\frac{\partial T}{\partial x} + v\frac{\partial T}{\partial y} = \alpha\frac{\partial^2 T}{\partial y^2} \tag{6.16}$$

$$u\frac{\partial u}{\partial x} + v\frac{\partial u}{\partial y} = -\frac{1}{\rho}\frac{\partial p}{\partial x} + v\frac{\partial^2 u}{\partial y^2} \tag{6.17}$$

Equation 6.11, together with Eq. 6.14, leads to the condition $\partial p/\partial y = 0$, which means the pressure is a function of x. This $p(x)$ can be obtained from other conditions and so we can treat p as known in Eq. 6.17.

We would now change Eq. 6.16 and 6.17 to a non-dimensional form. For this purpose, we define non-dimensional quantities or similarity parameters as

$$
\begin{aligned}
x^* &= x/L \\
y^* &= y/L \\
u^* &= u/u_\infty \\
v^* &= v/u_\infty \\
T^* &= (T - T_s)/(T_\infty - T_s)
\end{aligned}
$$

where L is the length of the plate and u_∞ is the free stream velocity as shown in Figure 6.2. Using these non-dimensional quantities, Eq. 6.16 and 6.17 can be written as

$$u^*\frac{\partial T^*}{\partial x^*} + v^*\frac{\partial T^*}{\partial y^*} = \frac{\alpha}{u_\infty L}\frac{\partial^2 T^*}{\partial y^{*2}} \tag{6.18}$$

$$u^* \frac{\partial u^*}{\partial x^*} + v^* \frac{\partial u^*}{\partial y^*} = -\frac{dp^*}{dx^*} + \frac{\nu}{u_\infty L} \frac{\partial^2 u^*}{\partial y^{*2}} \tag{6.19}$$

where $p^* = p/\rho u_\infty^2$ is a non-dimensional pressure and $p^*(x^*)$ depends on geometry in general. Defining Reynolds number Re_L as

$$\text{Re}_L = \frac{u_\infty L}{\nu} \tag{6.20}$$

and Prandtl number

$$\text{Pr} = \frac{\mu c_p}{k} \tag{6.21}$$

Eq. 6.18, 6.19 can be written as

$$u^* \frac{\partial T^*}{\partial x^*} + v^* \frac{\partial T^*}{\partial y^*} = \frac{1}{\text{Re}_L \, \text{Pr}} \frac{\partial^2 T^*}{\partial y^{*2}} \tag{6.22}$$

$$u^* \frac{\partial u^*}{\partial x^*} + v^* \frac{\partial u^*}{\partial y^*} = -\frac{dp^*}{dx^*} + \frac{1}{\text{Re}_L} \frac{\partial^2 u^*}{\partial y^{*2}} \tag{6.23}$$

From these equations, we can write symbolically

$$T^* = f(x^*, y^*, \text{Re}_L, \text{Pr}, dp^*/dx^*) \tag{6.24}$$

For a given geometry, contribution from the pressure gradient term is fixed and Eq. 6.24 can be written as

$$T^* = f(x^*, y^*, \text{Re}_L, \text{Pr}) \tag{6.25}$$

Let us now consider the definition of the convective heat transfer coefficient (Eq. 6.8), which can be written in terms of the non-dimensional temperature and distance as

$$h = \frac{-k_{fluid} \frac{\partial T}{\partial y} \big|_{y=0, \, in \, fluid}}{T_s - T_\infty} = \frac{k_{fluid}}{L} \frac{\partial \left(\frac{T-T_s}{T_\infty-T_s} \right)}{\partial \left(\frac{y}{L} \right)} \Big|_{y=0} = \frac{k_{fluid}}{L} \frac{\partial T^*}{\partial y^*} \Big|_{y^*=0} \tag{6.26}$$

which can be rewritten as

$$\frac{hL}{k_{fluid}} = \frac{\partial T^*}{\partial y^*} \Big|_{y^*=0} \tag{6.27}$$

Using this definition of h and using the functional form of the temperature solution (Eq. 6.25), we can write a functional form for the dependence of h as

$$\frac{hL}{k_{fluid}} = f(x^*, \text{Re}_L, \text{Pr}) \tag{6.28}$$

Thus, h should depend on position, in general. An average h can be obtained by integrating over the length of the plate as

$$h_L = \frac{\int_0^L h\,dx}{L} \tag{6.29}$$

This average is independent of x. Therefore,

$$\frac{h_L L}{k_{fluid}} = f(\text{Re}_L, \text{Pr}) \tag{6.30}$$

This is the final equation showing the functional relationship of the average heat transfer coefficient h_L as related to the two non-dimensional parameters Re_L and Pr. The non-dimensional quantity on the left of Eq. 6.30 is defined as the Nusselt number:

$$Nu_L = \frac{h_L L}{k_{fluid}} \tag{6.31}$$

From Eq. 6.27, we can see that the Nusselt number is a non-dimensional temperature gradient. Note the apparent similarity (Table 6.1) between the Nusselt number and the Biot number defined in earlier chapters. Both numbers involve heat transfer coefficient, length, and thermal conductivity. The Biot number relates to thermal conduction in a solid relative to the convection in a fluid in contact with a solid. It therefore involves solid thermal conductivity. The Nusselt number, on the other hand, compares thermal conduction in a fluid relative to convection in the same fluid. It relates entirely to a fluid and involves fluid thermal conductivity. The various non-dimensional parameters defined in this chapter are summarized in Table 6.1.

6.6 Calculation and Physical Implications of Convective Heat Transfer Coefficient Values

Conceptually, for every physical situation, the governing equation for convection can be solved to obtain an explicit form of the function f. This has been performed for many flow situations. Often, for complex situations, h is found experimentally. Such

Table 6.1: Non-dimensional parameters and their physical significance

Dimensionless number	Definition and physical significance
Reynolds number	$\mathrm{Re} = \dfrac{u_\infty L \rho}{\mu} = \dfrac{\text{Inertia force}}{\text{Viscous force}}$
Nusselt number	$\mathrm{Nu} = \dfrac{hL}{k_{fluid}} = \dfrac{\text{Diffusive resistance}}{\text{Convective resistance}}$
Prandtl number	$\mathrm{Pr} = \dfrac{\mu c_p}{k} = \dfrac{\mu/\rho}{\alpha} = \dfrac{\text{Viscous effect}}{\text{Thermal diffusion effect}}$
Biot number	$\mathrm{Bi} = \dfrac{hL}{k_{solid}} = \dfrac{\text{Internal diffusive resistance}}{\text{Surface convective resistance}}$
Grashof number	$\mathrm{Gr} = \dfrac{\beta g L^3 \Delta T}{\nu^2} = \dfrac{\text{Buoyancy force}}{\text{Viscous force}}$
Rayleigh number	$\mathrm{Ra} = \mathrm{Gr} \times \mathrm{Pr}$

formulas (correlations) for h for various physical situations are now discussed. For definition of Nu, Re, Gr, and Pr, see Table 6.1. In this table, the length L refers to a *characteristic length* that depends on the geometry of the surface over which flow takes place. The characteristic lengths for various geometries are introduced at appropriate places in this section. The correlations provided in this section are used in practice to find h for a given situation. For more details on these equations and to find correlations for situations not discussed below, see standard heat transfer texts, such Incropera and Dewitt (1996).

Before presenting the specific convection formulas, some general comments are made about them. As Re increases, the velocity and the thermal boundary layer thickness decreases (see, for example, Eq. 6.6 for velocity boundary layer thickness over a flat plate). Since the slower moving fluid in the boundary layer provides the thermal "insulation," reduced thickness will mean higher heat transfer and therefore larger values of h.

The heat transfer coefficient h varies considerably with location. Typical variation of h in a flat plate is shown in Figure 6.4. At the entrance ($x = 0$), h is very large due to the jump in temperature from T_s for the fluid on surface to T_∞ in the bulk fluid, such that

$$h = \frac{-k_{fluid} \left.\frac{\partial T}{\partial y}\right|_{y=0,\, \text{in fluid}}}{T_s - T_\infty} \rightarrow \infty \qquad \text{at } y = 0 \qquad (6.32)$$

since $\left.\frac{\partial T}{\partial y}\right|_{y=0,\, \text{in fluid}}$ goes to ∞ at the entrance. Thus, when using the correlations for h, one needs to be aware whether the correlations are for local h at a location or for average h over a certain length. In the equations described below, Nu_x is calculated from local h at a location x, whereas Nu_L or Nu_D are averaged over length or diameter, respectively.

Since the temperature varies across a thermal boundary layer, fluid properties such as viscosity, density, and thermal conductivity will also vary across the thermal boundary layer, influencing the heat transfer rate. This influence of temperature variation is generally treated in a simple way by evaluating the properties at an average temperature, called the *film temperature*, defined as

$$T_f = \frac{T_s + T_\infty}{2} \qquad (6.33)$$

where T_s is the surface temperature and T_∞ is the bulk fluid temperature. Thus, for all the correlations presented below, the properties are to be evaluated at this film temperature, T_f.

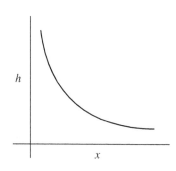

Figure 6.4: Variation of heat transfer coefficient on a flat plate. The edge of the plate is at $x = 0$.

6.6.1 Flat Plate, Forced Convection

The characteristic length for forced convection over a flat plate is the distance along the flow on the plate. As mentioned earlier, the plate surface heat transfer coefficient varies with distance along the flow. This variation is captured in a local Nusselt number, Nu_x ($= h_x x / k_{fluid}$), *at a location x along the flow*, and an average Nusselt number, Nu_L ($= h_L L / k_{fluid}$), *over a distance L along the flow*. Two sets of formulas are provided for each laminar and turbulent flow:

$$\mathrm{Nu}_x = 0.332 \mathrm{Re}_x^{\frac{1}{2}} \mathrm{Pr}^{\frac{1}{3}} \quad \text{for laminar } (\mathrm{Re}_x < 2 \times 10^5) \qquad (6.34)$$

$$\mathrm{Nu}_L = 0.664 \mathrm{Re}_L^{\frac{1}{2}} \mathrm{Pr}^{\frac{1}{3}} \quad \text{for laminar } (\mathrm{Re}_L < 2 \times 10^5) \qquad (6.35)$$

$$\mathrm{Nu}_x = 0.0288 \mathrm{Re}_x^{\frac{4}{5}} \mathrm{Pr}^{\frac{1}{3}} \quad \text{for turbulent } (\mathrm{Re}_x > 3 \times 10^6) \qquad (6.36)$$

$$\mathrm{Nu}_L = 0.0360 \mathrm{Re}_L^{\frac{4}{5}} \mathrm{Pr}^{\frac{1}{3}} \quad \text{for turbulent } (\mathrm{Re}_L > 3 \times 10^6) \qquad (6.37)$$

Note that the equations for average values, i.e., Nu_L, can be derived from the equations for point values, i.e., Nu_x. For example, for laminar flow, starting from Eq. 6.34,

$$\frac{h_x x}{k} = 0.332 \left(\frac{u_\infty x \rho}{\mu} \right)^{\frac{1}{2}} \text{Pr}^{\frac{1}{3}}$$

where h_x is h at location x. Rewriting in terms of h_x,

$$h_x = 0.332 \left(\frac{u_\infty \rho}{\mu} \right)^{\frac{1}{2}} \text{Pr}^{\frac{1}{3}} k x^{-1/2}$$

Using the definition of average h,

$$h_L = \frac{\int_0^L h_x dx}{L}$$

Substituting and performing the integration,

$$= \frac{1}{L} \left(0.332 \left(\frac{u_\infty \rho}{\mu} \right)^{\frac{1}{2}} \text{Pr}^{\frac{1}{3}} k \int_0^L x^{-1/2} dx \right)$$

$$= \frac{1}{L} \left(0.332 \left(\frac{u_\infty \rho}{\mu} \right)^{\frac{1}{2}} \text{Pr}^{\frac{1}{3}} k \left. (2x^{1/2}) \right|_0^L \right)$$

$$= \frac{1}{L} \left(0.332 \left(\frac{u_\infty \rho}{\mu} \right)^{\frac{1}{2}} \text{Pr}^{\frac{1}{3}} k 2 L^{1/2} \right)$$

Simplifying and rewriting,

$$\frac{h_L L}{k} = 0.664 \left(\frac{u_\infty \rho L}{\mu} \right)^{\frac{1}{2}} \text{Pr}^{\frac{1}{3}}$$

$$= 0.664 \text{Re}_L^{\frac{1}{2}} \text{Pr}^{\frac{1}{3}}$$

6.6.2   *Example: Heat Loss from Elephant Pinna*

The African elephant has developed the pinna or external ear (Figure 6.5) that functions as a radiator-convector (see page 102 for more discussion). The pinna can be modeled as a flat plate with air flow over it. The ambient air temperature is 20°C, the air velocity is 2 m/s, the surface area of one side of one ear is 0.84 m^2, and the average length of one side of the ear is 1 m. For properties of air, see Table C.8, page 568. 1) Find an equation for the convective heat loss from the pinna as a function of surface temperature of the pinna. Plot this equation as the surface temperature ranges from 21°C to 36°C. 2) If the rate of metabolic heat production for a 2000 kg elephant is 1650 W, what is the maximum percentage of the metabolic heat that can be lost by convection through the pinna? Remember that an elephant has two ears, and heat is lost equally from both sides of the ear.

Figure 6.5: African elephant showing its large ear (pinna). (iStock.com/Chris Fourie)

Solution

Understanding and formulating the problem *1) What is the process?* Heat flows from the warmer flat plate (the pinna) to the cooler air flowing over it. *2) What are we solving for?* Convective heat loss. Note that the convective heat transfer coefficient, h, is not provided, so we have to calculate that as well. *3) Schematic and given data:* A schematic is shown in Figure 6.6 with some of the given data superimposed on it. *4) Assumptions:* Air flow and heat transfer are steady.

Generating and selecting among alternate solutions *1) What solutions are possible?* For convective heat transfer, the solutions we have available are shown in the solution chart in Figure 6.21. *2) What approach is likely to work?* As we look at the solution chart in Figure 6.21, the solution is straightforward. This convection problem is steady state since we made the assumption that our air flow and heat transfer are steady. We are interested in heat transfer at the surface (of pinna) rather than inside the bulk air. For this surface, one simply needs to choose the formula for the geometry and the flow conditions. Of the four equations (Eqs. 6.34–6.37) for forced convection over a flat plate, since we are only interested in the total heat loss over a pinna and not its variation with location (i.e., from the leading to the trailing edge of pinna), we need the equations that provide average h over the plate, i.e., Eq. 6.35 or Eq. 6.37, depending on whether the flow is laminar or turbulent.

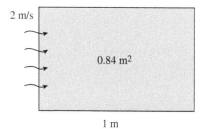

Figure 6.6: Schematic for Example 6.6.2.

Implementing the chosen solution The rate of convective heat loss over the pinna is given by

$$q = hA(T_s - T_a)$$

where h needs to be calculated. To choose between Eq. 6.35 or Eq. 6.37, we first need to find the Reynolds number to determine whether the flow is laminar or turbulent:

$$
\begin{aligned}
\text{Re} &= \frac{\rho u D}{\mu} \\
&= \frac{(1.2177 \; [\text{kg/m}^3])(2 \; [\text{m}])(1 \; [\text{m}])}{1.7985 \times 10^{-5} \; [\text{kg/m}^3]} \\
&= 1.354 \times 10^5
\end{aligned}
$$

Equation 6.35 applies since $\text{Re} < 2 \times 10^5$ (laminar flow). Plugging in values for this equation,

$$
\begin{aligned}
\text{Nu} &= 0.664 \text{Re}_L^{1/2} \text{Pr}^{1/3} \\
\frac{hL}{k} &= 0.664 \left(1.354 \times 10^5\right)^{1/2} (0.71)^{1/3} \\
\frac{h \; [\text{W/m}^2 \cdot \text{K}] \; 1 \; [\text{m}]}{0.02547 \; [\text{W/m} \cdot \text{K}]} &= 218 \\
h &- 5.55 \; [\text{W/m}^2 \cdot \text{K}]
\end{aligned}
$$

Thus, heat loss through one side of one ear is

$$
\begin{aligned}
q &= (5.55 \; [\text{W/m}^2 \cdot \text{K}])(0.84 \; [\text{m}^2])(36 - 20) \\
&= 74.59 \, \text{W}
\end{aligned}
$$

Heat loss through all four sides of the two ears is $74.62 \times 4 = 298.47$ W. Thus, the convective heat lost as a percentage of total metabolic heat production is $298.47/1650 \times 100\% \approx 18\%$.

Evaluating and interpreting the solution *1) Does the computed value (heat loss) make sense?* The metabolic heat generation for the elephant can be contrasted with what an average human male adult will produce at a typical pace of walking, which is 150 W. So the person needs to lose 150 W to maintain thermoneutrality. As mentioned in Section 1.3.3, about 40% of our required heat loss is from convection, which would be 40% of 150, or 60 W. The heat loss calculated above for an elephant, 298 W, is consistent in the sense that it is considerably higher, as it should be. *2) What do we learn?* The heat loss will vary with the environmental temperature and air flow conditions, of course. The remainder of the metabolic heat is lost by radiation and transepidermal evaporation.

6.6.3 Flat Plate, Natural Convection

During natural convection, fluid rises or sinks depending on its change in density compared to the surroundings, due to changes in temperature. The coefficient of thermal expansion, β, characterizes the change in density that causes the flow, and is given by

$$\beta = -\frac{1}{\rho}\frac{\partial\rho}{\partial T}\Bigg|_{P=\text{constant}} \qquad (6.38)$$

where ρ is the density, T is the absolute temperature, and P is the pressure. The quantity β is a thermodynamic property of the fluid. The manner in which β depends on temperature T varies with fluids. For an ideal gas (see Section 9.1.1 on page 317), $\rho = P/R_g T$ and

$$\beta = -\frac{1}{\rho}\frac{\partial\rho}{\partial T}\Bigg|_{P=\text{constant}} = \frac{1}{\rho}\frac{P}{R_g T^2} = \frac{1}{T} \qquad (6.39)$$

Equation 6.39 can be used to calculate β for ideal gases. For liquids, β is obtained experimentally, as shown in Table C.12 on page 572.

The velocity of the fluid arises from the density differences. Since there is no forced velocity, the concept of the Reynolds number is not that useful. Instead, we define another dimensionless parameter, the Grashof number, as

$$\text{Gr} = \frac{\beta g \rho^2 L^3 \Delta T}{\mu^2} \qquad (6.40)$$

where g is the acceleration due to gravity, L is the characteristic dimension, ΔT is the temperature difference (typically between a surface and the bulk fluid), and μ is the viscosity of the fluid. Another non-dimensional number Ra, called the Rayleigh number, is defined in terms of the Grashof number Gr and the Prandtl number Pr as:

$$\text{Ra} = \text{Gr} \times \text{Pr} \qquad (6.41)$$

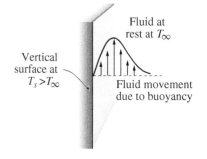

Figure 6.7: Natural convection over a heated vertical surface.

Since the Grashof number characterizes natural convection, the Rayleigh number also relates to natural convection.

Formulas for natural convection depend on the orientation of the surface since buoyancy drives the flow. In the vertical position (Figure 6.7), the air with difference in density can flow more easily and therefore the heat transfer coefficient is higher. In the extreme case of two heated horizontal plates (Figure 6.8), one facing upward and the other downward, the one with the heated face downward gives a much lower value. This results from the absence of flow for the latter situation due to the warmer and lighter fluid being at the top, which is its stable position.

Vertical surface For a heated vertical surface (Figure 6.7), the heat transfer coefficient is calculated from

$$\text{Nu}_L = \left(0.825 + \frac{0.387\text{Ra}_L^{1/6}}{\left[1 + (0.492/\text{Pr})^{9/16}\right]^{8/27}}\right)^2 \qquad (6.42)$$

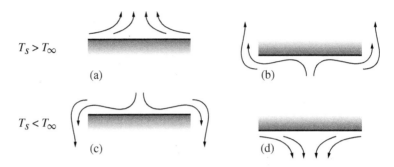

Figure 6.8: Natural convection over (a) hot surface facing up, (b) hot surface facing down, (c) cold surface facing up, and (d) cold surface facing down.

Here the characteristic length L is the height of the vertical surface.

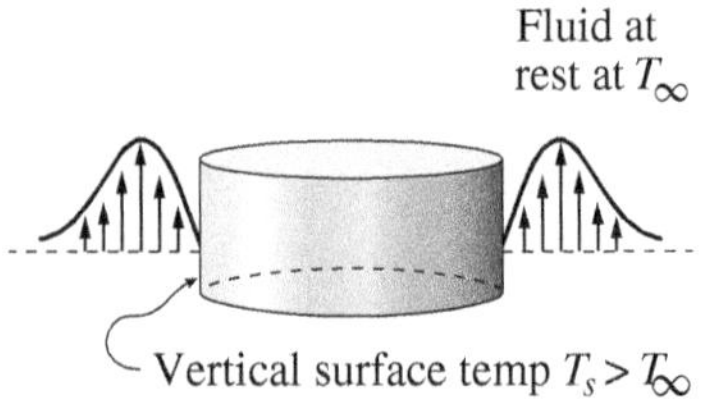

Figure 6.9: Natural convection over a heated vertical cylinder.

Horizontal surface   For a horizontal surface (Figure 6.8), with the hot side facing up or the cold side facing down, the heat transfer coefficients are calculated from

$$\text{Nu}_L = 0.54\text{Ra}_L^{\frac{1}{4}} \qquad 10^5 < \text{Ra}_L < 2 \times 10^7 \tag{6.43}$$

$$\text{Nu}_L = 0.14\text{Ra}_L^{\frac{1}{3}} \qquad 2 \times 10^7 < \text{Ra}_L < 3 \times 10^{10} \tag{6.44}$$

For a horizontal surface, with the cold side facing up or the hot side facing down, the heat transfer coefficients are calculated from

$$\text{Nu}_L = 0.27\text{Ra}_L^{\frac{1}{4}} \qquad 3 \times 10^5 < \text{Ra}_L < 10^{10} \tag{6.45}$$

The characteristic length in the equations for horizontal plates is calculated as $L = A/P$, where A is the surface area and P is the perimeter of the plate (see Figure 6.10).

6.6.4   Flow over Cylinder, Natural Convection

For a vertical cylinder (Figure 6.9), expressions for a plane surface can be used provided the diameter D is large enough compared to the length L; specifically, the following relationship has to be satisfied:

$$\frac{D}{L} \geq \frac{35}{\text{Gr}_L^{\frac{1}{4}}} \tag{6.46}$$

Figure 6.10: Surface area and perimeter of a horizontal plate.

Thus, the characteristic length would be the height of the cylinder, L. For natural convection over a horizontal circular cylinder (Figure 6.11), the characteristic length is the outside diameter of the cylinder. One correlation that is valid for this situation for a wide range of the Rayleigh number is

$$\mathrm{Nu}_D = \left(0.60 + \frac{0.387\mathrm{Ra}_D^{1/6}}{\left[1 + (0.559/\mathrm{Pr})^{9/16}\right]^{8/27}}\right)^2 \quad \text{for } 10^{-5} < \mathrm{Ra}_D < 10^{12} \quad (6.47)$$

6.6.5 Flow over Cylinder, Forced Convection

For forced convection over a cylinder (Figure 6.12), the characteristic length is the outside diameter. For both laminar and turbulent flow over such a cylinder, the following correlation is widely used:

$$\mathrm{Nu}_D = B\mathrm{Re}_D^n \mathrm{Pr}^{1/3} \quad (6.48)$$

where $\mathrm{Re}_D = u_\infty D/\nu$ is the Reynolds number based on the outside diameter D of the cylinder, and the values of B and n are given by

Re_D	B	n
0.4-4	0.989	0.330
4-40	0.911	0.385
40-4000	0.683	0.466
4000-40,000	0.193	0.618
40,000-400,000	0.027	0.805

6.6.6 Flow through Cylinder, Forced Convection

For internal flow through a cylindrical tube (Figure 6.13), the characteristic length is the inner diameter of the tube such that $\mathrm{Nu}_D = hD/k$ and $\mathrm{Re}_D = u_{av}D/\nu$, u_{av} is the mass average velocity of the fluid. The following correlations are provided for laminar and turbulent flow when the velocity profile is fully developed:

$$\mathrm{Nu}_D = 3.66 \text{ for } \mathrm{Re}_D \leq 2300 \quad (6.49)$$
$$\mathrm{Nu}_D = 0.023\mathrm{Re}_D^{0.8}\mathrm{Pr}^n \quad \text{for } \mathrm{Re}_D \geq 10000 \text{ and } L/D \geq 10 \quad (6.50)$$
$$0.6 \leq \mathrm{Pr} \leq 160$$
$$n = 0.3 \text{ for fluid being cooled and}$$
$$n = 0.4 \text{ for fluid being heated}$$

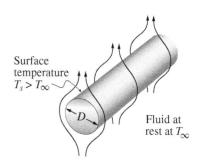

Figure 6.11: Natural convection over a heated horizontal cylinder.

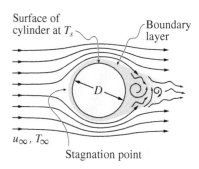

Figure 6.12: Forced convection over a heated cylinder.

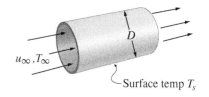

Figure 6.13: Forced convection through a cylindrical tube.

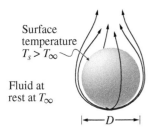

Figure 6.14: Natural convection over a heated sphere.

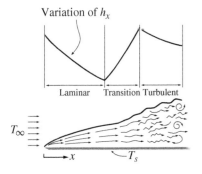

Figure 6.15: Variation of local heat transfer coefficient for flow over a flat plate as flow changes from laminar to turbulent.

6.6.7   Flow over Sphere, Natural Convection

For natural convection over a sphere (Figure 6.14), the following correlation is used:

$$\mathrm{Nu}_D = 2 + 0.43 \mathrm{Ra}_D^{\frac{1}{4}} \quad \text{for } 1 < \mathrm{Ra}_D < 10^5, \ \mathrm{Pr} \cong 1 \tag{6.51}$$

Here the characteristic length is the diameter of the sphere such that $\mathrm{Nu}_D = hD/k$ and $\mathrm{Gr}_D = \beta g D^3 \Delta T / \nu^2$.

6.6.8   Flow over Sphere, Forced Convection

Flow over a sphere is similar to that over a cylinder, discussed earlier. For forced convection over a sphere, the following correlation developed by Whitaker (1972) is widely recommended in various textbooks:

$$\mathrm{Nu}_D = 2 + \left(0.4 \mathrm{Re}_D^{1/2} + 0.06 \mathrm{Re}_D^{2/3}\right) \mathrm{Pr}^{0.4} \tag{6.52}$$

$$\text{for } 3.5 < \mathrm{Re}_D < 7.6 \times 10^4 \text{ and } 0.71 < \mathrm{Pr} < 380$$

Here the characteristic length is also the diameter of the sphere such that $\mathrm{Nu}_D = hD/k$ and $\mathrm{Re}_D = u_\infty D / \nu$.

6.6.9   Laminar vs. Turbulent Flow

As mentioned in Section 6.3, turbulence increases the bulk mixing and is more effective for heat transfer. This is shown schematically in Figure 6.15 and can be seen from the equations of h for laminar and turbulent flow. For example, for flow over a flat plate, Eqs. 6.35 and 6.37 can be used to show that the ratio

$$\frac{\mathrm{Nu}_{L,\text{turbulent}}}{\mathrm{Nu}_{L,\text{laminar}}} = 0.0542 \, \mathrm{Re}_L^{0.3} \tag{6.53}$$

is more than one in the high Reynolds number range. Thus, for turbulent flow, Nu or h values will be higher.

6.6.10   *Example: Convective Heat Transfer in an Incubator*

To increase their chances of survival, some neonates are cared for at their thermoneutrality in closed, convectively heated incubators soon after they are born. For the purpose of the problem, the baby is assumed to be a long cylinder and the incubator is simplified to the shape shown in Figure 6.16. The baby exchanges heat through radiation and convection. In this problem, we are concerned with only the convective heat loss from the body in the absence of any forced air flow and only along the curved surface of the cylinder (not the ends). The effect of the outer enclosure (incubator) can be ignored, i.e., it is not taking part in the convection process. The mean surface temperature of the clothed body is 35.5°C, the incubator air temperature is 32°C, the thermal conductivity of air is 0.0268 W/m·K, the density of air is 1.1507 kg/m³, the specific heat of air is 1006.6 J/kg·K, and the viscosity of air is 1.8785×10^{-5} kg/m·s. The volumetric expansion coefficient of air is given by $1/T$ where T is its absolute temperature. Calculate the convective heat transfer coefficient between the body surface and the air.

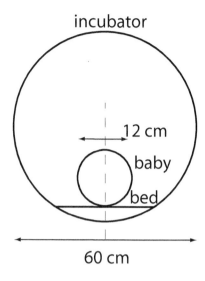

Figure 6.16: Schematic of an incubator with the baby placed in supine position.

Solution

Understanding and formulating the problem *1) What is the process?* Natural convection heat transfer from the warmer surface of the baby (approximated as a long horizontal cylinder) to the surrounding cooler air. The ends of the cylinder are not included. *2) What are we solving for?* Convective heat transfer coefficient between the baby's skin surface and air. *3) Schematic and given data:* A schematic is shown in Figure 6.16 with some of the given data superimposed on it. *4) Assumptions:* Process is steady (our interest is for the time period when fluid flow and heat transfer has become steady) and the incubator surface has no effect on the process, i.e., natural convection is the same as that in the case of open air.

Generating and selecting among alternate solutions *1) What solutions are possible?* For convective heat transfer, the solutions we have available are shown in the solution chart in Figure 6.21. *2) What approach is likely to work?* As we look at the solution chart in Figure 6.21, the convection problem is steady state since we made that assumption. We are interested in heat transfer at the (baby's skin) surface rather than in the bulk air. It is a natural convection problem since no velocity is provided (also, the volumetric expansion coefficient not being provided is a hint). One simply needs to use the formula for the appropriate geometry (cylinder) and conditions (natural convection).

Implementing the chosen solution From the list of formulas given in Section 6.6, we look for the appropriate geometry which is a cylinder, oriented horizontally and for natural convection over it. The only formula provided is Eq. 6.47, which is reproduced here:

$$\mathrm{Nu}_D \;=\; \left(0.60 + \frac{0.387 \mathrm{Ra}^{1/6}}{\left(1 + (0.559/\mathrm{Pr})^{9/16}\right)^{8/27}}\right)^{2}$$

Since $Nu_D = hD/k$, by plugging in values for the other non-dimensional numbers Ra and Pr, we can calculate h: The film temperature at which the properties of air are to be evaluated is calculated as follows:

$$
\begin{aligned}
T_s &= 35.5°C = 308.65 \text{ K} \\
T_\infty &= 32°C = 305.15 \text{ K} \\
T_f &= 0.5(T_s + T_\infty) = 0.5(308.65 + 305.15) = 306.9 \text{ K}
\end{aligned}
$$

$$
\begin{aligned}
\text{Ra} &= \text{Gr} \times \text{Pr} \\
\text{Gr} &= \frac{\beta g D^3 \Delta T \rho^2}{\mu^2} \\
&= \frac{0.00326[1/\text{K}] \times 9.8[\text{m/s}^2] \times 0.12^3[\text{m}^3] \times 3.5[\text{K}] \times 1.1507^2[\text{kg/m}^3]^2}{(1.8785 \times 10^{-5})^2[\text{kg/m} \cdot \text{s}]^2} \\
&= 725032
\end{aligned}
$$

where $\beta = 1/T_f = 1/306.9 = 0.00326 \ 1/K$, valid for a perfect gas (see Eq. 6.39) has been used.

$$
\begin{aligned}
\text{Pr} &= \frac{\mu c_p}{k} \\
&= \frac{1.8785 \times 10^{-5} \times 1006.6}{0.0268} \\
&= 0.7056
\end{aligned}
$$

Thus, as the Rayleigh number is calculated as

$$
\begin{aligned}
\text{Ra} &= \text{Gr} \times \text{Pr} \\
&= 725032 \times 0.7056 \\
&= 5.116 \times 10^5
\end{aligned}
$$

which is within the range of applicability ($10^{-5} < \text{Ra}_D < 10^{12}$) of this formula, the above formula is indeed valid for this problem. Plugging in, we get

$$
\begin{aligned}
\text{Nu} &= \left(0.60 + \frac{0.387 \times 511600^{1/6}}{\left[1 + (0.559/0.7056)^{9/16} \right]^{8/27}} \right)^2 \\
&= 12.05
\end{aligned}
$$

Since $\text{Nu}_D = hD/k$, we can calculate h as

$$\frac{hD}{k} = 12.05$$

$$\frac{(h\ [\text{W/m}^2\text{K}])(0.12\ [\text{m}])}{k\ [\text{W/m} \cdot \text{K}]} = 12.05$$

$$h = 2.69\ \text{W/m}^2 \cdot \text{K}$$

Evaluating and interpreting the solution *1) Does the computed value (heat transfer coefficient) make sense?* If we look at the typical range for heat transfer coefficients in Figure 6.19, the computed value here (2.69) is just outside the range for natural (free) convection in air. It makes sense that the value is on the very low end of this range since the temperature difference between the surface and bulk air is rather small (3.5°C). *2) What do we learn?* Convection is one of the modes for losing metabolic heat, other modes being radiation and evaporation.

6.6.11 Wind Chill Factor and Boundary Layer Thickness

During the winter, the weather report often provides us with a "wind chill." This wind chill temperature or wind chill factor is lower for higher wind speed but the same ambient temperature. Also, the lower the wind chill, the colder we feel. Thus, how cold we feel depends on the wind speed in addition to the temperature. What is its explanation from a heat transfer standpoint?

At higher wind speed, the boundary layer, which is a layer of relatively stagnant air providing insulation, becomes thinner (Figure 6.17). This increases the heat loss from the skin and tends to lower its temperature. As a result, a person or an animal feels colder. This is equivalent to being exposed to colder air, but without the wind. As shown in Figure 6.17, still air that has a thicker boundary layer can produce the same gradient (same heat loss) at the skin surface for a lower temperature.

To develop a relationship between wind chill and the ambient air temperature and velocity, note that they are equivalent, e.g., they lead to the same heat loss from the body. If the skin temperature is assumed constant at T_s, heat loss in terms of the ambient (true) temperature is $h_a(T_s - T_a)$ where h_a is the heat transfer coefficient for the ambient conditions. Heat loss in terms of the wind chill temperature T_c is $h_c(T_s - T_c)$ where h_c is the heat transfer coefficient without any wind blowing (under natural convection). Since these two heat losses are equivalent,

$$h_c(T_s - T_c) = h_a(T_s - T_a) \tag{6.54}$$

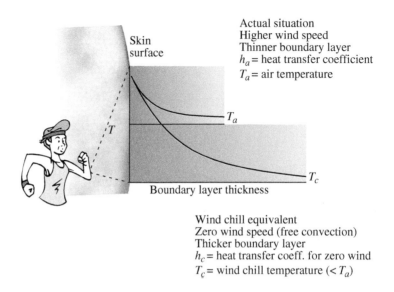

Figure 6.17: Boundary layer explanation of the wind chill factor. The wind chill temperature, T_c, is a fictitious temperature that can produce the same temperature gradient (and therefore the same heat loss) at the skin surface for a thicker boundary layer corresponding to still air.

Upon simplification, we obtain

$$T_c = \frac{h_a}{h_c} T_a + \left(1 - \frac{h_a}{h_c}\right) T_s \qquad (6.55)$$

Figure 6.18 shows the relationship between the wind chill temperature and the ambient temperature. We can try to explain the data in Figure 6.18 using Eq. 6.55. At 0 mph wind, $h_a = h_c$ and therefore $T_c = T_a$. As wind speed increases, the heat transfer coefficient increases. If we use the formula for a flat plate in laminar flow (Eq. 6.35) at low air speeds, $h_a \propto u^{0.5}$ and the slope of Eq. 6.55 increases with speed as $u^{0.5}$. Beyond a certain wind speed, the flow is likely to be turbulent and the relationship will change to $h_a \propto u^{0.8}$, as given by Eq. 6.37. Thus, when the flow changes to turbulent, h_a will increase faster with u, as seen in the rapidly increasing slope and intercept between 5 and 15 mph. Eventually, the turbulent relationship will continue and the curves become closely spaced.

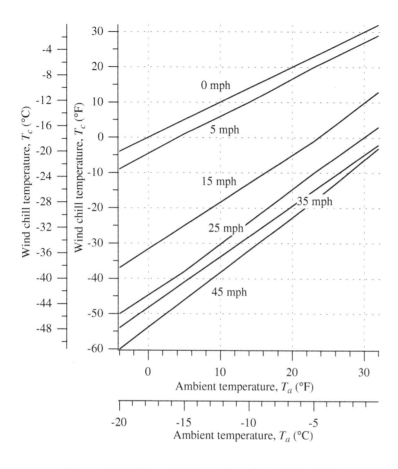

Figure 6.18: Wind chill chart. Data from Folk (1966).

6.7 Orders of Magnitude for Heat Transfer Coefficient Values

Order of magnitude estimates for the values of h for various physical situations are given in Figure 6.19. Note that this figure does not replace the equations described above, which are to be used for engineering calculations.

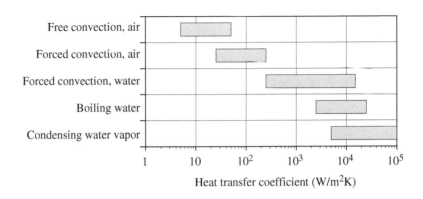

Figure 6.19: Order of magnitude estimates of the heat transfer coefficient h for various situations.

6.7.1   *BioConnect: Coefficients for Air Flow over Human Subjects*

As an example of a biological application, consider the convection of air over the human body. For this specific heat transfer situation, the convective coefficients for three different conditions are shown in Figure 6.20. Note that this plot is intended only to show the nature of variation. Significant variability is expected for any real situation. Such data are useful for designing indoor environments to achieve thermal comfort.

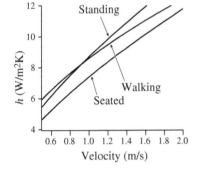

Figure 6.20: Typical values of the convective heat transfer coefficient for humans. For walking, the velocity corresponds to the speed of walking. Prediction equations are taken from Shitzer and Eberhart (1985).

6.8   Chapter Summary—Convective Heat Transfer

- **Temperature Profiles and Boundary Layers (page 185)**

  1. The convective resistance to heat transfer between a solid surface and a fluid flowing over it is restricted to a thin layer on the surface where the fluid moves relatively slowly. This layer is called a boundary layer.

- **Convective Heat Transfer Coefficient Calculations (page 193)**

  1. The convective heat transfer coefficient, h, represents the inverse of the thermal resistance between a surface and a fluid flowing over it.

  2. h includes the effect of conduction in the fluid and flow. It is therefore also a function of the flow parameter Re in addition to the thermal conductivity

k of the fluid. Thus h should not be confused with a material property such as the thermal conductivity, k.

3. The many significant variables such as the flow velocity that h depends on are grouped into dimensionless parameters as shown in Table 6.1.

4. Since h varies with the flow situation, its exact value depends on the flow velocity and other variables. The exact value of h for a situation is given by formulas in Section 6.6 (page 193). For forced convection, the flow parameter is the Reynolds number, Re. For natural convection, Re is replaced by the Grashof number, Gr.

6.9 Problem Solving—Convective Heat Transfer

The map in Figure 6.21 shows that of the many possible types of problems, in this book we are only dealing with steady convection over a fluid surface. Other situations such as unsteady state and convection and temperatures inside the fluid are also important—it is just that we are not covering them in this book. Here we calculate the heat transfer coefficient and convective flux of energy. The types of problems that we can solve are

▶**Calculate boundary layer thicknesses.**   This is done using Eq. 6.6 and 6.7.

▶**Calculate heat flow for a flat plate when h is changing with x.**   Obtain flux at x as $h_x(T_{surface} - T_\infty)$. This is flux over a strip at x. Integrate this heat flux over the entire area, i.e., all strips.

▶**Calculate h**   This is the most common type of question we will answer. Calculation steps for h can be as follows:

1. Choose geometry (flat plate, cylinder, sphere)

2. Find L based on dimensions provided (L is length along the flow)

3. Note natural or forced (velocity not provided may be natural)

4. h_x (at a location x) or h_L (average over L)?

5. Calculate Reynolds number to choose between laminar/turbulent

6. Choose equation from those provided

7. Properties should be evaluated at film temperature (average of surface and bulk). Often we end up using constant properties

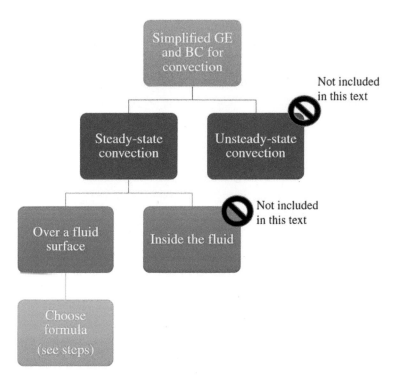

Figure 6.21: Problem solving in convection. For steps in using a convection formula, see "Calculate h" Section 6.9.

►**Calculate how h varies with different parameters**   This is calculated from the respective equation.

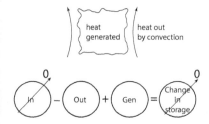

Figure 6.22: Steady state with generation.

►**Perform energy balance using h**   Once calculated, h can also be used to perform energy balance. Two possible scenarios are

- Steady state. As shown in Figure 6.22, by performing energy balance at steady state, we can write

$$\underbrace{hA(T_{\text{surface}} - T_\infty)}_{\text{Rate of energy out by convection}} = \text{Rate of energy generated}$$

If energy generated is known, h or other items can be calculated.

- Unsteady state. As shown in Figure 6.23, by performing energy balance at unsteady state (without generation in this case), we can write

Rate of energy in by convection $\quad=\quad$ Rate of change in storage

$$hA(T_{\text{surface}} - T_\infty) \quad = \quad \frac{d}{dt}(mc_p T_{av})$$

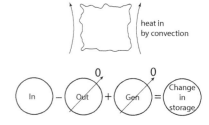

Figure 6.23: Steady state with generation.

6.10 Concept and Review Questions

1. What is meant by a velocity boundary layer? How is its thickness defined?

2. What is meant by a thermal boundary layer? How is its thickness defined?

3. Is the velocity boundary layer always thicker than the thermal boundary layer?

4. Distinguish between laminar and turbulent flow from a physical standpoint.

5. Define the film temperature and describe how it is used.

6. Define the Reynolds number and describe in words its physical significance.

7. Define the Grashof number and describe in words its physical significance.

8. Define the Prandtl number and describe in words its physical significance.

9. What are some of the advantages and disadvantages in using the heat transfer coefficient, h, in the analysis of convection heat transfer.

10. How does the Archimedes principle relate to natural convection?

11. What dimensionless group in natural convection substitutes for the Reynolds number in forced convection?

12. Describe the flow pattern for natural convection a) in a room having people with one wall exposed to cold outside, b) inside a stationary vertical cylindrical can of liquid being heated from all sides for sterilization.

13. Give examples of various kinds of convective heat transfer in a) plant system, b) mammalian system, c) the environment.

14. How is convection heat transfer related to conduction heat transfer?

15. Referring to boundary layers, 1) What is the significance of these boundary layers? 2) If the fluid velocity increases, how do they change?

Further Reading

Anantheswaran, R. C. and M. A. Rao. 1985. Heat transfer to model Newtonian foods in cans during end-over-end rotation. *Journal of Food Engineering* 4:1–19.

Chandra, S., P. Lindsey, and N. Bassuk. 1992. Measurement of the mass flow rate of water in trees. Presented at the National Heat Transfer Conference, Aug. 9–12, San Diego, CA.

Folk, G. E. Jr. 1966. *Introduction to Environmental Physiology.* Lea and Febiger, Philadelphia.

Fox, R. W. and A. T. McDonald. 1985. *Introduction to Fluid Mechanics.* John Wiley & Sons, New York.

Geankoplis, C. J. 1983. *Transport Processes and Unit Operations.* Allyn and Bacon, Inc., Boston.

Griffin, O. M. 1973. Heat, mass and momentum transfer during the melting of glacial ice in seawater. *Transactions of the ASME, J. Heat Transfer* 95:317 323.

Incropera, F. P. and D. P. Dewitt. 1996. *Fundamentals of Heat and Mass Transfer.* John Wiley & Sons, New York.

Oliveira, A. V. M., A. R. Gaspar, S. C. Francisco and D. A. Quintela. 2012. Convective heat transfer from a nude body under calm conditions: Assessment of the effects of walking with a thermal manikin. *International Journal of Biometeorology* 56(2):319–332.

Phillips, P. K. and J. E. Heath. 1992. Heat exchange by the pinna of the African elephant (Loxodonta africana). *Comparative Biochemistry and Physiology* 101A (4):693–699.

Prandtl, L. 1904. Fluid motion with very small friction (in German). Proceedings of the Third International Congress on Mathematics, Heidelberg. English translation available as NACA TM 452, March 1928.

Shitzer, A. and R. C. Eberhart. 1985. Heat generation, storage, and transport processes. In *Heat Transfer in Medicine and Biology* edited by A. Shitzer and R. C. Eberhart, Plenum Press, New York.

Whitaker, S. 1972. Forced convection heat transfer correlations for flow in pipes, past flat plates, single cylinders, single spheres, and for flow in packed beds and tube bundles. *AIChE Journal,* 18:361–371.

6.11 Problems

6.1 Thickness of Boundary Layer over Face

Consider a cold wintry breeze at a temperature of 7°C blowing parallel to your face, which has an average facial dimension of 5 cm along the direction of the air flow. The air velocity is 1 m/s (corresponding to 3.6 km/hour). Assume laminar flow. Air properties can be found in Appendix C.8. At the end of the 5 cm length, calculate the thickness of the 1) velocity boundary layer and the 2) thermal boundary layer. 3) If the air velocity is halved, calculate the new thickness of the thermal boundary layer and explain the physical significance of this change in the boundary layer thickness, i.e., whether you would feel any different, thermally speaking.

6.2 Average and Point Values of h

1) Starting from Eq. 6.34, the expression for the heat transfer coefficient at a location x for laminar flow over a flat plate, derive Eq. 6.35, the expression for the average heat transfer coefficient over the entire length L for the same plate. 2) Give one physical (real) situation where a point value of the heat transfer coefficient will be more useful and one where an average value will be more useful.

6.3 Understanding the Convective Heat Transfer Coefficient

Consider a cold fluid flowing over a warmer flat surface at a velocity low enough for the flow to be laminar. 1) Is the rate of heat removal the same at any location x, measured from the edge of the surface? Illustrate this by writing the equation for h as a function of x. 2) Plot (roughly) the h vs. x relationship and explain (qualitatively) the physical reasoning behind such variation. 3) The properties of air and water at 300 K are given in the table below. Assume plate temperature does not change. For the same bulk velocity and film temperature (300 K) of the fluids, how many times faster would water cool (remove heat from) the plate as compared to air? 4) Keeping all other parameters the same, if the bulk velocity doubles (still laminar flow), how does *average h* change?

Property	Air	Water
density [kg/m^3]	1.1421	995.8
viscosity [kg/m·s]	17.84×10^{-6}	8.6×10^{-4}
specific heat [J/kg·K]	1.0408×10^3	4.179×10^3
thermal conductivity [W/m·K]	0.0262	0.614

6.4   Heat Transfer Coefficient in a Motionless Fluid

The heat transfer coefficient, h, captures both heat conduction and flow effects. Thus, in the case when flow is negligible, this coefficient would measure only the effect of conduction. This can be demonstrated by solving a pure conduction problem. A heated sphere at temperature T_s is in a large, *motionless* fluid at temperature T_∞ far from the sphere. Find the expression for the heat transfer coefficient over this sphere surface in terms of its diameter, D, and the thermal conductivity of the fluid, k. *(Hint: Starting from the governing equation, set up the heat conduction problem (without bulk flow or convection) for the fluid in spherical coordinates with temperature T_s at a radius $D/2$ and temperature T_∞ at infinity.)*

6.5   Does Air Cool as Rapidly as Water?

Consider a cold fluid flowing over a warmer flat surface of a plate, at a velocity low enough for the flow to be laminar. Assume the plate temperature does not change. 1) Consider two fluids, water and air, whose properties are given in the appendix. For the same bulk velocity of the fluids and a film temperature of 350 K, how many times faster would water cool the plate (i.e., remove heat from the plate) as compared to air? 2) Keeping all other parameters the same for the same laminar flow, if the bulk velocity doubles (still laminar flow), how does *average h* change? 3) Can you provide physical reasoning for the answer you obtained in 1)?

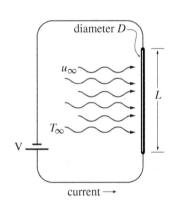

Figure 6.24: Schematic of hot wire anemometry.

6.6   Measuring Air Velocity

Hot wire anemometry depends on the convective heat transfer from forced flow across a small diameter wire, as shown in Figure 6.24. To maintain a known wire surface temperature of T_s when the known temperature of the flowing air is T_∞, an electric energy input of Q watts per meter length of wire is needed. Consider a length L of the wire and diameter D. In terms of the parameters given, write an expression for 1) the total energy input to the wire needed in watts, and 2) the total energy lost from the wire by convection in watts. 3) Under steady-state conditions, use the information from steps 1 and 2 to find an expression for the air velocity, u_∞. 4) For an air temperature of 30°C and heat input to a 0.15 mm diameter wire at 15 W/m, the wire temperature was measured to be 80°C. What is the air velocity, u_∞, in m/s?

6.7   Convection Heat Transfer and Exercise

Although it is well known that exercise increases the overall rate of heat generation in the body, it is also important to realize that the rate of heat loss must increase (by sweating and increased convection) during exercise to avoid large increases in the temperature of the body. Here we are interested in the convective heat loss from an individual walking on a treadmill. For simplicity, assume that the subject is not clothed, and that the mean skin temperature increases 1°C from rest during exercise. The surface area of the body is 1.7 m^2, the ambient air temperature is 20°C, and the skin temperature at rest is 33°C. The heat transfer coefficient at rest is 2.1 W/m$^2 \cdot$K and the heat transfer coefficient walking on a treadmill is $6.5(u)^{0.36}$ W/m$^2 \cdot$K, where u is the speed of the treadmill in m/s. Compare the convective heat loss when the subject is seated to the convective heat loss when the subject is walking on the treadmill at 6.4 kilometers per hour.

Figure 6.25: Running on a treadmill. (Diamond_Images–Sutterstock.com)

6.8 Heat Loss from Elephant Pinna

Moved to solved example.

6.9 Iceberg as a Source of Clean Water

Problem moved to solved problems

6.10 Wind Chill and Heat Loss

Wind chill charts used in practice often come from an empirical formula developed by Paul Siple in 1945 (Folk, 1966). He was an Eagle Scout who accompanied Richard E. Byrd on his second Antarctic expedition in 1933. His empirical formula was

$$q'' = \left(\sqrt{100u} + 10.45 - u \right) (33 - T_A) \ \frac{\text{kcal}}{\text{cm}^2\text{min}} \qquad (6.56)$$

where u is wind speed in m/s and T_A is the air temperature, in °C. A skin temperature of 33°C is assumed in the above equation. 1) Compare this equation to the equation for heat loss as a function of velocity for a 40 cm long flat plate with the same surface temperature of 33°C (in average film temperature of -50°C), and compare their results for the velocities of 5, 10, 20, and 30 km/h. Estimate the multiplying factor to make the flat plate formula approximately the same as the above formula over this range of velocities. 2) Using Eq. 6.56, calculate the ratio of the heat loss per unit area from the skin for a calm day (assume zero wind speed) to that for a windy day with a 40 km/h wind. 3) What air temperature on a calm day would produce the same heat loss occurring with the air temperature at -5°C on a windy day with a 30 km/h wind? 4) Is there a physical interpretation for the air temperature calculated in 3?

Figure 6.26: Paul Siple, co-developer of the wind chill factor. Source: Wikipedia.

6.11 Steady-State Temperature and Natural Convection

You left one of the heaters in your cooking range on by mistake as you hurried for your lectures. For simplicity, consider the heater as a flat disk of 15 cm diameter that continually receives an energy input of 500 W in the form of heat in a room whose air temperature stays at 25°C. Additionally, assume that the heater does not lose any heat from the bottom or side and unobstructed air surrounds the heater. 1) Discuss why the temperature of the heater would reach a steady state. 2) What would the steady-state temperature be? *(Hint: You need to assume a temperature of the disk to get started in the iterative process.)*

6.12 Forced Water Cooling of an Apple after Picking

Apples of diameter 12 cm are picked from a tree and cooled in a stream of cold water to reduce their metabolism so that they store longer. Consider a stream of water at 5°C and a velocity of 0.04 m/s flowing over an individual apple. The density of the apple is 740 kg/m^3 and other thermal properties of apple and water are given in Appendix C. If the initial uniform temperature of the apple was 25°C, calculate the time it takes for the center of the apple to reach 8°C. Also calculate the Biot number to confirm that both external convection and internal conduction are important in this problem.

Figure 6.27: Dumping of apples in water. The problem assumes running water over an apple. Figure source: http://www.fao.org/docrep/008/y4893e/y4893e0u.jpg

6.13 Heat Loss from a Free Water Surface

A rectangular cattle watering fountain has an open water surface with a size equal to 0.6 m × 0.3 m, with the longer side being parallel to the prevailing wind. Wind at −23°C is blowing over the surface at 1 m/s. An electrical heating element maintains the water surface and interior temperature at 10°C. Assume that the bottom and the sides of the fountain are perfectly insulated. Also, evaporative cooling, which can be significant, is ignored and the water surface is level with the top edge of the fountain. Assuming steady state, find the power needed for the heater.

Figure 6.28: A cattle watering fountain. (iStock.com)

6.14 Natural Convection Heating of Containerized Liquids

In food and biochemical processes, sometimes liquid to be sterilized is heated in a container without any kind of agitation, thus heating by natural convection. Such heating sets up flow patterns, as shown in Figure 6.29. Assume the average heat transfer coefficient, defined in relation to the average fluid temperature and wall temperature, reduces with time as

$$h = 200 \exp(-t/600) \qquad (6.57)$$

where t is the time of heating, in seconds, and h is in units of $W/m^2 \cdot K$. For simplicity, we will treat this problem as a lumped parameter situation (where the entire fluid is at one average temperature, analogous to solid in Section 5.1), and use Eq. 5.6. The mass average temperature of the fluid, T_{av}, will replace the temperature, T, in this equation. 1) Explain why it physically makes sense for the heat transfer coefficient to reduce with time. 2) Are there any changes in the properties of the fluid as it is being heated which would tend to increase the heat transfer coefficient? 3) Integrate Eq. 5.6 with a variable h given by Eq. 6.57 to obtain an expression for temperature as a function of time. 4) The fluid in the container has a mass of 0.2 kg, specific heat of 4190 J/kg · K, and the container has a surface area of 0.04 m². Find the time to heat a fluid to reach an average temperature of 115°C from an initial uniform temperature of 30°C when the surface of the container is kept at 121°C.

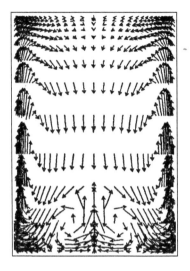

Figure 6.29: Natural convection patterns in an unagitated fluid.

6.15 Agitation to Improve Heat Transfer in Sterilization of Containerized Liquids

The rate of heating of a canned liquid food is increased substantially by agitating, an example of which is shown in Figure 6.30. For this type of agitation, the heat transfer coefficient is given by Anantheswaran and Rao (1985) as

$$Nu = 2.9Re^{0.436}Pr^{0.287} \tag{6.58}$$

where the characteristic diameter used is not the diameter of the can but the diameter of rotation ($= 2R_r$ in Figure 6.30) and the Reynolds number is given by $Re = D_r^2 N\rho/\mu$ where N is the speed of rotation, in revolutions per second. By what factor does the rate of heat transfer increase when the speed of rotation is doubled?

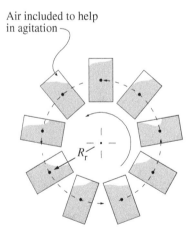

Figure 6.30: End-over-end agitation of a canned liquid to increase rate of heat transfer.

6.16 Heat Transfer between Fluid and Particles in Continuous Flow

Continuous sterilization (as opposed to canning) of foods containing particulates, such as chunky soups, is of major interest to the food industry as it provides much improved quality. Consider a spherical particle (such as a pea) flowing with a carrier liquid (such as a soup) in a tube. Whereas the liquid can be agitated to heat faster, heating of a solid particle is limited by heat conduction inside the particle and the heat transfer coefficient at the surface of the particle, which depends on the relative velocity between the fluid and the particle. A small relative velocity between the fluid and a particle would lead to a lower surface heat transfer coefficient, i.e., slower heating of the particle. Since sterilization processes must be designed to sterilize all particles, some researchers have suggested that the smallest possible heat transfer coefficient between the fluid and a particle, i.e., worst heating scenario, be used to determine heating time. 1) For a

Figure 6.31: Peas heated with water. (iStock.com/Milous Chab)

Figure 6.32: Water incident on body surface. (iStock.com)

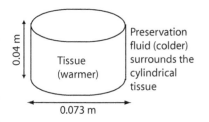

Figure 6.33: Schematic of tissue cylinder cooled by a surrounding fluid.

pea of 5 mm diameter that is in a carrier fluid of water (thermal conductivity of 0.5 W/m·K), what is this smallest value of the heat transfer coefficient? 2) What would be the disadvantage of deciding the heating time based on the smallest value of the heat transfer coefficient?

6.17 Local Heat Transfer Coefficient

You are taking a shower in cold water. Consider your back, where the water hits, to be a flat plate and the water flow to be laminar. How much colder would you feel (as defined by the ratio of rates of heat loss) at 1 cm from where the water hits your back than at 45 cm from the same point?

6.18 Heat transfer coefficient for flow over a flat plate

Same as 6.10.3–deleted.

6.19 Rate of Cooling for Two Different Cooling Fluids

The quality of preserved tissue can depend on its rate of cooling after harvest. We want to study the rate of cooling due to natural convection of a cylindrical tissue immersed vertically in two different fluids, perfluorocarbon (PFC) and University of Wisconsin (UW) solution. The tissue cylinder dimensions are shown in Figure 6.33. The vertical cylindrical surface of the tissue can be assumed to be a vertical plate, ignoring the curvature. The properties of these two fluids are given in the table below.   As the tissue

Fluid	Thermal conductivity W/m·K	Density kg/m^3	Specific heat J/kg · K	Viscosity Pa · s	Expansion coefficient K^{-1}
PFC	0.057	1917	1050	0.0051	0.00021
UW	0.6	1025	4180	0.0037	0.00021

cools, the temperature difference between the tissue and the fluid reduces, thus reducing the natural convection. We are only interested in the instantaneous h value between the tissue and the fluid at the start of cooling when the temperature of the tissue is the body temperature of 37°C and the preservation solution is at 23°C. Acceleration due to gravity is 9.81 m/s^2.

Calculate the heat transfer coefficient due to natural convection from the *vertical surface* of the cylinder for 1) the PFC fluid; 2) the UW fluid. 3) Explain briefly how the contrasting property values of the two fluids contribute to the difference in the heat transfer coefficient for the two fluids.

6.20   Wind Chill Temperature

Assume for a moment you are standing in such a way that wind is blowing on one of your cheeks while the other cheek is somehow shielded from air flow (Figure 6.34). Although the blowing air on one cheek and the still air on the other cheek are at the same temperature, one cheek feels colder. For this simple analysis, assume the temperatures on both cheek surfaces are 20°C while the air temperature is 10°C. 1) Why does one cheek feel colder? 2) Consider the cheeks to be flat plates 7 cm in length, along which air is blowing at 20 miles per hour (= 8.96 m/s). Air properties are in the appendix. Calculate the average heat transfer coefficient over the cheek length. 3) Calculate the average heat flux over the cheek length. 4) On the side that air is not blowing, there is still natural convection. Why? 5) Due to the natural convection, a smaller heat transfer coefficient of 10 W/m²K results. What imaginary air temperature on this side will lead to the same average heat flux as you just calculated in step 3 for the blowing side? 6) This imaginary temperature is what we call the wind chill temperature. If this wind chill temperature is significantly below zero while the air temperature is as provided, can your cheek freeze? Explain.

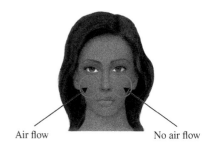

Air flow   No air flow

Figure 6.34: Heat loss from skin.

6.21   Convective Heat Transfer in an Incubator

Moved to example problem.

6.22   Convective Heat Transfer Coefficient over a Body in Water

The surface heat transfer coefficient for water flow over a body is important in predictions of heat loss and tolerance of humans in cold water. In an experiment using calorimetry, for nude subjects who stayed in a horizontal position, Eq. 6.59 was obtained from experimental data:

$$h_{expt} = 0.09(GrPr)^{0.275} \qquad (6.59)$$

We would like to compute the heat transfer coefficient first using available formulas for ideal cylindrical bodies for a water temperature of 20°C and a skin surface temperature that is assumed to stay at 34°C. Assume the diameter of the body to be 0.3 m. The density of water is 1000 kg/m³. The viscosity and thermal properties needed can be found in the appendix. 1) Compute the heat transfer coefficient using formulas for ideal bodies available to you. 2) Compute the heat transfer coefficient using Eq. 6.59

Figure 6.35: Human subject floating in water. (iStock.com)

Figure 6.36: A climate chamber with thermal manikin. *International Journal of Biometeorology*, "Convection heat transfer from a nude body under calm conditions: assessment of the effects of walking with a thermal manikin," pp 319-332, Oliveira, et al., with permission of Springer.

and compare it with the value computed in the previous step. Is it higher or lower? 3) To explain the discrepancy, we look for a number of factors that would differ in an experimentation with real bodies and living entities. First we compute, using the formula from step 1, the heat transfer coefficient from the arm that is 0.075 m in diameter. Does having this small diameter in different parts of the body explain the discrepancy? 4) Do you think it is reasonable to assume that the skin surface temperature will continue to be at 34°C as the body stays in cooler water? Would this explain the discrepancy? 5) Can shivering explain the discrepancy? 6) Does the heat transfer coefficient calculated in step 1 include heat conduction from the body into the water? Explain.

6.23   Convective Heat Transfer over Various Body Limbs

A research group is studying the effect of walking on the heat loss from individual body limbs, using a manikin, as shown in Figure 6.36. Measured data for h over the right leg during standing and walking (the leg can be considered as a long vertical cylinder) is shown in Figure 6.37. Ignore any air velocity in the chamber other than that due to walking. Steps are 0.7 m long and the average diameter and length for the right leg are 0.14 m and 0.4 m, respectively. The average air properties that can be used for convection calculations for viscosity, density, thermal conductivity, specific heat, and thermal expansion coefficient are 1.85×10^{-5} kg/m·s, 1.18 kg/m³, 0.0262 W/m·K, 1.006 kJ/kg·K and 1/300 1/K, respectively. The skin temperature and bulk air temperature are 34°C and 20°C, respectively.

1) Calculate the expected h value over the right leg when standing still and note whether it is close to the measured data in Figure 6.37. 2) Develop the equation for h over the right leg as a function of walking speed, substituting all other property and parameter data, and compare the trend provided by this equation with the measured trend. 3) The figure shows that h increases slightly with $T_{skin} - T_{air}$ for standing. Refer to the appropriate equation for h and mention *qualitatively* whether or not the observed trend is predicted by this equation.

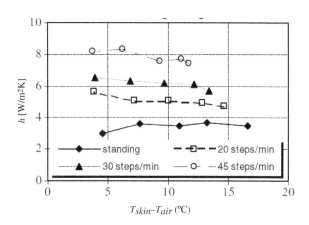

Figure 6.37: Measured convective heat transfer coefficient over the right leg. *International Journal of Biometeorology*, "Convection heat transfer from a nude body under calm conditions: assessment of the effects of walking with a thermal manikin," pp 319-332, Oliveira, et al., with permission of Springer.

Chapter 7

HEAT TRANSFER WITH CHANGE OF PHASE

CHAPTER OBJECTIVES

After you have studied this chapter, you should be able to

1. Explain the process of gradual freezing of a biomaterial over a temperature range, as opposed to one fixed temperature.

2. Explain the process of freezing in a cellular tissue and how the success of freezing depends on the rate of freezing.

3. Describe the changes in thermal and physical properties of biomaterials during freezing.

4. Calculate the time to freeze a plant or an animal tissue of given size.

KEY TERMS

- **heat of fusion or latent heat**
- **freezing point depression**
- **cryopreservation**
- **"mushy" zone in frozen biomaterials**

- **vapor pressure**
- **evaporation**
- **boiling point elevation**
- **evapotranspiration**

Energy Transfer

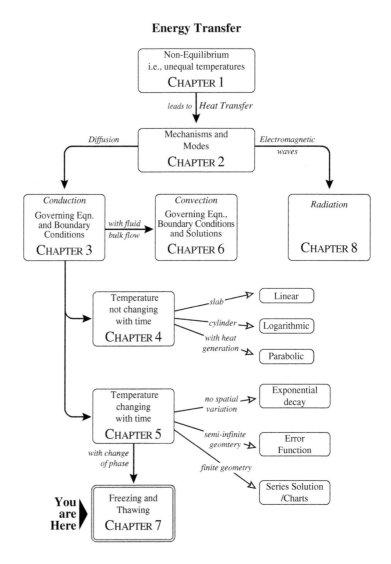

Figure 7.1: Concept map of energy transfer showing how the contents of this chapter relate to other chapters in the part on energy transfer.

This chapter deals with energy transport when a change of phase is involved. The first part of this chapter deals with freezing, the change of state from liquid to solid.

Figure 7.1 shows how this chapter relates to other chapters in energy transfer.

Phase change brings complications in studying heat transfer, particularly in bio-materials, due to drastic changes in property. One primary comment that can be made about phase change processes is that they are energy intensive; i.e., a high energy exchange occurs in the change from one state to another, as shown in Figure 7.2. The enthalpy difference due to change of phase from solid to liquid, occurring at the melting temperature, is termed the *latent heat of fusion*. Similarly, the enthalpy difference due to phase change from liquid to vapor, occurring at the boiling temperature, is termed the *latent heat of vaporization*. Figure 7.2 shows that, for water, the latent heat of vaporization is almost seven times the latent heat of fusion. Also, sensible heat or enthalpy change due to temperature difference is typically smaller than the latent heat.

7.1 Freezing and Thawing

Freezing and thawing is one of the most promising methods of preserving biological material. The term *cryopreservation* generally refers to the preservation of a living system at low temperatures in such a way that the system can be brought back to life. One of the ultimate goals, to preserve a whole human body reversibly by freezing and storage at low temperatures, is still at the science fiction stage. At present success in cryopreservation is limited to smaller mammalian systems, including blood, embryos, spermatozoa, culture cells, hepatocytes, bone marrow cells, pancreatic tissue and corneas. Other successfully frozen systems include protozoa and other microor-

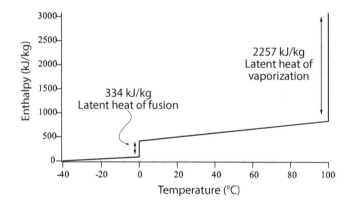

Figure 7.2: Enthalpy changes in water showing the latent heat involved for phase change from ice to water and from water to steam.

ganisms, insects, fish, plants, and algae.

Freezing is one of the most common methods of preserving large quantities of food materials. Biochemical spoilage reactions are reduced due to lower temperatures and the unavailability of water in liquid form. Frozen foods are considered the best alternative to fresh foods, as demonstrated by the $15 billion of frozen foods sold in the U.S. in 1990, which represents 7–8% of all foods sold in supermarkets. Intact food tissues, such as in fresh whole vegetables or large cuts of meat, have many similarities to living specimens, and the goal in food freezing is to preserve a material as closely as possible to its original state so that it is close to the fresh state upon thawing.

As well as preserving materials, freezing is sometimes used to purposely destroy living material. In cryosurgery a localized necrosis in diseased tissues or organs is produced by freezing. A cryosurgical probe is inserted into a tissue and suddenly brought to a cryogenic temperature. The diseased tissue in the vicinity of the probe undergoes a phase change and is destroyed.

Quantitative studies of the thermodynamic and heat transfer processes during freezing and thawing have resulted in significant progress in cryobiology. In industrial food processing, these studies have led to reduced energy use and better quality frozen food. In cryosurgery they help determine the optimum freezing protocol to minimize unwanted damage to surrounding tissues.

Since biomaterials are mostly water, the freezing process of pure water should provide significant information about the freezing of biomaterials. It makes sense, therefore, to study the basic aspects of the liquid–solid phase change of pure water. After an explanation of the pure water system, more complex systems such as solutions and tissues will be studied.

7.2   Freezing of Pure Water

7.2.1   Freezing Process

Several steps are involved in the freezing of pure water—supercooling, nucleation, crystal growth, and maturation. The temperature at which crystallization is initiated is uncertain. Ice does not always form at precisely 0°C. Water frequently cools to a sub-zero temperature before freezing, a phenomenon called *supercooling* or *undercooling*. Ultrapure water may not freeze until around −40°C. For any kind of crystal to grow, a stable seed that can act as a foundation is required. Such seeds are termed nuclei. At around −40°C, homogeneous nucleation takes place in ultrapure water—nuclei spontaneously generate, and the water freezes. However, homogeneous nucleation happens rarely in unpurified water because extraneous (heterogeneous) nuclei are usually present as impurities, and ice crystals can form easily around such nuclei.

Thus, heterogeneous nucleation takes place at temperatures much higher than −40°C, typically −2 to −3°C.

Crystal growth is possible once nucleation has taken place. As long as a stable ice crystal is present, further growth is possible. The rate of this growth is controlled in part by the rate of heat removal from the system. Rapid freezing tends to form small ice crystals, whereas slow freezing tends to produce large ice crystals. During storage, even at constant temperatures, a *maturation* process occurs in which smaller ice crystals decrease in size, while larger ice crystals enlarge. With time, the number of ice crystals decreases, and their average size increases, in part reversing the initial effects produced by rapid freezing. This increase in crystal size with time can be important. For example, ice cream stored for a long time in the freezer grows large ice crystals, which diminish its appeal.

7.2.2 Property Changes during Freezing

As the temperature of water decreases, the water's specific volume and density change, as shown in Figure 7.3. As it freezes, the water expands greatly, around 9%. This high degree of expansion has many consequences. Water pipes can rupture. In plant and animal cells expansion can rupture the cell membranes. Freezing in order to preserve food material can lead to cracks in the food. Stresses caused by expansion during a phase change are also a major problem in cryosurgery.

Soils or pavement in colder climates often heave from the expansion of water in the soil due to freezing. However, the heaves are much bigger than what the normal amount of water in soil can account for. As temperature decreases near the surface, the surface tension of the water increases. Water moves up from the warmer interior soil and groundwater due to capillary action. Freezing and expansion of this accumulated water lead to the larger heaves. The total water movement is partly due to a difference in vapor pressure from the warmer groundwater (higher vapor pressure) to the colder surface soil (lower vapor pressure). During the spring, when a substantial amount of ice melts and greatly increases the water content of the soil, the strength of the soil greatly decreases, and pavements can suffer serious structural damage. Measures to avoid such damage include lowering of the groundwater table, replacement of frost-susceptible soils, and use of impervious membranes.

The thermal conductivity of ice is more than three times that of water, as shown in Figure 7.4. Such a large change in thermal conductivity has implications, for example, in a thawing process. For a given material, thawing is slower than freezing. The layer near the surface of the material, which provides the thermal resistance to heat flow from the inside, has a higher resistance when it is thawed, due to lower thermal conductivity.

The specific heat of water drops considerably during freezing and continues to drop

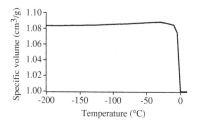

Figure 7.3: Increase in specific volume (decrease in density) of water during freezing.

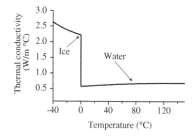

Figure 7.4: Changes in thermal conductivity with temperature showing a large increase in thermal conductivity for ice over water.

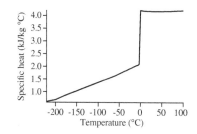

Figure 7.5: Specific heat of water and ice as a function of temperature.

as the temperature is lowered beyond the freezing point, as shown in Figure 7.5. An average value of the specific heat of water is 4.179 kJ/kg·K, and that of ice is 2.04 kJ/kg·K.

Dielectric properties measure the interaction of a material with electromagnetic waves such as microwaves. One of the two dielectric properties, dielectric loss, is a measure of the ability of a material to absorb electromagnetic energy and turn it into heat. The dielectric loss of water changes significantly during freezing. At the microwave frequency of 2450 MHz, the frequency used in most domestic microwave ovens in the U.S., the dielectric loss of water changes, as shown in Figure 7.6. Since the dielectric loss is a measure of microwave absorption, the figure shows that ice cannot absorb microwaves readily.

The lower absorption of microwaves by ice has many practical implications. For example, in microwave thawing of food, initially very few of the microwaves are absorbed. However, due to differences in heating rates and other factors, one region of the food will thaw first, and that region will selectively absorb more of the microwaves and may start to boil while other areas remain frozen! This non-uniformity of thawing is undesirable, and to minimize it, the thawing cycle in microwave ovens typically does not heat continuously. Instead it cycles between heating and rest periods, so that temperatures at warmer and colder areas can equilibrate by diffusion.

Mechanical properties such as the modulus of elasticity also change drastically during freezing and continue to change as the temperature drops, as in cryogenic freezing. The material becomes brittle at very low temperatures and may crack, depending on the rate of freezing, as explained later (see Figure 7.11).

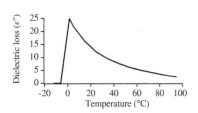

Figure 7.6: Dielectric loss of water and ice at a microwave frequency of 2450 MHz.

7.3   Freezing of Solutions and Biomaterials

7.3.1   Solutions

Dissolved solutes in water depress the freezing point of the solution below the freezing point of pure water. Solutions experience less supercooling and earlier nucleation than pure liquids. The depressed freezing point T_A of a solution is related to its concentration x_A by

$$\frac{\Delta H_f}{R_g}\left[\frac{1}{T_{A0}} - \frac{1}{T_A}\right] = \ln\, x_A \tag{7.1}$$

where ΔH_f is the latent heat of fusion of pure liquid A in J/mol, T_{A0} is the freezing point of pure liquid A in K, T_A is the freezing point of the solution in A in K, R_g is the gas constant equal to 8.314 kJ/kmol · K, and x_A is the mole fraction of A in the solution. This relationship is plotted in Figure 7.7 for water. Since nucleation occurs easily in solutions, they undergo little supercooling; i.e., solutions in fact freeze at their expected freezing point of T_A, given by Equation 7.1. For dilute solutions, the above equation can be simplified to

$$\frac{\Delta H_f}{R_g}\left[\frac{T_A - T_{A0}}{T_{A0}\,T_A}\right] = \ln\left(1 - x_B\right) = -x_B \quad \text{for } x_B \ll 1 \tag{7.2}$$

where x_B is the mole fraction of solute. Defining

$$\Delta T_f = T_{A0} - T_A$$

as the drop in freezing point or freezing point depression and simplifying for small changes in T_A as $T_A \cdot T_{A0} \cong T_{A0}^2$, we get

$$\frac{\Delta H_f}{R_g} \cdot \frac{\Delta T_f}{T_{A0}^2} = x_B \tag{7.3}$$

Note that Equation 7.3 provides a linear relationship between the drop in freezing point and mole fraction of solute for dilute solutions, as plotted in Figure 7.7. For dilute solutions the mole fraction can be approximated as below:

$$x_B = \frac{m_B/M_B}{\frac{m_A}{M_A} + \frac{m_B}{M_B}} = \frac{m_B}{M_B} \cdot \frac{M_A}{m_A} = \frac{m_B/M_B}{m_A} \cdot M_A$$

where m_A is the mass of water, m_B is the mass of solute, M_A is the molecular weight of water, and M_B is the molecular weight of solute. Defining $\mathcal{M}$ as the molality or moles of solute per unit mass of water (or other solvent), and substituting in Equation 7.3, we get the freezing point depression equation for a dilute solution:

$$\Delta T_f = \frac{R_g T_{A0}^2 M_A \mathcal{M}}{\Delta H_f} \tag{7.4}$$

which is sometimes written as

$$\Delta T_f = K_f \mathcal{M} \tag{7.5}$$

where K_f is known as the *cryoscopic constant* and depends only on the solvent.

This process of gradual freezing of a solution is shown as a schematic for a hypothetical system in Figure 7.8. The various regions in the pie chart represent the relative amounts of various components at a location in a solution. As the temperature is lowered, the amount of ice is zero until the initial freezing point of the solution is reached, depending on the initial concentration of the solution. As ice begins to form, the rest of the solution concentrates since the ice formed reduces available liquid water while

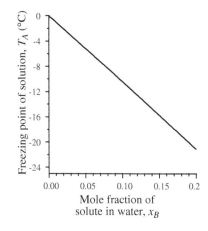

Figure 7.7: Freezing point depression of an aqueous solution as a function of concentration.

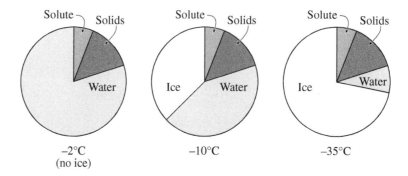

Figure 7.8: Schematic showing various components in gradual freezing of a solution.

the amount of solute remains the same. As temperature is continually lowered, this process involving freezing of pure water and the concentration of the remaining solution continues until all the water that can be crystallized as ice freezes, leaving only the solute and its water of hydration. At this solute concentration, the solution is termed a eutectic. Further reduction in temperature results in the solidification of the remaining solution as a whole, i.e., without formation of pure ice. The lowest temperature at which the solution remains liquid is referred to as the eutectic temperature.

7.3.2   Example: Percent Water Frozen at a Temperature for a Solution

Consider the freezing of grape juice. Assume that the initial weight percentage of water in grape juice is 85% and the effective molecular weight of solute is 183.61 kg/kmol. The latent heat of water is 334 kJ/kg and the molecular weight of water is 18.015 kg/kmol.

1) Assuming that the grape solute does not freeze, predict the percent of initial water that is frozen in grape juice when the temperature has been reduced to −6°C. 2) Repeat the calculations in step 1 when the temperature has been reduced to −15°C. 3) Would your answer change if we started from a more concentrated grape juice of, say, 50% water? Why?

Solution

Grape juice is a solution of water and other solutes. These solutes lower the freezing point of water in the solution. At the given temperature of $-6°C$ for the entire solution, enough water would have frozen such that the freezing point of the solution is $-6°C$. Since the amount of solute has stayed the same as the initial amount, the solvent (water) amount has reduced (frozen water is no longer available) so that the solution is concentrated. From the solvent concentration at $-6°C$, obtained from the freezing point depression formula, we can find the solvent amount at $-6°C$. By subtracting from the initial solvent amount, we can find the amount of the solvent (water) that must have changed to ice.

1) From the freezing point depression equation (Eq. 7.1), we get, using x as the gram of water per gram of juice at a temperature T,

$$\frac{\Delta H_f}{R_g}\left[\frac{1}{T_{water,o}} - \frac{1}{T_{water}}\right] = \ln x_{water}$$

$$\frac{6013.4\,\text{J/mol}}{8.314\,\text{J/mol}\cdot\text{K}}\left[\frac{1}{273} - \frac{1}{267}\right] = \ln\frac{\overbrace{x/18}^{\substack{\text{moles of}\\\text{water}}}}{\underbrace{(0.15/183.61)}_{\substack{\text{moles of}\\\text{solute}}} + \underbrace{(x/18)}_{\substack{\text{moles of}\\\text{water}}}}$$

$$\frac{x/18}{(0.15/183.61) + (x/18)} = 0.94220$$

$$(1 - 0.94220)\frac{x}{18} = \frac{(0.94220)(0.15)}{183.61}$$

from which we get

$$x = 0.2398$$

Since the initial (before freezing) value of x was 0.85,

$$\text{Fraction frozen at } -6°C = \frac{0.85 - 0.2398}{0.85} = 0.718$$

which means percent of water frozen is 71.8.

2) At a temperature of $-15°C$, we write Eq. 7.1 as

$$\frac{6013.4\,\text{J/mol}}{8.314\,\text{J/mol} \cdot \text{K}} \left[\frac{1}{273} - \frac{1}{258} \right] = \ln \frac{x/18}{(0.15/183.61) + (x/18)}$$

from which we solve for percent frozen as 89.6.

3) Equation 7.1 implies that a given solution temperature, T_A, corresponds to one concentration value, x_A. Thus, at $-6°C$, the solution concentration (solvent mole fraction) would have to be 0.94221. If we started with a more concentrated solution, the percent of water frozen at $-6°C$ would be lower.

7.3.3 Cellular Tissues

The process of freezing in plant tissues is considerably more complicated than in a solution, primarily due to the presence of cellular structures (Figure 7.9). Cells possess semi-permeable membranes that selectively control the passage of molecules in and out of the cells. During tissue freezing, the most fundamental physical change taking place is the transport of water across a cell's semi-permeable membrane. The cell walls are generally considered freely permeable to water. As cooling proceeds, the intracellular as well as the extracellular fluid reaches the freezing temperature. Depending on the extent of supercooling, the random process of ice nucleation can be initiated preferentially in the intracellular or extracellular fluid. The cell has been considered devoid of nucleating agents, and thus it is normally assumed that ice forms in the extracellular fluid first. With the formation of ice, the remaining extracellular fluid becomes more concentrated. The cell contents remain unfrozen and supercooled, presumably because the plasma membrane blocks the growth of ice crystals into the cytoplasm. This difference in concentration between the inside and outside develops osmotic pressure, and water flows out of the cell. The expelled water freezes in the extracellular space. The extent of cellular dehydration is important, because the rehydration process depends on the amount and rate of water loss. In addition, the presence of too much intracellular ice may rupture the cell membrane.

7.3.4 Cooling Rates and Success of Freezing

The rate of cooling has microscopic as well as macroscopic consequences. At the cellular level, the osmotic water migration mentioned above is higher for a slower cooling process, leading to dehydration and a large change in the cell size, as shown in Figure 7.10. Fast cooling, on the other hand, will cause the least amount of migration

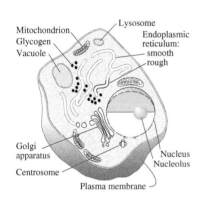

Mitochondrion
Glycogen
Vacuole

Lysosome
Endoplasmic
reticulum:
— smooth
— rough

Golgi
apparatus
Centrosome

Nucleus
Nucleolus

Plasma membrane

Figure 7.9: Hypothetical composite cell.

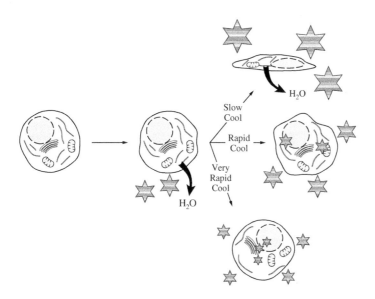

Figure 7.10: Schematic showing possibilities during cell freezing. Faster freezing reduces the transport of water out of the cell.

and will leave the cell closer to its original shape and size. For this and other reasons, fast freezing is often practiced in cryopreservation and is the recommended procedure for industrial food-freezing applications.

Fast freezing can lead to damage, however. At the cellular level, formation of intracellular ice can cause membrane damage. At the macroscopic level, for specimens as small as a few centimeters, cracks due to stress from thermal expansion can occur. Cracking is more likely for larger sizes. Such thermal stresses arise because all the material is not at the same temperature. Due to the difference in temperature, all the material is not expanding to the same rate. This differential expansion causes stresses in the material. Cracks have been reported for organs and for food systems such as fish, potatoes, and beans frozen in liquid N_2. As shown in Figure 7.11, potato material can crack significantly when dipped in liquid N_2.

In Figure 7.11, slower freezing at $-40°C$ does not crack the potato cylinder while faster freezing, at $-200°C$, does lead to cracks. Cracking at $-200°C$ occurs in a lower temperature range (0 to $-20°C$) where most of the ice formation and related expansion takes place. However, ice formation alone is not the cause of cracks since the net amount of expansion due to ice formation and later contraction of the solid, starting

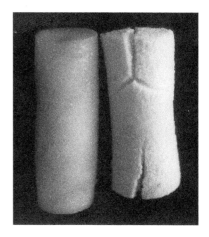

Figure 7.11: Cracks in cylindrical potato sample (1.78 cm dia) dipped in liquid nitrogen.

from water phase, is approximately the same for boundary temperatures of $-40°C$ and $-200°C$. Thus, it is the combination of faster cooling and large expansion from ice formation that cause the material to crack. Using fracture mechanics, this is explained as follows: a material's ability to resist cracks, measured by a property called fracture toughness, drops drastically as the material freezes (more ice forms) and becomes more brittle. During ice formation, this ability to resist cracks can be exceeded by the higher strain energy release rate at the crack tip for faster cooling. Thus, slow freezing that reduces the strain energy release rate during the ice formation, combined with fast freezing after most of the ice has formed, can avoid the fracture even when the temperature is eventually lowered to $-200°C$.

7.4 Temperature Profiles and Freezing Time

7.4.1 Freezing Time for an Infinite Slab of Pure Liquid

Since the freezing process involves latent heat in addition to sensible heat, analysis of this very simple process is rather complicated. A very simplified solution (known as Plank's solution) that is used in practice and that preserves some of the essential physics of the process is described below.

Consider symmetric (both surfaces at T_∞) freezing of a slab of pure liquid, as illustrated in Figure 7.12. Due to the symmetry, one needs to consider only the half thickness of the slab. The latent heat evolved at the interface of frozen and unfrozen regions is removed through the frozen layer. Although the thermal conductivity of the frozen layer is higher than that of the unfrozen layer, it is still small in absolute terms. The rate of heat transfer in the frozen layer is slow enough to be regarded as in pseudo-steady-state condition. Thus, although the temperature profile changes with time, this happens slowly enough that it approaches close to a steady-state profile at any time. In addition, the following assumptions are made:

1. Initially all material is at freezing temperature T_m but unfrozen.

2. All material freezes at one freezing point.

3. Thermal conductivity of the frozen part is constant.

With the above assumptions, heat loss through a frozen layer whose thickness is x at any time t is

$$q = kA\frac{T_m - T_s}{x}$$

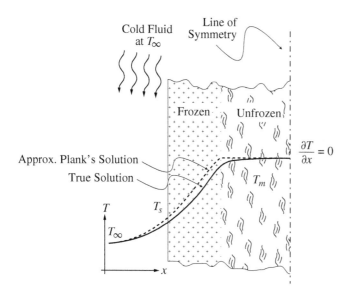

Figure 7.12: Freezing of a pure (stationary) liquid showing two distinct regions, frozen and unfrozen. Plank's assumption of a linear temperature profile is compared with the "true" profile.

This energy comes from the latent heat given off at the freezing front. The rate at which latent heat is given off is related to the rate at which the freezing front advances. If ΔH_f is the latent heat of fusion of the material per unit mass in J/kg,

$$q = \Delta H_f A \rho \frac{dx}{dt}$$

Equating the two,

$$\Delta H_f A \rho \frac{dx}{dt} = k A \frac{T_m - T_s}{x} \tag{7.6}$$

and solving for t, time to freeze as a function of x, the depth frozen,

$$\frac{k \left(T_m - T_s\right)}{\Delta H_f \rho} \int_0^t dt = \int_0^x x \, dx$$

$$t = \frac{\Delta H_f \rho}{k\,(T_m - T_s)}\frac{x^2}{2} \tag{7.7}$$

Freezing of the slab is complete when the freezing front reaches the midpoint, or the line of symmetry, for the slab. If the slab has a total thickness of $2L$, the freezing time for the slab is obtained by substituting $x = L$ in the above equation:

$$t_{slab} = \frac{\Delta H_f \rho}{k\,(T_m - T_s)}\frac{L^2}{2} \tag{7.8}$$

Two observations can be made from Equation 7.8. First, it shows that the depth frozen is related to time as

$$x \propto \sqrt{t}$$

Let's try to provide some physical explanations for this result. At any given time, a certain thickness of material is frozen. For another layer to freeze, the latent heat given off by that layer must be conducted through the frozen layer. The speed of freezing is determined by the rate at which heat can be conducted out. But as freezing progresses, the rate at which heat can be conducted out drops because the temperature gradient drops (same temperature difference over an increasing distance). Thus, the rate of advancement of the freezing layer slows.

Second, in the thawing process, the outer layer will be the unfrozen layer. The unfrozen material has lower thermal conductivity, and Eq. 7.8 clearly demonstrates that thawing of the same slab will take longer (although not by the same factor, in reality).

A more common boundary condition is convection at the surface instead of a specified temperature. For this case additional convective resistance $1/hA$ must be added to the conductive resistance x/kA in Eq. 7.6 with the total temperature difference becoming $T_m - T_\infty$ instead of $T_m - T_s$, leading to

$$\frac{T_m - T_\infty}{\frac{x}{kA} + \frac{1}{hA}} = \Delta H_f \rho A \frac{dx}{dt} \tag{7.9}$$

$$\frac{T_m - T_\infty}{\frac{x}{K} + \frac{1}{h}} = \Delta H_f \rho \frac{dx}{dt}$$

$$\frac{T_m - T_\infty}{\Delta H_f \rho}\int\limits_0^{t_F} dt = \int\limits_0^{L}\left(\frac{x}{k} + \frac{1}{h}\right) dx$$

$$\frac{T_m - T_\infty}{\Delta H_f \rho} t_F = \frac{L^2}{2k} + \frac{L}{h}$$

$$t_F = \frac{\Delta H_f \rho}{T_m - T_\infty}\left[\frac{L^2}{2k} + \frac{L}{h}\right] \tag{7.10}$$

Note that Eq. 7.8 becomes identical to Eq. 7.10 for large values of h, which is equivalent to a surface-temperature-specified boundary condition.

7.4.2 Example: Time for Cryosurgery Using a Cylindrical Probe

We would like to estimate the time needed in cryosurgery to freeze tissue using a long cylindrical cryosurgical probe of length (see cross-section in Figure 7.13) you can assume a length of L). To make things easier, you will be guided through the setup of the governing equation. Consider pseudo-steady-state heat transfer where the tissue touching the probe is at the liquid nitrogen temperature of $T_p = -196°C$. Consider the tissue to be initially unfrozen, but already at the constant freezing point of $T_m = -2°C$. Ignore any sensible heat involved.

1) Let us say the outer frozen layer increases by a value of dr in time dt. Write the expression for the rate of the heat flow that must be released for this additional layer of material to freeze. 2) The heat flow in step 1 has to be conducted through the frozen layer into the cryoprobe. Write an expression for heat flow due to conduction through the frozen layer at steady state, considering an appropriate expression for resistance. 3) Equate the expressions in step 1 and step 2 to set up the equation from which we can solve for the radius of frozen tissue at time t. 4) Complete the integration using the fact that $\int r \ln r\, dr = r^2(\ln r - 1/2)/2$. Do not try to simplify to get r as a function of t. 5) Instead, provide an approximate plot of r as a function of time t, based on your physical understanding of the problem.

Solution

Understanding and formulating the problem *1) What is the process?* This is freezing in a cylindrical geometry, outward in the radial direction, starting from the coldest point at the probe surface. *2) What are we solving for?* We need to calculate the time it takes for the freezing front to reach a particular radius. *3) Schematic and given data:* A schematic is shown in Figure 7.14 with some of the given data superimposed on it. *4) Assumptions:* Although it is freezing of water in a tissue, we are considering the freezing of pure water with nothing dissolved in it, so we can develop a simple formula for freezing.

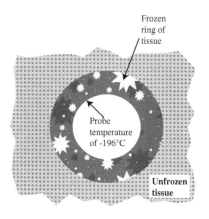

Figure 7.13: Schematic of iceball development around a cryoprobe.

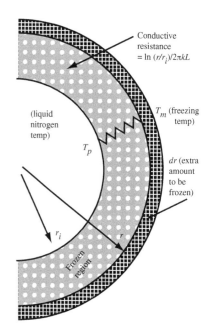

Figure 7.14: Freezing of a pure (stationary) liquid in cylindrical coordinates.

Generating and selecting among alternate solutions *1) What solutions are possible?* For freezing, the solutions we have available are shown in the solution chart in Figure 7.20. *2) What approach is likely to work?* As we look at the solution chart in Figure 7.20, since we are not using a numerical solution technique, we need to assume we are freezing pure water in the tissue, i.e., no dissolved solutes. Since this problem involves freezing from the inside out and it is in a cylindrical geometry, it does not match the solution derived in Section 7.4. The required formula, therefore, needs to be derived from scratch (the problem also asks for this derivation). Since the problems in slab, cylinder, and sphere differ by the geometry alone, we can use the solution process in Section 7.4 as an analogy and substitute quantities corresponding to cylindrical geometry, i.e., we will consider heat balance over a thin cylindrical shell instead of a thin slab.

Implementing the chosen solution We start from energy balance and apply this to the thin cylindrical shell in Figure 7.14 of thickness Δr that freezes in time Δt.

$$\text{In} \quad - \quad \text{Out} \quad + \quad \text{Gen} \quad = \quad \text{Change in storage}$$

$$0 \quad - \frac{T_m - T_p}{\left(\frac{\ln(r/r_i)}{2\pi kL}\right)}\Delta t + \quad 0 \quad = \quad -\Delta H_f 2\pi r \Delta r L\rho \quad (7.11)$$

where the individual terms are calculated as follows: the amount of heat removed in freezing a shell of thickness Δr is its mass, Δm, times the latent heat of freezing, given by

$$\begin{aligned}\text{Change in storage} \quad &= \quad \Delta H_f \Delta m \\ &= \quad \Delta H_f 2\pi r \Delta r L\rho\end{aligned}$$

The amount of heat conducted out of the frozen layer over time Δt is

$$\frac{\text{Temperature difference}}{\text{Thermal resistance}}\Delta t \quad = \quad \frac{T_m - T_p}{\left(\frac{\ln(r/r_i)}{2\pi kL}\right)}\Delta t$$

Rewriting Eq. 7.11 for $\Delta r \to 0$ and $\Delta t \to 0$, we get the governing differential equation

$$\frac{T_m - T_p}{\left(\frac{\ln(r/r_i)}{2\pi kL}\right)} \quad = \quad \Delta H_f 2\pi r L\rho\frac{dr}{dt}$$

from which we solve for r as a function of t as follows (note that freezing starts from the inner radius, r_i, while the outer radius, r, of the boundary of frozen material, changes with time, t):

$$\frac{(T_m - T_p) k \, dt}{\Delta H_f \rho} = \ln(r/r_i) \, r \, dr$$

$$\int_0^t \frac{(T_m - T_p) k}{\Delta H_f \rho} \, dt = \int_{r_i}^r \ln(r/r_i) \, r \, dr$$

$$\frac{(T_m - T_p) k}{\Delta H_f \rho} \, t = \int (\ln r - \ln r_i) \, r \, dr$$

$$= \int_{r_i}^r r \ln r \, dr - \ln r_i \int_{r_i}^r r \, dr$$

$$= \frac{r^2}{2}\left(\ln r - \frac{1}{2}\right)\Bigg|_{r_i}^r - \ln r_i \frac{r^2}{2}\Bigg|_{r_i}^r$$

$$= \frac{r^2}{2}\left(\ln r - \frac{1}{2}\right) - \frac{r_i{}^2}{2}\left(\ln r_i - \frac{1}{2}\right) - (\ln r_i)\left(\frac{r^2}{2} - \frac{r_i{}^2}{2}\right)$$

$$= \frac{r^2}{2}\left(\ln r - \frac{1}{2}\right) + \frac{r_i{}^2}{4} - \frac{r^2}{2}\ln r_i$$

$$= \frac{r^2}{2}\left(\ln \frac{r}{r_1} - \frac{1}{2}\right) + \frac{r_i{}^2}{4}$$

from which the expression for time to freeze up to a radius r is

$$t = \frac{\Delta H_f \rho}{k(T_m - T_p)}\left[\frac{r^2}{2}\left(\ln \frac{r}{r_i} - \frac{1}{2}\right) + \frac{r_i^2}{4}\right] \qquad (7.12)$$

Evaluating and interpreting the solution *1) Does the computed variation of freezing time with radius make sense?* Based on the physical understanding of the problem, as the frozen radius increases, time should increase more than proportionally with radius since there is rapidly increasing volume ($\sim \pi r^2 h$) of material to freeze. If we look at Eq. 7.12, the r^2 term is likely to dominate, i.e., time required to freeze will increase faster than the radius, so, indeed, it is consistent with our qualitative understanding. As an example, for $r_i = 0.002$ m, $k = 0.4416$ W/m $\cdot$ K, $\rho = 1200$ kg/m^3, $\Delta H_f = 335 \times 10^3$ kg/m^3, and $T_m - T_p = 196$ K, the time to freeze with radius is shown in Figure 7.15. *2) What do we learn?* If we look at the formulas for freezing time calculation in Table 7.1, for inward freezing of a cylinder (second column), with no external

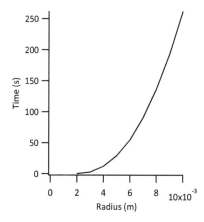

Figure 7.15: Time to freeze outward from a cylindrical probe of 2 mm radius.

resistance ($h \to \infty$), the time to freeze varies as the square of the radius. Our solution above for outward freezing is somewhat similar in the sense that freezing time also varies approximately as the square of the radius.

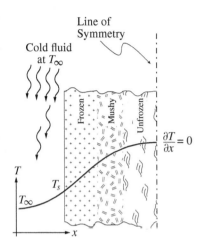

Figure 7.16: Freezing of a biological material showing a partially frozen or mushy zone.

7.4.3   Freezing Time for Biomaterials

The previous calculations of freezing time for a pure liquid will not work for even a simple solution where just one solute is dissolved. For one dissolved solute there will be two freezing fronts or interfaces (why?), and our calculations (Eq. 7.10) will not work. For every solute added, there will be an added interface to keep track of. The water in a biomaterial has many components dissolved in it whose many interfaces must be tracked, and this procedure would be hopeless for such materials. The effect of one or more dissolved components is a "mushy" zone in the material, where the material is only partially frozen, as shown in Figure 7.16.

Like spatial variation, time variation of temperature during freezing of a biomaterial would also be substantially different from what was assumed for the simple solution in the previous section. To have a qualitative understanding of the time variation of temperature in a biomaterial freezing process, consider the schematic given by Figure 7.17. Initially, the solid is at 10°C. Cold fluid is flowing over the surface to maintain the surface temperature at −10°C. The temperatures at the two locations A and B would follow approximately the curves shown. The plateau near 0°C is characteristic of a freezing process and occurs due to the need to remove latent heat for the entire region from the surface up to the location. The plateau would also occur near the temperature where most of the freezing takes place (which is 0°C here) and therefore most of the contribution from latent heat happens. The plateau would not be present in a simple cooling process where the material does not go through change of phase.

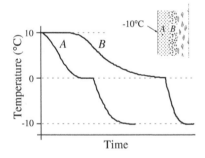

Figure 7.17: Schematic showing typical time–temperature history during a freezing process.

The gradual freezing of biomaterial due the effect of the "mushy" region can be successfully captured by following an alternative procedure for calculating freezing time. To develop this, we rewrite the heat equation as

$$\frac{\partial}{\partial x}\left(k\frac{\partial T}{\partial x}\right) = \rho c_{pa}\frac{\partial T}{\partial t} \tag{7.13}$$

where c_{pa} is an apparent specific heat that includes the latent heat. Thus, c_{pa} is defined as

$$c_{pa} = \frac{\partial H}{\partial T} \tag{7.14}$$

where H is the heat content, or enthalpy (i.e., the sum of sensible and latent heat) of a material per unit mass. The relationship between H and T

$$H = f(T) \tag{7.15}$$

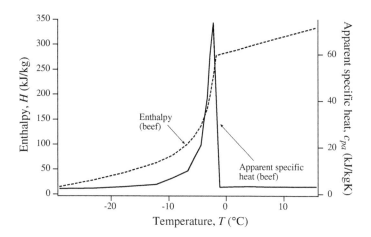

Figure 7.18: Experimental enthalpy (H) vs. temperature (T) relationship for beef muscle tissue around freezing temperature. The apparent specific heat (c_{pa}) is calculated from this data using Eq. 7.14.

is available from experimental data on the material. An example of such experimental data for beef muscle tissue is shown in Figure 7.18. There are also simple empirical equations to estimate H for a material, as shown in this figure, although such equations have considerable error in them. Equations 7.13 and 7.15 together can provide temperature as a function of time as needed for the freezing time calculation. Note that Eq. 7.13 automatically includes the phase change effects due to the use of the apparent specific heat, c_{pa}. Thus, even if the cellular structure and other complications are ignored for a biomaterial, its freezing process is more complex than that of a pure liquid.

It is apparent that the calculation of freezing time using Eqs. 7.13 and 7.15 can be involved. For practical use often the simplified formula (Eq. 7.10) derived in the previous section for pure materials is used to obtain an approximate value for freezing time. Note the assumptions mentioned in deriving the simple formula. Depending on the situation, the estimate can have considerable error. Also the latent heat in Eq. 7.10 should be that of the biomaterial, which is related to its water content by

$$\Delta H_{f,biomaterial} = \Delta H_{f,water} \times \left(\begin{array}{c} \text{mass fraction} \\ \text{of water in biomaterial} \end{array} \right) \qquad (7.16)$$

7.5   Evaporation

At any temperature above absolute zero, a molecule has kinetic energy. Consider the molecules of a liquid near its surface. At any temperature, some molecules have more than the average kinetic energy, while some molecules have less. While they are moving randomly, some molecules happen to be moving upward and may have enough energy to break out of the liquid. They can leave the liquid and join molecules of the gas above it. This change of state from the surface of a liquid to gas is called evaporation. When the change of state is from a solid to a vapor, the process is termed sublimation. Evaporation and sublimation are spontaneous processes driven by the energy of the molecules, i.e., by their temperature. Since the movement of the vapor molecules is random, some vapor molecules come back to the liquid (in the case of evaporation). Initially, more molecules leave the liquid for the vapor phase. However, when the vapor phase has enough vapor, the number of vapor molecules reentering the liquid phase is the same as the number leaving it. An equilibrium is reached at this point. The pressure of the vapor in equilibrium with the liquid (called *vapor pressure*) is a measure of the relative ease of evaporation. Vapor pressure increases with temperature, because the average molecule has more energy to leave the liquid at higher temperature. Vapor pressure of water as a function of temperature is shown in Figure 7.19.

As an example, at a temperature of 60°C, the vapor above water exerts a pressure of 0.02 MPa (≈ 0.2 atm). If the total pressure above the water is one atmosphere, 0.8 atmospheres will be exerted by air. From Figure 7.19 we can see that vapor pressure increases almost exponentially with temperature. At higher temperatures, more vapor can be formed. When the vapor pressure at a given temperature is equal to the external pressure, the evaporation that occurs is termed *boiling*. Because the vapor pressure above water at 100°C is one atmosphere, water will boil when it reaches 100°C if the external pressure is one atmosphere. At slightly less than 100°C and under one atmosphere, water will evaporate vigorously but will not be boiling, strictly speaking.

Both heat and mass are transferred when a liquid evaporates. The rate at which the liquid evaporates depends on the rate at which energy can be supplied to, and the rate at which vapor (mass) can be removed from, the interfacial region. Continued evaporation is possible only if vapor can be removed from the interfacial region. As a liquid evaporates, the more energetic molecules leave the liquid. Thus, the energy of the remaining liquid, i.e., its temperature, is lowered. This cooling process is termed evaporative cooling.

Evaporation is a very important process in biological systems. Evaporation from plant leaves, skin (sweating), soil surfaces, and free water surfaces are all important biological processes.

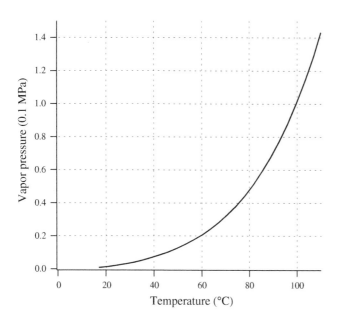

Figure 7.19: Vapor pressure of water at different temperatures.

7.5.1 Evaporation from Wet Surfaces

Evaporation from wet surfaces is frequent: from wet soils, food and biological materials, lakes, or falling droplets. Here, the surface vapor concentration is assumed to be the saturation value at the surface temperature. Thus, the rate of transfer of water vapor (*rate of evaporation*) per unit area at surface temperature T is given by

$$\begin{array}{c} \text{Flux of} \\ \text{water vapor} \end{array} = h_m(c_{\text{sat at } T} - c_{\text{vapor at } \infty}) \qquad (7.17)$$

where h_m stands for the mass transfer coefficient and c stands for the concentration of water vapor. (Use of the mass transfer coefficient is discussed in detail on page 513 under convective mass transfer.) At any given temperature, evaporation is limited by the rate at which vapor can be convected away, as given by Eq. 7.17. Thus, evaporation is a special case of convective mass transfer from a surface. This mass flux also removes energy and lowers the temperature (the evaporative cooling process). The energy flux due to evaporation is given by

$$\begin{pmatrix} \text{Flux of energy} \\ \text{from evaporation} \end{pmatrix} = \Delta H_{vap} \begin{pmatrix} \text{Flux of} \\ \text{water vapor} \end{pmatrix} \qquad (7.18)$$

where ΔH_{vap} is the latent heat of vaporization.

Evaporation in the context of various biological systems is now discussed qualitatively.

Plant systems (plant and soil surfaces)   *Transpiration* in a leaf is the passage outward of water vapor through stomatal pores. Transpiration is generated by photosynthesis (see Example 12.2.1 on page 425). *Evapotranspiration* is the flux of water vapor just above a plant canopy, which includes transpiration from the leaves and evaporation from the soil surface.

$$\text{Evapotranspiration} = \text{Evaporation} + \text{Transpiration} \qquad (7.19)$$

Evaporation from soil and the surface of small water bodies, and transpiration from vegetation, are difficult to separate and are therefore referred to in the single term called evapotranspiration. For fairly dense vegetation, evapotranspiration is appreciable, amounting to 60–90% of the flux of water vapor from a layer of open water at the same ambient air temperature (Nobel, 1974). Evapotranspiration is calculated using empirical formulas.

Penman's equation, a specialized equation for evaporation from a sufficiently wet soil surface, where the contacting air is fully saturated, can be seen in soil science textbooks such as Jury et al. (1991). As expected, Penman's equation to estimate potential evaporation is a function of the velocity and other conditions of the wind, just as it should be for any mass transfer coefficient. Thus, appropriate mass transfer coefficients as described in Section 14.5.4 can be used to find the potential evaporation. Note that the estimate will be for potential evaporation. Even when it is well supplied with water near the soil surface or at the roots, an evaporating vegetation cannot, in general, be considered wet, except after rainfall or dew formation. Thus, the humidity at the surface of the foliage elements is likely to be somewhat smaller than the saturation value at the corresponding temperature, and equations such as Penman's need to be modified as described in Brutsaert (1982).

Evaporation in animal systems (evaporative cooling)   As discussed under thermoregulation (page 97), animals routinely use panting and cutaneous evaporation, which are evaporative cooling processes. Evaporation accounts for about 20% of the body heat loss for humans, as discussed on page 13 under human and animal comfort.

Evaporative cooling is used to reduce building air temperature in the U.S. in poultry houses and sometimes in swine and dairy housing. During evaporative cooling, the evaporation of water removes energy from the surrounding air, thus cooling it. Evaporative cooling works best in areas of low relative humidity. In a pad system air is blown through a wet porous pad. High pressure fogging systems that create a fine aerosol are preferred in broiler houses. Direct spray systems where nozzles produce mist or spray droplets that wet animals directly are used during hot weather in swine or dairy confinement facilities with solid concrete or slotted floors.

7.5.2 Evaporation inside a Solid Matrix

Inside an unsaturated porous matrix, the rate of evaporation is related to temperature, the amount of pore space available for evaporation, and permeability (ease with which vapor can escape). As liquid evaporates and more vapor forms, pressure rises. The extent to which it rises depends on how much pressure can be released (measured by permeability). This vapor generation and flow is part of all drying (soil, foods, biomaterials) and similar processes (baking and frying of food, for example). Intensive heating such as the use of microwaves tends to produce higher pressures. A common assumption is that liquid and vapor are at equilibrium in a pore (at any point). Thus, pressure and temperature are related as shown in Figure 7.19. In a typical drying process, moisture is evaporated internally and the vapor either diffuses or is blown to the surface and convected away. Additionally, any liquid on the surface can evaporate.

7.5.3 Evaporation of Solutions

Evaporation is a very common method of concentrating food and other biological solutions. It is an important operation in the processing of liquid food materials, to improve their storability or taste and to reduce bulk in transit. In this context evaporation is really boiling; i.e., it occurs at the boiling point of the solution. A product is concentrated by boiling off the solvent, generally water. Because products such as food materials are often heat sensitive, i.e., because they degrade at higher temperatures, they are boiled at lower temperatures. A lower boiling point is achieved by lowering the pressure of the enclosure where boiling takes place. Figure 7.19 shows that boiling will take place at any given temperature, if the pressure over the water is equal to the vapor pressure at that temperature. Details of the evaporation process for concentrating food and biological solutions can be found in texts on food process engineering or in Minton (1987).

As mentioned earlier, the evaporation of water is an energy-intensive process, and therefore other means of concentrating food products such as freeze concentration are often used. In freeze concentration the product is partially frozen and the ice is separated out. Because the product is not heated, quality is not degraded as much as in evaporation.

In heating a solution, the boiling point of the solution is higher than that of pure liquid. The reasons are the same as those giving rise to freezing point depression in the freezing of solutions, discussed in Section 7.3. Dissolved solutes elevate the boiling point above that of pure liquid. The elevated boiling point, T_b, of a solution is related to its concentration x_A by

$$\frac{\Delta H_{vap}}{R_g} \left[\frac{1}{T_b} - \frac{1}{T_{b0}} \right] = \ln x_A \qquad (7.20)$$

where ΔH_{vap} is the latent heat of vaporization of pure liquid A in J/mol, T_{b0} is the boiling point of pure liquid A in K, R_g is the gas constant, and x_A is the mole fraction of A in solution. For a small increase in concentration x_A, the increase in T_b is linear.

7.6   Chapter Summary—Heat Transfer with Change of Phase

• Freezing of Water (page 226)

1. As water freezes, its density and specific heat decrease while its thermal conductivity increases. Microwave absorption of ice is much lower than that of water.

• Freezing of Solutions and Biomaterials (page 228)

1. As water freezes in a solution, the rest of the solution is more concentrated and its freezing point lowers. This decrease in freezing point is given by Eq. 7.1.

2. Water in biomaterials has dissolved solutes. As pure water freezes, it concentrates the rest of the solution, continuously lowering its freezing point. This makes biomaterials freeze over a range of temperatures.

3. In a slow freezing process of a tissue, the solution outside a cell (extracellular fluid) freezes first and concentrates, leading to the transport of intracellular water to the outside.

4. Higher cooling rates minimize the transport of intracellular water and are often preferred for biomaterial freezing.

• Freezing Time Calculations (page 234)

1. Approximate formulas for the freezing time for a slab and cylinder are given by Eq. 7.10 and Problem 7.3, respectively.

2. A more accurate procedure for calculating the freezing time of biomaterials is provided in Section 7.4.3

• Evaporation (page 243)

1. The spontaneous change of molecules from liquid to vapor state is termed evaporation.

2. Vapor pressure is the partial pressure of water vapor in equilibrium with the liquid, at any given temperature. Vapor pressure increases with temperature.

3. At any temperature, if the external or surrounding pressure is equal to the vapor pressure, the evaporation at that temperature is termed boiling.

4. The evaporation rate from a very wet surface is often limited by the rate at which vapor can be taken away. The rate of vapor removal is given by the mass transfer equation 7.17.

7.7 Problem Solving—Freezing

In freezing, there are primarily two types of problems that we can solve in this chapter—freezing point calculation and freezing time calculation, of which the second type is the significant one:

▶ **Calculate concentrations and temperatures related to freezing point depression.**

- Calculate the freezing point of a solution of given concentration. This is straightforward use of Eq. 7.1.

- Calculate the concentration of a solution for a given freezing point. This is the opposite of the question above. This question can also be put as what is the concentration at a particular temperature (it is implied that this temperature is the freezing point of the solution at the given concentration since no more freezing occurs).

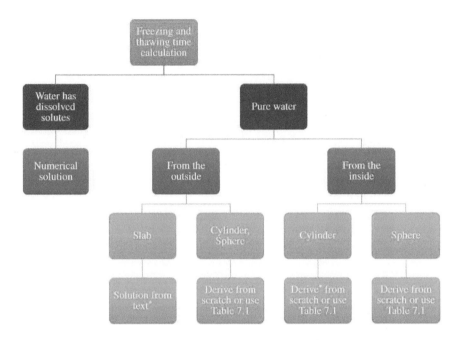

Figure 7.20: Problem solving in freezing. *Derivations for slab and cylinder are in Section 7.4 and Example 7.4.2, respectively.

- Calculate the fraction of water frozen at a particular solution temperature. From the previous step, we know what the concentration must be at a particular temperature. Knowing the concentration allows us to calculate how much water is in solution form. If we know the initial (unfrozen solution) amount of water, the initial water minus what is currently in solution form is the water that is frozen.

▶**Calculate freezing (or thawing time).** Figure 7.20 shows there are primarily two types of approaches to freezing time calculation—simple, analytical solutions that ignore the effect of dissolved solute such as freezing point depression, and a more comprehensive approach (e.g., Eq. 7.13) that requires numerical solution. Analytical solutions are divided into freezing from outside in and inside out. Some of the possibilities are:

- Calculate freezing/thawing time of a slab/cylinder/sphere using solutions already available. This is straightforward use of formulas in Table 7.1 (note that the sensible heat is ignored in all these formulas; the effect of surface convection, h, is also ignored in this table—you may need to derive these as needed). The steps are provided below:

1. Select the characteristic length, L. Formulas we have covered are in one dimension. For an arbitrary object, depending on the dimensions, it has to be approximated to a one-dimensional freezing (i.e., a thin cylinder will be approximated as a slab, with characteristic length as half thickness of the slab whereas a long cylinder will be just that, with the characteristic length as the radius). Also, note the characteristic length is half thickness when freezing is symmetric while, for a layer such as a lake, it is the thickness being frozen.

2. Calculate the appropriate latent heat $\lambda_{tissue} = \lambda_{water} \times$ (Water content in fraction), as it is only the water that is being frozen.

3. Select the appropriate thermal conductivity for freezing vs. thawing. Why?

4. Choose the appropriate formula from Table 7.1 (or their more complete versions that include the effect of h) and plug in.

- Derive the solutions for freezing/thawing of a cylinder/sphere, analogous to a slab. While resistance for a slab was L/kA, resistance for a cylindrical and spherical shell are given in Table 4.2 on page 92. See Problem 7.3 for a cylinder and Problem 7.4 for a sphere.

7.8 Concept and Review Questions

1. Why do biomaterials freeze over a temperature range?

2. Would the freezing of pure water lead to a "mushy" zone?

3. Does Plank's formula to calculate freezing time consider this "mushy" zone?

4. Do biomaterials start to freeze at $0°C$? Why?

5. When a tissue is frozen slowly, cells often shrink. Why?

6. Is faster freezing always preferable? Explain.

7. What are the ways to minimize cracking from faster freezing?

8. What would be the problems in cryopreservation of a whole human body?

9. In thawing a frozen food, the microwave oven is programmed to cycle on and off, instead of staying on continuously. Explain why.

Table 7.1: Freezing time formulas for a pure liquid for various geometries (slab, cylinder, sphere) when freezing from outside in (with a finite h) and when freezing from inside out.

Geometry	Time to freeze (frozen outside in, with finite h)	Time to freeze (frozen inside out starting at radius r_i)
	$t = \frac{\Delta H_f \rho}{T_m - T_s} \left(\frac{L^2}{2k} + \frac{L}{h} \right)$	
	$t = \frac{\Delta H_f \rho}{T_m - T_s} \left(\frac{R^2}{4k} + \frac{R}{2h} \right)$	$t = \frac{\Delta H_f \rho}{k(T_m - T_s)} \left[\frac{r^2}{2} \left(\ln \frac{r}{r_i} - \frac{1}{2} \right) + \frac{r_i^2}{4} \right]$
	$t = \frac{\Delta H_f \rho}{T_m - T_s} \left(\frac{R^2}{6k} + \frac{R}{3h} \right)$	$t = \frac{\Delta H_f \rho}{k(T_m - T_s)r_i} \left[\frac{r^3}{3} + \frac{r_i^3}{6} - \frac{r_i r^2}{2} \right]$

10. A given solution can be concentrated by partially freezing it and removing the ice. However, the same solution can also be concentrated by boiling it. Which process would you choose, and why?

11. Why do you expect the thawing time for a biomaterial to be longer than the freezing time for the same material?

Further Reading

Freezing

Bischof, J., N. Merry, and J. Hulbert. 1997. Rectal protection during prostrate cryo-surgery: Design and characterization of an insulating probe. *Cryobiology*, 34:80–92.

Cleland, A. C. and R. L. Earle. 1984. Assessment of freezing time prediction methods. *Journal of Food Science*, 49:1034–42.

Fahy, G. M., J. Saur, and R. J. Williams. 1990. Physical problems with the vitrification of large biological systems. *Cryobiology*, 27:492–510.

Fennema, O., W. D. Powrie, and E. Marth. 1973. *The Low Temperature Preservation of Foods and Living Matter*. Marcel Dekker, New York.

Laverty, J. 1991. Physio-chemical problems associated with fish freezing. In *Food Freezing: Today and Tomorrow*. W. B. Bald, ed. Springer-Verlag, London.

McGrath, J. J. 1985. Preservation of biological material by freezing and thawing. In *Heat Transfer in Medicine and Biology* (Vol. 1), edited by A. Shitzer and R. C. Eberhart. Plenum Press, New York.

Meryman, H. T. 1966. *Cryobiology*. Academic Press, New York.

NEMC. 2000. Cryosurgical Ablation of the Prostate (CSAP) for Patients with Prostate Carcinoma. WWW document at http://www.nemc.org/urology/cryo.htm accessed on June 29.

Shi, X., A. K. Datta, and Y. X. Mukherjee. 1998. Thermal stresses from large volumetric expansion during freezing of biomaterials. *Journal of Biomechanical Engineering, Transactions of the ASME*, 120:720–726.

Shi, X., A. K. Datta, and J. Throop. 1998. Mechanical property changes during freezing of a biomaterial. *Transactions of the ASAE*, 41(5):1407–1414.

Evaporation

Brutsaert, W. 1982. *Evaporation into the Atmosphere: Theory, History, and Applications*. D. Reidel Publishing Co., Dordrecht, Holland.

Dowdy, J. A., R. L. Reid, and E. T. Handy. 1986. Experimental determination of heat- and mass-transfer coefficients in aspen pads. *ASHRAE Transactions*, 92(2A):60–69.

Hagiwara, T. and R. W. Hartel. 1996. Effect of sweetener, stabilizer, and storage temperature on ice recrystallization in ice cream. *J. Dairy Sci.*, 79:735–744.

Jones, F. E. 1991. *Evaporation of Water*. Lewis Publishers, Inc., Chelsea, Michigan.

Jury, W. A., W. R. Gardner, and W. H. Gardner. 1991. *Soil Physics*. John Wiley & Sons, New York.

Kung, S. K. J. and T. S. Steenhuis. 1986. Heat and moisture transfer in partly frozen nonheaving soil. *Soil Science Soc. Am. J.*, 50(5):1114–1122.

Minton, P. E. 1987. *Handbook of Evaporation Technology*. Noyes Publications, Park Ridge, New Jersey.

Nobel, P. S. 1974. *Biophysical Plant Physiology*. W.H. Freeman and Company, San Francisco.

7.9   Problems

7.1   Freezing of a Solution

Moved to solved problems

7.2   Storing at Subzero Temperature without Freezing

A large piece of biological tissue has to be kept at a temperature as low as possible without freezing (ice formation). The water in the tissue has several solutes dissolved in it and the effective molecular weight of the dissolved solutes is 50. The weight of all dissolved solutes combined is 10% of the weight of water. The latent heat of fusion of water is 335 kJ/kg and the molecular weight of water is 18. What is the lowest temperature that this tissue can be stored at, without ice formation?

7.3   Freezing of an Infinitely Long Cylinder

You are trying to freeze a long cylinder (length much greater than the radius) of pure material of radius R. The cylinder is at an initial temperature T_m, which is close to its freezing temperature, so that any sensible heat change may be ignored. The cylinder is exposed to an air blast in a freezer with an air temperature of T_∞ and a heat transfer coefficient of h at the cylinder surface. Ignore any convection of water in the unfrozen material. The material has density ρ, latent heat ΔH_f, and thermal conductivity k. Obtain an expression for the time to freeze this cylinder. *(Note: $\int r \ln r \, dr = r^2(\ln r - 1/2)/2$.)*

7.4   Freezing of a Sphere

You are trying to freeze a sphere of pure material of radius R. The sphere is at an initial temperature T_m, which is close to its freezing temperature, so that any sensible heat change may be ignored. The sphere is exposed to air blast in a freezer, maintaining the sphere surface at T_s. Ignore any convection in the unfrozen material. The material has density ρ, latent heat ΔH_f, and thermal conductivity k. Obtain an expression for the time to freeze this sphere.

7.5   Thawing of Blood

Thawing is the change of phase from solid to liquid. Consider the thawing of two frozen rectangular packets of blood, one having a thickness twice that of the other. Assume the thickness is much smaller than the other two dimensions. Consider the blood to be at the freezing temperature so that only the change in latent heat is to be considered. How would the thawing times compare?

7.6   Freezing vs. Thawing Time

We want to compare the freezing time vs. thawing time for one of the packets of blood in Problem 7.5. Assume no convection or bulk flow in the blood. The thermal conductivity of frozen blood is 1.5 W/m·K and that of unfrozen blood is 0.6 W/m · K. For the same temperature difference between the cooling or heating medium and the blood, and ignoring any convective resistance outside the packet, how does the freezing time compare with thawing time?

Figure 7.21: Filled bag of blood. (iStock.com/Wessel du Plooy)

7.7   Freezing of Hamburger Patty

Freezing of food materials for preservation is one of the most important applications of biomaterials freezing. Consider freezing of a hamburger patty, as shown in Figure 7.22, that has a thickness of 2 cm and diameter of 8 cm. For simplicity, the patty can be approximated as a slab. Cold air at $-50°C$ blows at 0.5 m/s over both sides of the patty. The initial temperature of burgers is $-5°C$, which is the temperature at which we will consider all freezing to take place. The patty contains 75% water. The thermal conductivity of the frozen patty is 1.4 W/m · K and the density of the patty is 900 kg/m^3. The latent heat of fusion of pure water is 335 kJ/kg. 1) The properties of air at the relevant temperature are: $\rho = 1.3947$ kg/m^3, $\mu = 1.596 \times 10^{-5}$Pa · s, $k = 2.23 \times 10^{-2}$ W/m · K, and Pr = 0.72. Calculate the heat transfer coefficient over the top surface. 2) Calculate the time to freeze the patty (i.e., when the center plane

Figure 7.22: Hamburger patty. (iStock. com)

just freezes), when the top and bottom surfaces have the same value of heat transfer coefficient.

7.8 Inadvertent Thawing

During transportation and storage of food, unexpected and undesired variations in the surrounding temperature can pose health hazards if the temperature rises or lowers to the range for optimum bacterial growth. Consider one of the frozen hamburger patties from Problem 7.7, left on the countertop by mistake. Assume that it is initially frozen, with a uniform temperature of $-5°C$, and that there is no heat transfer from the bottom of the patty. The room air temperature is at $30°C$ and the heat transfer coefficient at the top of the patty from natural convection is $10 \ W/m^2 \cdot K$. The unfrozen patty has 75% water and a thermal conductivity of $0.4 \ W/m \cdot K$, a density of $1040 \ kg/m^3$, and a freezing temperature of $-5°C$. The latent heat of fusion of pure water is $335 \ kJ/kg$. The dimensions of the patty are given by a thickness of 2 cm and a diameter of 8 cm. Neglecting the sensible heat change, how long would it take for the hamburger to completely thaw from its initial frozen state?

7.9 Freezing of Meat (Use of Characteristic Length)

Consider two pieces of sausage being frozen in a cylindrical form in an air blast freezer. The diameter for each cylinder is 7.5 cm. The height of one cylinder is 10 cm, and that of the other is 5 cm. The meat has already been cooled to a uniform initial temperature of $-4°C$, which is the initial freezing temperature. The air temperature in the freezer is $-30°C$, and the blowing leads to a high heat transfer coefficient of $200 \ W/m^2 \cdot K$. The meat has 70% moisture and, when frozen, it has a density of $950 \ kg/m^3$ and a thermal conductivity of $1.4 \ W/m \cdot K$. The latent heat of fusion of pure water is 335 kJ/kg. Assuming 1D heat transfer, calculate the freezing times for the two cylinders. *(Hint: Be careful to select the right formula based on characteristic dimension.)*

Figure 7.23: A sausage to be frozen. (iStock.com)

7.10 Depth of Freezing of a Lake

Consider the freezing of a large stagnant lake during a winter that lasts for 60 days. The average air temperature for this period is $-10°C$ with an average heat transfer coefficient of $20 \ W/m^2 \cdot K$ over the lake surface. Neglect any sensible heat effect by assuming that the lake temperature stays at $0°C$ during the season. The latent heat for freezing of water is 335 kJ/kg, the density of water is $1000 \ kg/m^3$, the thermal conductivity of water is $0.61 \ W/m K$, and the thermal conductivity of ice is $2.3 \ W/m K$. 1) Determine the depth to which this lake will be frozen at the end of the winter. 2) In the springtime, the lake begins to thaw. Assume the entire frozen layer is at $0°C$ when thawing begins. The air temperature for spring is 10°C and it has the same surface heat transfer coefficient value. Calculate the time it would take for the frozen layer (that you just calculated) to thaw in the springtime. 3) Explain the reasons for the thawing time being longer or shorter.

7.11 Depth of Frost Penetration in Soil

In regions where ground surface temperatures fall below the freezing point of water, freezing of the pore water in the soil is possible. A cold front which freezes the pore water in the subsoil (layer beneath the topsoil) is generally defined as the frost front, and the depth to which the soil freezes is called the depth of frost penetration. Consider a uniform region of homogeneous silt soil, with a water content equal to 20% and a density of 1750 kg/m^3. Assume the soil water freezes at 0°C. The thermal conductivity of the frozen soil is 1.4 W/m·K. The latent heat of water is 335 kJ/kg. Calculate the depth of frost penetration for a winter lasting 5 months (about 30 days per month), if the temperature at the soil surface remains constant at −13°C.

7.12 Time for Cryosurgery

Moved to solved examples.

Figure 7.24: Freezing of ground. (iStock.com)

7.13 Time for Cryosurgery: Spherical Ball

We would like to estimate the time needed in cryosurgery where tissue is frozen using a spherical cryosurgical probe inserted into the tissue. To make things easier, you will be guided through the setup of the governing equation. Consider pseudo-steady-state heat transfer where the tissue touching the probe is at the liquid nitrogen temperature of $T_p = -196°C$. Consider the tissue to be initially unfrozen, but already at the constant freezing point of $T_m = -2°C$. Ignore any sensible heat involved.

1) Let us say the outer frozen layer increases by a value of dr in time dt. Write the expression for the rate of heat flow that must be released for this layer of material to freeze. 2) The heat flow in step 1 has to be conducted through the frozen layer into the cryoprobe. Write an expression for heat conduction through the frozen layer at steady state, considering an appropriate expression for thermal resistance of a spherical shell. 3) Equate the expressions in step 1 and step 2 to set up the equation from which we can solve for the radius of frozen tissue at time t. 4) Integrate from radius r_i (at time 0) to any radius r at time t. Do not try to simplify to get r as a function of t. 5) Instead, provide an approximate plot of r as a function of time t, based on your physical understanding of the problem.

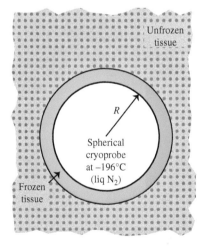

Figure 7.25: Schematic of a cryoprobe inserted in tissue.

7.14   Freezing of Vaccines

Vaccines are sometimes frozen prior to storage or transport. To design the packaging, you are trying to decide between two rectangular blocks, one that is twice as thick as the other. Assume the thickness is much smaller than the other two dimensions for both slabs. 1) By what factor would the freezing time change for the thicker slab compared with that for the thinner slab? 2) Suppose that, instead of a vaccine, the material being frozen is biological tissue having cellular structure. Explain (qualitatively) what, if any, effect choosing a thicker (vs. a thinner) slab of tissue to freeze would have on the tissue. *(Hint: think of transport of water across a cell membrane.)*

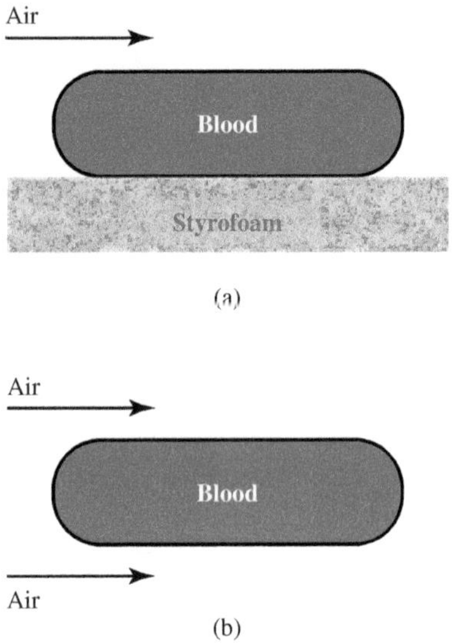

7.15   Thawing with Changes in Boundary Conditions

We would like to compare the differences in thawing time for packaged frozen blood in two different scenarios, as shown in the schematic in Figure 7.26. For the package on the left, the bottom surface, placed on styrofoam, can be considered insulated. For the package on the right, the bottom is on a wire mesh such that the same convection occurs over the bottom and top surfaces. The thickness of the packages is equal to 2.5 cm and the other two dimensions of the packages are much greater than the thickness, so the heat transfer can be considered to be one dimensional along the thickness. The surface heat transfer coefficient is 20 W/m²·K. The thermal conductivity of the blood is 0.55 W/m·K. 1) What is the ratio (left/right) of the times that it would take to thaw the packages entirely? 2) Explain why the time decreased or increased, going from the right scenario to the left.

7.16   Relative Durations for Thawing of Blood

We would like to compare the differences in thawing time for packaged frozen blood in two different scenarios, as shown in Figure 7.26: a) The bottom surface of the package, placed on styrofoam, can be considered insulated. b) The bottom is on a wire mesh such that the same convection occurs over the bottom surface and top surface.

The thickness of the packages is equal to 2.5 cm and the other two dimensions of the package are much greater than the thickness, so the heat transfer can be considered to be one dimensional along the thickness. The initial temperature of the blood is at its freezing temperature of –0.5°C. For simplicity, consider the entire volume of blood to freeze at –0.5 °C, i.e., no freezing point depression. The air temperature is 15 °C. The surface heat transfer coefficient is 20 W/m² · K. The thermal conductivity of the blood material is 0.55 W/mK and its density is 950 kg/m³. The latent heat of fusion for blood is 335 kJ/kg. What are the thawing times for the two conditions shown?

Figure 7.26: Thawing of frozen blood.

7.17   Freezing and Shapes

Freezing rapidly stops metabolic and other reactions occurring in cells and therefore is an important procedure in studying kinetics of cellular processes. Consider freezing of muscle tissue in cylindrical bundles that is very soft (and therefore pliable). They are frozen using a surface temperature of –20°C. The water content of the tissue is 78% and the freezing point can be assumed as 0°C for simplicity. From the literature, a quick estimation of the thermal conductivity and density of unfrozen tissue is 0.5 W/m·K and 1050 kg/m^3, respectively, and that of frozen tissue is 1.4 W/m·K and 970 kg/m^3, respectively. The latent heat of water is 335 kJ/kg. 1) Explain which set of data (frozen or unfrozen tissue) is the appropriate one to use for freezing time calculations. 2) What is the freezing time for a long tissue bundle of diameter 0.2 mm? 2) What is the freezing time for a long tissue bundle that has a diameter 5 times larger? 3) For the needed kinetic study, if the freezing time can be no longer than the time you calculated for the 0.2 mm diameter bundle, how can we achieve this for a 1 mm diameter soft tissue, *without changing the surface temperature*? Any ideas are welcome.

7.18   Time to Freeze Dry

Freeze drying is a method of drying food, tissue, blood plasma, and pharmaceuticals without destroying their physical structure. In a freeze drying process, instead of the ice going to water as in thawing, ice sublimates (goes from solid to vapor) since heating is performed at a very low pressure. For a slab being dried symmetrically, as shown in Figure 7.27, sublimation occurs at the interface between the frozen and dried layers and the vapor diffuses out of the slab through the dried region without any significant resistance.

A frozen slab of thickness $2L$ that is initially at the freezing temperature, T_m, is being dried symmetrically from both sides. Heat is transferred from both surfaces through convective heat transfer with a coefficient h by air at temperature T_∞. A pseudo-steady-state assumption can be made for heat transfer in the dried layer. Enthalpy of sublimation (enthalpy needed for going from solid to vapor) is given by ΔH_s in J/kg. Starting from known heat flow equations, derive the formula for the time to freeze dry the slab of thickness $2L$. Define any symbols you use with appropriate units.

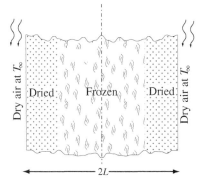

Figure 7.27: Schematic of a slab during freeze drying symmetrically.

7.19   Freezing Time for a Sphere with a Central Core and Convection on the Surface

We would like to compute the freezing time for a spherical geometry, as shown in Figure 7.28, where the inner core has no water, i.e., does not take part in freezing.

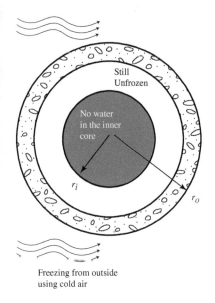

Freezing from outside
using cold air

Figure 7.28: Cross-section of a spherical
geometry showing region to be frozen.

Figure 7.29: Common warts on
the surface of skin to be frozen.
(https://en.wikipedia.org/wiki/Wart#/
media/File:Wart_ASA_animated.gif)

Consider a cranberry with a seed having such a geometry. The outside of the cranberry has a convective heat transfer coefficient of h. The frozen part has a thermal conductivity of k, latent heat of fusion of λ, density of ρ, freezing temperature of T_m (which is the temperature for all unfrozen material), and bulk air temperature of T_a.

1) Write an expression for the rate of heat flow that must be released, in joules per second, from an unfrozen shell of thickness dr for it to freeze. 2) The heat amount in step 1 must be conducted out through the frozen layer and the outside air. Write an expression for this rate of heat flow at steady state. 3) Using your answers from steps 1 and 2, write the differential equation for the radius to which the tissue would be frozen at any time t. 4) Integrate to solve for the time it takes to freeze the entire material. 5) To freeze more rapidly, the spherical fruit was cut in half and placed with the cut side on a thermally insulating surface while the air flow and all other parameters stay the same. Explain to what extent the freezing time will change.

7.20   Freezing Time for a Wart

Warts on the skin are treated by freezing (both the epidermal and dermal layer) to ensure complete cell death using an ultra cold nitrous oxide probe. We can approximate a wart as a slab where a distance of 2 mm from the surface needs to be frozen. The initial temperature of the skin is assumed to be at its freezing temperature of $-2°C$. The probe can be assumed to maintain a temperature of $-190°C$ with a heat transfer coefficient of $500,000$ W/m^2·K between the probe and skin surface. To perform a simplified analysis, you can make the major assumption that all tissue freezes at $-2°C$. The thermal conductivity and density of tissue to be frozen are 0.4416 W/m · K and 1200 kg/m^3, respectively. The latent heat of fusion for water is 335 kJ/kg and the mass fraction of water in the tissue is assumed to be 61.5%.

1) Calculate the time needed to freeze 2 mm of skin. 2) If the skin is drier, having 80% of the mass fraction of water *mentioned above*, what is the ratio of the new freezing time to the freezing time calculated in question 1? Explain why the ratio is higher or lower. 3) If the average specific heat of the tissue is 3480 J/kg·K, calculate the enthalpy (per unit area) that must be released to reach the freezing temperature of $-2°C$, starting from a normal skin surface temperature of $32°C$. 4) Have you considered this type of enthalpy (in question 3) removal in your freezing time calculation in question 1? Explain. 5) Calculate the enthalpy of phase change per unit area for the wart (with 61.5% water) in joules per unit area. 6) Based on your computed values for steps 3 and 5, would the actual freezing time (time to reach $-2°C$ starting from $32°C$) be lower or higher than what you calculated in step 1? Why?

7.21 Freezing Porcine Liver with Glycerol

Consider a jet of highly filtered dry air at −73°C flowing over a piece of porcine liver (approximated as a sphere) at 15 m/s. We would like to compute the freezing time for this sphere. The diameter of the sphere is 3 cm. The liver has been soaked in glycerol and a phosphate buffer to aid freezing and it has already been cooled down to the freezing temperature of −31°C.

1) The viscosity, specific heat, density, and thermal conductivity, respectively, of air are 1.44×10^{-5} kg/m · s, 1.0059×10^3 J/kg · K, 1.5kg/m^3, and 0.0198W/m · K. Calculate the heat transfer coefficient over the liver that would be needed to compute the freezing time. 2) What temperature should have been used to read the air properties used in step 1? 3) We need to derive the freezing time for a sphere. (Relax, you will only be setting up the problem). Let us say the frozen layer increases inwardly by a value of dr in time dt. Write the expression for the *rate of heat flow* in J/s that must be released for this additional material to freeze. Leave in terms of symbols but define each one, including their units. 4) The heat flow in step 2 has to be conducted through the frozen shell and through convection outside. Write an expression for the *rate of heat flow* in J/s through the frozen shell and the air at steady state, considering the thermal resistance of a spherical shell. Again, leave in terms of symbols but define each one. 5) *Set up the integral, including limits* from which we can solve for the radius of unfrozen tissue as a function of time t. You do not need to integrate.

7.22 Freezing Time for a Layer of Biomaterial

1) Derive the freezing time for a layer of water containing biomaterial that is covered with a plastic wrap (Figure 7.31) with convection of air over the plastic wrap (h is less than infinity), i.e., starting from energy balance, derive how the thickness of a frozen layer changes with time. Assume every parameter that you need and define all the symbols that you use very precisely. For simplicity, consider the water in the biomaterial having no dissolved components. 2) Refer to the formula you derived and explain how the plastic layer will affect the time to freeze.

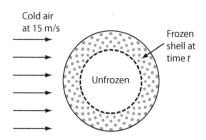

Figure 7.30: Schematic of porcine liver with air flow over surface.

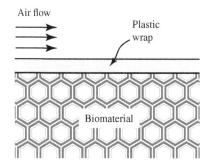

Figure 7.31: Cross-section of a biomaterial slab being frozen from the surface.

Figure 7.32: Three hamburger patties stacked. (iStock.com)

7.23 Freezing Time and Size

By what factor would the freezing time decrease if, instead of the freezing in the stack of three hamburger patties shown in Figure 7.32, they are frozen as single patty. Show your calculations.

Chapter 8

RADIATIVE ENERGY TRANSFER

After you have studied this chapter, you should be able to

1. Explain the characteristics of electromagnetic radiation as it is incident on a biological or environmental surface or emitted by the surface.

2. Calculate the magnitude and spectral quality of electromagnetic radiation emitted by a surface.

3. Calculate the radiative energy exchange between two bodies.

Key Terms

- **reflection, absorption, and transmission of waves**

- **blackbody**

- **emission and emissivity**

- **solar radiation**

- **greenhouse effect**

- **radiative exchange**

- **view factor**

There are only two fundamental mechanisms of energy transfer—diffusion and electromagnetic radiation (convection refers to the addition of bulk flow to these modes). This chapter deals with transport of energy by the mechanism of electromagnetic radiation. No medium is necessary in transporting energy using electromagnetic radiation, in contrast to diffusion and convection which, by definition, require a medium to be present, i.e., cannot occur in vacuum.

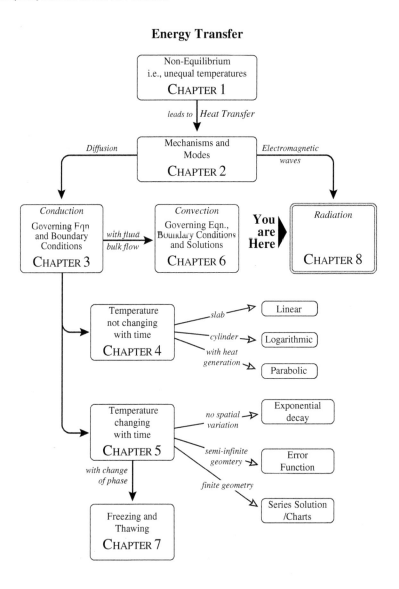

Figure 8.1: How the contents of this chapter relate to other chapters in energy transfer.

8.1 The Electromagnetic Spectrum

Unlike conductive or convective heat transfer, no medium is required for the radiative exchange between bodies since the exchange is through electromagnetic waves. The magnitude of the radiative exchange depends significantly on temperature level and not just temperature difference. The quality of thermal radiation also depends on the temperature level.

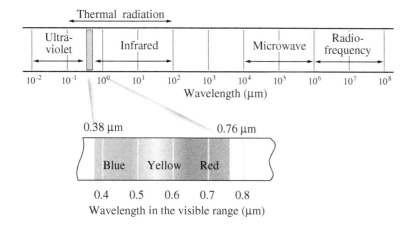

Figure 8.2: The electromagnetic spectrum.

Electromagnetic radiation travels at the speed of light, c, of $3 \times 10^8 \text{m/s}$. The wavelength and frequency of electromagnetic radiation are related by

$$\left(\begin{array}{c} \lambda \\ \text{Wavelength} \end{array} \right) \left(\begin{array}{c} \nu \\ \text{Frequency} \end{array} \right) = \left(\begin{array}{c} c \\ \text{Speed of Light} \end{array} \right) \qquad (8.1)$$

In the electromagnetic spectrum (Figure 8.2), thermal radiation generally refers to the range 0.1–100 microns (1 micron = 10^{-6} m) and includes some ultraviolet and all infrared radiation, as well as the entire visible spectrum of 0.38–0.76 microns. It is interesting to note, that in the visible spectrum, the sensitivity of the human the eye retina varies with wavelength, as shown in Figure 8.3. Microwaves typically refer to wavelength (in air) of 1–100 cm and radiofrequency waves are 1–100 m. Microwaves and radio frequencies are efficiently absorbed by many materials, including biomaterials. This will be discussed later.

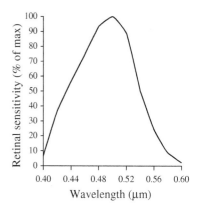

Figure 8.3: Sensitivity of the human eye to various wavelengths in the visible region.

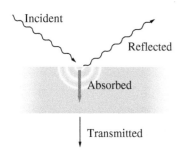

Figure 8.4: Reflection, absorption, and transmission of electromagnetic waves at a surface.

8.2 Reflection, Absorption, and Transmission of Waves at a Surface

Being part of the electromagnetic spectrum, infrared, microwave, and radiofrequency waves incident on a surface are reflected, absorbed, and transmitted similar to visible light. Figure 8.4 shows reflection, absorption, and transmission of electromagnetic waves at a surface. The reflection can be specular or diffuse. For a specular reflection, the angle of incidence is equal to the angle of reflection. For diffuse reflection, which is the case for most materials, waves are reflected in all directions.

Thus, the following relation would apply at the surface:

$$\left(\begin{array}{c} \rho \\ \text{Reflectivity} \end{array} \right) + \left(\begin{array}{c} \alpha \\ \text{Absorptivity} \end{array} \right) + \left(\begin{array}{c} \tau \\ \text{Transmissivity} \end{array} \right) = 1 \qquad (8.2)$$

All three properties, reflectivity, absorptivity, and transmissivity, are functions of wavelength. In addition, reflectivity and absorptivity are directional, i.e., they vary with the direction of the incident wave. In this text, only directional averages of reflectivity and absorptivity will be used.

8.2.1 Transmissivity of a Leaf and Photosynthesis

Figure 8.5 shows representative values for the reflectivity, absorptivity, and transmissivity of a leaf. This has implications in plant growth. For example, a shaded leaf in a canopy will get only solar radiation between 0.7 and 2 μm since that is where the leaves have high transmissivity. Solar radiation in this region is not very useful for photosynthesis. As illustrated in Figure 8.6, green chlorophyl (responsible for photosynthesis) absorbs strongly in the red (0.660 μm), blue (0.435 μm), and some of the lower wavelength portions of the visible region. Moreover, the stomatal opening in a leaf (see Example 12.2.1 on page 425) is partially closed at the lower illumination levels, which increases the resistance to CO_2 exchange by the leaf and further decreases photosynthesis.

8.2.2 Transmissivity of the Atmosphere: Greenhouse Effect

The atmosphere traps heat. As seen from space, the earth radiates energy at wavelengths and intensities characteristic of a body at $-18°C$. Yet the average temperature at the surface is about 33°C higher. Heat is trapped between the surface and the level, high in the atmosphere, from which radiation escapes. The ability of the earth's atmosphere to trap the heat from the sun is known as the greenhouse effect.

The greenhouse effect is a consequence of the spectral dependence of the transmissivity of the atmosphere. Some of the gases in the atmosphere, CO_2, methane, and

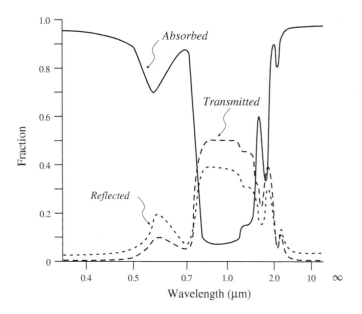

Figure 8.5: Representative spectral values for the reflectivity, absorptivity, and transmissivity of a leaf.

chlorofluorocarbons are relatively transparent to sunshine (smaller wavelengths) but opaque to the longer-wavelength infrared radiation released by the earth. Thus, the radiation from earth is absorbed by the atmosphere, increasing its temperature. Gases such as CO_2, methane, and chlorofluorocarbons are called greenhouse gases. The relationship between CO_2 concentration and the rise in temperature of the earth was discussed in Chapter 1.

8.2.3   Albedo—Reflection from Soil

The overall fraction of incident radiation that is reflected from celestial surfaces such as water, soil, etc., is termed albedo. Albedo is dependent on, among other factors, the position of the sun and the surface composition and roughness. An example of variations in albedo for different surfaces can be seen in Figure 8.7.

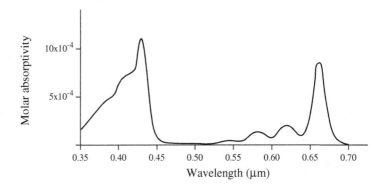

Figure 8.6: Visible spectra of green chlorophyl.

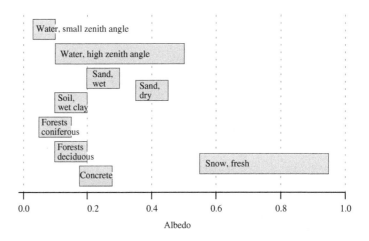

Figure 8.7: Albedo of various surfaces, data from Guyot (1998).

8.2.4   Absorption and Transmission in Biomaterials

As the waves are transmitted in a material, they are also absorbed. The decay in the intensity due to the absorption is often characterized by the simple Beer–Lambert law given by

$$F = F_0 e^{-x/\delta} \tag{8.3}$$

where F_0 is the flux in W/m^2 entering a material at its surface (incident flux minus the reflected flux) and F is the flux at a location inside the material at a distance x from the surface, as illustrated in Figure 8.8. The penetration depth, δ, defined by Eq. 8.3, is the distance at which the flux becomes $1/e$ or approximately 37% of its value at the surface, since using $x = \delta$ in Eq. 8.3 we get

$$F = F_0 e^{-\delta/\delta} = \frac{F_0}{e} \qquad (8.4)$$

A smaller penetration depth means a quicker decay of the flux or good absorptivity and therefore poor transmissivity. Table 8.1 compares the penetration depth for infrared, microwave, and radiofrequency, which are the three types of waves normally used in heating applications such as drying, baking, tempering, sterilizing, etc. The low penetration depth in the infrared region signifies that most solids are opaque to infrared radiation. Infrared provides primarily surface heating, whereas radiofrequency can provide uniform heating to a much greater depth.

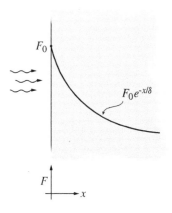

Figure 8.8: Illustration of the decay in the intensity of a wave as it is transmitted through a material, as given by Eq. 8.3.

Table 8.1: Heating characteristics of various types of electromagnetic waves for water that often makes up the bulk of a biomaterial

Type of electromagnetic wave	Frequency (typical) in MHz	Wavelength in air (cm)	Penetration depth, δ, in biomaterials
Infrared	70,000,000	0.0004	few microns
Microwave	2,450	12	few centimeters
Radiofrequency	27	1100	hundred centimeters

8.3 Thermal Radiation from an Ideal (Black) Body

To describe radiation emitted from a surface, several terms are first defined. The total emissive power of a surface, E [W/m^2], is defined as the total rate of thermal energy emitted via radiation in all directions and at all wavelengths per unit surface area. The emissive power of an ideal (black) body, E_b, is the largest possible E for a given temperature.

The monochromatic emissive power E_λ is defined similarly to total emissive power but restricted to one wavelength λ. Thus, the monochromatic emissive power is the energy emitted between wavelengths λ and $\lambda + d\lambda$, such that

$$dE = E_\lambda d\lambda$$

The energy emitted between two wavelengths λ_1 and λ_2 is given by

$$\int_{\lambda_1}^{\lambda_2} E_\lambda d\lambda$$

The total energy emitted by a body can be written as

$$E = \int_0^\infty E_\lambda d\lambda$$

The fraction of total energy emitted between two wavelengths is

$$F = \frac{\int_{\lambda_1}^{\lambda_2} E_\lambda d\lambda}{\int_0^\infty E_\lambda d\lambda} \tag{8.5}$$

To describe the radiation characteristics of real surfaces, the concept of radiation from an ideal surface is introduced. This ideal surface neither reflects nor transmits any radiation, i.e., it absorbs all radiation incident upon it. Since no light is considered to be reflected, this body will appear black and hence it is referred to as a blackbody. A blackbody is only an idealization and it is a standard against which radiation from real surfaces is compared. An example of a blackbody can be a cavity with a small hole. Radiation entering this cavity through the small hole is continuously reflected inside the cavity and is almost entirely absorbed before it can come out. This approximates blackbody behavior.

Radiation emitted by a blackbody has a spectral distribution. The amount of radiation emitted by a blackbody at temperature T between two wavelengths λ and $\lambda + d\lambda$ (monochromatic emissive power $E_{b,\lambda}$) is given by Planck's law of radiation:

$$E_{b,\lambda} = \frac{2\pi c^2 h \lambda^{-5}}{\exp(ch/\kappa\lambda T) - 1} \tag{8.6}$$

where c is the speed of light, h is Planck's constant and has the value 6.625×10^{-34}J·s, κ is Boltzmann's constant and has the value 1.380×10^{-23}J/K, and T is the absolute temperature in K. Equation 8.6 is plotted in Figure 8.9 for different temperatures. Note that the vertical axis in the figure is power flux per unit wavelength.

Important features of Planck's law can be seen from Figure 8.9 and Equation 8.6 as

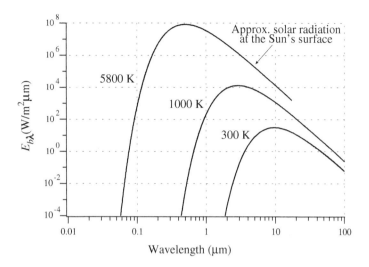

Figure 8.9: Spectral blackbody emissive power. Note that the y-axis has the dimension of power per unit area per unit wavelength, $W/m^2 \cdot \mu m$.

1. The emitted radiation varies continuously with wavelength.

2. At any wavelength the magnitude of the radiation increases with increasing temperature.

3. The spectral region in which the radiation is concentrated depends on temperature, with comparatively more radiation appearing at shorter wavelengths as the temperature increases. The relationship between the peak of emission λ_{max} and temperature is given by

$$\lambda_{max} T = 2897.6 \; \mu mK \tag{8.7}$$

As a consequence of Equation 8.7, a significant fraction of the radiation emitted by the sun, which may be approximated as a blackbody at 5800 K, is in the visible region of the spectrum. In contrast, for $T \leq 800$ K, emission is predominantly in the infrared region of the spectrum and not visible to the eye.

Using Equation 8.6, the total energy emitted between two wavelengths λ and $\lambda + d\lambda$ is given by $E_{b,\lambda} d\lambda$. To obtain the total energy emitted at all wavelengths, this emitted energy is integrated over all wavelengths to yield the Stefan–Boltzmann law:

$$E_b = \int_0^\infty E_{b,\lambda} d\lambda = \frac{2\pi^5 \kappa^4 T^4}{15 c^2 h^3} = \sigma T^4 \tag{8.8}$$

where $\sigma = 5.670 \times 10^{-8} \mathrm{W/m^2 \cdot K^4}$ is the Stefan–Boltzmann constant.

8.4 Fraction of Energy Emitted by an Ideal Body in a Given Range of Wavelengths

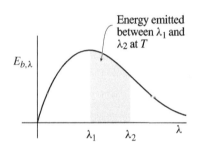

Figure 8.10: Illustration of energy emitted by a blackbody between two wavelengths.

It is often quite useful to estimate the amount of energy emitted by a body between any two wavelengths or in a certain band. For a black or an ideal body at temperature T, the energy emitted between two wavelengths λ_1 and λ_2 is given by

$$\int_{\lambda_1}^{\lambda_2} E_{b,\lambda} d\lambda$$

as illustrated in Figure 8.10. This emitted energy can be written as a fraction of the total energy emitted by the same blackbody to be

$$\frac{\displaystyle \int_{\lambda_1}^{\lambda_2} E_{b,\lambda} d\lambda}{\displaystyle \int_0^\infty E_{b,\lambda} d\lambda} \tag{8.9}$$

Note that the denominator in Eq. 8.9 representing the total energy emitted is given by Eq. 8.8. Substituting this value in the denominator and separating the integral in the numerator,

$$= \frac{\displaystyle \int_0^{\lambda_2} E_{b,\lambda} d\lambda - \int_0^{\lambda_1} E_{b,\lambda} d\lambda}{\sigma T^4}$$

$$= \int_0^{\lambda_2} \frac{E_{b,\lambda}}{\sigma T^4} d\lambda - \int_0^{\lambda_1} \frac{E_{b,\lambda}}{\sigma T^4} d\lambda$$

$$= \int_0^{\lambda_2 T} \frac{E_{b,\lambda}}{\sigma T^5} d(\lambda T) - \int_0^{\lambda_1 T} \frac{E_{b,\lambda}}{\sigma T^5} d(\lambda T) \tag{8.10}$$

Substituting Eq. 8.6 for the integrand in the two integrals above, we get

$$\frac{E_{b,\lambda}}{\sigma T^5} = \frac{2\pi c^2 h (\lambda T)^{-5}}{\sigma \left(\exp(ch/\kappa \lambda T) - 1 \right)} \tag{8.11}$$

which shows that the integrand is exclusively a function of λT (the rest are constants). If the integrals are denoted by F such that

$$F_{0-\lambda T} = \int_0^{\lambda T} \frac{E_{b,\lambda}}{\sigma T^5} d(\lambda T) \qquad (8.12)$$

the $F_{0-\lambda T}$ values, representing the fraction of total energy emitted by a blackbody at temperature T between wavelengths 0 and λ, are illustrated in Figure 8.11. Then the fraction of energy between the two wavelengths λ_1 and λ_2 is given by

$$\text{Fraction of energy between } \lambda_1 \text{ and } \lambda_2 \text{ at temp } T = F_{0-\lambda_2 T} - F_{0-\lambda_1 T} \qquad (8.13)$$

as illustrated in Figure 8.12. The $F_{0-\lambda T}$ values calculated using Eq. 8.12 are given in Table 8.2.

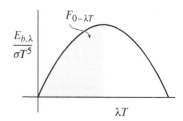

Figure 8.11: Graphical illustration of $F_{0-\lambda T}$.

8.4.1 Example: What Fraction of Solar Energy Is in the Visible Region (0.38–0.76 μm)?

As mentioned earlier, solar energy can be assumed to radiate from a blackbody at the surface temperature of 5800 K. Thus, a fraction of energy between 0.38 and 0.76 μm at 5800 K can be calculated using Eq. 8.13 as

$$
\begin{aligned}
\text{Fraction of energy between } 0.38 \text{ and } 0.76 \ \mu m \text{ at } 5800 \text{ K} &= F_{0-(0.76)(5800)} - F_{0-(0.38)(5800)} \\
&= F_{0-4408} - F_{0-2204} \\
&= 0.5491 - 0.1013 \quad \text{(from Table 8.2)} \\
&= 0.4478
\end{aligned}
$$

Thus, 44.78% of the sun's energy is in the visible region. Note that this value does not include the effect of atmospheric absorption and therefore is true for solar radiation just outside the atmosphere.

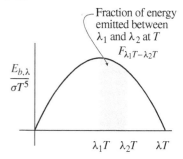

Figure 8.12: Graphical illustration of $F_{0-\lambda_2 T} - F_{0-\lambda_1 T}$.

8.5 Thermal Radiation from a Real Body: Emissivity

The emissive powers of all real bodies are less than that for the ideal body and depend on surface condition, among other factors. *Total emissivity* ϵ compares the emissive power E of an actual body to the emissive power of the ideal (black) body:

$$\epsilon = \frac{E}{E_b} \quad \text{where } 0 \le \epsilon \le 1 \qquad (8.14)$$

Table 8.2: Blackbody radiation functions

λT [μmK]	$F_{0-\lambda T}$	λT [μmK]	$F_{0-\lambda T}$	λT [μmK]	$F_{0-\lambda T}$
0	0	10200	0.91823	20400	0.986375
200	7.01016e-27	10200	0.91823	20600	0.986733
400	2.20179e-12	10400	0.92197	20800	0.98707
600	9.96926e-08	10600	0.925491	21000	0.987413
800	1.70667e-05	10800	0.928807	21200	0.987735
1000	0.000328258	11000	0.931932	21400	0.988047
1200	0.00216717	11200	0.9344881	21600	0.988349
1400	0.00787475	11400	0.937665	21800	0.98864
1600	0.0198754	11800	0.942781	22000	0.988922
1800	0.0395777	12000	0.945133	22200	0.989195
2000	0.0670396	12200	0.94736	22400	0.989459
2200	0.10126	12400	0.94947	22600	0.989715
2400	0.140672	12600	0.95147	22800	0.989962
2600	0.183564	12800	0.953368	23000	0.990202
2800	0.228346	13000	0.955169	23200	0.990434
3000	0.273688	13200	0.95688	23400	0.990659
3200	0.318549	13400	0.958506	23600	0.990877
3400	0.362168	13600	0.960052	23800	0.991088
3600	0.404021	13800	0.961523	24000	0.991293
3800	0.443775	14000	0.962924	24200	0.991492
4000	0.481246	14200	0.964258	24400	0.991685
4200	0.51636	14400	0.96553	24600	0.991872
4400	0.549119	14600	0.966743	24800	0.992053
4600	0.579582	14800	0.9679	25000	0.99223
4800	0.607839	15000	0.969004	25200	0.992401
5000	0.634007	15200	0.97006	25400	0.992567
5200	0.658212	15400	0.971068	25600	0.992728
5400	0.680584	15600	0.972032	25800	0.992885
5600	0.701254	15800	0.972954	26000	0.993038
5800	0.720351	16000	0.973836	26200	0.993186
6000	0.737997	16200	0.974681	26400	0.99333
6200	0.754307	16400	0.97549	26600	0.99347
6400	0.769389	16600	0.976265	26800	0.993606
6600	0.783343	16800	0.977008	27000	0.993738
6800	0.796263	17000	0.977721	27200	0.993867
7000	0.808234	17200	0.978405	27400	0.993993
7200	0.819335	17400	0.979061	27600	0.994115
7400	0.829637	17600	0.979691	27800	0.994234
7600	0.839206	17800	0.980297	28000	0.994349
7800	0.848101	18000	0.980879	28200	0.994462
8000	0.856379	18200	0.981438	28400	0.994572
8200	0.864089	18400	0.981976	28600	0.994678
8400	0.871276	18600	0.982494	28800	0.994783
8600	0.877981	18800	0.982992	29000	0.994884
8800	0.884244	19000	0.983471	29200	0.994983
9000	0.890098	19200	0.983933	29400	0.995079
9200	0.895575	19400	0.984378	29600	0.995173
9400	0.900703	19600	0.984807	29800	0.995265
9600	0.90551	19800	0.98522	30000	0.995354
9800	0.910019	20000	0.985619		
10000	0.914252	20200	0.986004		

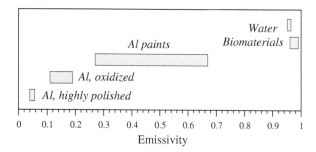

Figure 8.13: Emissivity of biomaterials as compared to other materials such as aluminum and water.

Figure 8.13 shows the total emissivity values for a number of surfaces. Emissivities of highly polished metallic surfaces are low, those of non-metallic surfaces are higher, and those of biomaterials are in the range 0.9–0.95. For example, the low emissivity value of polished aluminum is exploited on the outer surface of space suits to reduce radiative heat loss. Emissivity typically varies with wavelength. *Monochromatic emissivity* is defined as the ratio of the monochromatic emissive power of a body to the monochromatic emissive power of a blackbody at the same wavelength and temperature. Thus

$$\epsilon_\lambda = \frac{E_\lambda}{E_{b,\lambda}} \tag{8.15}$$

For example, Figure 8.14 shows the typical spectral dependence of monochromatic emissivity for human skin. An average value of 0.98–0.99 for human skin monochromatic emissivity is often used. When the monochromatic emissivity is independent of the wavelength, the body is described as a *gray body*.

Emissivities of soil surfaces range from 0.9 for dry quartz sand to 0.98 or higher depending on organic matter content, mineral composition, and moisture content. Between a wet and dry soil it can vary by almost 0.04. Differences in soil emissivity are not likely to be significant in the energy balance of bare soils (having a negligible effect on surface temperature). The emissivity differences are important for thermal infrared imagery.

Note that the emission discussed here is passive emission, i.e., it happens automatically at a rate dependent on the temperature level. This is to be distinguished from emission studied in spectroscopy. In spectroscopy, an external radiation field is introduced which is reflected, absorbed, or transmitted. The absorption causes molecules to make a transition to an energetically excited state. The transition from the excited

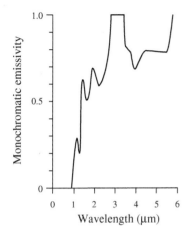

Figure 8.14: The spectral emissivity of human skin at room temperature. The corresponding total emittance is 0.993. Most authors use values between 0.98 and 0.99. Adapted from R. Elam, D. W. Goodwin, and K. L. Williams. 1963. Optical properties of human epidermis. *Nature*. 198(4884):1001–1002. Reproduced with permission from MacMillan Publishers Ltd.

state back to the lower state is accompanied by emission of radiation. This emission has information about the molecules, to obtain which is the purpose of emission spectroscopy.

8.5.1 *BioConnect: Emission from Human Bodies—Infrared and Microwave Thermography for Clinical Uses*

As mentioned earlier, all bodies above a temperature of absolute zero emit radiation naturally over a range of wavelengths. Thus, a human body would also emit radiation. At a temperature of approximately 300 K, which is near the comfortable human skin surface temperature of 306 K (92°F), the spectral curve of blackbody radiation is shown in Figure 8.15. As shown in this figure, human radiation will cover infrared as well as microwave regions.

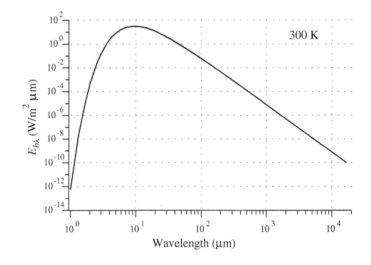

Figure 8.15: Radiation emitted from a blackbody at 300 K showing the approximate radiation emitted from humans.

Several medical problems, including cancer and vascular occlusions, are accompanied by local changes in temperature of the order of 1°C. Infrared and microwave thermography are passive, non-invasive techniques that can detect and diagnose pathologic conditions in which there are disease-related temperature differentials (Carr, 1989). These radiometric techniques, for the most part, have been directed at the early detection and diagnosis of breast cancer. Early detection can dramatically increase the survival rate. Present detection techniques other than radiometry require that the tumor have mass and contrast with respect to the surrounding tissue (i.e., physical examination, mammography, ultrasonography, etc.). The average tumor diameter when first detected and diagnosed using these techniques is still quite large, and radiometric techniques that depend on thermal activity rather than mass could provide still earlier detection.

At the human body temperature of approximately 300 K, the maximum intensity of radiative emission (according to Eq. 8.7) occurs at a wavelength of approximately 10 μm. This has led to the operation of infrared thermographs at wavelengths near 10 μm.

At a typical microwave frequency of 3 GHz (approximately 12 cm or $1.2 \times 10^5 \mu$m) wavelength), the intensity of radiative emission is reduced considerably. However, microwave radiometers, developed primarily for radio astronomy, can detect radiation with this intensity. These radiometers can detect intensity changes corresponding to emitter temperature changes of less than 1°C. At 3 GHz, microwaves penetrate approximately 5 cm in fat and 0.8 cm in muscle or skin. Thus, it would be possible to get tissue information from a depth of several centimeters. Of course, there are many complications, and microwave thermography is still at the research stage (Carr, 1989).

8.6 Solar, Atmospheric, and Earth Surface Radiation

8.6.1 Solar Radiation—Magnitude and Spectral Distribution

Outside the atmosphere, solar radiation reaches an approximate intensity E_{sc} of 1353 W/m^2. An approximation to the solar constant can be made by considering radiative exchange between two spheres, sun and earth, at their distance (see Example 8.4). Depending on the location on earth (Figure 8.16), the intensity of radiation just outside the earth's atmosphere (extraterrestrial radiation) can be calculated from

$$E_{et} = E_{sc} \cdot \beta \cdot cos\theta$$

where E_{et} is the extraterrestrial solar irradiation in W/m^2, E_{sc} is the solar constant of 1353 W/m^2, β is a correction factor to account for the eccentricity of the earth's orbit about the sun with $0.97 \leq \beta \leq 1.03$, and θ is the zenith angle, shown in Figure 8.16. The spectral distribution of solar radiation is compared with that emitted from a blackbody at 5800 K (Figure 8.17). Just outside the atmosphere this solar radiation is incident with an intensity as shown in Figure 8.17. The significant discrepancies which occur in the ultraviolet region (around 0.5 μm) are mainly due to electronic transitions which occur in the overlying gases of the sun. The various atmospheric processes

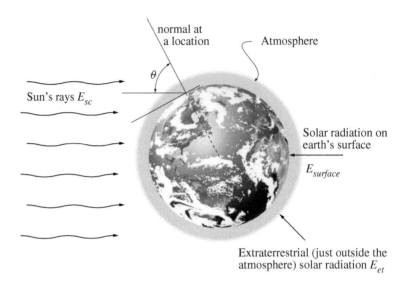

Figure 8.16: Relation of solar radiation on the earth's surface to the extraterrestrial solar radiation and the solar constant.

operate to change the spectral distribution as the radiation traverses to the earth's surface through the atmosphere to approximately that shown in Figure 8.17. The main absorption is by water vapor, which is responsible for the several strong bands in the infrared region (most of Figure 8.17), and by high-altitude atmospheric ozone, which effectively limits radiation which reaches the ground in appreciable quantities to wavelengths greater than about 0.30 μm.

8.6.2 Emission from Earth's Surface

The radiation emitted by the earth's surface is important since the earth has to lose (and thus balance) the energy it receives from the sun. The energy emitted by the earth is given by

$$E_{earth} = \epsilon_{earth}\sigma T_{earth}^4 \tag{8.16}$$

The average temperature of the earth's surface is considered to be 14°C (57°F). The emissivity of the earth's surface will be made up of that of water, vegetation, soil, etc. As shown earlier, these emissivities are close to unity.

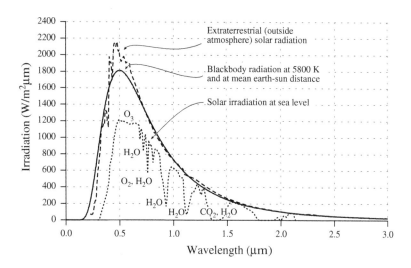

Figure 8.17: Spectral distribution of solar radiation outside the earth's atmosphere and on the earth's surface. The absorption of the radiation over various bands of wavelength occurs due to the atmospheric gases, ozone, oxygen, water vapor, and carbon dioxide, as noted. Data from Gast (1965).

8.6.3 Atmospheric Emissions

The atmospheric gases such as H_2O and CO_2 emit considerable radiation. The effective concentration of these atmospheric gases can vary from time to time. For example, clouds contain much water in the form of vapor, droplets, and/or crystals. The effective sky temperature can be substantially lower for a cold, clear sky than for a warm, cloudy sky. Atmospheric emissions are the only source of radiation at night. The radiation from these gases is estimated in practice (when measured data is not available) using an apparent sky emissivity as

$$G_{atm} = \epsilon_{sky}\sigma T_a^4 \qquad (8.17)$$

where T_a is the air temperature near the ground. The apparent sky emissivity, ϵ_{sky}, is an empirical constant that can be estimated from equations such as (Brunt, 1940)

$$\epsilon_{sky} = 0.55 + 1.8\left(\frac{p_{H_2O}}{P}\right)^{1/2} \qquad (8.18)$$

for a clear sky. Here p_{H_2O} is the partial pressure of water vapor in the atmosphere and P is the total atmospheric pressure.

8.6.4   Global Energy Balance

The approximate energy contributions from various sources (e.g., solar radiation, emissions from earth, atmospheric emissions, etc., discussed earlier) averaged over the entire earth's surface for annual mean conditions is shown in Figure 8.18. The solar radiation value, for example, is obtained by dividing the total radiation incident on earth by the total surface area of the earth.

Using some of the approximate energy contributions, a simple model of the greenhouse effect of the earth's atmosphere can be built—see Problem 8.18.

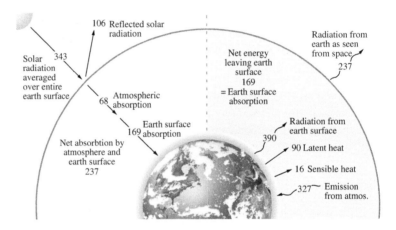

Figure 8.18: The global energy balance for annual mean conditions. Data from Ramanathan (1987). All quantities are fluxes, in W/m^2, averaged over the total surface area of earth.

8.7   Radiative Exchange between Bodies

When two bodies exchange thermal radiation, the net radiative exchange between them will depend on their size, shape, and the relative orientation of their radiating surfaces. The size, shape, and orientation factors are lumped in a parameter called the *configu-*

ration factor or *view factor*, denoted by F_{1-2}, and stands for

$$
F_{1-2} = \text{fraction of radiation leaving surface 1}
$$
$$
\text{that is intercepted by surface 2} \quad (8.19)
$$

The view factor, F_{1-2}, is dimensionless. Since radiation travels along a straight line in the absence of strong electromagnetic fields, the view factor from a surface to itself will be zero unless the surface "sees" itself. Thus, $F_{1-1} = 0$ for plane or convex surfaces and $F_{1-1} \neq 0$ for concave surfaces, as illustrated in Figure 8.19. The value of the view factor ranges between zero and one. The limiting case of $F_{1-2} = 0$ represents the situation when two surfaces do not have a direct view of each other, and thus radiation leaving surface 1 cannot reach surface 2 directly. The other limiting case of $F_{1-2} = 1$ represents the situation where surface 2 completely surrounds surface 1 so that all radiation leaving surface 1 reaches surface 2.

The view factor has been calculated and is available (see, for example, Siegel and Howell, 1981) for a large number of physical situations involving different combinations of size, shape, and orientation. As an illustration, the view factor between two aligned parallel rectangles of the same size can be derived as

$$
F_{1-2} = \frac{2}{\pi \bar{X} \bar{Y}} \left\{ \ln \left[\frac{(1 + \bar{X}^2)(1 + \bar{Y}^2)}{1 + \bar{X}^2 + \bar{Y}^2} \right]^{1/2} \right.
$$
$$
+ \bar{X}(1 + \bar{Y}^2)^{1/2} \tan^{-1} \frac{\bar{X}}{(1 + \bar{Y}^2)^{1/2}} + \bar{Y}(1 + \bar{X}^2)^{1/2} \tan^{-1} \frac{\bar{Y}}{(1 + \bar{X}^2)^{1/2}}
$$
$$
\left. - \bar{X} \tan^{-1} \bar{X} - \bar{Y} \tan^{-1} \bar{Y} \right\} \quad (8.20)
$$

where X and Y are the sides of the rectangles, D is the distance between them, and $\bar{X} = X/D$ and $\bar{Y} = Y/D$. Equation 8.20 is plotted in Figure 8.20 and shows how the view factor F_{1-2} increases as either the size of the rectangles is increased or the distance between the rectangles is decreased so that a greater fraction of radiation leaving surface 1 is intercepted by surface 2. Figure 8.20 can be used, for example, to find the view factor between the two parallel surfaces of a roof and a floor for a structure.

The simplest situation of radiative exchange occurs when two blackbodies are exchanging radiation. We recall that blackbodies absorb all incident radiation and therefore do not reflect any. Energy is only emitted from a blackbody. The net radiative exchange between two such bodies is given by

$$
q_{1-2} = \sigma A_1 F_{1-2} \left(T_1^4 - T_2^4 \right) \quad (8.21)
$$

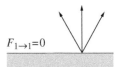

$F_{1 \to 1} = 0$

Plane surface

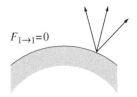

$F_{1 \to 1} = 0$

Convex surface

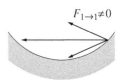

$F_{1 \to 1} \neq 0$

Concave surface

Figure 8.19: Illustration of view factors showing the view factor is zero for a surface to itself for a plane or a convex surface, but non-zero for a concave surface.

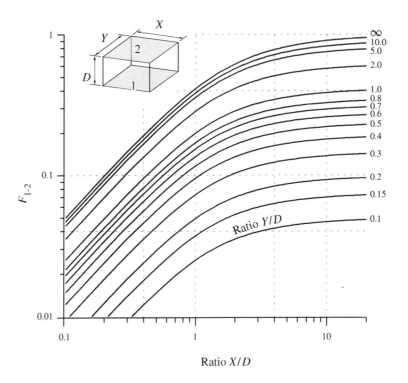

Figure 8.20: View factor between two aligned parallel rectangles of equal size.

or

$$q_{1-2} = \sigma A_2 F_{2-1} \left(T_1^4 - T_2^4 \right) \tag{8.22}$$

where q_{1-2} is the net energy transfer between bodies 1 and 2, in W, A_1 and A_2 are the areas of the bodies in m², F_{1-2} is the view factor between bodies 1 and 2, while F_{2-1} is the view factor between bodies 2 and 1, and T_1 and T_2 are the absolute temperatures of the respective surfaces, in K. Note that either Eqs. 8.21 or 8.22 can be chosen depending on whether F_{1-2} or F_{2-1} is easier to calculate.

If we have an enclosure made up of several black surfaces all at different temperatures, the net radiative heat transfer between a surface i and all other surfaces is given by

$$q_i = \sum_{k=1, k \neq i}^{n} \sigma A_i F_{i-k} \left(T_i^4 - T_k^4 \right) \tag{8.23}$$

where n is the total number of surfaces.

Instead of blackbodies, when two *gray* bodies are exchanging radiation, the situation becomes more complex since these bodies are also reflecting part of the radiation. For the special case when two gray surfaces form a complete enclosure or close to it, as illustrated in Figure 8.21, the net radiative exchange is given by

$$q_{1-2} = \frac{\sigma \left(T_1^4 - T_2^4\right)}{\dfrac{1-\epsilon_1}{\epsilon_1 A_1} + \dfrac{1}{A_1 F_{1-2}} + \dfrac{1-\epsilon_2}{\epsilon_2 A_2}} \tag{8.24}$$

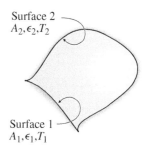

Surface 2
A_2, ϵ_2, T_2

Surface 1
A_1, ϵ_1, T_1

Figure 8.21: Schematic of two gray surfaces that form a complete enclosure.

where q_{1-2} is the net energy transfer between bodies 1 and 2, in W, ϵ_i is the emissivity of body i, A_i is the area of body i in m^2, F_{1-2} is the view factor between bodies 1 and 2, dimensionless, and T_i is the absolute temperature of surface i, in K. For example, radiative exchange between two large (infinite) parallel planes, as illustrated in Figure 8.22, would be given by a special case of Eq. 8.24 for $A_1 = A_2 = A$ and $F_{1-2} = 1$:

$$q_{1-2} = \frac{\sigma A \left(T_1^4 - T_2^4\right)}{\dfrac{1-\epsilon_1}{\epsilon_1} + 1 + \dfrac{1-\epsilon_2}{\epsilon_2}} \tag{8.25}$$

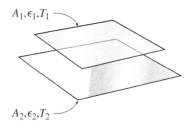

A_1, ϵ_1, T_1

A_2, ϵ_2, T_2

Figure 8.22: Schematic of two gray infinite parallel planes.

When a relatively small object 1 is completely enclosed by a much larger surface 2, such as a person inside a large room, $A_1/A_2 \approx 0$ and $F_{1-2} = 1$, and Eq. 8.24 simplifies to

$$q_{1-2} = \epsilon_1 \sigma A_1 \left(T_1^4 - T_2^4\right) \tag{8.26}$$

8.7.1 Radiative Heat Transfer Coefficient

For a special case of radiative heat exchange, when temperatures T_1 and T_2 are close, Eq. 8.21 can be written as

$$q_{1-2} = \sigma A_1 F_{1-2} \, 4T_1^3 \left(T_1 - T_2\right) \tag{8.27}$$

$$= F_{1-2} A_1 h_r \left(T_1 - T_2\right) \tag{8.28}$$

where

$$h_r = 4\sigma \, T_1^3 \tag{8.29}$$

can be termed a radiative heat transfer coefficient analogous to the convective heat transfer coefficient. Note that the unit for h_r is W/m$^2 \cdot$ K, the same as that for the convective heat transfer coefficient. Since Eq. 8.29 is an approximation, in some cases, h_r

may be determined experimentally. For example, in radiative heat exchange between humans and the environment (Iberall and Schindler, 1973), the radiative heat transfer coefficient h_r was measured experimentally as

$$h_r = 3.5[1 + 0.0055(T_s + T_a)] \qquad (8.30)$$

where T_s is the average skin temperature and T_a is the environmental temperature, both in °C. For the range of environmental temperatures $0 < T_a < 50°$ C, h_r varies between 3.8 and 5.1 W/m$^2 \cdot$ K.

8.7.2 *BioConnect: Radiative Exchange Between a Leaf and Surroundings*

Radiative heat loss from an unshaded leaf (Figure 8.23) has important consequences. As will be discussed later, on a cloudless night, the effective temperature of the sky can be quite low. If this leaf is radiating to the sky, it can get to very low temperatures and freeze even when the surrounding air is above freezing. Note that the air temperatures in such cases are considerably warmer than the effective sky temperature. Therefore, convection or wind actually warms the plant. In fact, radiative freeze typically occurs when the wind is calm or still. In some orange groves in the state of Florida in the United States, where such radiative freeze frequently causes considerable damage, large fans are employed to blow air when such freezing is imminent. Also, since the uppermost leaves in a tree are more exposed to the sky, such leaves will tend to have the lowest temperatures and frost tends to form there first.

In the daytime, with large amounts of radiative heat gain from the sun, the surrounding air typically cools the leaf. This also has very important consequences, as this lowering of temperature keeps it in the range for optimal photosynthesis. Also, the lower the leaf temperature, the lower is the concentration of water vapor in the pores of the cell walls of the mesophyll cells, and consequently less water would then tend to be lost in transpiration (see mass transfer). This is an important consideration in arid and semi-arid regions.

8.7.3 *Example: Radiative Cooling of Orange Trees*

Consider a flat orange leaf as shown in Figure 8.23 of surface area 10 cm^2 parallel to the ground. The top surface of the leaf radiates to clear sky and the effective radiation temperature of the sky is assumed at $-133°$C. The bottom surface of the leaf radiates to the ground and the ground temperature is 30°C. Both the top and bottom surfaces are also exchanging heat with the air through convection, with a convective heat transfer coefficient of 10 W/m$^2 \cdot$K. The air temperature is 5°C. Consider the leaf to be very thin so that the entire leaf (including the top and the bottom surfaces) is at one temperature. 1) What is the leaf temperature at steady state? 2) Suppose we forget to include the convective heat transfer from the bottom, how much of an error would we make in the leaf temperature?

Solution

Understanding and formulating the problem *1) What is the process?* Leaf surfaces (top and bottom) exchange energy by radiation with the soil and the sky, and by convection with air. *2) What are we solving for?* Leaf temperature at steady state. *3) Schematic and given data:* A schematic is shown in Figure 8.23 with some of the given data superimposed on it. *4) Assumptions:* Since the leaf has little temperature variation across its thickness (it is very thin), we can treat the heat transfer as lumped (see Section 5.1 on page 129).

Generating and selecting among alternate solutions *1) What solutions are possible?* For radiative heat transfer, the solutions we have available are shown in the solution chart in Figure 8.27. *2) What approach is likely to work?* Following the solution chart in Figure 8.27, we see this is not an enclosure problem, but a problem involving pairs of surfaces; thus, we stay on the left side. For two surface problems, the gross simplification of $h_r = \sigma 4 T^3$ should not be applied unless we know for sure that the temperatures are really close. Here we have large variations in temperatures (30°C to −133°C) and therefore should not use the gross simplification. We therefore use the two surface exchange formulas with energy balance.

Implementing the chosen solution 1) We choose the entire leaf as the domain, although energy exchange is only at the surface. Energy balance, with the leaf as the domain, is written as

$$
\underset{\text{In}}{\text{Energy}} - \underset{\text{Out}}{\text{Energy}} + \underset{\underset{\substack{\text{radiation absorbed} \\ \text{at surface only}}}{\underbrace{\text{Generation}}}}{\text{Energy}}^{\,0} = \underset{\underset{\text{steady state}}{\underbrace{\text{Storage}}}}{\text{Energy}}^{\,0} \tag{8.31}
$$

The energy generation term can be non-zero if some of the radiation incident on the leaf penetrates into the leaf. Here it is assumed that all radiation incident on the leaf surface is absorbed at the surface, leading to zero heat generation in the domain. The relevant two-surface exchange formula for black surfaces is (Eq. 8.21)

$$
q_{1-2} = \sigma A_1 F_{1-2} \left(T_1^4 - T_2^4 \right)
$$

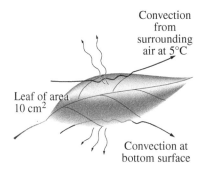

Figure 8.23: Schematic of radiation exchange of a leaf with the sky and the ground.

We now follow the numbered steps shown in Section 8.9 for performing energy balance. Let T_l be the leaf temperature in K (one temperature for the entire domain). The various terms are now calculated as

Radiative heat exchange between the leaf and the sky

$$= \sigma A F_{leaf-sky} \quad \underbrace{(T_l^4 - T_{sky}^4)}_{\text{energy from leaf to sky}}$$

Radiative heat exchange between the leaf and the ground

$$= \sigma A F_{leaf-ground} \quad \underbrace{(T_l^4 - T_{ground}^4)}_{\text{energy from leaf to ground}}$$

Convective heat exchange between both surfaces of the leaf and air

$$= 2hA \quad \underbrace{(T_l - T_{air})}_{\text{energy from leaf to air}}$$

Note that all three exchange formulas written above are heat flows and they have their implied directions. Since they are all from the leaf, they are *out* terms. The view factor $F_{leaf-sky} = 1$ since it is assumed that all of the radiation coming from the leaf's top surface would reach the sky. Likewise, $F_{leaf-ground} = 1$ since it is assumed that all of the radiation coming from the leaf's bottom surface would reach the ground. Substituting in Eq. 8.31, we get

$$\underbrace{0}_{\text{Energy in}} - \underbrace{(5.676 \times 10^{-8})(10 \times 10^{-4})(T_l^4 - 140^4)}_{\text{Energy out, radiative (to sky)}}$$

$$- \underbrace{(5.676 \times 10^{-8})(10 \times 10^{-4})(T_l^4 - 303^4)}_{\text{Energy out, radiative(to sky)}}$$

$$- \underbrace{2(10)(10 \times 10^{-4})(T_l - 278)}_{\text{Energy out, convection}} = 0$$

There is no term corresponding to "Energy In" in Eq. 8.31. This is due to the formula for *net* heat exchange being used here for the radiative heat loss term. The leaf temperature T_l can be solved from the above equation by trial and error, and its value is 271.96 K or −1.04°C.

2) If convective heat exchange is considered from only one surface, Eq. 8.31 is modified to

$$0 \quad - \quad (5.676 \times 10^{-8})(10 \times 10^{-4})(T_l^4 - 140^4)$$
$$- \quad (5.676 \times 10^{-8})(10 \times 10^{-4})(T_l^4 - 303^4)$$
$$- \quad (10)(10 \times 10^{-4})(T_l - 278) \ = \ 0$$

This leads to a leaf temperature of $-4.22°C$.

Evaluating and interpreting the solution *1) Does the temperature make sense?*
When convective heat transfer is considered from only one surface, the leaf becomes colder, since it is unable to receive as much convective heat from the air (i.e., air is warming it). This tells us temperatures are in the right direction. *2) What do we learn?* The area term drops out since it is a steady-state problem (no transient term with heat capacity) and both convective and radiative exchanges are over the same area.

8.7.4 *Example: Radiative Heat Transfer in an Incubator (Relates to Example 6.6.10 on Page 202)*

Recall the problem described in Example 6.6.10 (see Figure 8.24 with additional information). Although the baby is 40 cm long and the incubator is 70 cm long, assume that the view factor formula provided below, true for radiative exchange from an infinitely long cylinder (baby) to a non-concentric infinitely long cylindrical surface (incubator), is applicable for this situation. Thus, exchange from the cylinder ends is ignored. Ignore any exchange with the region labeled "bed." The average emissivity of the body surface is 0.95 and that of the incubator surface is 0.94. The mean surface temperature of the body is 35.5°C and the incubator internal surface temperature is 25°C. 1) The radiative heat transfer coefficient between the baby and the incubator internal surface is defined by $q_{1-2} = h_r A_1 (T_1 - T_2)$. Obtain the radiative heat transfer coefficient. 2) Obtain the ratio of the radiative heat transfer coefficient to the convective heat transfer coefficient calculated in Example 6.6.10 and comment on the relative importance of the two modes of heat transfer in this situation.

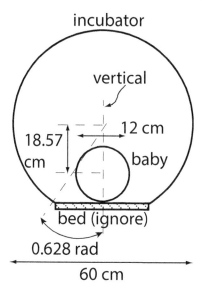

incubator

vertical

18.57 cm

12 cm

baby

bed (ignore)

0.628 rad

60 cm

Figure 8.24: Schematic of an incubator with the baby placed in supine position.

$$F_{1-2} = \frac{1}{2\pi} \left\{ \frac{\alpha_2 - \alpha_1}{2} + \tan^{-1} \left[\left(\frac{1+E}{1-E} \right) \tan \frac{\alpha_2}{2} \right] - \tan^{-1} \left[\left(\frac{1+E}{1-E} \right) \tan \frac{\alpha_1}{2} \right] \right\}$$

where, as shown in Figure 8.25, α_1 and α_2 are in radians, and $E = e/R$.

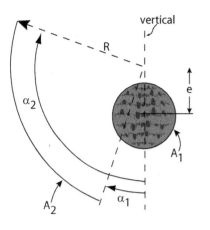

Figure 8.25: Schematic for view factor from an infinitely long cylinder to non-concentric cylindrical enclosure (Feingold and Gupta, 1970).

Solution

Understanding and formulating the problem *1) What is the process?* Radiative heat exchange between the baby's skin surface and the incubator's inside surface. Temperatures of the radiative surfaces are provided. *2) What are we solving for?* Radiative heat transfer coefficient, which is a simplified parameter for describing the net radiative heat exchange. *3) Schematic and given data:* A schematic is shown in Figure 8.24 with some of the given data superimposed on it. *4) Assumptions:* The two surfaces are infinitely long, perfect cylinders that match the geometry for the available view factor formula given by Figure 8.25.

Generating and selecting among alternate solutions *1) What solutions are possible?* For radiative heat transfer, the solutions we have available are shown in the solution chart in Figure 8.27. *2) What approach is likely to work?* As we follow the solution chart in Figure 8.27, since the portion labeled "bed" is to be ignored, the problem can be treated as an enclosure problem with the heating surface enclosing the baby's skin surface. Another hint for using an enclosure solution as opposed to two surfaces is that the two-surface problem dealt with in this chapter are for black only, whereas the surfaces in this problem are grey (emissivity provided). Thus, it is a two-surface enclosure problem (to do multi-surface, all surfaces need to be black to use the formulas we have available). We have not been told to make the assumption of $A_1 \gg A_2$, so we cannot make it.

Implementing the chosen solution The two-surface enclosure formula is given by

$$q_{bi} = \frac{\sigma \left(T_b^4 - T_i^4 \right)}{\frac{1-\varepsilon_b}{\varepsilon_b A_b} + \frac{1}{A_b F_{bi}} + \frac{1-\varepsilon_i}{\varepsilon_i A_i}}$$

From the given data,

$$
\begin{aligned}
\varepsilon_b &= 0.95 \\
\varepsilon_i &= 0.94 \\
T_b &= 35.5°C = 308.5 \text{ K} \\
T_i &= 25°C = 298\text{K} \\
A_b &= \pi D_b L = \pi \,(0.12)\,(0.4) \text{ m}^2 \\
&= 0.151 \text{ m}^2 \\
A_i &= \pi D_i L = \pi \,(0.6)\,(0.7) \text{ m}^2 \\
&= 1.319 \text{ m}^2
\end{aligned}
$$

For view factor F_{b-i}, we have the equation provided in the problem description. To calculate it, first we find $E = 18.57 \text{ cm}/30 \text{ cm} = 0.619$ and plug other geometry data into the equation:

$$
\begin{aligned}
F_{b-i} &= \frac{1}{2\pi} \left\{ \frac{\alpha_2 - \alpha_1}{2} + \tan^{-1} \left[\left(\frac{1+E}{1-E} \right) \tan \frac{\alpha_2}{2} \right] - \tan^{-1} \left[\left(\frac{1+E}{1-E} \right) \tan \frac{\alpha_1}{2} \right] \right\} \\
&= \frac{1}{2\pi} \left\{ \frac{5.655 - 0.628}{2} + \tan^{-1} \left[\frac{1+0.619}{1-0.619} \tan \frac{5.655}{2} \right] \right. \\
&\quad \left. - \tan^{-1} \left[\frac{1+0.619}{1-0.619} \tan \frac{0.628}{2} \right] \right\} \\
&= \frac{1}{2\pi} \left\{ 2.5135 + \tan^{-1} \left[4.249 \, \tan (2.827) \right] - \tan^{-1} \left[4.249 \tan (0.314) \right] \right\} \\
&= \frac{1}{2\pi} \left\{ 2.5135 + \tan^{-1} \left[-1.3826 \right] - \tan^{-1} \left[1.3798 \right] \right\} \\
&= \frac{1}{2\pi} \left\{ 2.5135 + \tan^{-1} \left[-\tan (0.9446) \right] - \tan^{-1} \left[1.3799 \right] \right\}
\end{aligned}
$$

Since α_2 is in the fourth quadrant, $\alpha_2/2$ is in the second quadrant. Using this information for the first inverse,

$$
F_{b-i} = \frac{1}{2\pi} \left\{ 2.5135 + \tan^{-1} \left[\tan (\pi - 0.9446) \right] - \tan^{-1} \left[1.3799 \right] \right\}
$$

$$
F_{b-i} = \frac{1}{2\pi} \left\{ 2.5135 + \pi - 0.9446 - 0.9437 \right\} = 0.6
$$

Plugging in,

$$
\begin{aligned}
q_i &= \frac{5.67 \times 10^{-8} \frac{W}{m^2 K^4} \left[(308.5)^4 - (298)^4 \right] K^4}{\frac{1-0.95}{0.95 \times 0.151 m^2} + \frac{1}{0.6 \times 0.151 m^2} + \frac{1-0.94}{0.94 \times 1.319 m^2}} \\
&= 5.80 \text{ W}
\end{aligned}
$$

The radiative heat transfer coefficient is defined analogously to the convective heat transfer coefficient,

$$
q_{b-i} = A_b h_r (T_b - T_i)
$$

Using this definition of h_r,

$$
\begin{aligned}
h_r &= \frac{q_{bi}}{A_b (T_h - T_i)} = \frac{5.80 \text{ W}}{0.151 \text{ m}^2 \times (308.5 - 298) \text{ K}} \\
&= 3.66 \frac{W}{m^2 K}
\end{aligned}
$$

2) The ratio of the radiative heat transfer coefficient calculated here to the convective heat transfer coefficient between the baby's surface and surrounding air calculated in Example 6.6.10 on page 202 is 3.66/2.69 = 1.36. Thus, radiative heat transfer is more important in this situation than convective heat transfer.

Evaluating and interpreting the solution *1) Does the computed value (radiative heat transfer coefficient) make sense?* If we go by Figure 1.9 on page 13, we see that radiative heat loss is generally comparable to convective heat loss. The fact that we found the radiative heat transfer coefficient value similar to the convective heat transfer coefficient gives us some confidence in our computation. *2) What do we learn?* As previously stated, radiative heat loss is an important mechanism for our thermoneutrality.

8.7.5 Radiative Exchange between a Human (or Animal) and Its Surroundings

As mentioned in Chapter 4, body heat loss by radiation is an important mechanism for maintaining a thermal balance and comfort. Figure 8.26 shows the (shaded) zone for human thermal comfort as related to surrounding wall temperatures. The mean radiant temperature (MRT) is an average temperature of the walls that produces an approximately similar radiative effect. The higher the MRT value, the lower the corresponding air temperature required for comfort. Note that, as the MRT increases, a lower air temperature is required to increase convective heat loss and thus maintain thermal comfort.

8.8 Chapter Summary—Radiative Energy Transfer

- ### Electromagnetic Radiation (page 263)

  1. Electromagnetic waves travel at the speed of light. Waves having a wavelength of 0.1–100 microns are referred to as thermal radiation.

  2. The waves are reflected, absorbed, and sometimes transmitted at a surface. The reflectivity, absorptivity, and transmissivity are material properties.

- ### Radiation Emitted by a Body (pages 267 and 271)

  1. All bodies above absolute zero emit radiation. The total radiation emitted by an ideal body called a blackbody is given by Eq. 8.8. The radiation emitted in a given range of wavelength is given by Eq. 8.13.

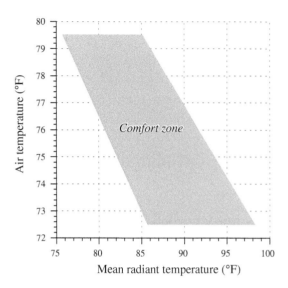

Figure 8.26: Human thermal comfort and mean radiant temperature. Data for lightly clothed subjects engaged in sitting activities. Relative humidity is 50% and air velocities range from about 15 to 60 fpm (0.076 to 0.305 m/s) for the data. Adapted from EGAN, CONCEPTS IN THERMAL COMFORT., 1st Ed., ©1976. Reprinted by permission of Pearson Education, Inc., New York, New York.

2. Emission from a real body is less than that from an ideal body. It is calculated by multiplying the emission from an ideal body by a surface property called the emissivity of the real body. A polished body has a lower emissivity. Biomaterials with lots of water have emissivity close to 1.

- ## Solar and Atmospheric Radiation (page 275)

  1. The spectral distribution of solar radiation is given by Figure 8.17. The peak solar radiation is in the visible region. Ozone in the atmosphere attenuates the harmful ultraviolet rays before they reach the earth's surface.

  2. Earth's atmosphere allows shortwave radiations from the sun to reach the earth's surface but blocks the longwave radiation emitted by the earth's surface from leaving the atmosphere. This trapping of energy is termed the greenhouse effect.

- **Radiative Exchange (page 278)**

    1. When two bodies are exchanging radiation, only a fraction of radiation from one body reaches the other and vice versa. The net exchange of energy is given by equations such as Eq. 8.21 and 8.24.

    2. Radiation is a major mechanism for losing body heat to maintain thermoregulation in mammals.

8.9   Problem Solving—Radiation

Types of problems we can solve:

▶ Calculate emission and absorption at a surface

- Total radiation emitted by a body (*not* exchanged with other bodies) $= \epsilon \sigma T^4$. For an ideal body, $\epsilon = 1$.

- The fraction of emission between wavelengths λ_1 and λ_2 is given by $F_{0-\lambda_2 T} - F_{0-\lambda_1 T}$; This is true for a blackbody having emissivity of 1 or for any surface as long as its emissivity does not change with wavelength. The F value is read from Table 8.2.

- The wavelength at which the maximum emission for a blackbody at temperature T occurs is calculated from $\lambda_{\max} T = 2897.6 \ \mu$mK.

- $\rho + \alpha + \tau = 1$ and $\alpha = \epsilon$ are always true; when $\tau = 0$, $\rho + \alpha = 1$ and thus if α is given, since $\epsilon = \alpha$, all properties are known.

- If absorption is not all on the surface, information on the decay constant must be provided that you can use to consider exponential decay. Exponential decay is considered in problems where temperature as function of position is desired.

▶ Calculate energy exchange by radiation

- Problems in radiative exchange are considered as being either between two surfaces not forming an enclosure or they are form an enclosure. Figure 8.27 shows how the radiative exchange formulas covered in this book are related. Use this to decide the right formula to use.

- View factors needed in formulas are available from different sources.

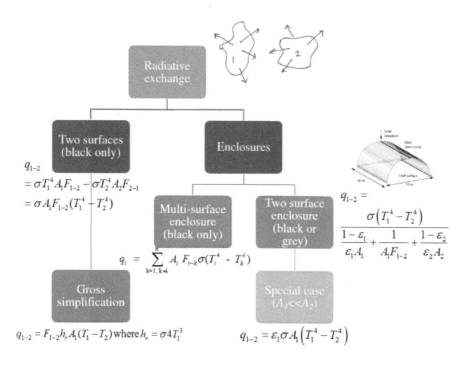

Figure 8.27: Problem solving in radiative exchange.

►**Peform energy balance that includes radiative exchange.**   Below is one step by step approach to do energy balance problems including radiation and convection:

1. Define the domain over which to perform energy balance.

2. Set up energy balance In − Out + Gen = Change in Storage. Generally, these are steady–state problems, making change in storage zero.

3. Note how many distinct convective and radiative exchange surfaces are present within the domain.

4. For each radiating surface, identify the surfaces it exchanges radiation with and write the exchange formula, noting whether they are in or out (amount $q_{12} = \sigma A_1 F_{1-2} \left(T_1^4 - T_2^4 \right)$ is going surface 1 to surface 2, i.e., it is out for surface 1 while in for surface 2).

   • Which exchange formula to use can be decided based on the flow chart in Figure 8.27.

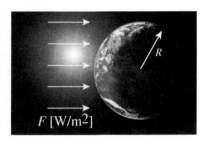

Figure 8.28: Total energy incident on a curved surface. Total energy incident on a sphere [W] when radiation is coming from one direction is equal to $F[\text{W/m}^2]\pi R^2[\text{m}^2]$. (Alexey Repka—Shutterstock.com)

- Obtain the view factor (from this book or elsewhere), if needed.

5. Include any energy incident on a surface that is not already considered using exchange formulas. On a curved surface, effective incident energy is the flux multiplied by the projected area (see Figure 8.28)

6. Write the amount of convective exchange from *each* surface, making sure of proper direction (amount $q_{1-2} = h(T_1 - T_2)$ is going 1 to 2, i.e., it is out for 1 while in for 2).

7. Include the radiative and convective exchange amounts in the energy balance equation in Step 2.

8. Solve the energy balance equation for the required parameter

8.10   Concept and Review Questions

1. Electromagnetic radiation is characterized by (a) frequency, (b) wavelength, and (c) propagation velocity. Define these characteristics and indicate which ones are independent of the substance through which the radiation is transmitted.

2. What is solar radiation? Does it contain components of microwave and radio-wave radiation?

3. List several important biological effects of (a) visible light and (b) ultraviolet radiation.

4. Explain how it is possible for an individual to get a sunburn on a cloudy day, even though the skin actually feels cool.

5. What are microwaves and what role do they play in heat transfer processes that occur in microwave ovens?

6. Examine a microwave oven door and explain how it is possible to see through the window during the cooking process without being harmed by the microwave radiation.

7. Explain why the sky appears blue and the sun appears yellow.

8. What are the main factors responsible for attenuating the irradiation from the sun that reaches the surface of the earth?

9. What is the solar constant?

10. How does thermal radiation differ from other types of electromagnetic radiation?

11. What is the Stefan–Boltzmann law?

12. What is a gray body?

13. What is meant by the radiation shape factor?

14. What is meant by the atmospheric greenhouse effect?

Further Reading

Albright, L. D. 1990. *Environmental Control of Animals and Plants*. American Society of Agricultural Engineers, St. Joseph, Michigan.

Bohm, M., A. Browen, and O. Noren. 1991. Thermal environment in cabs. Presented at the International Meeting of ASAE, Dec. 17–20, Chicago, IL.

Brunt, D. 1940. Radiation in the atmosphere. *Quarterly Journal of the Royal Meteorological Society* 66(Suppl.):34–40.

Carr, K. L. 1989. Microwave radiometry: Its importance to the detection of cancer. *IEEE Transactions on Microwave Theory and Techniques* 37(12):1862–1868.

Coulson, K. L. 1975. *Solar and Terrestrial Radiation*. Academic Press, New York.

Feingold, A. and Gupta, K. G., 1970, New analytical approach to the evaluation of configuration factors in radiation from spheres and infinitely long cylinders, *J. Heat Transfer* vol. 92, no. 1, pp. 69–76.

Garzoli, K. V. and J. Blackwell. 1987. An analysis of the nocturnal heat loss from a double skin plastic greenhouse. *Journal of Agricultural Engineering Research* 36:75–85.

Gast, P. R. 1965. Solar electromagnetic radiation. In *Handbook of Geophysics and Space Environments*. Edited by S. L. Valley. McGraw-Hill, Inc., New York.

Guyot, G. 1998. *Physics of the Environment and Climate*. Praxis Publishing, Chichester, UK.

Hejazi, S., D. C. Wobschall, R. A. Spangler, and M. Anbar. 1992. Scope and limitations of thermal imaging using multiwavelength infrared detection. *Optical Engineering* 31(11):2383–2392.

Iberall, A. S. and A. M. Schindler. 1973. On the physical basis of a theory of human thermoregulation. *Trans. ASME, J. Dynamic Systems Measurement and Control* 95:68–75.

Incropera, F. P. and D. P. Dewitt. 1996. *Fundamentals of Heat and Mass Transfer.* John Wiley & Sons, New York.

Jacob, D. J. 1999. *Introduction to Atmospheric Chemistry.* Princeton University Press, Princeton, NJ.

Jones, B. S., W. F. Lynn and M. O. Stone. 2001. Thermal modeling of snake infrared reception: Evidence for limited detection range. *Journal of Theoretical Biology* 209(2): 201–211.

Metaxas, A. C. 1996. *Foundations of Electroheat: A Unified Approach.* John Wiley & Sons, Chichester, UK.

Myers, P., N. L. Sadowsky, and A. H. Barrett. 1979. Microwave thermography: Principles, methods, and clinical applications. *Journal of Microwave Power* 14(2): 105–115.

Ramanathan, V. 1987. The role of earth radiation budget studies in climate and general circulation research. *Journal of Geophysical Research* 92(D4):4087–4095.

Roussy, G. and J. Pearce. 1995. *Foundations and Industrial Applications of Microwaves and Radio Frequency Fields.* John Wiley & Sons, New York.

Shitzer, A. and R. C. Eberhart (ed) 1985. *Heat Transfer in Medicine and Biology* (2 volumes). New York: Plenum Press.

Siegel, R. and J. R. Howell. 1981. *Thermal Radiation Heat Transfer.* McGraw-Hill, New York.

Smith, D. M. S., I. R. Noble, and G. K. Jones. 1985. A heat balance model for sheep and its use to predict shade-seeking behaviour in hot conditions. *Journal of Applied Ecology* 22(3):753–774.

Steketee, J. 1973. Spectral emissivity of skin and pericardium. *Phys. Med. Biol* 18(5):686–694.

ten Berge, H. F. M. 1990. Heat and water transfer in bare topsoil and the lower atmosphere. Center for Agricultural Publishing and Documentation, Wageningen, the Netherlands.

Wathen, P., Mitchell, J. W., Porter, W. P., 1971. Theoretical and experimental studies of energy exchange from jackrabbit ears and cylindrically shaped appendages. *Biophysical Journal* 11(12), 1030.

8.11 Problems

8.1 Solar Energy Flux

Assume an average surface temperature of the sun to be 5800 K. What is the energy flux at the surface of sun? Compare this to the average solar flux just outside the earth's atmosphere, 1353 W/m^2, and explain the difference.

8.2 Spectral Emission from the Sun

The sun can be treated as a blackbody at a temperature of 5800 K. What is the wavelength λ_1 below which the lower 10% of the solar emission is concentrated? What is the wavelength λ_2 above which the upper 10% of the solar emission is concentrated? Determine the maximum spectral emission power of the sun and the wavelength at which this emission occurs.

8.3 Fraction of Solar Energy for Photosynthesis

Solar radiation between the wavelengths 0.35 μm and 0.70 μm is useful for photosynthesis. Consider the sun to be a blackbody at 5800 K. 1) What fraction of total energy radiated from the sun is useful for photosynthesis? 2) Is this fraction the same as the fraction of the solar energy incident on the earth's surface that is useful for photosynthesis?

8.4 Extraterrestrial Solar Radiation

Assume the sun to be a blackbody. Estimate the solar energy flux incident just outside the earth's atmosphere from knowing the sun's surface temperature of 5800 K, the sun's radius of 0.695×10^6 km, and the sun-to-earth distance of 1.5×10^8 km.

8.5 Simple Energy Balance of Earth

Consider the earth to be a sphere of average radius 6.38×10^6 m. The solar radiation just outside the earth's atmosphere is 1353 W/m^2 and is incident *only on the plane of the earth facing the sun, in a direction normal to the earth's surface*. About 30% of this energy is reflected from the atmosphere; the rest of it is transmitted into the atmosphere and absorbed by the earth's surface. The radiation from earth (outside of atmosphere) to outer space seems to be from an average surface temperature of 15°C. The emissivity of the earth's surface is approximately 0.97. 1) Calculate the total radiation absorbed by the earth's surface from the sun, in watts. 2) Calculate the total energy lost to outer space by means of radiation from the earth. 3) Compare the total radiation received by the earth from the sun with the total radiation lost from the earth by finding the ratio of the two quantities you just calculated.

8.6   Human Radiation

The radiation emitted from the human body has been used for a variety of applications ranging from military night vision devices to diagnostic medical imaging. For thermoregulation, i.e., maintaining body temperature, human beings have to lose energy. Consider a standing person (do not consider clothes) in the center of a *large* room. Approximate the person as a 1.8 m tall vertical cylinder with a 0.3 m diameter. Heat loss from the ends of the cylinder can be neglected. The average surface temperature of the person is 33°C, the room surface temperature is 20°C, and the emissivity of the skin surface is 0.95. Consult Appendix C.8 (page 568) for the properties of air. 1) Calculate the net radiative heat loss from this person. 2) What is the wavelength, λ_{max}, at which the maximum amount of energy will be radiated from the person? 3) What region of the electromagnetic spectrum does the wavelength calculated in step 2 primarily fall in? 4) If a detector is available that can detect only in the wavelength range $\lambda_{max} \pm 5$ μm, what fraction of the total energy from the human being will this detector be sensitive to? 5) Calculate the convective heat loss from this person to the room due to natural convection. *(Note: For the vertical surface of the cylinder, formulas for a vertical plate can be used.)* 6) Compare the convective heat loss with the radiative heat loss (from step 1) and state if they are close.

8.7   Radiative Heating in a Greenhouse

The degree of heat delivery from a steam heating system can vary based on different conditions of the surface that lead to different emissivities. Consider a steam heating pipe, having a surface temperature of 95°C, in a large greenhouse. The surface has been painted with an aluminized paint having an emissivity of 0.45. 1) Calculate the radiant flux leaving the heated surface of the pipe. 2) How much would the flux change if the pipe were repainted with a layer of oil-based paint having an emissivity of 0.95? (Albright, 1990)

8.8   Heat Loss from Animals in a Barn

Consider a pig alone in a large barn exchanging thermal radiation with its inside walls. The surface area of the pig can be approximated to be 2 m^2, with a skin temperature of 35°C and a skin emissivity of 0.90. The mean radiant temperature of the inside walls of the barn is 5°C. 1) Calculate the *net* radiative heat exchange (loss) from the pig. 2) Calculate the thermal radiation emitted by the pig and explain the difference between it and what is calculated in step 1 (Albright, 1990).

8.9   Radiative Cooling of Orange Groves (Variation of Example 8.7.3)

Consider a flat orange leaf of surface area 10 cm^2 parallel to the ground. The leaf absorbs all incident radiation and its emissivity is 1. The top surface of the leaf radiates to the clear night sky, and the effective radiation temperature of the sky is assumed to be 140 K. The bottom surface of the leaf radiates to the ground and the ground temperature is 7°C. Both the top and bottom surfaces are also exchanging heat with the air through convection. The air temperature is 5°C. Consider the leaf to be very thin so that the entire leaf (including the top and the bottom surfaces) is at one temperature. 1) By making an energy balance on the leaf involving the convective and the radiative heat transfer, write an equation from which you can solve for the steady-state leaf temperature. 2) For a convective heat transfer coefficient of 25 W/m^2·K, calculate the leaf temperature. *(Note: Solution requires iteration.)* 3) Compare your leaf temperature with that obtained in Example 8.7.3 and comment on what would be the simplest or least expensive solution in the orange grove of Example 8.7.3 to prevent the leaf from freezing.

8.10   Heat Balance on Greenhouse Glass

Due to the selective absorption by the glass roof and walls of a greenhouse, as shown in Figure 8.30, more of the solar radiation comes inside than leaves the greenhouse, effectively trapping the energy and raising the inside temperature. Some gases in the atmosphere around the earth have a similar effect as the glass and lead to an increase in the temperature of the earth's surface, thus deriving the name "greenhouse gases." In this problem, we would like to calculate the temperature of the greenhouse glass reached at steady state by the greenhouse glass due to a balance of all energy exchanges.

Assume a section of glass of a greenhouse has the property that it absorbs no radiation (transmissivity of 1) for $\lambda < 1$ μm and absorbs all of the radiation (with zero reflectivity) for $\lambda > 1$ μm, as shown in Figure 8.29. Its emissivity is 1 in the wavelength

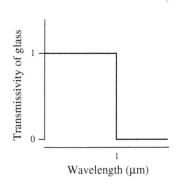

Figure 8.29: Transmissivity of glass for Problem 8.10.

region $\lambda > 1$ μm. This glass receives solar radiation at 1000 W/m^2, atmospheric radiation at 200 W/m^2, and radiation emitted from the interior surfaces (plants, walls, etc.) at 500 W/m^2. The solar radiation can be assumed to be that from a blackbody at 5800 K. The radiation from the atmosphere, the interior surfaces, and the glass are concentrated in the far infrared region ($\lambda > 8$ μm) of the spectrum, i.e., no energy is emitted below 8 μm. Both sides of the glass also exchange energy via convection (net energy loss) with the surrounding air. The heat transfer coefficient on the outside surface and the outside air temperature are 50 W/m^2·K and 20°C, respectively. The heat transfer coefficient on the inside surface and the greenhouse interior air temperature are 15 W/m^2·K and 25°C, respectively. In order to perform an energy balance on the glass, 1) calculate the energy coming into the glass. 2) Now, assuming that both sides of the glass are at the same temperature, write an expression for the energy leaving the glass. 3) Calculate the glass temperature at steady state. *(Note: Solution requires iteration.)*

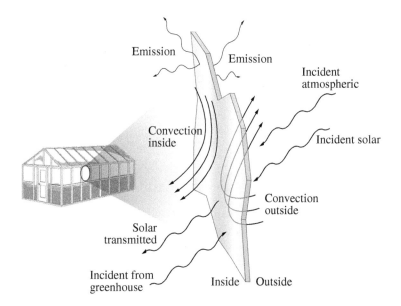

Figure 8.30: Schematic of various energy exchanges on a greenhouse glass for Problem 8.10.

8.11 Steady-State Glass Temperature

Consider solar radiation incident on a glass plate with air on both sides. The solar radiation can be assumed to be that from a blackbody at 5800 K temperature. The glass absorbs all radiation between 0.3 μm and 3 μm. Outside this range of wavelengths, all radiation is transmitted through the glass without any absorption. The radiation incident to the glass is 700 W/m^2 and the ambient air temperature is 20°C. One side of the glass exchanges radiation with the ground (also at 20°C) and all radiation from this glass surface reaches the ground. The other side of the glass sees the sky but the radiation *from the sky* is ignored. 1) Calculate the solar energy that is absorbed in the glass, in W/m^2. 2) Write an energy balance for the glass at steady state, from which you can calculate the steady-state temperature. 3) For a convective heat transfer coefficient of 50 W/m$^2 \cdot$ K, what is the steady-state temperature? *Note: Solution requires iteration.*

8.12 Thermocouple Measurement Error Due to Radiation

A thermocouple, a temperature measuring device, can introduce measurement errors due to its radiative exchange with the surroundings. Consider a cylindrical thermocouple sheath that is 4 mm in diameter and with a surface emissivity of 0.5. It is kept horizontal during the measurement in a large room where the air can move only by natural convection. The air temperature is 25°C and the mean radiant temperature of the wall surfaces is 35°C. For heat transfer coefficient calculations only, assume a thermocouple temperature of 30°C. Air properties are provided in Appendix C.8 (page 568). 1) What temperature will the thermocouple indicate, i.e., what is the temperature of the thermocouple? 2) What is the measurement error, i.e., the difference between indicated temperature and true air temperature?

8.13 Solar Energy Received by a Shaded Leaf

Consider a small leaf completely shaded by a large one, as shown in the schematic in Figure 8.31. The idealized absorptivity of the large leaf is shown in Figure 8.32. 1) What fraction of solar radiation (sun's surface temperature is approximately 5800 K) reaches the bottom leaf? 2) Comment on how useful the solar radiation reaching the bottom leaf is for the purpose of photosynthesis (see absorptivity of chlorophyll in Figure 8.6 in text). 3) If the temperature of the large leaf is 35°C, what is the total amount of radiation emitted from the large leaf? *(Note that it is the total leaving and* **not** *the net amount leaving due to exchange with other bodies.)*

Figure 8.31: Quality of and quantity of energy reaching a shaded leaf.

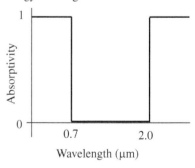

Figure 8.32: Idealized spectral absorptivity

8.14 Combined Radiation and Natural Convection inside a Greenhouse

Side view

Sunshine

45°

Natural
Convection

Top view

Sunshine

Projection
plane

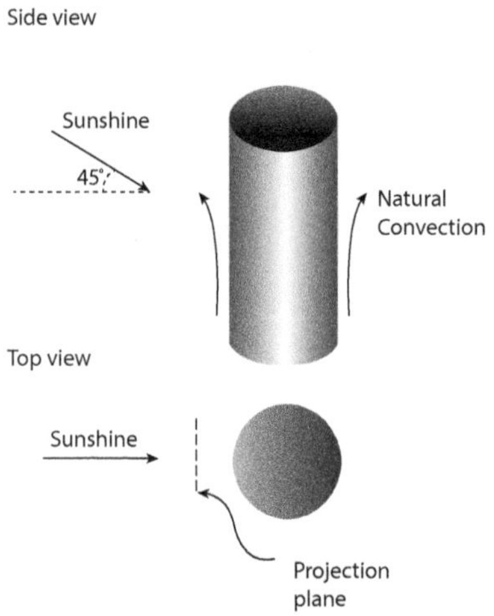

Figure 8.33: Schematic for Problem 8.14.

You are standing vertically surrounded by glass in a nice greenhouse on an otherwise cold (but sunny) wintry day. Approximate yourself as a vertical cylinder with a diameter of 30 cm and a height of 1.8 m. Sunshine (see schematic in Figure 8.33) is providing radiative energy at 700 W/m² (outside the greenhouse) at an angle 45 degrees to the horizontal onto one side of the surface of your body, and only the horizontal component of the sunshine on the projected area is useful for energy balance. Glass absorbs all energy in the wavelengths above 1 μm. The solar radiation can be assumed to come from a blackbody at 5800 K. The absorptivity and emissivity of your body and the glass surfaces are 1.0, i.e., they are blackbodies at the given temperatures. Inside the greenhouse the air is moving a bit due to natural convection. Assume $D/L > 35/\mathrm{Gr}_L^{1/4}$ is true for formulas in natural convection. Your body surface temperature is 35°C and the glass temperature surrounding you is −10°C. The inside air temperature is 25°C. Neglect any heat exchange through the ends of the cylinder. Other needed data are: thermal expansion coefficient of 0.0033/K, density of air of 1.18 kg/m³, viscosity of air of 1.85×10^{-5} kg/m · s, thermal diffusivity of 2.25×10^{-5} m²/s, and thermal conductivity of air of 0.0265W/m · K. By doing an energy balance on the body surface, find the net heat exchange from the body surface in watts. Don't forget to indicate if this is a net loss or gain.

8.15 Radiative Energy Transfer in an Incubator

Radiative heat transport can be an important component in an infant incubator. Consider an infant lying on its back with its back surface being warmed by a radiant heater. Both the top of the infant (stomach side) and radiant heater surface can be approximated as long and flat parallel surfaces, as shown in the schematic below. Assume the emissivity of both the radiant heating surface and the baby surface is 1, i.e., they are black bodies. The formula for the shape factor calculation is below. Air in the incubator has only a slight movement, leading to a heat transfer coefficient of 5 W/m²K. Assume the infant surface has a temperature of 35°C. The radiative heating surface has an effective temperature of 90°C, and the air temperature in the incubator is 23°C. The width of the heating surface and baby surface are 30 cm and 15 cm, respectively. 1) What is the radiative heat flux (*not the net*) from the top plane surface of the infant? 2) If we want to have a *net radiative heat* flux of 200 W/m² into the infant top surface from the incubator, what should the position (distance L) of the radiative surface be? You do not need to solve for L, but clearly show an equation in which L is the only unknown and units of all quantities are properly considered. 3) What is the convective heat flux from the infant top surface into the surrounding air? 4) Considering both convection and radiation, what is the *net* heat flux into the infant top surface? 5) What important mode of heat transfer is being ignored here?

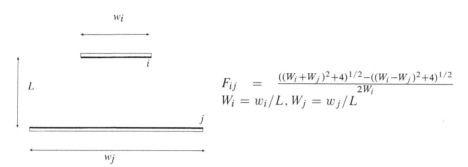

$$F_{ij} = \frac{((W_i+W_j)^2+4)^{1/2}-((W_i-W_j)^2+4)^{1/2}}{2W_i}$$
$$W_i = w_i/L, \, W_j = w_j/L$$

Figure 8.35: View factor equation for parallel plates with midlines connected by perpendicular

Figure 8.34: Baby in an incubator. (Sayanjo65—Shutterstock.com)

8.16 Greenhouse Energy Balance

A greenhouse that has its bottom surface (floor) covered with leafy plants is shown as a schematic in Figure 8.36. The glass dome of the roof is semi-circular and the greenhouse *can be considered long* in the horizontal direction (actual length 30 m). Incident vertical solar radiation is 1000 W/m^2. Solar radiation can be assumed to be coming from a blackbody of 5800 K. Ignore any other atmospheric radiation. Transmissivity of the glass as a function of wavelength is given by Figure 8.37 and has no directional dependence. Reflectivity of the glass is considered zero at all wavelengths. The energy exchange between the leafy plants and the inside surface of the glass is given by the equation for a two-surface enclosure. For this particular exchange, the glass is effectively a blackbody. The leafy plants are arranged in such a way that the area of the plants can be approximated to be the same as the area of the greenhouse floor. The emissivity of the surface of the plants is 0.95. The inside surface of glass has a heat transfer coefficient of 5 W/m$^2 \cdot$ K while the outside glass surface has a heat transfer coefficient of 10 W/m$^2 \cdot$ K. The plant surface is at 30°C, the glass is at 45°C, and the outside air is at 25°C. 1) Calculate the amount of radiative heat exchange between these two bodies—glass and the leafy plant floor. Note that the glass property data means that it is effectively a blackbody for this exchange. 2) In order to find the greenhouse air temperature, first write the equation for energy balance for the glass surface, making sure all modes of heat transfer are included. 3) Calculate the greenhouse air temperature from the energy balance.

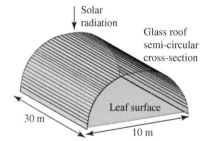

Figure 8.36: Schematic of a greenhouse.

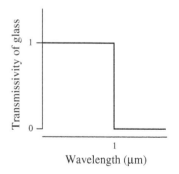

Figure 8.37: Transmissivity of the greenhouse glass.

8.17   Energy Balance on a Surface

A certain manufacturer of a non-stick pan says the non-stick coating will not decompose or release any gases unless heated to above 430°C. Consider the extreme situation when the temperature can get quite high when an empty pan is left on a burner unattended. The burner continuously supplies 500 W to the pan, which sits horizontally on top of it. The diameter of the pan is 20 cm, and its characteristic length, L, is the ratio of its surface area to perimeter. The radiative heat loss from the top surface of the pan can be simplified by assuming that the pan surface and the surroundings are blackbody surfaces with $\epsilon = 1$, and that there is a view factor of 1 between them. The surroundings are at the same temperature as the room air temperature, which is at 25°C. The volumetric expansion coefficient of air is given by $1/T$ where T is the absolute temperature. Other air properties can be found in the appendix.

1) Write the equation for energy balance for the pan from which the temperature, T, of the pan can be calculated. Plug in all known quantities. 2) This problem requires iteration to get the final solution, and you have to assume a surface temperature to get started. Show the first iteration and provide clearly the steps for continuing the iterations without carrying them out.

8.18   Modeling Earth's Greenhouse Effect

We will build a simple model of the greenhouse effect of the atmosphere. Here the atmosphere will be treated as an isothermal layer at temperature T_1 at some small distance from the earth's surface whose temperature is T_0. The solar flux outside the atmosphere is F_s W/m^2 and is incident only on the plane facing the sun. The spectral property of the atmosphere is assumed as follows: For solar radiation, the atmospheric layer has a reflectivity of ρ (includes reflection from earth's surface), and zero absorptivity, both properties averaged over all wavelengths. For radiation emitted by the earth's surface that is incident on the atmosphere, the absorptivity of the atmosphere is α, averaged over all wavelengths. The spectral property of the earth is assumed as follows: For solar radiation, the earth's surface has an absorptivity of 1. Reflectivity of solar radiation from the earth's surface is included in ρ, so no additional reflection needs to be considered. The radius of the earth and the atmosphere can be considered the same (say R). Those interested in learning more can see Jacob (1999).

1) Perform a radiative energy balance of the earth + atmosphere system at the outer layer of atmosphere (at the dashed line as shown in Figure 8.38), considering this system to be at steady state with no other sources of heat generation or transport (conduction or convection). *(Hint: What radiation enters this system (dashed line)? How much radiation from this system leaves the dashed line?)* 2) Perform a radiative energy balance of the atmosphere only, assuming it to be at steady state. *(Hint: How*

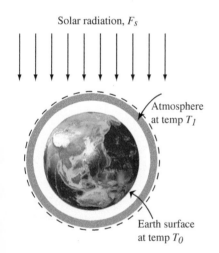

Solar radiation, F_s

Atmosphere at temp T_1

Earth surface at temp T_0

Figure 8.38: A simple schematic of earth and atmosphere system (not to scale) with atmosphere approximated as a layer above earth's surface. (Globe illustration by u3d-Shutterstock.com)

much radiative energy is absorbed by the atmosphere? How much radiative energy is released by the atmosphere?) 3) Combining 1) and 2), calculate the earth's surface temperature when the solar flux is $F_s = 1353$ W/m^2, $\alpha = 0.75$, and $\rho = 0.28$. 4) What would happen to the calculated earth's surface temperature if the absorptivity of the atmosphere, α, increases?

8.19 Effect of Shading of Leaves on Photosynthesis

Figure 8.39a shows two leaves, one completely shaded by the other by being exactly under and within a short distance of it. Radiation in the range 0.4–0.7 μm is useful for photosynthesis; an idealized spectral curve for the leaves is shown in Figure 8.39b. Assuming the sun is a blackbody at 5800 K, what fraction of radiation from solar rays useful for photosynthesis will reach the bottom leaf?

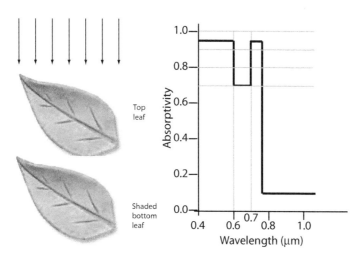

Figure 8.39: a) A leaf completely shading another below it. b) An idealized spectral absorptivity curve for a leaf.

8.20 Radiative Heat Transfer in an Incubator (relates to Example 6.6.10)

Moved to solved problems

8.21   Equation for Thermal Comfort

We would like to make a very simple model to describe the process of thermoregulation whereby the convective and radiative heat losses keep the body at steady state and in thermal comfort. Assume a person is shopping in a mall (walking very slowly), leading to a heat generation of 85 watts per m^2 of body surface. The clothed body has an assumed emissivity of 1 and view factor of 1 with the surrounding walls. The body surface temperature is not very far from the surrounding wall temperature such that the radiative exchange between the body and walls can be written in terms of a radiative heat transfer coefficient. Ignore evaporative heat loss. The average surface area of the body is 1.8 m^2, and the temperature of the clothed body surface is 33°C. The convective heat transfer coefficient is 11.5 W/m$^2 \cdot$ K. 1) Considering only the convective and radiative heat losses from the body, derive the equation that shows the needed air temperature as a function of wall temperature for maintaining the same level of comfort (i.e., losing the same amount of heat) for this activity. Substitute numerical values and simplify as much as possible. 2) Compare the slope of this line (air temperature as a function of wall temperature) with the slope of the line in Figure 8.40 for human thermal comfort zone, as used in practice. 3) What condition would the dashed line signify (qualitative answer only)?

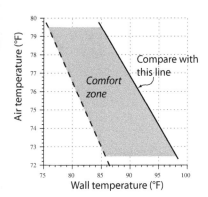

Figure 8.40: Relation between mean wall temperature and surrounding air temperature for human thermal comfort.

8.22   Emission of Earth in the Microwave Band

Assuming the earth is a blackbody with an average surface temperature of 14°C, what is its emissive power (in W/m$^2 \cdot \mu$m) at the mid microwave band of $10^5 \mu$m? If the wavelength band for microwaves is 10^4–$10^6 \mu$m, *very approximately* how much (in W/m^2 of earth's surface) microwaves would earth emit? *(Hint: The blackbody radiation function in the text does not have tabulated values for this wavelength range.)*

8.23   Radiative Exchange of a Leaf with Forest Fire

A forest fire is advancing as a plane, shown as a wall of fire in Figure 8.41, and a leaf is exchanging radiant energy with this wall of fire. On the side of the leaf facing the fire, we will ignore heat exchange with the sky (the wall of fire is very tall). The view factor between a small area and an infinite plane, as shown in Figure 8.42, is $F_{dA1-A2} = (1/2)(1 + \cos \phi)$. The other side of the leaf (facing away from the fire) is exchanging radiation with trees, etc., whose approximate radiant temperature is the same as the surrounding air temperature. The leaf absorbs all electromagnetic radiation emitted by the fire as well as that from the surroundings (i.e., it behaves as a blackbody). Both sides of the leaf are also exchanging convective energy with the surrounding air. The convective heat transfer coefficient for the situation is approximated as 150

W/m$^2 \cdot$ K. The surrounding air temperature is 150°C, and the temperature of the fire wall is approximated as 800°C. A leaf will catch fire when its temperature reaches its ignition temperature, which is 300°C for the particular type of leaf. Assume the leaf to be at steady state.

1) Calculate the maximum angle of the leaf from the plane of the wall of fire, ϕ (degrees), as shown in Figure 8.42, for which it will catch fire. 2) How do you interpret $F_{dA1-A2} = 1/2$ (for $\phi = 90°$) in the view factor equation provided above?

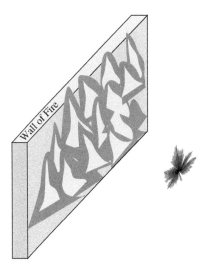

Figure 8.41: Wall of fire and a leaf.

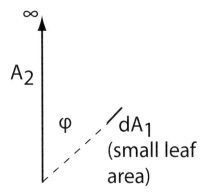

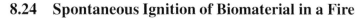

Figure 8.42: Differential element of any length to semi-infinite plane. The plane containing the element and receiving the semi-infinite plane intersect at angle ϕ at the edge of the semi-infinite plane.

8.24 Spontaneous Ignition of Biomaterial in a Fire

Figure 8.43 shows a highly simplified schematic of a wall of fire facing a dense row of trees with leaves on them, separated by a small creek flowing in between. We will consider both the fire and the leaf surfaces facing the fire to be walls of different heights that are vertical, long, and parallel to each other, and assumed to be blackbodies. The walls are close enough that we can ignore radiation with the creek and the sky. Possible view factor formulas are provided in Figure 8.44. The "wall of leaves" is completely surrounded by other leaves (at 80°C) on the side away from the fire, with which it exchanges radiative energy. There is also convective heat exchange of the leaves (both sides) with the surrounding air. The convective heat transfer coefficient between the leaf and the air is approximated at 100 W/m$^2 \cdot$ K. The surrounding air temperature is 100°C and the temperature of the wall of fire is approximated at 800°C. A leaf catches fire when its temperature reaches the ignition temperature of 300°C.

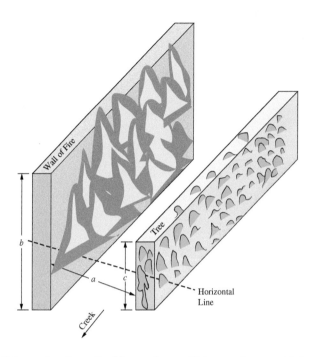

Figure 8.43: Schematic of a wall of fire and a parallel row of trees with leaves on them.

1) Develop the equation for the minimum distance a, in m, to avoid the leaves catching fire when $b = 50$ m and $c = 30$ m. You do not need to solve for the numerical value but your equation should have a as the only unknown. 2) Explain whether or not sufficient wind can prevent the leaves from catching fire. 3) If the wall of fire and the trees are of equal height, everything else staying the same, explain (using the appropriate equation) whether the chances of the leaf catching fire increase or decrease as the tree height (and consequently the wall of fire) increases.

Some relevant formulas for view factor are given by (see Figure 8.44; Here $H = h/w$, $B = b/a$, and $C = c/a$)

$$F_{1-2} = F_{2-1} = \sqrt{1 + H^2} - H \quad \text{Figure 8.44a} \tag{8.32}$$

$$F_{1-2} = \frac{1}{2B}\left(\sqrt{(B+C)^2 + 4} - \sqrt{(C-B)^2 + 4}\right) \quad \text{Figure 8.44b (8.33)}$$

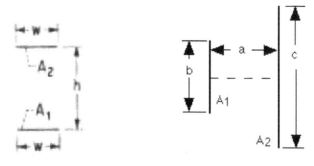

Two infinitely long, directly opposed parallel plates of the same finite width

Two infinitely long parallel plates of different widths; centerlines of plates are connected by perpendicular between plates

(a)

(b)

Figure 8.44: Two view factor formulas that may be relevant. Source: www.thermalradiation.net

8.25 Changing Position of a Jackrabbit Ear Can Make It Feel Different

A jackrabbit uses its ears to increase heat transfer by changing position or location to either encounter warmer air or increase solar radiation. We will make several gross simplifications to perform a simple steady-state heat transfer analysis on a jackrabbit ear (see Figure 8.46). Model the ear surface as a flat plate of dimensions 15.24 cm × 8.16 cm with 0.5 cm thickness where the entire ear volume is at one temperature. Both top and bottom ear surfaces behave as blackbodies. At rest, a jackrabbit ear generates heat at the rate of 1015 W/m^3. At some point during the day, 1000 W/m^2 of solar radiation is incident on the ear at an angle of 72° to the normal to the surface (only the normal component of solar radiation is involved in radiation exchange). Both the top and the bottom surfaces of the ear exchange radiation with both the sky and the ground. View factor information is provided in Figure 8.46. Exchange with the other ear or the body is ignored for simplicity. The ambient air at 21°C is flowing over the plate (and parallel to the long end) at a velocity of 1 m/s. The effective sky temperature is 24°C and the ground temperature is 40°C.

1) For calculations of the convective heat transfer coefficient for flow over the jackrabbit ear, explain what temperature you would ideally use to look up air proper-

Figure 8.45: A jackrabbit showing the ears. (Ingrid Curry–Shutterstock.com)

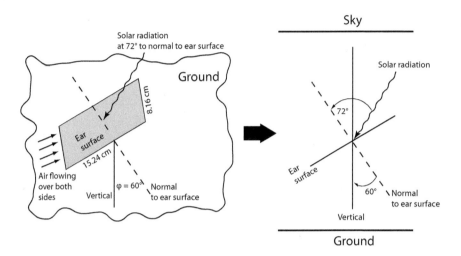

Figure 8.46: The ground and the sky are both considered plane surfaces that are parallel to each other. When the normal to the lower surface of the ear is at an angle of ϕ with the vertical, as shown on the left figure, the view factor between the bottom surface of the ear and the ground is $(1/2)(1 + \cos\phi)$ and the view factor between the bottom surface of the ear and the sky is $(1/2)(1 - \cos\phi)$. The view factors for the top surface of the ear are treated the same way, with ϕ as the angle that the normal to the top surface makes with the vertical.

ties. Using air properties given in the appendix and assuming a reasonable temperature for properties, find this heat transfer coefficient. 2) Perform an energy balance for one ear, making sure you include all forms of energy exchange. 3) From the energy balance, solve for the ear temperature. 4) What would be the new ear temperature when the sun is at an angle of $10°$ to the normal to the top ear surface? 5) Comment on whether the change in ear temperature from step 4 makes sense, i.e., *would there be any biological factor that would work toward reducing such drastic change?*

8.26 Prey Detection by Snakes using Infrared

Snakes sense infrared (often for prey detection) using facial pit organs (Figure 8.47) (Jones et al., 2001). Assume the prey has a surface temperature of 25°C and can be assumed to be a blackbody. One study assumed the atmosphere to be totally transparent to infrared (no absorption) in the 9–11 μm bandwidth and complete absorption (no transmission) outside this bandwidth. Based on this, they estimated the detection range of prey to be only 5 cm, i.e., the snake receives enough infrared from prey up to 5 cm away. Another study disagreed with this and claimed the totally transparent region of the atmosphere to be wider, 8–12.5 μm. Assume the infrared sensitivity of the pit organ is uniform over the entire bandwidth (8–12.5 μm). If the snake's detection range increases proportionally to the amount of infrared radiation the snake is receiving, what will the new prey detection range be, based on the 8–12.5 μm bandwidth of radiation that would be received?

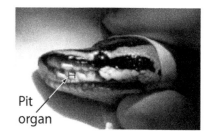

Figure 8.47: Pit organ on a snake that senses infrared radiation.

8.27 Heat Load on a Grazing Sheep

The growth of animals can depend significantly on their ability to balance heat. Consider a grazing sheep approximated as a cylinder, as shown in Figure 8.49a. We would like to estimate the temperature variation along the wool surface by partitioning the sheep body surface (outside layer of the cylinder) into small strips, as shown in Figure 8.49a. We ignore sweating and assume all evaporation occurs in the respiratory tract (this simplification was used in the research paper as the goal was to predict water usage by the sheep for cooling). The wool surface temperature reaches steady state by balancing incoming solar energy with radiative exchange with the ground and effective sky, convection of heat to the surrounding air, and conduction of heat to the body core. All radiative heat exchange surfaces can be considered as planes that are blackbodies. We ignore heat conduction between strips and along the length of the strips. We are interested only in the shaded quadrant of the cylinder shown in Figure 8.49b, i.e., develop the equations below only for this region on the surface.

Figure 8.48: A grazing sheep exchanging thermal radiation. (Simon Annable—Shutterstock.com)

The incident solar radiation is 1360 W/m², and we are considering the instant when solar radiation is at a 45° angle to the vertical. Note that only the component of solar radiation normal to the strip surface is relevant. The view factor between a strip and a plane is given in terms of the angle θ the strip makes with the plane (see Figure 8.49)—for ground, it is $(1/2)(1 + \cos\theta)$ and for sky, it is in terms of the corresponding angle, $(180 - \theta)$. Conduction from the wool surface into the body can be estimated as $6.4(T - T_b)$ W/m² where T is the wool surface temperature and $T_b = 39$ is the body core temperature, both in °C. The surrounding air temperature is 30°C and the convective heat transfer coefficient is 50 W/m² · K.

1) Perform an energy balance on a strip on the cylinder, as shown in Figure 8.49, by including all the relevant terms for energy transport (do not plug in numbers yet). 2) Plug in all the numerical values in appropriate units and write the equation from which the wool surface temperature, T, can be obtained as a function of the angle, θ, over the shaded region. Our interest is in the region with θ varying between 90 and 180 degrees (shaded quadrant of the cylinder in the figure).

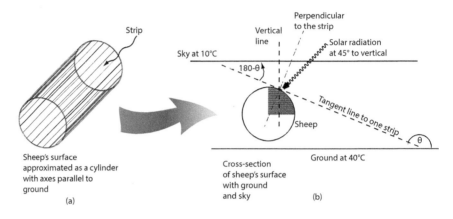

Figure 8.49: a) Approximation of a sheep body as cylinder, and b) schematic showing a section through the body and the ground and sky. *Only the shaded quadrant is of interest in this problem.*

8.28 Radiative and Convective Heat Transfer in a Greenhouse

In the idealized greenhouse in Figure 8.51, the solar radiation (the sun is assumed to be a blackbody at 5800 K) is incident on glass surface 1 at an angle of 30° to the vertical. Only the vertical component is involved in radiative exchange. The floor surface is covered with plants whose surface temperature is 20°C and can also be considered ideal or black for radiative exchange. Ignore reflection of solar or any other radiation from the glass surfaces. The three surfaces form an equilateral triangle, i.e., the respective areas are equal. View factors between these three surfaces can be found in Figure 8.51c. The glass temperature of surface 1 is 30°C while that of surface 2 is 15°C (ignore any variation of glass temperature over its thickness). The outside air temperature is 10°C while the inside air temperature is unknown and what we need to find. The outside heat transfer coefficient is 5 W/m$^2 \cdot$ K. The thermal expansion coefficient is equal to $1/T$.

1) Since you do not know the inside air temperature, you can initially assume (for this step only) the inside air temperature to be 20°C. Calculate the heat transfer coefficient for the inside of glass surface 1, assuming natural convection and the surface being vertical. 2) The radiative properties provided in the problem statement and in Figure 8.51b for glass are equivalent to glass being treated as a blackbody, for radiation exchange inside the enclosure (not counting direct solar radiation). Explain why this is so, using numerical calculations to justify your answer. 3) Write the energy balance equation for the left side glass (surface 1) where the only unknown should

Figure 8.50: A greenhouse and its idealization in Figure 8.51. (iStock.com)

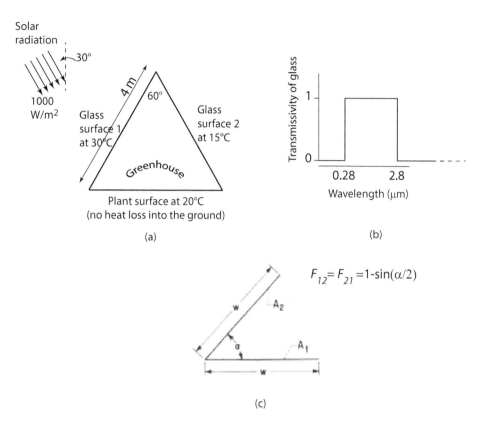

Figure 8.51: a) An idealized greenhouse with plants covering the floor while the other two surfaces are glass. b) Glass radiative property is as shown. Ignore reflection from glass surface. c) A view factor formula that could be useful.

be the inside air temperature. Radiation to the sky can be ignored. 4) Calculate the inside air temperature. 5) If your calculated air temperature is different from the assumed temperature for step 1, how would you proceed to improve your air temperature calculation? *Answer without any calculations.*

Part II

Mass Transfer

Chapter 9

EQUILIBRIUM, MASS CONSERVATION, AND KINETICS

CHAPTER OBJECTIVES

After you have studied this chapter, you should be able to

1. Explain equilibrium between various phases—solid, liquid, and gas.

2. Understand the various components needed for setting up conservation of mass for a single mass species.

3. Understand simple rates of reaction in the decay or generation of a mass species.

4. Understand the increase in reaction rate with temperature.

KEY TERMS

- **mass species**
- **mass concentration**
- **mass conservation**
- **equilibrium**
- **Henry's law**

- **isotherms**
- **partition coefficient**
- **reaction rate constant**
- **zeroth– and first–order reaction**

We have concluded the *first part* of the text, which dealt with energy transfer. This is the first chapter in the *second part* of the text on mass transfer. Figure 9.1 is a concept map which shows the relationship of this chapter to the other chapters on mass transfer. Like Chapter 1, this first chapter in the part on mass transfer discusses concepts of equilibrium, with non-equilibrium in concentration being the driving force.

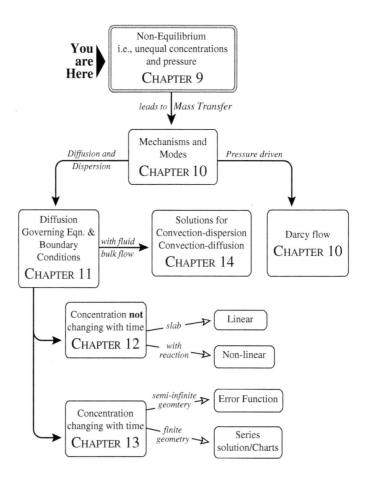

Figure 9.1: Concept map of mass transfer showing how the contents of this chapter relate to other chapters in mass transfer.

9.1 Concentration

The subject of mass transfer involves movement of one mass species through one or more other species. Therefore, we will always be dealing with systems consisting of two or more species, i.e., multi-species systems. The quantity of a certain species is described by its concentration, which is defined as the amount of substance per unit mass or volume. In a multi-species system having many mass species, the concentration of any one species can be expressed in many different ways. We will primarily use two different measures of concentration:

$$\rho_A = \text{mass concentration} = \frac{\text{mass of A}}{\text{unit volume}} = \frac{\text{kg of A}}{\text{m}^3}$$

$$c_A = \text{molar concentration} = \frac{\text{moles of A}}{\text{unit volume}} = \frac{\text{mol of A}}{\text{m}^3}$$

Thus, the two concentrations can be interchanged using the relation

$$\rho_A = c_A M_A$$

where M_A is the molecular weight of component A. For simplicity, however, we will not keep track of two separate symbols of concentration corresponding to two different units (mass and molar). Instead, the symbol c_A will be used to denote concentration of species A in either units, except when both units are used simultaneously. Although mass concentration is analogous to the amount of energy in heat transfer, it should be noted that in heat transfer only one variable, temperature, was sufficient, while in mass transfer, we have to keep track of the concentration of each component. Generally, this results in increased complexity.

9.1.1 Concentrations in a Gaseous Mixture

In a mixture of perfect gases, the concentrations of individual gases can be calculated from their partial pressures. Using the ideal gas law for a mixture of gases of total volume V at temperature T, we can write for the component A

$$p_A V = n_A R_g T \tag{9.1}$$

where p_A is the partial pressure of gas A for which there are n_A moles are present and R_g is the gas constant. Using this equation, the molar concentration of gas A can be written as

$$c_A = \frac{n_A}{V} = \frac{p_A}{R_g T} \tag{9.2}$$

Thus, concentration of any gas species i can be written in terms of its partial pressure as

$$c_i = \frac{p_i}{R_g T} \qquad (9.3)$$

The total concentration c of all the species is related to the total pressure, P. Using $PV = nRT$ where n is the total number of moles ($= \sum n_i$), the total concentration c is written as

$$c = \frac{P}{R_g T} \qquad (9.4)$$

9.1.2 Example: Concentration of Individual Gases in Air

Consider an air–water vapor mixture at a total pressure of 1 atm and a temperature of 60°C having 20% water vapor, 17% oxygen, and 63% nitrogen. The percentages refer to the ratio of partial pressures to total pressure. For simplicity, we are ignoring the other constituent gases in air. Calculate the molar and mass concentrations of each of the three gases in air.

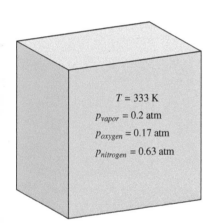

$T = 333$ K

$p_{vapor} = 0.2$ atm

$p_{oxygen} = 0.17$ atm

$p_{nitrogen} = 0.63$ atm

Figure 9.2:   Schematic for Example 9.1.2.

Solution

Understanding and formulating the problem *1) What is the process?* Mixture of gases at equilibrium at 333 K where the partial pressure of each gas is given. *2) What are we solving for?* Molar and mass concentrations of each gas. *3) Schematic and given data:* A schematic is shown in Figure 9.2 with some of the given data superimposed on it. *4) Assumptions:* Ideal gas law applies to the mixture.

Generating and selecting among alternate solutions This step is skipped since it is a straightforward application of Eq. 9.3.

Implementing the chosen solution Using Eq. 9.3, the concentrations are calculated as

$$c_{vapor} = \frac{p_{vapor}}{R_g T} = \frac{0.2 \times 1.013 \times 10^5 \text{ N/m}^2}{8.315 \text{ J/mol·K} \times 333 \text{ K}} = 7.32\frac{\text{mol}}{\text{m}^3}$$

$$c_{oxygen} = \frac{p_{oxygen}}{R_g T} = \frac{0.17 \times 1.013 \times 10^5 \text{ N/m}^2}{8.315 \text{ J/mol·K} \times 333 \text{ K}} = 6.22\frac{\text{mol}}{\text{m}^3}$$

$$c_{nitrogen} = \frac{p_{nitrogen}}{R_g T} = \frac{0.63 \times 1.013 \times 10^5 \text{ N/m}^2}{8.315 \text{ J/mol·K} \times 333 \text{ K}} = 23.05\frac{\text{mol}}{\text{m}^3}$$

The mass concentrations are calculated by multiplying the molar concentrations by the respective molecular weights as

$$\rho_{vapor} = 7.32 \times 18 \times 10^{-3} \frac{kg}{m^3} = 0.1317 \; \frac{kg}{m^3}$$

$$\rho_{oxygen} = 6.22 \times 32 \times 10^{-3} \frac{kg}{m^3} = 0.1990 \; \frac{kg}{m^3}$$

$$\rho_{nitrogen} = 23.05 \times 28 \times 10^{-3} \frac{kg}{m^3} = 0.6454 \; \frac{kg}{m^3}$$

The total molar concentration of gases is given by

$$c = c_{vapor} + c_{oxygen} + c_{nitrogen}$$

$$= 7.32 + 6.22 + 23.05 \; \frac{mol}{m^3} = 36.59 \; \frac{mol}{m^3}$$

and the total mass concentration of the gases is given by

$$\rho = \rho_{vapor} + \rho_{oxygen} + \rho_{nitrogen}$$

$$= 0.1317 + 0.1990 + 0.6454 = 0.9761 \frac{kg}{m^3}$$

Note that the total mass concentration, ρ, is another name for the density of the gas mixture.

Evaluating and interpreting the solution *1) Do the mass concentrations make sense?* Since we know the density of air is of the order of 1 kg/m³, the calculated values are probably in the ballpark.

9.2 Species Mass Balance (Mass Conservation)

Like total energy, total mass is conserved. In Section 3.3 (page 41) for energy transfer, we wrote a balance for thermal energy. Other forms of energy can convert into thermal energy and vice versa. Likewise, one species of mass can convert to other species and vice versa, for example, through chemical reactions. By analogy to the thermal energy conservation equation (Eq. 3.3), consider mass conservation of a species for an arbitrary control volume shown in Figure 9.3. A word equation for mass conservation of a species can be written from this figure as

$$\underbrace{\begin{array}{c}\text{Rate of}\\\text{Mass In}\end{array}}_{\text{species } i} - \underbrace{\begin{array}{c}\text{Rate of}\\\text{Mass Out}\end{array}}_{\text{species } i} + \underbrace{\begin{array}{c}\text{Rate of}\\\text{Mass Generation}\end{array}}_{\text{species } i} = \underbrace{\begin{array}{c}\text{Rate of}\\\text{Change in Mass Storage}\end{array}}_{\text{species } i} \quad (9.5)$$

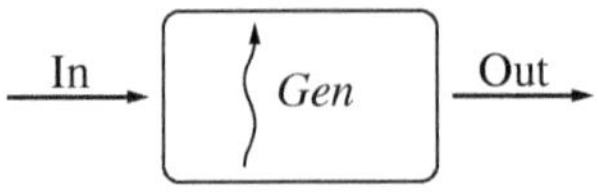

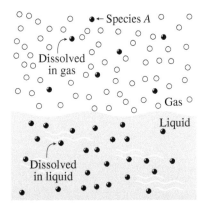

Figure 9.3: A control volume for mass conservation showing different components.

Increase in storage is the increase in concentration of the species. The mass of a species can be generated or utilized in chemical reactions or biological processes. Note that this is a transformation of one mass species into another. For example, under suitable growth conditions, the microbial mass in a fermentor increases. At the same time there is utilization of resources like O_2 in the fermentor in order to sustain growth.

9.3 Equilibrium

The transfer of mass within a phase or between two phases requires a departure from equilibrium. One needs to study interphase equilibrium in order to describe interphase mass transfer.

9.3.1 Equilibrium between a Gas and a Liquid

An example of equilibrium between a gas and a liquid is water saturated with dissolved oxygen in contact with air, as shown in Figure 9.4. Here the oxygen in water and the oxygen in air are considered to be in equilibrium with each other. Such equilibrium between a gas and a liquid phase can be described by Henry's law:

$$p_A = H \, x_A \tag{9.6}$$

where p_A is the partial pressure of species A in gas phase at equilibrium, x_A is the concentration of species A in liquid phase at equilibrium, and H is Henry's constant. Equation 9.6 is illustrated in Figure 9.5. One can think of Henry's law as the pressure dependence of solubility of a gas in a solution. It says that gas solubility should be directly proportional to pressure.

$$x_A = \frac{p_A}{H} \tag{9.7}$$

Figure 9.4: Schematic of equilibrium between a gas and a liquid phase.

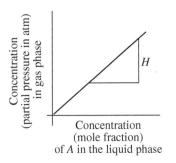

Henry's constants for several gases in water are given in Figure 9.6. In this figure, H increases with temperature, thus showing a decrease in solubility of gases in water at higher temperatures, as given by Eq. 9.7.

Figure 9.5: Equilibrium relationship between a gas phase and a liquid phase following Henry's law.

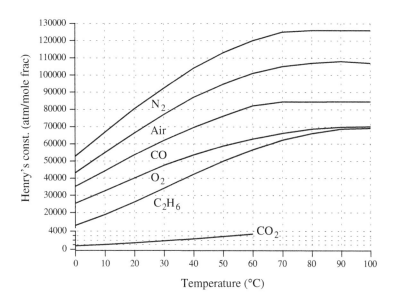

Figure 9.6: Henry's constant, H, for gases slightly soluble in water. Data abridged from Foust et al. (1960).

9.3.2   *BioConnect: Bends from Diving in Deep Water*

There are many biological applications of this pressure dependence of solubility. A condition known as caisson disease or bends or compressed air sickness results from a rapid change in atmospheric pressure from high to normal, causing nitrogen bubbles to form in blood and body tissues. Divers can get the bends if they rise too rapidly.

9.3.3   *Example: Oxygen Concentration in Water in Equilibrium with Air*

Compared to air, there is little oxygen (dissolved) in water, even under the best of conditions. Thus, fish or other aquatic animals must either pump large quantities of water over their respiratory surface to obtain reasonable amounts of oxygen or else be limited to relatively low metabolic rates. This relatively low amount of dissolved oxygen is further reduced at higher temperature, as a direct consequence of the Henry's law just studied.

Water at 5°C is in contact with a large volume of ordinary air at a total pressure of 1 atmosphere. 1) How much oxygen is dissolved in the water in mg O_2/liter of water? Suppose this water got warmed up to 25°C. 2) What is the new amount of oxygen dissolved in the same units? 3) What happens if water at 25°C and having initially an amount of oxygen 10 mg O_2/liter of water is brought into contact with ordinary air at the same temperature?

Both liquid and gas are at 5°C or 25°C

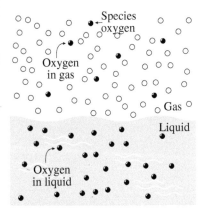

Figure 9.7: Schematic for Example 9.3.3.

Solution

Understanding and formulating the problem *1) What is the process?* Oxygen in water is in equilibrium with oxygen in air. *2) What are we solving for?* The amount of oxygen dissolved in water at equilibrium. *3) Schematic and given data:* A schematic is shown in Figure 9.7 with some of the given data superimposed on it. *4) Assumptions:* The gas–liquid system is at one temperature and has been in contact for a long enough time to have reached equilibrium. The volume of air is very large, so the oxygen content in it does not change. Ordinary air has 21% oxygen.

Generating and selecting among alternate solutions We skip this step as this is a simple application of Henry's law (Eq. 9.6).

Implementing the chosen solution 1) Using Henry's law (Eq. 9.6) to find the concentration of oxygen in the water, x_{O_2}, at equilibrium at 5°C,

$$x_{O_2} = \frac{p_{O_2}}{H}$$

Here H, Henry's law constant, is found from Figure 9.6 to be 2.9×10^4 and 4.45×10^4 atm/mole fraction of oxygen in the water at temperatures 5°C and 25°C, respectively. Since dry air contains 21% oxygen, the partial pressure of oxygen is given by $p_{O_2} = 0.21$ atm.

$$
\begin{aligned}
x_{O_2} &= \frac{0.21 \text{ atm}}{2.9 \times 10^4 \text{ atm/mole fraction}} \\
&= 7.241 \times 10^{-6} [\text{mol } O_2/\text{mol solution}] \\
&\approx 7.241 \times 10^{-6} [\text{mol } O_2/\text{mol water}] \\
&= 7.241 \times 10^{-6} \times \frac{32}{18} [\text{g } O_2/\text{g water}] \\
&= 12.87 [\text{mg } O_2/\text{liter of water}]
\end{aligned}
$$

where the mass of water with and without oxygen is considered approximately the same. Also 1 g $\approx$ 1 cm^3 = 10^{-3} liter has been used for water.

2) Similarly, the amount of oxygen dissolved in water at 25°C is

$$x_{O_2} = \frac{p_{O_2}}{H}$$
$$= \frac{0.21 \text{ atm}}{4.45 \times 10^4 \text{ atm/mole fraction}}$$
$$= 4.719 \times 10^{-6} [\text{mol } O_2/\text{mol solution}]$$
$$= 8.39 [\text{mg } O_2/\text{liter of water}]$$

3) Since the final equlibrium concentration of oxygen in water at 25°C is 8.39 mg O_2/liter of water, water having initially 10 mg O_2/liter of water will *lose* oxygen until it comes to the final equilibrium concentration of 8.39 mg O_2/liter of water.

Evaluating and interpreting the solution *1) Does the calculated concentration make sense?* From available data (web), oxygen solubility in fresh water is ~ 10 mg/liter so our answer is probably reasonable. *2) Comments:* Since Henry's constant increases with temperature, as shown in Figure 9.6, the solubility of oxygen decreases with temperature, and fish and aquatic life are expected to have greater difficulty in obtaining the needed oxygen at higher water temperature. Actually, the fish are in double jeopardy because at high water temperatures their metabolic rates increase, hence their physiologic demand for oxygen increases.

9.3.4 Equilibrium between a Gas and a Solid (with Adsorbed Liquid)

The common example of a solid coming into equilibrium with a gas is the potato chip typically becoming soggy when left out of the package. The moisture in the potato chip (solid phase) comes to equilibrium with the typically higher amount of water vapor in the outside air than inside the package, thus absorbing more moisture from the air and becoming soggy. Any wet solid, when brought into contact for a sufficient time with a large amount of air having a certain concentration of moisture (water vapor), will reach a final moisture content that is the *equilibrium moisture content* of the solid for the particular concentration of moisture in the air. This equilibrium between moisture in the solid and moisture in the gas is illustrated in Figure 9.8. If the solid initially (before coming in contact with the air) had more moisture than the equilibrium moisture content corresponding to the concentration of moisture in air, it will reach equilibrium with the air by losing the excess moisture. Conversely, if the solid had less moisture

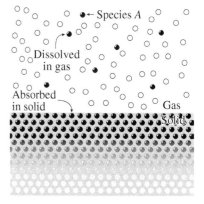

Figure 9.8: Schematic of equilibrium between a gas and a solid phase.

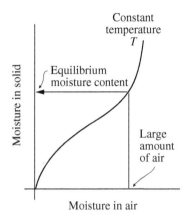

Figure 9.9: Typical equilibrium moisture isotherm for a solid. Superimposed on it is a process where the moisture in the solid comes to equilibrium with the moisture in the gas, due to an excess of the gas.

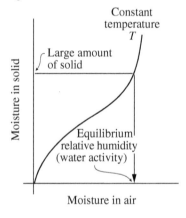

Figure 9.11: Typical equilibrium moisture isotherm for a solid. Superimposed on it is a process where the moisture in the gas comes to equilibrium with the moisture in the solid, due to an excess of the solid.

than the corresponding equilibrium moisture content, it will gain moisture in coming to equilibrium with the air.

The relationship between the concentration[1] of moisture in air and the corresponding equilibrium moisture content of a solid can be expressed as an equilibrium moisture curve. A typical curve is shown in Figure 9.9. Equilibrium moisture content is also a function of temperature. Thus the curve in Figure 9.9 refers to a specific temperature, which is why the curve is often called an isotherm. An example of an empirical equation (Henderson, 1952) used to represent the equilibrium moisture content curve is

$$1 - RH = b_1 \exp(-b_2 w) \qquad (9.8)$$

where RH is the relative humidity (in fraction), b_1 and b_2 are constants, and w is the equilibrium moisture content (%). There are many such empirical equations. Equilibrium moisture content of a solid will also be typically different depending on whether moisture is being added or removed as the material comes to equilibrium with the surrounding air, i.e., whether it is a drying or a wetting process.

Examples of equilibrium moisture isotherms are shown in Figure 9.10 (see next page) for wood and soil. The moisture contents in this figure are on a wet weight basis, i.e., kg of water per kg of total. To convert to a dry basis, use the relationship $M_d = M_w/(1 - M_w)$ where M_w is the moisture content in fraction on a wet basis and M_d is the moisture content in fraction on a dry basis. Note that the moisture content in air is given in this figure in terms of relative humidity (see footnote).

Hygroscopic and non-hygroscopic solids

To understand the relative degree to which water is freely available in a solid, consider the same information as shown in Figure 9.9 except that now, instead of a large amount of air coming in contact with a solid, the solid is sealed inside a container together with a small amount of air. The moisture in air would come to equilibrium with moisture in the solid, as illustrated in Figure 9.11. This equilibrium relative humidity of the air is also called the *water activity* of the solid at its moisture content. This water activity of the solid is a measure of the relative ease at which water is available in the solid and is extremely important for biological activities such as growth of microorganisms.

---

[1]Concentration of moisture in air is typically measured in terms of humidity or relative humidity. Humidity denotes the absolute amount of moisture (water vapor) in air, in kg of water/kg of dry air. Relative humidity, on the other hand, refers to the amount of moisture in air relative to the maximum capacity of the air to carry water vapor. Symbolically, $RH = 100 p_v/p_{vs}$ where RH is the relative humidity, p_v is the partial pressure of water vapor and p_{vs} is the partial pressure of water vapor at saturation at the same temperature. Thus, when air is carrying the maximum possible amount of moisture (it is saturated), the relative humidity is 100%. Conversion between humidity and the relative humidity can be done using the psychrometric chart shown in Appendix C.9 on page 569.

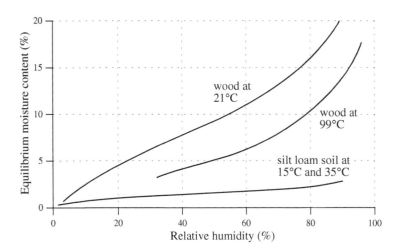

Figure 9.10: Equilibrium moisture isotherms for wood and soil. Wood data courtesy of USDA Forest Service and cited by Siau (1984). Soil data adapted from Berge (1990).

The equilibrium moisture isotherms in Figure 9.10 show that the relative humidity and therefore the vapor pressure of air surrounding a solid in a closed system are a function of both moisture content and temperature of the solid. Solids showing such characteristics are called hygroscopic. When the vapor pressure surrounding a solid is a function of temperature only and is thus independent of the solid moisture level at all times, the solid is called non-hygroscopic. Examples of non-hygroscopic solids are sand, crushed minerals, polymer particles, and some ceramics. Water on the surface of these solids behaves as if the solid were not there, and the vapor pressure above the surface as a function of temperature is given by the Clausius–Clapeyron equation:

$$\frac{d \ln p_v}{d T} = \frac{\Delta H_{vap}}{RT^2} \tag{9.9}$$

where p_v is the vapor pressure at absolute temperature T and ΔH_{vap} is the enthalpy of vaporization. The amount of *physically bound* water is negligble in a non-hygroscopic solid, while there can be a large amount of physically bound water in a hygroscopic solid. Once the hygroscopic solid is saturated with water, it can no longer bind more water and begins to behave like a non-hygroscopic solid. Thus, even in a hygroscopic solid, above a level of moisture content the vapor pressure is a function of temperature

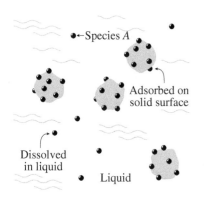

Figure 9.12: Schematic of equilibrium between dissolved and surface adsorbed species.

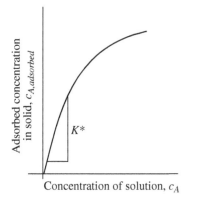

Figure 9.13: Equilibrium relationship between dissolved and surface adsorbed quantity, known as Freundlich isotherm.

only. A non-hygroscopic solid does not shrink during drying. In a hygroscopic solid, shrinkage often occurs during drying due to the large amount of physically bound water. Many biological solids are hygroscopic.

9.3.5   Equilibrium between Solid and Liquid in Adsorption

Consider a solid surface in contact with a liquid that has a dissolved species. As illustrated in Figure 9.12, some of the dissolved solute will bind to the solid surface, instead of remaining in solution. This phenomenon is called adsorption. An example is the adsorption of pollutant chemicals to a soil surface when water with dissolved pollutants is flowing through it. The dissolved species in the liquid reaches an equilibrium with the same species adsorbed to solid surfaces. Such equilibrium relationships are represented as adsorption isotherms and are often empirical. A common adsorption isotherm is the Freundlich isotherm shown in Figure 9.13. Other possible isotherms are Langmuir and BET, which are described well in numerous soil science and other textbooks.

The isotherm shown in Figure 9.13 can be represented mathematically as

$$c_{A,adsorbed} = K^* c_A^n \tag{9.10}$$

where $c_{A,adsorbed}$ is the concentration of adsorbed solute A, c_A is the concentration of solute A in solution, and K^* and n are empirical constants. Note that, even though the relationship expressed in Figure 9.13 is non-linear, it can be treated as linear in the ranges of small concentration. In the linear region of small concentrations (here $n = 1$), the slope of the curve in Figure 9.13 is sometimes referred to as the *partition coefficient* or *distribution coefficient*.

Important note about K^*

Note that K^*, defined as a distribution coefficient in Eq. 9.10, is different from another definition to come in Eq. 13.23. In both situations, K^* carries the same idea of concentrations being different in the two phases in contact, it being the ratio of the two concentrations. How we form the ratio, however, varies in the literature depending on which two phases (solid/liquid/gas) are in contact and the application area. For example, when we are dealing with solid–liquid systems in adsorption, as in the example of a soil matrix containing water with pollutants dissolved in it (Section 14.3), we will use Eq. 9.10 for K^*. For a fluid flowing over a solid, as in the example of airflow over a solid being dried, we will use Eq. 13.23 as the definition for K^*. Thus, the discussion in this section, which also refers to the Heisler charts on pages 555–557, would use K^* defined by Eq. 13.23.

Table 9.1: Multiple definition of K^*, the distribution coefficient or partition coefficient

	K^*	
Where defined	Eq. 9.10	Eq. 13.23
How defined	$K^* = \dfrac{c^{solid}}{c^{fluid}}$	$K^* = \dfrac{c^{fluid}}{c^{solid}}$
Where else used	Dialysis membrane (Section 12.1.3 and Example 12.1.4)	Drying of a solid (Example 13.3.1 on page 454)
What phases	Liquid–solid	Gas–solid

9.3.6 Example: How to Obtain K^* for Solid–Gas Equilibrium: Moisture in a Paper–Air System

As in solid–liquid equilibrium, gas–solid equilibrium information in Figure 9.10 can also be linearized. As an example, consider the isotherm of paper shown in Figure 9.14. This linear portion can, likewise, be defined by a K^* as

$$K^* = \left. \frac{c_{\text{vapor in air}}}{c_{\text{water in solid}}} \right|_{\text{surface}} \tag{9.11}$$

The concentration of vapor in air, $c_{\text{vapor in air}}$, depends on the conditions (temperature and humidity) of the air. Properties of air corresponding to a temperature and humidity are noted in a psychrometric chart (Appendix C.9). From this chart, for the given condition of air (25°C and 20% RH), moisture content of air is 0.004 kg of water/kg of dry air and the specific volume of air is 0.85 m³/kg of dry air. Thus,

$$c_{\text{vapor in air}} = \frac{0.004}{0.85} \frac{\text{kg water/kg dry air}}{\text{m}^3/\text{kg dry air}}$$
$$= 0.0047 \text{ kg water/m}^3$$

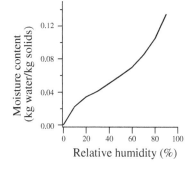

Figure 9.14: Equilibrium moisture content of paper. Moisture content is given in kg of water per kg of solids. Data from ICT, 1933.

The concentration of water in a solid (paper) that is in equilibrium with this water vapor in air, $c_{\text{water in solid}}$, is obtained from the equilibrium moisture content curve for paper in Figure 9.14. However, in the figure, the units of this concentration are in kg of water/kg of dry solids, which has to be converted to kg water/m³ of solid. For this, the bulk density of paper (as in a compost pile) is needed, which is assumed to be 700 kg dry solid/m^3. Thus,

$$c_{\text{water in solid}} = \left(0.03\frac{\text{kg water}}{\text{kg dry solids}}\right)\left(700\frac{\text{kg dry solids}}{\text{m}^3}\right)$$

$$= 21\frac{\text{kg water}}{\text{m}^3}$$

Substituting into Eq. 9.11,

$$K^* = \left.\frac{c_{\text{vapor in air}}}{c_{\text{water in solid}}}\right|_{\text{surface}}$$

$$= \frac{0.0047\ \text{kg water}/\text{m}^3}{21\ \text{kg water}/\text{m}^3}$$

$$= 2.238 \times 10^{-4}$$

Note that K^* is a dimensionless quantity signifying the ratio of concentrations at the two phases in equilibrium.

9.4 Chemical Kinetics: Generation or Depletion of a Mass Species

So far in this chapter we have discussed the mechanisms of mass movement. In systems undergoing chemical changes, mass concentration at a location can also change due to a chemical reaction, independent of any diffusive or convective transport. For example, a chemical transporting through soil can also degrade simultaneously. In heat transfer the term Q was used to denote the generation of heat within the system. Similarly with mass transfer, the term r_A is used to denote the rate of generation or depletion of a mass species A.

Chemical Kinetics is the study of the rate and mechanism by which one chemical species is converted to another species. The mechanism of a chemical reaction is the sequence of individual chemical events whose overall result produces the observed reaction. It is not always necessary to know the detailed mechanism of a reaction in order to quantify the observed rate of the overall reaction. A satisfactory equation for the intrinsic rate of the reaction is sufficient. Reaction mechanisms are reliably known for only a few systems. We will not deal with any mechanisms in this book.

Thermodynamics tells us the maximum extent to which a reaction can proceed, since the equations are correct only for equilibrium conditions. The maximum possible conversion, found at equilibrium, is important as a standard for evaluating the actual performance of the reaction. For practical use, we also need to know the time required for the reaction to proceed. The rate of a chemical reaction is defined by

$$\text{Rate} = \frac{\text{Mass of product produced or reactant consumed}}{\text{(Unit volume)(Time)}} \tag{9.12}$$

with the units being $kg/m^3 \cdot s$. A few simple relationships describing rates of reactions are now discussed.

9.4.1 Rate Laws of Homogeneous Reactions

Rate laws describe the dependence of reaction rates on concentration. Consider an irreversible reaction:

$$s_A A + s_B B \rightarrow s_C C + s_D D \tag{9.13}$$

where s_A, s_B, s_C and s_D are stoichiometric coefficients, respectively. The rate of disappearance of A may be written in terms of concentration c_A and c_B of A and B, respectively, as

$$r_A = -\frac{dc_A}{dt} = k'' c_A^\alpha c_B^\beta \tag{9.14}$$

where α is the order of reaction with respect to A, β is the order of reaction with respect to B, $\alpha + \beta$ is the overall order of reaction, and k'' is the *reaction rate constant*. Since r_A measures rate of disappearance, in the transport equation r_A will have a (+) or (-) sign depending on whether A is generated or depleted, respectively.

The reaction rate constant k'' includes the effects of all variables other than concentration. While many variables may affect the reaction, temperature is the most important. However, if the reaction requires a catalyst, k'' might depend on the nature and concentration of the catalytic substance. We will now consider several special cases of Eq. 9.14.

9.4.2 Zeroth-Order Reaction

A reaction is of zeroth order when the rate of reaction is independent of the concentration of the species:

$$r_A = -\frac{dc}{dt} = k'' \tag{9.15}$$

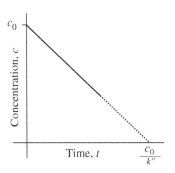

Figure 9.15: Linear change in concentration for a zeroth-order reaction.

where the concentration of species A, c_A has been replaced by c for simplicity. Integrating, and noting that c can never become negative:

$$c_0 - c = k''t \quad for \; t \le \frac{c_0}{k''} \tag{9.16}$$

Equation 9.16 is plotted in Figure 9.15. The dashed line in the figure signifies that the zeroth-order reaction may not continue all the way. As a rule, reactions are only zeroth order in certain high concentration ranges, because the small change in the amount of material caused by the reaction is negligible compared to the amount of material present. If the overall concentration of the material is decreased to a certain point, the reaction will usually become dependent on the concentration level, at which point the reaction order will increase from zero.

An example of a zeroth-order reaction is the study of oxygen transport in human tissue. Augusu Krough won a Nobel prize by studying this mass transfer process. By considering a tissue cylinder surrounding each blood vessel, he proposed the diffusion of oxygen away from the blood vessel into the annular tissue was accompanied by a zeroth-order reaction. This reaction was necessary to explain the metabolic consumption of oxygen to produce carbon dioxide.

9.4.3 First-Order Reaction

A reaction is first order when the rate of reaction is linearly related to the concentration of the species; thus,

$$r_A = -\frac{dc}{dt} \;\; = \;\; k''c \tag{9.17}$$

$$-\frac{dc}{c} \;\; = \;\; k'' \, dt$$

$$-\int_{c_0}^{c} \frac{dc}{c} \;\; = \;\; \int_{o}^{t} k'' \, dt$$

$$-\ln \frac{c}{c_0} \;\; = \;\; k''t$$

$$c \;\; = \;\; c_0 \, e^{-k''t} \tag{9.18}$$

Equation 9.18 is plotted in Figure 9.16. The exponential decay of concentration is characteristic of the first-order reaction. The concentration approaches zero asymptotically, or, in other words, it never actually reaches zero even though it becomes very small.

Half-life for a first-order reaction is a convenient alternative description of reaction rate that is used widely in practice. It is defined as the time to change concentration by

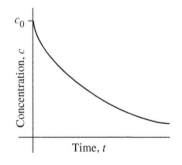

Figure 9.16: Exponential change in concentration for a first-order reaction.

50%. If $t_{1/2}$ denotes half-life, substituting in Eq. 9.18, we get

$$
\begin{aligned}
0.5c_0 &= c_0 e^{-k''t_{1/2}} \\
-k''t_{1/2} &= \ln(0.5) \\
&= -0.693 \\
k'' &= \frac{0.693}{t_{1/2}}
\end{aligned}
\tag{9.19}
$$

Equation 9.18 can be rewritten in terms of half-life as

$$
c = c_0 e^{-\frac{0.693}{t_{1/2}}t}
\tag{9.20}
$$

An example of a first-order reaction is the degradation of certain pesticides in soil. Representative values of $t_{1/2}$ of some pesticides in soil are shown in Table 9.2.

A larger half-life means a lower reaction rate. For example, in the case of a pesticide, a longer half-life means the pesticide remains in the soil longer. Using the half-life data from Table 9.2 and with an assumed initial concentration of 1 unit, the concentrations are plotted in Figure 9.17 to show the relative rates of decay. It can be seen from the curves that a concentration of 0.5, half of the initial concentration,

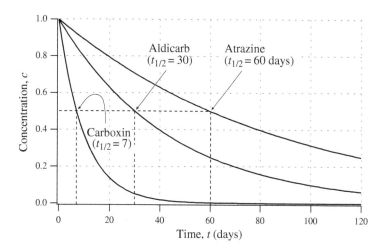

Figure 9.17: Concentration changes with time (due to decay) for the pesticides listed in Table 9.2.

Table 9.2: Half-life of some important pesticides. Here the health advisory level is the concentration level of a substance which is considered safe to be consumed daily throughout a person's lifetime.

Pesticide	$t_{1/2}$ in soil (days)	Health advisory level (μg/L or ppb)
Aldicarb	30	10
Atrazine	60	3
Carboxin	7	700

is reached for each chemical at times equal to their half-life. During food processing, the thermal destruction of microorganisms, most nutrients, and quality factors (texture, color, and flavor) generally follow first-order kinetics, although the exact mechanisms of these reactions are often complex and are still subjects of research. For example, bacterial death by heat is often attributed to the loss of reproductive ability due to the denaturation of one gene or a protein essential to reproduction. Data on kinetics of first-order reactions in food systems is provided in Table D.7 (page 580).

9.4.4 *nth*-Order Reaction

When the mechanism of a reaction is not known, we often attempt to fit the data with an *nth*-order rate equation of the form

$$r_A = -\frac{dc}{dt} = k'' c^n \tag{9.21}$$

Integrating,

$$\int_{c_0}^{c} \frac{dc}{c^n} = -\int_{0}^{t} k'' \, dt$$

$$\frac{c^{-n+1}\big|_{c_0}^{c}}{-n+1} = -k'' t$$

Putting the limits and rearranging,

$$c^{1-n} - c_0^{1-n} = (n-1)\, k'' t \,, \quad n \neq 1 \tag{9.22}$$

which provides concentration c as a function of time, t, for a *nth*-order reaction.

9.4.5 Effects of Temperature

Higher temperature speeds up most reactions. Arrhenius in 1889 (Smith, 1970) and others later showed that, for most chemical reactions, the temperature dependence of the reaction rate is of the form

$$k'' = k_0'' e^{-E_a/(R_g T)} \qquad\qquad (9.23)$$

where k_0'' is called the frequency factor, E_a is called the activation energy (J/mole), R_g is the gas constant (J/mole·K), T is the absolute temperature in K. The above equation is often referred to as the Arrhenius law of temperature dependency. It is an empirical equation that has been found to represent many chemical and biological processes. Attempts to arrive at this expression from a theoretical standpoint have had some success (e.g., the collision theory), but only for very idealized gaseous reactions. More details about the Arrhenius law can be found in most textbooks on physical chemistry or chemical kinetics.

9.4.6 *Example: Heating for a Shorter Time at Higher Temperatures Can Keep More Nutrients in Food*

Destruction of bacteria and nutrients occurring during heat sterilization of food is often assumed to follow first-order reactions. Consider heat sterilization of a food, in which the entire food mass is at one temperature during heating, i.e., heating can be considered lumped. The bacterial death reaction is given by $k_0'' = 4.669 \times 10^{38}$/s and $E/R = 35,751$ K, and vitamin C destruction is given by $k_0'' = 8.885 \times 10^{11}$/s and $E/R = 14,300$ K. 1) Calculate the fraction of initial concentrations of bacteria and nutrients that would be retained by a heating process that maintains the food temperature at 120°C for 10 minutes. 2) If the food is maintained at 140°C instead, how long should it be heated to destroy the same fraction of the initial bacterial population as in step 1? 3) For the duration of heating calculated in step 2 at the temperature 140°C, what would be the fraction of the nutrients retained? 4) Would you prefer to heat the food at 120°C or 140°C? Why?

Solution

Our solution is a straightforward application of first-order kinetics and the Arrhenius temperature dependency.

1) Combining Eq. 9.18 and Eq. 9.23, we get

$$c = c_0 e^{-k_0 e^{-E/RT} t}$$

For bacteria, the fraction of initial concentration that would be retained is

$$
(c/c_0)|_{\text{bacteria}} = e^{-\left(4.669\times10^{38}\frac{1}{[\text{s}]}e^{-\frac{35751}{393}}\right)(10\times60)\,[\text{s}]}
$$
$$
= 1.54 \times 10^{-38}
$$

For nutrients, the fraction of initial concentration that would be retained is

$$
(c/c_0)|_{\text{nutrients}} = e^{-\left(8.885\times10^{11}[\frac{1}{\text{s}}]e^{-\frac{14300}{393}}\right)10\times60\,[\text{s}]}
$$
$$
= 0.919
$$

2) When the food is heated at 140°C instead, the time required to destroy the same fraction of initial bacterial population is

$$
1.54 \times 10^{-38} = e^{\left(-4.669\times10^{38}e^{-35751/413}\right)t\,[\text{s}]}
$$
$$
= e^{-11.881871t}
$$

from which we get

$$
t = 7.33 \text{ s}
$$

3) For 7.33 s of heating at 140°C, the fraction of nutrients retained is

$$
(c/c_0)|_{\text{nutrients}} = e^{-\left(8.885\times10^{11}[\frac{1}{\text{s}}]e^{-\frac{14300}{413}}\right)7.328\,[\text{s}]}
$$
$$
= 0.994
$$

which is higher than the fraction retained at 120°C, 0.919.

4) Since the food heated at higher temperature retains more nutrients, it would generally be preferred. This conclusion of higher nutrient retention when heated at a higher temperature for the same bacterial destruction may not always be true. Our analysis assumed that the entire food is at one temperature (there was only one T) at any time, i.e., it is a lumped system. When spatial variation in temperature inside the food is significant, this conclusion may not be true and would require a different analysis.

9.5 Chapter Summary—Equilibrium, Mass Conservation, and Kinetics

- **Equilibrium**

    1. Equilibrium between a gas and a liquid is described by Henry's law (page 320).

    2. Equilibrium between a gas and a solid is described by isotherms (page 323).

    3. Equilibrium between a solid and a liquid in absorption (page 326).

- **Kinetics of Chemical Reactions (page 328)**

    1. A reaction where the rate is constant is called a zeroth-order reaction. Concentration change for this is given by Eq. 9.16.

    2. A reaction is called first order whenever the reaction rate is proportional to the concentration of a single reactant. Concentration change for this is given by Eq. 9.18. Many reactions are often approximated as first-order reactions.

    3. Higher temperature speeds up most reactions. This temperature dependency of reaction rate for most reactions is given by Eq. 9.23.

9.6 Problem Solving—Equilibrium, Mass Conservation, and Kinetics

▶**Obtaining concentration in one phase in equilibrium with another**   Table 9.3 shows how concentration in one phase can be obtained when the same is provided in another phase.

▶**Performing mass conservation to find various quantities**

1. Define the domain over which to perform mass balance.

2. Set up mass balance as

$$\text{In} - \text{Out} + \text{Gen} = \text{Change in Storage} \qquad (9.24)$$

where the terms can be all either rate or amount over time.

Table 9.3: Equilibrium relations for various pairs of phases

Phases	Typically referred to as	Concentration relations are given by
Liquid–Gas	Henry's law	$x_A = p_A/H$
Solid–Gas	Water activity relations	Figure 9.10 and others
Solid–Liquid	Distribution coefficient	$c_{A,liquid} = \dfrac{c_{A,adsorbed}}{K^*}$

3. Calculate each quantity, **In** and **Out**, making sure of the direction of mass flow, as given by the equations used. For example, the convection equation (introduced in the following chapter), $n_A = h_m A(c_1 - c_2)$, is the amount of mass n_A going from 1 to 2. Whether this mass transport is **In** or **Out** depends on which body we are referring to as 1 and which one as 2. One does not need to be concerned about which body has a higher concentration and which has a lower.

4. Consider the **Generation** term based on zero, first-order, or other reaction kinetics provided. If the species is being produced, the **Generation** term is positive. Conversely, if the species is being consumed (degraded or used up), the **Generation** term is negative. See Table 9.4 for the expression to be included.

5. Since mass content of a volume V is Vc, **Change in storage** is $V\Delta c$, where Δc is change in concentration during a chosen period, i.e., $c_{end} - c_{begin}$.

6. Making sure of directions of mass flow in **In** and **Out** terms, as done in step 3, and the appropriate sign for **Generation**, plug in all the terms in the above equation. Typically, this will lead to an algebraic equation or a differential equation for concentration.

7. All the terms in the equation need to be in the same units, i.e., they can be in terms of mass amount, kg, or they can be rates as mass per unit time, kg/s.

▶**Using equilibrium to perform mass balance**　Equilibrium decides the final concentration of a species. You can use this final concentration to determine how much of a species will be lost or gained—this will be needed in mass balance. An example of this is Problem 9.11.

Table 9.4: Generation term to be included in mass balance equation

	Zeroth order	First order
Generation [kg/m$^3 \cdot$ s]	k''	$k'' c$
Depletion or decay [kg/m$^3 \cdot$ s]	$-k''$	$-k'' c$

▶**Finding time for a reaction or equivalent time–temperature combinations**

- Given the rate of a reaction and the initial concentration, you can find concentration at any time, for either a zero-order reaction (Eq. 9.16) or a first-order reaction (Eq. 9.18).

- The same reaction at two different temperatures will proceed at two different rates. For the final concentration to be the same,

$$\frac{c_{final}}{c_0} = e^{-k_0 e^{(-E/RT_1)} t_1} = e^{-k_0 e^{(-E/RT_2)} t_2} \qquad (9.25)$$

From this equation, we can find (T_1, t_1) and (T_2, t_2) that are equivalent temperatures and times; given three of these four quantities (T_1, t_1, T_2, t_2), we can find the fourth using the above equation.

- Similarly, for a zeroth-order reaction, equivalent time–temperature combinations are related by

$$c_{final} - c_0 = k_0 e^{-E/RT_1} t_1 = k_0 e^{-E/RT_2} t_2 \qquad (9.26)$$

9.7 Concept and Review Questions

1. Why are zeroth-order reactions not encountered often?

2. How is equilibrium different from steady state?

3. Why is equilibrium important in the context of transport?

4. At higher temperature, does the solubility of gases increase or decrease?

5. How long does it take for a process to come to complete equilibrium with the surroundings?

6. When considering the transport of pollutants through the soil to groundwater, is a higher partition coefficient more desirable?

Further Reading

Foust, A. S., L. A. Wenzel, C. W. Clump, L. Mans, and L. B. Anderson. 1960. *Principles of Unit Operations*. John Wiley & Sons, New York.

Henderson, S. M. 1952. A basic concept of equilibrium moisture. *Agric. Eng.* 33 (1): 29–31.

ICT. 1933. International critical tables of numerical data, physics, chemistry and technology. Volume II. Published for the National Research Council of the United States of America by McGraw-Hill, New York.

Kasting, G. B. and N. D. Barai. 2003. Equilibrium water sorption in human stratum corneum. *Journal of Pharmaceutical Sciences*. 92(8): 1624–1631.

Masters, G. M. 1998. *Introduction to Environmental Engineering and Science*. Prentice Hall, New Jersey.

Siau, J. F. 1984. *Transport Processes in Wood*. Springer-Verlag, New York.

Smith, J. M. 1970. *Chemical Engineering Kinetics*. McGraw-Hill Inc., New York.

Smith, L. S. 1982. *Introduction to Fish Physiology*. T.F.H. Publications, Inc., Neptune City, NJ.

ten Berge, H. F. M. 1990. Heat and water transfer in bare topsoil and the lower atmosphere. Center for Agricultural Publishing and Documentation, Wageningen, the Netherlands.

Welty, J. R., C. E. Wicks, R. E. Wilson, and G. Rorrer. 2001. *Fundamentals of Momentum, Heat, and Mass Transfer*. John Wiley & Sons, New York.

Yang, W. J. 1989. *Biothermal-Fluid Sciences: Principles and Applications*. Hemisphere Publishing Corporation, New York.

9.8   Problems

9.1   Concentration in Various Units

Calculate the molar concentration, mass concentration, mole fraction, and mass fraction of nitrogen and oxygen in standard air (1 atmosphere, 20°C) assuming the air is made up of 79% nitrogen and 21% oxygen (percentages refer to partial pressure in relation to total pressure). Assume also that air can be treated as an ideal gas.

9.2   Mass Conservation: Salt Balance in an Agricultural Field

The salt content of the soil in a field is an important parameter in a crop's success. The minerals in these salts provide nutrients for the growth of the plants, but too much salt can be detrimental to the plants' health as well. 1) For the conditions given below, find the annual *change* in salt concentration per hectare (10,000 m^2) in the soil. 2) If the maximum salt concentration a particular field could tolerate is 5,000 kg/ha, has the field reached this limit?

  Rainfall occurs entirely in winter and amounts to 30 cm, with a total salt concentration of 40 ppm (kg salt/10^6 kg water). Capillary rise from shallow, saline groundwater occurs in the spring and fall, and amounts to 10 cm at a concentration of 1,000 ppm. During the summer growing season, 90 cm of irrigation water is applied with a concentration of 400 ppm. Drainage occurs during the summer and amounts to 20 cm, with a soluble salt concentration of 800 ppm. Fertilizers and soil amendments add an additional 120 g/m^2 of soluble salt, while the crops use up 100 g/m^2 of salts. Assume that the salt concentration at the beginning of the year was 1,000 kg/hectare and that there is no salt generation or decomposition within the soil. The density of water is 1000 kg/m^3.

9.3   Mass Conservation: Indoor Air Quality and Smoking

Tobacco smoke is often considered the most harmful and widespread contaminant of indoor air. Scientists have identified over 4,000 different chemical compounds in tobacco smoke of which at least 50 are known carcinogens. Long-term exposure to tobacco smoke has been linked to heart disease, cancer, respiratory illness in young children, and retardation of growth and development of fetuses during pregnancy.

  Consider an enclosed space with a volume of 12 m $\times$ 12 m $\times$ 4 m that has 30 persons smoking two cigarettes every hour. One of the gases coming out of the cigarettes is formaldehyde, and each cigarette may be assumed to emit 1.35 mg of formaldehyde. Conversion of formaldehyde to carbon dioxide can be assumed to be first order, and the rate of reaction is 0.40 per hour. Outside (fresh) air enters the enclosed space at the rate of 800 m^3 per hour. Assume the smoke becomes completely mixed with air and this smoke-mixed air leaves the enclosed space at the same rate that the fresh air enters. 1) What is the concentration of formaldehyde, in mg/m^3, in the room air at steady state? 2) The threshold for eye irritation due to formaldehyde is 0.05 ppm (parts per million). At the assumed temperature of 25°C and pressure of 1 atm for the enclosed space, the conversion for ppm uses the formula ppm $=$ mg/m$^3 \times 24.45/M$, where M is 30, the molecular weight of formaldehyde. Considering this threshold, what minimum flow rate of fresh air needs to be maintained for the enclosed space? (Adapted from Masters, 1991).

9.4   Effect of Altitude on Blood Oxygen Level

Consider a mountain climber at 10,000 m above sea level where the total air pressure is about 0.25 atm. Assume the air surrounding blood vessels, etc., has the same composition as that at sea level, i.e., 21% oxygen and 79% nitrogen. What fraction of oxygen (compared to that at sea level) will be dissolved in the blood at this lower pressure?

9.5   Equilibrium of O$_2$ in Blood

One of the primary functions of the blood is to carry oxygen from the lungs to cells in all parts of the body. One mechanism of transport is to dissolve the oxygen in the plasma as the blood passes through the lungs. The oxygen concentration equilibrates between the blood and the inhaled air. Assume that the concentration of oxygen in the inhaled air is not affected by the transport of oxygen into the blood plasma. In addition, the blood plasma is well mixed so the concentration of dissolved oxygen is constant throughout the vessel. Assume that blood plasma has the same material properties as water, and that the mole fraction of O$_2$ at 1 atm in 38°C air is 0.145, Henry's constant is 5.21×10^4 atm/mole fraction of O$_2$ in water, molecular weight of O$_2$ is 32, and molecular weight of H$_2$O is 18.

1) Find the concentration (percent by mass) of dissolved oxygen in the blood plasma. 2) Does this concentration seem reasonable? 3) What are the other factors involved in the transport of oxygen to the cells?

9.6   The Deep Diver and the Bends

Consider a diver in water down to a level where the total pressure is 3 atm. Consider air in tissue and in areas surrounding the blood vessels of the diver to be still at the same composition as the air before entering the water, i.e., 21% oxygen and 79% nitrogen. Use the appropriate chart in the book for the needed parameter. 1) How much more nitrogen will be dissolved in the blood at this higher pressure of 3 atm than in blood at 1 atm? Leave your answer in terms of mole fractions. 2) What happens to this dissolved nitrogen as the diver starts to go up rapidly?

9.7   Oxygen Availability for Fish in Aquaculture

Consider an aquaculture system where water containing fish is surrounded by air. Calculate the fraction of original oxygen that will be retained in the water as the temperature of the water and the air increases from 5°C to 15°C.

9.8   Dissolved Oxygen in Water

The amount of oxygen dissolved in water is an important parameter for industrial waste treatment. Often it is desired to know the dissolved oxygen level in the efflux. Determine the saturation concentration of oxygen (in mg/liter) in pure water that is exposed to dry air (0.21 mole fraction O_2) at 1 atm and 20°C.

9.9   The "Zing" in Soft Drinks

In the making of carbonated soft drinks, dissolved CO_2 is an important ingredient which provides the necessary "zing" to the beverage. We need to maintain a desirable level of CO_2 concentration in the drink. Henry's constant for CO_2 in water at 10°C is 1040 atm/mole fraction. At a temperature of 10°C, what partial pressure of CO_2 (in Pa) must be kept in the gas surrounding the beverage to keep the CO_2 concentration in the water at 0.9×10^{-4} kg CO_2/kg of water?

9.10   Moisture Equilibration during Storage

A potato that has reached equilibrium with drying air at 60°C and a humidity of 0.04 kg of moisture per kg of dry air is packaged in a bag that is impermeable to moisture. The bag initially had air at 60°C and 0.008 kg of moisture per kg of dry air. Consider the bag to have 95% by weight of potato (dry potato + water) and 5% by weight of air (dry air + water). The equilibrium moisture content of potato at 60°C is shown in Figure 9.18. We would like to know the final moisture content of the potato in the sealed bag. 1) Formulate the problem, i.e., write the equation from which the final moisture content of the potato in the package can be calculated. *(Hint: Set up a mass conservation equation, where moisture lost from the potato is gained by the air.)* 2) Solve for the final moisture content in step 1.

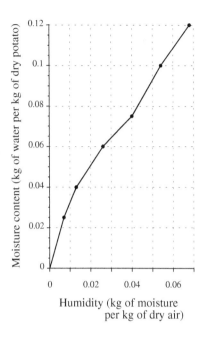

Figure 9.18: Equilibrium moisture content of potato.

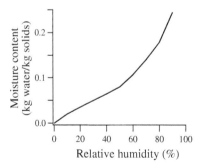

Figure 9.19: Equilibrium moisture content of soap.

9.11   Moisture Loss During Storage

Consider Figure 9.19 on the equilibrium moisture content of soap at approximately 25°C. Note that the moisture content is given on a dry solids basis. Consider a packaged bar of soap initially of weight 100 g (of which 40 g is water). The soap was left unpackaged in your bathroom during the summer. The air temperature in the bathroom stayed pretty much the same at approximately 25°C. However, the relative humidity in the summer was about 60% in the bathroom while in the winter it was down to 10%. 1) Determine the weight loss in soap after it was unpackaged in summer and left for several days. 2) Determine the weight loss in the soap from the summer to the winter.

9.12   Change in Moisture Content in a Hygroscopic Material during Storage

The equilibrium moisture content of a type of wood is given in Figure 9.10, where moisture content is given by water content per unit of total (water + dry wood) weight. Calculate the moisture that will be gained per unit of dry weight of a certain wood when stored at 21°C if the relative humidity changes from 20% to 80%.

9.13   Half–Life in a First-Order Reaction

If a first-order reaction has an activation energy of 25,000 cal/mol and, in the equation $k'' = k_0'' e^{-E_a/RT}$, k_0'' has a value of 5×10^{13}/s, at what temperature will the reaction have a half-life of (1) 1 min, and (2) 1 month (30 days)?

9.14   Decay of Pesticides in Soil

Consider two pesticides, aldicarb and atrazine, in the soil degrading independently as a first-order reaction given by $c = c_0 e^{-k''t}$. They were initially applied in equal amounts to the soil. The soil was tested 30 days after application and the relative concentration $c_{aldicarb}/c_{atrazine}$ was found to be 0.70. Calculate the half-life of aldicarb in the soil if the half-life of atrazine in the soil is 60 days.

9.15   Heating Shorter at Higher Temperatures Can Keep More Nutrients in Food

Moved to solved problems.

9.16 Adjustment of Sterilization Time

Sterilization of foods and biomaterials involves killing of bacteria that is often assumed to follow first-order reaction kinetics. A particular food material was to be heated at 121°C for 6 minutes to reduce the bacterial concentration to a very low value. The ratio of activation energy to the universal gas constant, E/R, for bacterial death is 35751 K. Due to the failure of the heating system, the food could only be heated at 111°C. To reduce the bacterial concentration to the same low final value, how many minutes should the food be heated at this lower temperature?

9.17 Indoor Air Quality

Consider the problem of smoking in an enclosed room. Instead of the usual steady-state assumption, we would like to consider the transient build-up of formaldehyde levels in the room starting with a zero value. The relevant data is as follows: a) Volume of the room is 576 m^3; b) 30 persons are smoking 2 cigarettes an hour, each cigarette producing 1.35 mg of formaldehyde; c) Conversion of formaldehyde to CO_2 is first order with a rate of reaction of 0.4/hour; d) Outside air comes in at 800 m^3/hour with no formaldehyde in it; e) The smoke becomes completely mixed with air and this smoke-mixed air leaves the enclosed room at the same rate that the fresh air enters. 1) Write a mass balance for smoke over time Δt in which the concentration changes by Δc. 2) Make $\Delta t \to 0$ and write the differential equation for c. 3) Solve the equation. 4) Plot approximately the transient solution.

9.18 Oxygen Concentration in a Crowded Room

We know our classroom can get fairly warm during the course of the lecture. It would be nice to check if, in addition, the O_2 level in the room also goes down, adding to the poor indoor air quality. Consider the schematic in Figure 9.20. The airflow rates in and out are given by Q. The concentration of oxygen coming into the room is that corresponding to outside air. Consider this concentration of oxygen to be also the initial concentration in the room. There is enough mixing of air in the room such that the entire room is at one concentration c at any time t. The rate of oxygen consumption is r amount of oxygen mass per person per unit time. The relevant data are as follows: The volume of the room is 5 m × 3 m × 3 m. The air flow rates in and out are 0.01 m^3/s. The outside and the room air temperature are at 25°C and at a pressure of one atmosphere. The outside air has 21% by volume of oxygen such that the partial pressure of oxygen in outside air is 0.21 atm. There are 80 persons in the room. A generally accepted value for constant rate of oxygen consumption in an adult human is 550 liters of pure oxygen per day. 1) Calculate the concentration of oxygen in

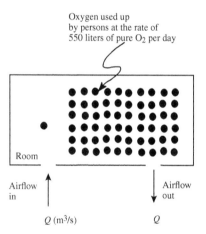

Oxygen used up by persons at the rate of 550 liters of pure O$_2$ per day

Room

Airflow in

Q (m^3/s)

Airflow out

Q

Figure 9.20: Schematic for Problem 9.18.

air, in kg/m^3, that is coming into the room. 2) Convert the oxygen consumption data provided into its mass units, in kg/min per person. 3) Write a mass balance for oxygen over time Δt, using symbols only (no data, so it looks clean). 4) Convert this mass balance to a differential equation for concentration as a function of time. 5) Solve the differential equation to obtain concentration as a function of time. 6) Plug in all the numerical values and comment whether the concentration reaches a steady state and if that happens during the class period of 50 minutes.

9.19   Quantifying Thermal Burn as a Zeroth-Order Reaction

Burns have been modeled as zeroth-order reactions. Intuitively, we can see that the same burn can happen due to different combinations of temperature and time. Experimentally, it is known that a second-degree burn occurs for either a 70°C exposure for 1 s or for a 60°C exposure for 5 s. What duration of exposure in seconds would cause a second-degree burn at 50°C? *(Hint: Think of the second-degree burn as reaching a final concentration value of the reaction.)*

9.20   Equilibrium between Liquid and Gas Phase

Although the data is kept confidential by the manufacturers, quite likely there are different amounts of CO_2 in different sodas that contribute to the individual flavor, mouthfeel, and anti-microbial effects. For the sake of this problem, suppose the ratio of CO_2 dissolved in Pepsi is 1.25 times that dissolved in 7-UP. Considering the space above the liquid in the cans to consist of only CO_2, how does the gas pressure in a can of Pepsi compare with that in a can of 7-UP?

9.21   Time–Temperature Equivalence in Hyperthermia Treatment Planning

In planning a hyperthermia treatment, a widely used relationship between thermal exposure time and temperature is given by

$$t_{43} = t\,(t_R)^{43-T} \tag{9.27}$$

where t_{43} is the time of heating at 43°C to achieve an equivalent heating of an injury (called isoeffect) as would be achieved by an actual time of heating of t minutes at a temperature of T°C for the same injury. Here T is assumed to be close to 43°C. The quantity t_R is the ratio of times that produce equivalent heating when temperature is increased by 1°C. Considering thermal injury to be a zeroth-order reaction, derive the above relationship. *(Hint: First use the two times for temperature 1°C apart. Next do* *the same for temperatures $43°C$ and T. Use the approximation $e^{-E/RT_1T_2} \sim e^{-E/RT_1^2}$ when T_1 and T_2 are close. Be careful about when temperatures are in $°C$ and when they are in K.)*

9.22 Drying Out in Air

After a class demonstration, a professor had left some of the potato slices out in the classroom. The equilibrium moisture curve for potato is shown in Figure 9.21. Knowing initially they had 4 kg water/kg dry solids, state approximately how much water/kg of dry solids the potato slices would lose after a long time, when the room relative humidity (in fraction) stays at 0.5.

9.23 Mass Balance in an Aquaculture System

A tank-based intensive aquaculture system has a tank volume of V m^3. Fresh water is coming in at a rate of F m^3/s, which is also the rate at which water is leaving the tank. The oxygen concentration in water at the inlet is c_i. Water in the tank is circulating enough such that the oxygen concentration, c [g/m^3], can be assumed uniform throughout the tank. Oxygen is consumed by the fish at a rate of k_1 g per m^3 per second. Oxygen is produced by the plants in the tank at a rate of k_2 g per m^3 per second. Air is bubbled through the tank and the oxygen transfer from the air to the tank is given by $k_3(c_b - c)$, where k_3 is the rate constant in 1/s, and c_b is the appropriate concentration of oxygen in air in the bubble (a constant value).

1) Write a mass balance for the tank for change in oxygen concentration, Δc, over a time, Δt. 2) Develop the differential equation for concentration as a function of time. 3) Solve the differential equation, assuming an initial concentration of c_i. 4) Plot the concentration versus time. 5) What is the steady-state concentration? 6) What is the steady-state concentration if there is no bubbling of air?

9.24 Time–Temperature Combination for Burns

Burns have been modeled as a zeroth-order reaction. Intuitively, we can see that the same burn can happen due to different combinations of temperature and time. Experimentally, it is known that a second-degree burn occurs for either a $70°C$ exposure for 1 s or for a $60°C$ exposure for 5 s. In a burn situation, where the pathologist confirms that it is a second-degree burn and we happen to know the duration of exposure to high temperature to be 10 s, what must have been the temperature of exposure? *(Hint: Think of the second-degree burn as reaching a final concentration value of the reaction.)*

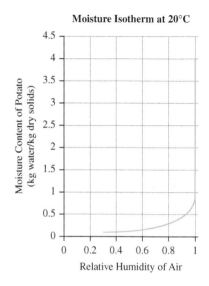

Figure 9.21: Equilibrium moisture curve (isotherm) for potato at the temperature needed.

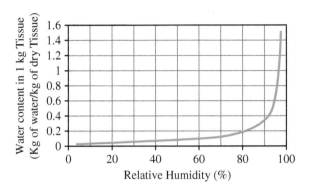

Figure 9.22: Equilibrium moisture content of skin as a function of relative humidity of air. Data from Kasting and Barai (2003).

9.25   Temperature Effect on Dissolved Gases

Assuming the partial pressure of oxygen does not change across seasons, what is the ratio between the dissolved oxygen concentration in the surface of the water in Cayuga Lake in winter (0°C) and the concentration in summer (20°C)?

9.26   Equilibrium between Skin Surface and Air

Based on Figure 9.22, how much moisture, in g of water per g of total weight of tissue (dry tissue plus water), would your skin surface lose as the weather changes from 20% to 90% relative humidity?

9.27   Mass Balance for Chloroform in the Air over an Indoor Swimming Pool

Although chlorine is the most widely used biocide for water disinfection due to it being inexpensive, one of the byproducts formed in the disinfection process, chloroform, can cause cytotoxicity and therefore have adverse health effects. Consider a swimming pool of floor space (all covered with water) 50 m × 25 m and a room height of 12 m. Clean (chloroform-free) air is blown into the room and the room air can be considered well mixed. We would like to compute the flow rate of clean air into the room needed to maintain the chloroform level in the room air below a certain level.

1) The source of chloroform in the room air is that coming from the air–water interface at the pool's surface, through convective mass transfer (this is analogous to convective heat transfer, i.e., mass flux = $h_m(c_{surface} - c_\infty)$, as will be studied later.

Here $c_{surface}$ and c_∞ refer to concentrations in the air at the air–water interface and bulk, respectively). The amount of chloroform in the water is 10 μg/l (a liter, l, is 1000 cm^3), Henry's law constant for chloroform in the air–water system is 304 Pa $\cdot$ m^3/mol, and the mass transfer coefficient for the airflow present is 0.03×10^{-2} m/s. The molecular weight of chloroform is 119.5. Water and air can be assumed to be at 20°C. Calculate this source term in g/s, in terms of concentration of chloroform in the air. 2) Write a mass balance for the chloroform in air in terms of its concentration change, Δc, over a time, Δt. 3) Develop the differential equation for concentration as a function of time. 4) Solve the differential equation assuming an initial concentration of c_i (the concentration before air flow was started). 5) Sketch a plot of concentration versus time assuming the initial concentration is higher than the steady-state concentration. 6) If we want to maintain the steady-state concentration of chloroform below 0.375 μg/l, what air flow rate (in m^3/s) is necessary?

9.28 How Blood Alcohol Changes with Time

Figure 9.23 shows measured blood alcohol concentration over time after 1–4 drinks taken rapidly. We want to develop a similar model for how blood alcohol changes with time after consumption. Do not worry about comparing your units with those in Figure 9.23, as we are trying to capture the qualitative picture in this simple model. Let c be the concentration of alcohol (cm^3 of ethyl alcohol/cm^3 of blood) at any time, t (in hr), V be the blood volume in the body in cm^3, and k (in 1/hr) be the rate constant for approximate first-order decay (metabolization) of alcohol in blood (a major assumption as the decay has also been modeled as reactions of various other rates). No alcohol leaves the blood volume.

1) We will greatly simplify the entry of alcohol into the bloodstream from the stomach as $m = m_f(1 - e^{-at})$ where m_f is the amount of alcohol (in ml, same as cm^3) in one or more drinks taken rapidly and a has the unit of 1/hr. Write the expression for the rate of alcohol entry into the bloodstream (the time derivative of m) that will be needed for setting up a mass balance in step 2. 2) By setting up an appropriate mass balance, derive the differential equation for concentration of alcohol in the blood stream as a function of time. ***Important: Be careful in making sure the units of all the terms are consistent.*** 3) Solve for concentration, c, as a function of time assuming no initial alcohol in the blood. The following solution is provided for your convenience: Solution to

$$\frac{dc}{dt} + Pc = Q \qquad (9.28)$$

where P and Q do not have to be constants, is

$$c = \left(e^{-\int P dt} \int Q e^{\int P dt} dt\right) + D e^{-\int P dt} \qquad (9.29)$$

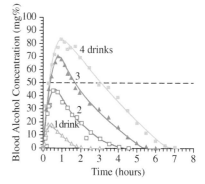

Figure 9.23: Blood alcohol concentration (BAC) after the rapid consumption of different amounts of alcohol by eight adult fasting male subjects.

where D is the constant of integration. 4) Assume one drink has 18 ml of alcohol. An average individual's body has about 4.8 liters of blood. The rate constants a and k are 4.2/hr and 0.693/hr, respectively. Assume an average individual appears sober when the concentration of alcohol decays to 0.002 cm^3 of ethyl alcohol/cm^3 of blood. It is commonly assumed that it requires 1 hour per drink to become sober. Test if this assumption is reasonably supported by this model by checking the alcohol concentration for two situations—1 and 3 drinks. *General warning: the value of 0.002 is not to be confused with those in everyday use.* 5) At what time after consumption is the alcohol content in blood at its highest value for one drink?

9.29 Dissolved Oxygen in a Bioreactor

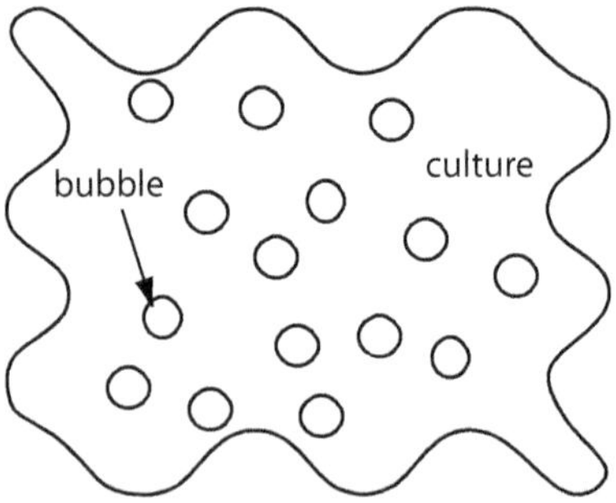

Figure 9.24: Schematic of a bacterial culture with air bubbling through it.

The cell productivity of bacteria depends on the dissolved oxygen concentration in the medium. A well-mixed culture of *E. coli* has a volume, V, at a concentration of N [bacteria per cm^3] that consumes oxygen at a constant rate of k [cm^3 O$_2$ per bacteria per second]. Oxygen is also bubbled through at a rate of $k_2(c_0 - c)$, where c_0 is the concentration of oxygen in the bubble, c is the concentration of oxygen in the culture, and k_2 has the units of 1/s. Let c_i be the amount of O$_2$ initially in the culture when the bubbling starts. All concentrations have units of cm^3 O$_2$ per cm^3 of total culture volume.

1) Develop the needed differential equation and initial condition that can be solved to obtain the concentration of O$_2$ as a function of time. 2) Solve the equation to obtain the concentration of O$_2$ as a function of time. 3) Provide an equation for the steady-state O$_2$ concentration in the culture. 4) To double the steady-state bacterial concentration, how would k_2 in the rate of bubbling need to be changed?

9.30 Predicting Phosphorous Concentration in Cayuga Lake

Cayuga Lake is listed on the list of NYS DEC impaired waters for having too high levels of phosphorous in its southernmost end. Phosphorous is ecologically harmful in high concentrations and leads to eutrophication and excessive algal bloom. We would like to relate the phosphorous level to flow and treatment conditions.

The total inflow for the lake can be approximated as a stream of $\dot{V}$ m^3/s with a phosphorus concentration of c_{pi} kg/m^3. For now, the lake does not evaporate significantly but empties at the same rate of $\dot{V}$ m^3/s. Thus, its current volume M m^3 is constant. For this analysis, the lake can be assumed to be well mixed. Phosphorous is degraded by absorption to sediment in the lake bottom by a first-order kinetic reaction with rate constant k.

1) Perform a mass balance on the lake over time to obtain a differential equation for the phosphorus concentration, c_p, as a function of time, t. 2) Under steady-state

conditions, solve for k that would satisfy the EPA standard phosphorous concentration of $2 \times 10^{-5} \text{kg/m}^3$ in the lake (i.e. concentration in the lake is equal to this value), for the situation when $\dot{V} = 2 \text{ m}^3/\text{s}$, $c_{pi} = 0.00024 \text{ kg/m}^3$, and $M = 339800 \text{ m}^3$. 3) Explicitly solve for c_p as a function of t for an initial concentration c_{pi}. 4) How would you rewrite the equation *(Important—do not solve)* in step 1 if you decided to include the evaporative loss of water, estimated as $h_m A(c_s^{vapor} - c_\infty^{vapor})$ kg/s, where c_s^{vapor} and c_{vapor}^∞ refer to the concentration of water vapor at the water surface and in the bulk air, respectively? Note that M is no longer a constant. 5) Write the differential equation for change in M over t *(Important—do not solve)* from which you can obtain M for the scenario in step 4).

9.31 Destruction of Pathogenic Bacteria

In pasteurization of milk, many disease-producing microorganisms are destroyed by applying heat. For example, Salmonella is one of the many microorganisms that pasteurization targets. Destruction of microorganisms is often modeled as first-order reactions. Heating time–temperature combinations for milk pasteurization can be one of the following, under the assumption that any time–temperature combination will lead to the same level of destruction of the microorganism.

Temp (°C)	Duration of heating (s)
89	1
100	0.01

1) Compute E/R for the microorganism. 2) Instructions for home pasteurization say "Milk must receive a minimum heat treatment of 66°C continuously for 30 minutes." Are we safe if we follow this instruction, i.e., would the microorganism concentration be reduced to the required level if this instruction were followed? *(Hint: Calculate the time needed at 66°C to achieve the same level of destruction of the microorganism and compare with the time given in the instruction.)*

9.32 Estimating Absorption of an Orally Administered Drug

Human drug absorption in the gastrointestinal tract (GI) can be modeled by considering the stomach, small intestine, and blood plasma as three compartments (Figure 9.25) and performing a mass balance of drug in each of the three compartments separately. Assume that there is no drug initially in the intestine and blood plasma, and that the drug is well mixed in each compartment.

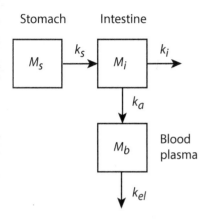

Stomach   Intestine

Figure 9.25: Absorption–disposition kinetic model. Symbols M_s, M_i, and M_b represent the total amount of drug at any time in that compartment.

1) The amount of drug, taken orally, that initially enters the stomach is M_o. This drug is emptied from the stomach (going to the small intestine) as a first-order process with a rate constant of k_s 1/s, i.e., as $k_s M_s$. Write the mass conservation differential equation for the amount of drug in the stomach, M_s, and solve for M_s as a function of time. 2) The drug leaving the stomach, $k_s M_s$, enters the small intestine. In the small intestine, drug is lost to intestinal transit at the rate of k_i and it also leaves to go to the blood plasma at the rate of k_a, both to be treated as first-order processes, like the amount leaving in step 1. Solve for the concentration of drug in the small intestine, M_i, as function of time. For your convenience, the solution to

$$\frac{dc}{dt} + Pc = Q \tag{9.30}$$

where P and Q do not have to be constants, is

$$c = \left(e^{-\int P \, dt} \int Q e^{\int P \, dt} \, dt \right) + D e^{-\int P \, dt} \tag{9.31}$$

where D is a constant of integration. 3) The drug enters the blood plasma as a fraction F of the amount $k_a M_i$ from the intestine (noted in step 2). In the plasma, the drug is also eliminated at the rate of k_{el} (first-order process). Write *(do not solve)* the differential equation from which you can calculate the amount of drug in blood plasma with time.

Chapter 10

MODES OF MASS TRANSFER

CHAPTER OBJECTIVES

After you have studied this chapter, you should be able to

1. Explain the process of molecular diffusion and its dependence on molecular mobility. Explain the process of capillary diffusion.

2. Explain the process of dispersion in a fluid or in a porous solid.

3. Understand the process of convective mass transfer as due to bulk flow added to diffusion or dispersion.

4. Explain saturated flow and unsaturated capillary flow in a porous solid.

5. Have an idea of the relative rates of the different modes of mass transfer.

6. Explain osmotic flow.

KEY TERMS

- **diffusion and diffusivity**
- **dispersion coefficient**
- **capillarity, osmotic flow**
- **mass and molar flux**

- **Fick's law**
- **Darcy's law**
- **Convective mass transfer**

In this chapter, we will study the fundamental ways mass can be transferred. Figure 10.1 shows how the contents of this chapter relate to other chapters on the subject of mass transfer. We study the processes of molecular and capillary diffusion, dispersion, and bulk flow or convection. Of these processes, diffusion and dispersion can be formulated the same way and are treated together. These two processes together are also the subject of later chapters. Hydraulic or Darcy flow in a porous medium is introduced in this chapter as a description for bulk flow through such media.

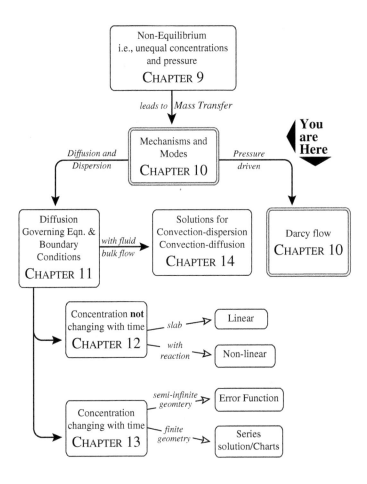

Figure 10.1: Concept map showing how the contents of this chapter relate to other chapters in mass transfer.

10.1 A Primer on Porous Media Flow

Before we introduce the three modes of mass transfer— diffusion, dispersion, and convection in the following sections, it is important to note that the convection mode of mass transfer requires information about the velocities that we can obtain from studying pressure-driven flow in a course on fluid mechanics. The important case of pressure-driven flow through porous media is not typically studied in an undergraduate course. Flow through a porous medium is introduced here due to its importance in applications as varied as pollutant transport through soil in a field and water transport in a food material during drying. The introduction to porous media covered here will provide the foundations for studying transport due to capillary diffusion and convection–dispersion through such a porous medium.

Movement of liquid in a porous material can be described by Darcy's law, written as

$$n^v = -K \frac{\partial \mathcal{H}}{\partial s} \tag{10.1}$$

where n^v is the volumetric flux (also called specific discharge) in $m^3/m^2 \cdot s$, $\mathcal{H}$ is the hydraulic or water potential in m that is causing the flow, and s is distance along flow in m. The proportionality constant K is termed hydraulic conductivity. The units of K can be calculated as

$$[K] = \frac{[n^v]}{[\partial \mathcal{H}/\partial s]} = \frac{m/s}{m/m} = m/s \tag{10.2}$$

Hydraulic potential is the sum of pressure and matric potential h, and gravitational potential z:

$$\mathcal{H} = h + z \tag{10.3}$$

The matric potential measures the physical forces, such as capillarity, which bind the water to the porous matrix. The retention of water is a result of attractive forces between the solid and liquid phases. In a soil, for example, these matric forces enable it to hold water against such forces as gravity, evaporation, and uptake by plant roots. In soil, there are three mechanisms for binding of water to the solid matrix: direct adhesion of water molecules to solid surfaces by London–van der Waals forces, capillary binding, and osmotic binding in double layers. This matric potential is important as a driving force for flow in unsaturated soil and other systems such as in the cell walls of root cortex and leaf mesophyll tissue. Matric potential is always negative or zero. The pressure potential is due to hydrostatic or pneumatic pressure applied to water. The gravitational component of the water potential is simply due to a difference in depth z in the vertical (parallel to g) direction from a reference point, usually taken as soil surface or surface of a water table.

Like Fourier's and Fick's laws, Darcy's law (Eq. 10.1) is an empirical relationship. The hydraulic conductivity, K, represents the ease with which fluid can be transported through a porous matrix, and is discussed more fully in Section 10.1.2. The units for the volumetric flux n^v in Eq. 10.1 are $m^3/m^2 \cdot s$ or m/s. Since the units for n^v are the same as those for velocity, it is also called Darcy velocity. Note that, even though n^v has the units of velocity, it is not the true average velocity of fluid through the pores, as flow takes place through only the porous part of the cross-sectional area A. Let ϕ be the volumetric porosity, defined as the ratio of volume of void space (pore volume) to the bulk volume of a porous medium. Average areal porosity—that is, the ratio of void area to the total area at a cross section—can be considered the same as average volume porosity ϕ. Thus, the true average velocity of fluid through the pores is

$$v_{average} = \frac{n^v}{\phi} \qquad (10.4)$$

10.1.1 Example: Flow of Water through a Column of Packed Sand

Consider the flow of water through a column of packed sand. The column has a cross-sectional area of 250 cm² and a length of 150 cm. The packed sand has a porosity of 0.30 and a hydraulic conductivity of 15 m/day. For a pressure drop of 200 cm through the column, calculate 1) volumetric flux (also called *specific discharge*) n^v, 2) volumetric flow rate, 3) average velocity of fluid in the pores.

Solution

1. Volumetric flux or specific discharge is calculated using Darcy's law:

$$
\begin{aligned}
n^v &= K \frac{\Delta \mathcal{H}}{\Delta x} \\
&= 15 \left[\frac{m}{day} \right] \frac{200}{150} \left[\frac{m}{m} \right] \\
&= 0.023 \ cm/s
\end{aligned}
$$

2. The volumetric flow rate can be calculated from the volumetric flux as

$$
\begin{aligned}
n^v A &= 0.023 [cm/s] \times 250 [cm^2] \\
&= 5.75 \ cm^3/s
\end{aligned}
$$

3. The *average velocity of fluid in the pores* is given by

$$v(\text{average velocity}) = \frac{n^v}{\text{porosity}} = \frac{0.023}{0.3} = 0.077 \ cm/s$$

10.1.2 Physical Interpretation of Hydraulic Conductivity K and Permeability k

Since hydraulic conductivity K relates the flux to the potential causing the flow, it must depend on both matrix and fluid properties. As a first approximation, a porous medium can be considered a bundle of tubes of varying diameter embedded in the solid matrix, as shown by the schematic in Figure 10.2. Using this simple approximation, we explicate the physical meaning of K. From fluid mechanics, the volumetric flow rate Q_i in a tube of uniform radius r_i (Poiseuille's flow) is

Figure 10.2: Idealization of a porous medium as bundle of tubes of varying diameter and tortuosity.

$$Q_i \quad = \quad -\frac{\pi r_i^4 \rho g}{8\mu}\frac{\partial h}{\partial s}$$

If ω_i represents the number of pores (tubes) in the ith pore size class with radius r_i, and Q_i is the discharge rate per pore in that group, then the total volumetric flux, n^v, through the collection of pores is given by

$$
\begin{aligned}
n^v &= \frac{\sum_i \omega_i Q_i}{A} \\
&= -\frac{\rho g}{8\mu} \frac{\sum_i \pi \omega_i r_i^4}{A} \frac{\partial h}{\partial s}
\end{aligned}
$$

where A is the total cross-sectional area. All pores do not run parallel to the flux direction. To account for the lengthened distance, a tortuosity factor τ is introduced that is the ratio of the actual roundabout path along the pore to the apparent, or straight, flow path. In soils, for example, the value of τ is typically between 1 and 2. Introducing tortuosity, the above equation is written as

$$
n^v = -\frac{\rho g}{8\mu\tau} \frac{\sum_i \pi \omega_i r_i^4}{A} \frac{\partial h}{\partial s} \tag{10.5}
$$

If $\Delta\beta_i$ is the volume fraction of pores with radius r_i, $\Delta\beta_i$ can be approximated as an area fraction

$$
\Delta\beta_i = \frac{\omega_i \pi r_i^2}{A} \tag{10.6}
$$

Using Eq. 10.6, the volumetric flux n^v is written as

$$
n^v = -\underbrace{\frac{\rho g}{8\mu\tau} \sum_i \Delta\beta_i r_i^2}_{K} \frac{\partial h}{\partial s} \tag{10.7}
$$

Comparing Eq. 10.7 with Darcy's law (Eq. 10.1), we can write

$$
K = \frac{\rho g}{8\mu\tau} \sum_i \Delta\beta_i r_i^2 \tag{10.8}
$$

which can be rearranged as

$$
K = \underbrace{\frac{\rho g}{\mu}}_{\substack{\text{fluid} \\ \text{property}}} \underbrace{\frac{1}{8\tau} \sum_i \Delta\beta_i r_i^2}_{\substack{\text{matrix} \\ \text{property}}} \tag{10.9}
$$

Equation 10.8 shows that, as expected, K depends on both fluid and matrix properties. The relevant fluid properties are density ρ and viscosity μ. The relevant solid matrix properties, as given by Eq. 10.8 are pore size distribution, shape of pores, porosity, and tortuosity. The sole effect of the matrix property can be included in *permeability* or *intrinsic permeability* k such that

$$k = \frac{1}{8\tau} \sum_i \Delta\beta_i r_i^2 \tag{10.10}$$

so that

$$K = \frac{k\rho g}{\mu} \tag{10.11}$$

The ratio ρ/μ represents the effect of fluid properties. To show the units of the various quantities in the above equation, we can write

$$K\left[\frac{m}{s}\right] = \frac{k\,[m^2]\,\rho\left[\frac{kg}{m^3}\right]g\left[\frac{m}{s^2}\right]}{\mu\left[\frac{kg}{ms}\right]}$$

Using this definition of permeability, k, Darcy's law can be rewritten as

$$\begin{aligned} n^v &= -K\frac{\partial h}{\partial s} \\ &= -\frac{k}{\mu}\frac{\partial(\rho g h)}{\partial s} \\ &= -\frac{k}{\mu}\frac{\partial P}{\partial s} \end{aligned} \tag{10.12}$$

since pressure is related to the head as $P = \rho g h$. Equation 10.12 is the alternate form of Darcy's law written in terms of gradient in pressure, P, instead of head, h. The units of k can be seen from Eq. 10.10 to be m^2.

When the solid is saturated with liquid, as in saturated soil where pores are completely filled with water, both the matrix contribution and the fluid contributions are constant. This leads to a constant hydraulic conductivity of a given solid saturated by a particular fluid. For an unsaturated solid, the hydraulic conductivity can vary dramatically, as described in Section 10.1.11.

10.1.3 *Example: Flow Due to Pressures Generated from Evaporation*

Small explosions can occur in rapid heating of water-containing materials (e.g., see Rebolleda et al., 1999). Consider a very simplified situation involving laser heating of a tissue where all the laser energy is absorbed in a region 10 mm below the skin

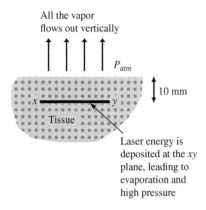

All the vapor
flows out vertically

P_{atm}

10 mm

x ———— y

Tissue

Laser energy is
deposited at the xy
plane, leading to
evaporation and
high pressure

Figure 10.3: Schematic of laser energy
deposited and resulting evaporation at a
depth of 10 mm from surface.

surface. All of this absorbed energy is assumed to turn to vapor instantly, and no
heat goes into diffusion. All of this vapor generated is transported through the 10 mm
layer into the atmosphere due to pressure-driven flow (with the tissue as the porous
media). The process can be considered at a steady state. The permeability of the tissue
is $10^{-14} m^2$, laser flux (fluence) is 13000 W/m², viscosity and density of vapor are
1.7×10^{-5} Pa · s and 1.2 kg/m³, respectively, and latent heat of vaporization of water
is 2.26×10^6 J/kg.

1) Write an expression for heat balance to relate the laser flux, F, to the mass flux
of vapor (don't plug in numbers). 2) Calculate the pressure that would be generated in
the tissue to maintain the constant mass flux of vapor calculated in step 1. 3) *Answer
qualitatively* if you would expect the tissue to explode under some conditions.

Solution

1) Since all the laser energy goes into evaporation, equating this energy to the mass
flux of evaporation, $\dot{m}$ kg/m² · s, over an area, A,

$$FA \quad = \quad \dot{m} A \lambda$$

or

$$F \quad = \quad \dot{m} \lambda$$

where λ is the latent heat of evaporation.

2) The pressure gradient developed due to this evaporation makes the vapor flow out
into the atmosphere. The flow rate would be related to the pressure gradient by Darcy's
law (Eq. 10.12)

$$\dot{m} = \rho \left(-\frac{k}{\mu} \frac{P_{atm} - P}{x} \right)$$

where x is the distance between the evaporation plane and the surface open to the
atmosphere, and ρ is the density of the vapor that converts from the volumetric flow
rate given by the Darcy's law. Equating the two,

$$F = \dot{m} \lambda$$
$$= \left(\frac{\rho k}{\mu} \frac{P - P_{atm}}{x} \right) \lambda$$

Solving for P,

$$P - P_{\text{atm}} = \frac{Fx}{\lambda} \frac{\mu}{\rho k}$$

$$= \left(\frac{13000 \text{ W/m}^2 \times 0.01 \text{ m}}{2.26 \times 10^6 \text{ J/kg}} \right) \frac{1.7 \times 10^{-5} \text{ Pa} \cdot \text{s}}{10^{-14} \text{ m}^2 \times 1.2 \text{ kg/m}^3}$$

$$= 81489.68 \text{ Pa}$$

$$= 0.0815 \text{ MPa}$$

3) The higher the laser flux, the higher the pressures developed from evaporation. Pressure causes mechanical stresses in the tissue and if the failure stress for the tissue is exceeded, it can rupture. Such "popping" has been reported in the literature.

10.1.4   Capillarity and Unsaturated Flow in a Porous Medium

Capillary flow is due to the difference between the relative attraction of the molecules of the liquid for each other and for those of the solid. As an example, such a difference causes the rise of water in an open tube of small cross-section (Figure 10.4). For the column of water of height h at equilibrium shown in this figure, the hydrostatic pressure, $P(= \rho g h)$, matches the capillary attraction given by $2\gamma/r$, where γ is the surface tension. Equating the two pressures results in the following relationship for the height of capillary rise:

$$h = \frac{2\gamma}{\rho g r} \qquad (10.13)$$

Data on surface tension of water is given on page 581 in Appendix D.8. As the radius becomes very small, capillary rise increases significantly. Capillarity is the reason, for example, that the soil does not get completely drained by gravity. Capillarity is also the mechanism by which water can rise to the top of a tree as tall as 30 m and be available for transpiration. The capillaries in a tree which are formed by numerous interstices of the cell wall of the xylem vessels aid the tree in transporting water from the roots and base to the upper branches and leaves. A representative radius of these channels in the cell was estimated as 5×10^{-9} m (Nobel, 1974). Using Eq. 10.13, it can be seen that a capillary of this size could support a water column 3 km in height.

In a porous solid, capillarity will cause the liquid to be attracted more strongly or held more tightly when there is less of it, i.e., at lower concentrations of the liquid. Conversely, the liquid will be held less tightly when there is more of it. This sets up a situation where differences in capillary action result in a flow of liquid from higher concentration (relatively loosely held) to lower concentration (more tightly held). This

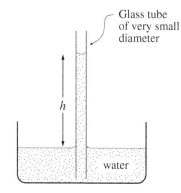

Figure 10.4: Capillary attraction between the tube walls and the fluid causes the fluid to rise.

is referred to as unsaturated flow. Unsaturated flow is extremely important, for example, in drainage of soils or drying of food materials. In an unsaturated solid, capillarity can be the primary mode of transport for the liquid. In contrast to the positive pressures (applied or gravitational) of liquid in a saturated solid, water pressure is negative or the water is attracted to the solid. As an example, Figure 10.5 shows this negative matric potential due to capillarity and other effects in a soil as a function of moisture content. This difference in the negative pressures drives the flow when the soil is unsaturated. As will be shown in Section 10.1.11, this capillary flow can be treated mathematically as analogous to molecular diffusion.

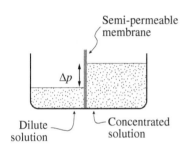

Figure 10.6: Osmotic flow from a dilute to a concentrated solution through a semi-permeable membrane.

10.1.5 Osmotic Flow in a Porous Medium

Osmotic flow, or osmosis, is the transport of a solvent from a region of low solute concentration to a region of high solute concentration through a semi-permeable membrane. A semi-permeable membrane is a porous structure that allows some of the species (here the solvent) to go through but stops the passage of other species (here the solute). Consider two solutions separated by a semi-permeable membrane, as shown in Figure 10.6. The solvent moves from low solute concentration to high solute concentration until a hydraulic pressure of magnitude Π is developed that exactly opposes the osmotic flow, stopping further flow. The pressure magnitude (Π) is called the osmotic pressure and it depends on the concentration c of the solute. This relationship is given by the Van't Hoff law

$$\Pi = cRT \qquad (10.14)$$

where c is the total concentration of the solutes in a solution, T is the absolute temperature, and R is the gas constant. Note the similarity of this equation to the ideal gas law (Section 9.1.1). Thus, osmosis is in the direction to reduce the solute concentration gradient, i.e., toward equalizing the concentration. Since cell membranes are permeable to water, water can flow through them. An example of osmotic flow is how a cell can shrink during freezing of a tissue, as explained in Figure 7.10 on page 233.

Since osmotic pressure is equivalent to hydraulic pressure, as shown in Figure 10.6, Darcy's law for flux can be generalized to include both hydraulic and osmotic pressures in a porous medium as

$$n^v = -\frac{k}{\mu}\frac{\partial}{\partial s}(P - \Pi) \qquad (10.15)$$

In Eq. 10.15 for osmotic flux, if the applied or hydraulic pressure P does not change,

$$n^v = \frac{k}{\mu}\frac{\partial \Pi}{\partial s} \qquad (10.16)$$

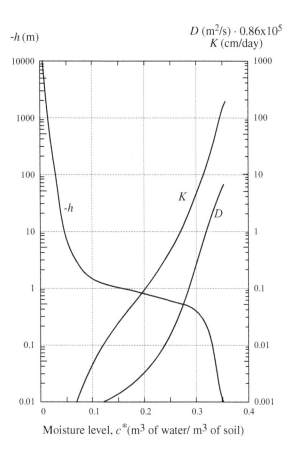

Figure 10.5: Typical relationship of capillary diffusivity to the moisture content in porous medium. Shown is data for soil where moisture diffusivity D is calculated from soil water characteristics (matric potential) h and hydraulic conductivity K. Here moisture content, c, in kg of water per m^3 of soil is related to c^* above as $c \approx 1000c^*$. Reprinted from Adapted from Rijtema, P.E. 1969. Soil moisture forecasting. Report 513. ICW, Wageningen. Cited in *Elements of Soil Physics* by P. Koorevaar, G. Menelik and C. Dirksen, with permission from Elsevier, New York. 1983.

In biological literature, this equation is often written in terms of an osmotic pressure difference $\Delta\Pi$ as

$$n^v = L_p \Delta\Pi \tag{10.17}$$

where L_p is the membrane permeability that lumps together the effects of the porous structure of the membrane, the thickness of the membrane, and the fluid properties. Note that membrane permeability, L_p (unit m/Pa · s), is different from hydraulic permeability k (unit m^2), although the term permeability is used in either case. The direction for the flux in Eq. 10.16 is from low solute concentration (low osmotic pressure) to high solute concentration (high osmotic pressure).

10.1.6 Molecular Diffusion

In a material with two or more mass species whose concentrations vary within the material, there is a tendency for mass to move. Diffusive mass transfer is the transport of one mass component from a region of higher concentration to a region of lower concentration. Examples of diffusive mass transfer are plentiful. For instance, perfume from one corner of a room can eventually be smelled from everywhere in the room, even if there is not much air flow in the room.

Diffusive mass transfer is analogous to diffusive heat transfer, described in Section 2.1. However, it is inherently more complicated since it deals with a mixture with at least two species. Diffusion is a natural, dynamic molecular process which tends to equilibrate the differences in concentration. There is net movement of a species from higher to lower concentration, simply due to the random molecular movement. Note the similarity between the mechanisms of molecular and thermal diffusion.

For molecular diffusion whose driving force is a concentration gradient, the rate at which mass is transported per unit area (diffusive flux) is related to the concentration gradient as

$$\underbrace{j_{A,x}}_{\text{Diffusive flux}} = -\underbrace{D_{AB}}_{\text{Diffusivity}} \underbrace{\left(\frac{dc_A}{dx}\right)}_{\substack{\text{Concentration} \\ \text{gradient}}} \tag{10.18}$$

where $j_{A,x}$ is the diffusive flux in $kg/m^2{\cdot}s$ of species A at point x, c_A is the concentration of A in kg/m^3, x is distance in m, and D_{AB} is the mass diffusivity of A in B in m^2/s, which is also called the *diffusion coefficient*. This is known as *Fick's law of diffusion*. In molar units, $j_{A,x}$ is in $kmol/m^2{\cdot}s$ and c_A is in $kmol/m^3$. The units for mass diffusivity D_{AB} are m^2/s, as can be seen from

$$[D_{AB}] = \frac{[j_{A,x}]}{\left[\dfrac{dc_A}{dx}\right]} = \frac{\left[\dfrac{kg}{m^2s}\right]}{\left[\dfrac{kg}{m^3m}\right]} = \left[\frac{m^2}{s}\right]$$

Physical interpretation of diffusivity

A solute introduced into a fluid at a certain point diffuses out. If we consider a 1-D diffusion process along x, then the resulting concentration profile is a Gaussian distribution (see Figure 10.7) in x at any time, with some solute molecules having moved further out from the origin than others. Since the molecules would move in the positive as well as negative direction, an average of the square of the displacements, $<x^2>$, can be defined as a measure of the movement of the molecules. This average is termed the mean-square displacement.

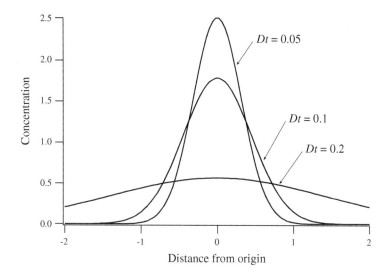

Figure 10.7: Concentration profiles at different times from an instantaneous source placed at zero distance. Here units of Dt and x are arbitrary, but related so that Dt/x^2 is dimensionless.

From the Gaussian distribution, the mean-square displacement $<x^2>$ for 1D diffusion can be evaluated with respect to the diffusivity by the relation (see, e.g., Barrow, 1981)

$$D = \frac{<x^2>}{2t} \tag{10.19}$$

The diffusivity is thus one half of the mean-square displacement per unit time for 1D diffusion. Using the square root of the mean-square displacement, an average diffusion velocity can be defined as

$$u_{diff} = \frac{\left(<x^2>\right)^{1/2}}{t} = \sqrt{\frac{2D}{t}} \qquad (10.20)$$

Note that the velocity u_{diff} is many orders of magnitude smaller than the molecular velocity. The molecules' movement in random directions is known as Brownian motion. Although molecular velocities are very large, any molecule encounters a large number of collisions per unit time. As a result, the net distance a molecule moves is much smaller, as shown by the diffusion velocity. For example, the root mean square displacement of an N_2 molecule at atmospheric pressure and 25°C is about 0.56 cm in one second. However, the total distance traveled along a zigzag path during this one second is about 475 m!

The mass diffusivity defined above is completely analogous to the thermal diffusivity defined earlier in the case of heat transfer (see Chapter 2). As thermal diffusivity was the proportionality constant between heat flux and energy gradient, mass diffusivity is the proportionality constant between diffusive mass flux and concentration gradient. Diffusional mass transfer is generally a slower process than diffusional heat transfer. To see this, compare mass diffusivities in Figure 10.8 with thermal diffusivities in Figure 2.7.

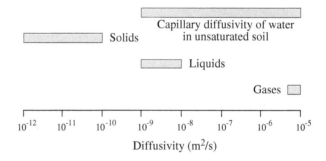

Figure 10.8: Typical ranges of diffusivity values.

The mass diffusivity depends on the pressure, temperature, and composition of the system. Since the mass diffusivity is a measure of molecular mobility, it is expected to be higher for gases than for liquids or solids. The typical ranges of diffusivities are given in Figure 10.8. Most available diffusivity values are experimental. Theoretical expressions for very idealized systems are discussed in the next section.

10.1.7 *Example: Diffusive Transport of the Protein Myoglobin*

At 20°C the diffusivity of the protein myoglobin (diameter 3.5 nm) in water is $11.3 \times 10^{-11} m^2/s$. What is the mean time required for a myoglobin molecule to diffuse a distance of 10 μm?

Solution

Understanding and formulating the problem *1) What is the process?* A molecule diffuses through a liquid. *2) What are we solving for?* Time required to move a particular distance when its diffusivity is provided. *3) Schematic and given data:* A schematic is shown in Figure 10.9 with some of the given data superimposed on it. *4) Assumptions:* None.

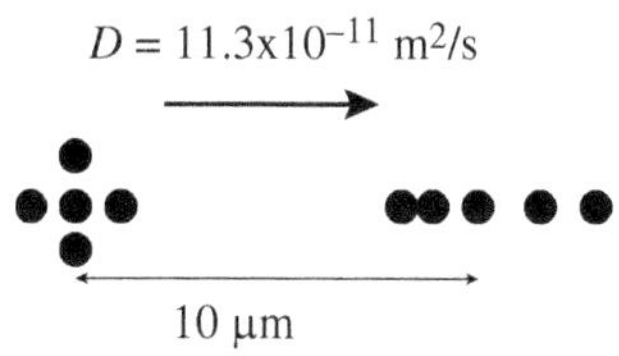

Generating and selecting among alternate solutions This step is skipped since it is a simple application of Eq. 10.19.

Implementing the chosen solution Using Eq. 10.19,

$$
\begin{aligned}
t &= \frac{<x^2>}{2D} \\
&= \frac{(10 \times 10^{-6})^2 m^2}{2(11.3 \times 10^{-11}) m^2/s} \\
&= 0.44 \text{ s}
\end{aligned}
$$

In spite of their large size, macromolecules like myoglobin (mol wt 17,000) diffuse rapidly in water. Note, however, to move a distance of 1 cm, it will take

$$
\begin{aligned}
t &= \frac{<x^2>}{2D} \\
&= \frac{(1 \times 10^{-2})^2 m^2}{2(11.3 \times 10^{-11}) m^2/s} \\
&= 5.12 \text{ days}
\end{aligned}
$$

Thus, for macroscopic distances, molecular diffusion is a slow process.

Evaluating and interpreting the solution *1) Does the calculated time make sense?* This is hard to say without comparing with experimental data but since diffusivity in liquid is a lot lower than that in gases, the molecule in liquid is expected to take a while to move.

$D = 11.3 \times 10^{-11} m^2/s$

10 μm

Figure 10.9: Schematic of 1D diffusion for Example 10.1.7.

Diffusivity for gases

Here, we discuss the diffusive movement of one gas inside another gas. For binary mixtures of low density gases that are non-polar and non-reacting, diffusivity can be estimated using the kinetic theory of gases. One such estimation of diffusivity is

$$D_{AB} = \frac{0.001858T^{3/2}\,(1/M_A + 1/M_B)^{1/2}}{p\sigma_{AB}^2\Omega_{D,AB}} \tag{10.21}$$

where D_{AB} is the diffusivity of A through B, T is the absolute temperature, M_A and M_B are molecular weights, p is absolute pressure in atm, σ_{AB} is the collision diameter in Å, and $\Omega_{D,AB}$ is a dimensionless function of the temperature and the intermolecular potential. As expected, Eq. 10.21 predicts an increase in diffusivity with temperature due to increased energy and a decrease in diffusivity for larger molecules.

Diffusivity for liquids

Here, we discuss the diffusivity of liquids or of dissolved solutes in other liquids. When studying diffusivity of liquids, it is useful to make a distinction between electrolytes and non-electrolytes. The diffusional coefficient of non-electrolytes can be predicted using hydrodynamical theory. A rigid sphere translating through a viscous medium experiences a frictional force, with the frictional coefficient f given by the Stokes equation as

$$f = 6\pi\mu r \tag{10.22}$$

where μ is the medium viscosity and r is the radius of the sphere. Einstein derived a relation between the macroscopic diffusion coefficient, D, and the frictional coefficient f. This relation is known as the Stokes–Einstein relation and is expressed as

$$D = \frac{\kappa T}{f} \tag{10.23}$$

where κ is the Boltzmann constant, and T is the absolute temperature. This relation has been used to estimate the radius of macromolecules, such as proteins. However, a certain amount of solvent is usually associated with a macromolecule in solution (termed solvation), which increases the effective radius of the molecule and therefore increases the frictional coefficient. Even so, the Stokes–Einstein relation is the most common basis for estimating diffusion coefficients in liquids. Substituting Eq. 10.22 in Eq. 10.23, we get

$$D = \frac{\kappa T}{6\pi\mu r} \tag{10.24}$$

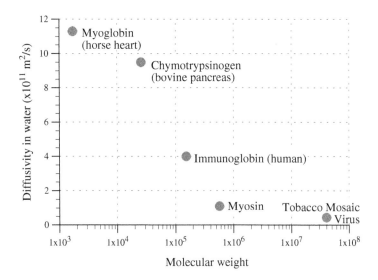

Figure 10.10: Diffusivity of some proteins in water at 20°C, as related to their molecular size.

Thus, the diffusivity is related to the solute molecule's mobility, i.e., the net velocity of the molecule. The reduction of diffusivity with size can be seen in the data for proteins in Figure 10.10.

Diffusivity for solids

Diffusion in solids can be quite complex. Also, diffusion of solids in solids is very slow (Figure 10.8) and does not seem to play a major role in biological or environmental systems. Thus, we discuss here only the diffusion of gases and liquids in solids. For simplicity, we can divide the solids into two groups: porous and non-porous. In the non-porous solids, the liquid or the gas is considered dissolved in the solid and diffuses through the solid. This type of diffusion is described by Fick's law (Eq. 10.18). In a porous solid, liquid movement is primarily due to capillarity and other forces. This is not molecular diffusion, although there can be some analogy in special cases, as discussed in Section 10.1.11. The rest of this section considers only gaseous molecular diffusion in a porous solid.

Diffusion of gases in a porous solid

The diffusion of gases in a porous solid plays an important role in biological systems. Most biological materials, such as a cell membrane or a tissue, and agricultural material, such as soil, can be described as capillary porous materials. Capillarity refers to small pore sizes. For example, Figure 10.11 shows such pores in wood. Liquid

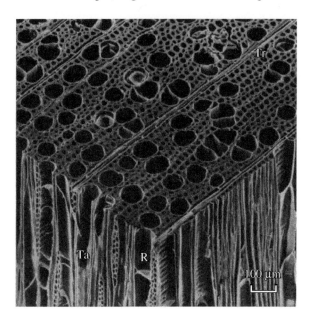

Figure 10.11: A scanning electron micrograph of wood showing pores with diameters ranging from 50 to 100 μm. Shown is a diffuse-porous hardwood Liriodendron tulipifera. Photograph courtesy of N. C. Brown Center for Ultrastructural Studies, College of Environmental Science and Forestry, State University of New York at Syracuse.

water transport in these materials is by capillarity and is described in Section 10.1.4. The vapor transport is typically by diffusion through the air in the pores. In soil, the metabolic activity of plant roots and microorganisms consumes oxygen and generates carbon dioxide. Oxygen, carbon dioxide, and water vapor all diffuse into and out of soil. A sufficient concentration of oxygen (adequate aeration) needs to be maintained in the soil for the absorption of nutrients by roots and for the beneficial activity of microorganisms.

Diffusion of gases in a porous solid is more complex than diffusion in a homogeneous non-porous solid, as the structure of the material plays a role. For example, as

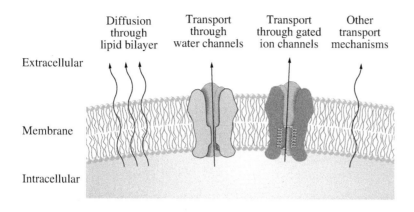

Figure 10.13: Schematic diagram of the cross-section of a cell membrane showing some of the transport mechanisms.

the pores become more tortuous, it takes longer for molecules to move between two given locations. This longer path decreases diffusivity. In addition, when the pores in the solid have diameters comparable to, or smaller than, the molecular mean free path, molecules collide with the walls of the pores (Figure 10.12). Such molecule-wall collisions change the nature of diffusion, and this is known as Knudsen diffusion. The effective diffusivity (D_{eff}) values for this type of diffusion can be significantly different, although Eq. 10.18 is still used to describe the process.

10.1.8 BioConnect: Transport in a cell membrane

A cell membrane is another example of a porous material. Cellular membranes are of fundamental importance to the cell because membranes act as selective barriers, controlling the intracellular contents and providing the cell with the capacity to produce internal compartments with specialized functions. Transport across cell membranes is quite a complex process, as illustrated in Figure 10.13 and passive molecular diffusion (subject of this text) is only one of several different transport mechanisms. Other mechanisms such as convective transport through water channels, transport through gated ion channels, carrier-mediated transport, and ion pumps are very important for transport, and the reader is referred to the excellent treatise by Weiss (1996) for details of these transport mechanisms. Using the mechanism of diffusion, some solutes dissolve in and diffuse through the lipid bilayer portion of a membrane (Figure 10.13). One of the important molecules that diffuses through the lipid bilayer is water.

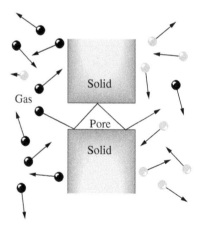

Figure 10.12: Schematic of Knudsen diffusion of a gas in a solid.

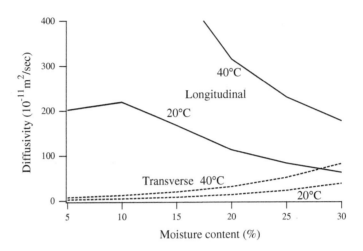

Figure 10.14: Directional and temperature variation of diffusivity in a biomaterial. Shown is data for wood at a oven dry specific gravity of 0.5, calculated using experimental data adapted from Siau (1984).

10.1.9 *BioConnect: Anisotropy in a biomaterial*

Complexities in structures of biomaterials and soils can lead to *anisotropy* of their transport properties. Anisotropy refers to when the material's physical properties vary along different directions. As an example, Figure 10.11 shows the pore structure in a wood that has the pores aligned in one direction. Therefore, in wood, resistance to diffusion in the longitudinal direction (along the pith) is expected to be different from resistance in the transverse direction (perpendicular to the pith). This is shown in Figure 10.14, where the diffusivity of moisture in wood is much higher in the longitudinal direction than in the transverse direction. Note that, for the moisture content range in this figure, the pith has mostly vapor, as opposed to pure liquid, which would move by capillarity. So this data refers to diffusion of vapor through the air in the pith and through the cell walls. The difference in diffusivities is explained by the much larger values in air than in solid cell wall material and by the presence of more of cell wall in the transverse direction.

Effect of temperature and moisture content

Several other characteristics of diffusion can be illustrated from Figure 10.14. Diffusivities generally increase with moisture content as more moisture is available for diffusion. This is seen for the transverse component. However, for the longitudinal component, the effect of the moisture isotherm between air and wood makes it decrease. All diffusivities increase with temperature as molecules become more mobile.

10.1.10 *BioConnect: Transport through Skin—Medication without Injection or Ingestion*

The pharmaceutical industries would like to send medication through skin, instead of via the traditional route of an injection or ingestion. Not having to use needles would simplify delivery, and it would avoid worries of intestinal degradation. Such topical application would offer a direct route for medication. A knowledge of transport properties through skin is required for such applications.

Substances can penetrate skin by two major routes (see Figure 5.16 on page 152 for a cross section of skin). The main path is direct diffusion through the stratum corneum (dried epidermal cells). Substances can also bypass the stratum corneum and penetrate via sweat ducts, hair follicles, or perhaps through minor breaches in the stratum corneum. However, for the most part, the effect of this second mode of transport is minimal owing to their relatively small fractional area (10^{-2} to 10^{-3} per unit skin area) and to the fact that they have significant diffusional resistance themselves. Nonetheless, for very hairy skin and for substances that penetrate the stratum corneum slowly, diffusion through these pathways can be significant. Diffusion through these pathways appears particularly important for highly polar, pharmacologically potent molecules that can act quickly on local structures of the skin. The permeability of intact stratum corneum to most electrolytes is extremely low and, for all practical purposes, these shunts provide the only means of access for such substances.

10.1.11 Capillary Diffusion

Capillarity in a porous solid was studied in Section 10.1.4 using Darcy's law (Eq. 10.1). Recall that the hydraulic potential, $\mathcal{H}$, is composed of matric and gravitational potentials. When a porous solid is unsaturated, i.e., pores are not completely filled with liquid, the matric potential, h, due to the binding of water from capillary and attractive forces, is negative and much stronger than the gravitational potential. Under such conditions, gravitational potential can be ignored and the capillary flux (volumetric) of liquid, n^v, can be written as

$$n^v = -K \frac{\partial h}{\partial s} \tag{10.25}$$

where s is distance along the flow. In terms of mass flux, this equation can be written as

$$n = -\rho_{liquid} K \frac{\partial h}{\partial s} \tag{10.26}$$

where n has the units of kg/m$^2 \cdot$s. Let c^* be the volume fraction of liquid in m^3 of liquid per m^3 of solid and c be the concentration of liquid in kg of liquid per m^3 of solid, such that

$$c = c^* \rho_{liquid} \tag{10.27}$$

Using this relationship, Eq. 10.26 can be transformed as

$$
\begin{aligned}
n &= -\rho_{liquid} K \frac{\partial h}{\partial c} \frac{\partial c}{\partial s} \\
&= -\frac{K}{\frac{\partial (c/\rho_{liquid})}{\partial h}} \frac{\partial c}{\partial s} \\
&= -\frac{K}{\left(\frac{\partial c^*}{\partial h}\right)} \frac{\partial c}{\partial s} \\
&= -\underbrace{D_{cap}}_{\substack{\text{capillary}\\\text{diffusivity}}} \underbrace{\frac{\partial c}{\partial s}}_{\substack{\text{concentration}\\\text{gradient}}}
\end{aligned}
\tag{10.28}
$$

where D_{cap} is given by

$$D_{cap} = \frac{K}{\left(\frac{\partial c^*}{\partial h}\right)} \tag{10.29}$$

Since the second term in Eq. 10.28 is a concentration gradient, D_{cap} can be interpreted as "diffusivity." It is important to note that, although the term "diffusivity" is being used here for movement of a liquid (water) through a solid, the mechanism of flow is primarily capillarity, not molecular diffusion. It so happens that Darcy's law for capillary flow can be cast in a form similar to molecular diffusion, as shown in Eq. 10.28. Thus, diffusivity D_{cap} here is really capillary diffusivity. However, in practice, often it is referred to simply as diffusivity and the subscript cap is dropped.

The term $\partial c^*/\partial h$ is termed differential water capacity. The relationship between the matric potential $h(c^*)$ and the volumetric water content c^* is a moisture characteristic curve that is used in soil science literature. Both $h(c^*)$ and permeability $K(c^*)$ are obtained from experiments. Figure 10.5 shows an example of such relationships for a particular type of soil. In this figure, $\partial c^*/\partial h$ is the slope of the soil moisture characteristic curve at any particular value of moisture level, c^*.

Thus, capillary diffusivity, D, can be shown to be a ratio of a transport coefficient K and differential capacity $\partial c^*/\partial h$. Note the similarity of this relationship (Eq. 10.29) with other diffusivities such as thermal diffusivity:

$$
\alpha \begin{pmatrix} \text{thermal} \\ \text{diffusivity} \end{pmatrix} = \frac{k_{thermal}}{\rho c_p}
$$

$$
= \frac{\text{thermal transport parameter}}{\text{thermal capacity}} \qquad (10.30)
$$

The capillary diffusivity $D(c^*)$ is very much material specific and depends on $h(c^*)$ and $K(c^*)$. It can be calculated from experimentally measured $h(c^*)$ and $K(c^*)$, and its value drops significantly (as illustrated in Figure 10.5) as the concentration of water decreases or the material dries. Such capillary diffusivity of water as a function of water content is also needed for engineering study of other important biological processes such as drying of foods, although detailed moisture variation of diffusivity is typically hard to find (Kiranoudis et al., 1994).

10.2 Dispersive Mass Transfer

Dispersion is the spreading of a mass component from higher concentrations to lower concentrations, for example, the spreading of smoke in air from a chimney, shown in Figure 10.15. Dispersion has qualitative effects similar to diffusion, but it is a different effect. Dispersion depends on *flow*, being generally an effect of turbulent flow. Molecular diffusion, due to the thermal-kinetic energy, is always present. In a turbulent flow, the random mixing effect, in addition to molecular diffusion, is defined to be dispersion (Cussler, 1997).

Hydrodynamic dispersion or dispersion in fluid systems is the spreading of a liquid or gas from the path it would follow due to the bulk flow or the convective hydraulics of the system. At the microscopic level, dispersion is not understood in detail. The lateral spreading of a plume from a smokestack (Figure 10.13) relative to its bulk flow in the wind direction is due to turbulence in air.[1] In surface-water regimes, dispersion

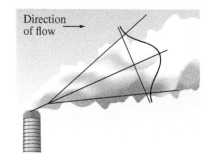

Figure 10.15: Convection–dispersion of smoke from a chimney.

---

[1]Spreading of a plume in air pollution is a convection–dispersion process (dispersion, combined with bulk flow in the wind direction), analogous to processes in liquid systems discussed in Section 14.1. However, spreading of a plume is inherently three-dimensional and its analysis is more complex. For further details, the reader is referred to books on air pollution, such as Stern (1976).

is also primarily due to turbulence. The net result is that some of the water molecules travel more rapidly than the average linear velocity, and some travel more slowly.

Mechanical dispersion in porous media, on a microscopic scale, is caused by three mechanisms (see Figure 10.16). The first occurs in individual pore channels because molecules travel at different velocities at different points in the channel due to the drag exerted on the fluid by the roughness of the pore surfaces. The second mechanism is due to the difference in pore sizes along the flow paths followed by the water molecules. Because of the differences in surface area and roughness relative to the volume of water in individual pore channels, different pore channels have different bulk velocities. The third mechanism leading to a dispersive process is related to the tortuosity, branching, and inter-fingering of pore channels.

Dispersion is discussed in the context of diffusion since the two processes can usually be described with very similar mathematics. Dispersive mass flux is written analogously to diffusion, defined in Eq. 10.18:

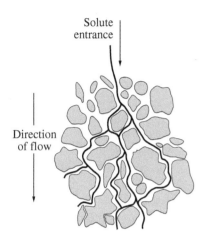

Solute entrance

Direction of flow

Figure 10.16: Convection–dispersion of fluid in a porous medium.

$$\underbrace{j_{A,x}}_{\text{Dispersive flux}} = -\underbrace{E_x}_{\substack{\text{Dispersion} \\ \text{coefficient}}} \underbrace{\left(\frac{dc_A}{dx} \right)}_{\substack{\text{Concentration} \\ \text{gradient}}} \tag{10.31}$$

where c_A is the mass concentration of species A being dispersed and the diffusivity is now replaced by a dispersion coefficient E_x. The dispersion coefficient has the units of m^2/s, which is the same as those of diffusivity, as can be seen by substituting the units in the above equation:

$$[E_x] = \frac{[j_{A,x}]}{\left[\frac{dc_A}{dx} \right]} = \frac{\left[\frac{kg}{m^2 s} \right]}{\left[\frac{kg}{m^3 m} \right]}$$

$$= \left[\frac{m^2}{s} \right] \tag{10.32}$$

Unlike diffusivity, the dispersion coefficient is not a strong function of the chemistry, e.g., the molecular weight or structure. Instead, as the subscript x in E_x indicates, the dispersion coefficient is a strong function of position. This is unlike diffusivity, which typically does not strongly depend on position. The dispersion coefficient is almost always measured experimentally.

Although dispersion is defined to be the random mixing component in addition to the molecular diffusion, unless the velocities are quite low, the dispersion effect is much stronger than diffusion. For example,[2] a crop canopy loses water vapor or

---

[2]See Nobel (1974).

absorbs carbon dioxide due to turbulent airflow (leading to dispersion) over the canopy. The dispersion coefficient for the rate of dispersive transport of water vapor or carbon dioxide between a plant and the bulk air can be 10,000 to 100,000 times larger than molecular diffusivities. Thus, in practice, if a dispersion coefficient is available, it generally includes the small or insignificant contribution from molecular diffusion. Moreover, it is nearly impossible to separately quantify the two processes.

10.3 Convective Mass Transfer

Convective mass transfer is studied the same way as convective heat transfer. It is the added effect of bulk flow on diffusion or dispersion. Convective mass transfer is the movement of mass through a medium as a result of the net motion of a material in the medium (e.g., air flow over a water surface which carries the water vapor away). This text will cover two scenarios of convection: 1) convection–diffusion over a surface and 2) convection–dispersion in a fluid or porous media.

10.3.1 Convection–Diffusion Mass Transfer over a Surface

Convection–diffusion mass transfer over a surface is completely analogous to Section 2.2 for convective heat transfer and is therefore characterized by a *convective mass transfer coefficient*. This is illustrated in Figure 10.17. As in heat transfer, we can have forced or free convection. Forced convection is due to an external force such as a fan, while free convection is driven by a density difference in the fluid (created by concentration or temperature differences, analogous to Figure 2.8 on page 31 in heat transfer). Convective mass transfer over a surface is described by

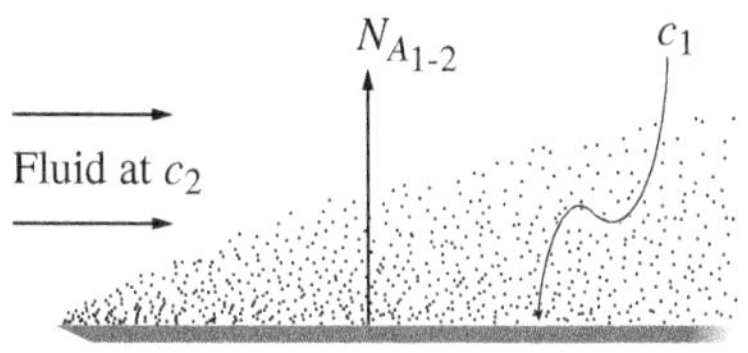

Figure 10.17: Schematic of convection–diffusion over a surface.

$$
\underbrace{N_{A_{1-2}}}_{\substack{\text{Mass flow rate}}} = \underbrace{h_m}_{\substack{\text{Convective} \\ \text{coefficient}}} \underbrace{A}_{\substack{\text{Area}}} \underbrace{(c_1 - c_2)}_{\substack{\text{Concentration} \\ \text{difference}}} \qquad (10.33)
$$

where $N_{A_{1-2}}$ is the mass flow rate *from* 1 to 2, A is the area normal to the direction of mass flow, $c_1 - c_2$ is the concentration difference between surface and fluid, h_m is the convective mass transfer coefficient, also called the film coefficient. Equation 10.33 is not a law but a defining equation for h_m, like the equation for convective heat transfer coefficient h. The units of h_m can be shown to be m/s:

$$
[h_m] = \frac{[N_{A_{1-2}}]}{[A]\,[(c_1 - c_2)]} = \frac{\left[\frac{\text{kg}}{\text{s}}\right]}{[\text{m}^2]\left[\frac{\text{kg}}{\text{m}^3}\right]} = \left[\frac{\text{m}}{\text{s}}\right]
$$

The convective mass transfer coefficient h_m includes the effects of diffusion and bulk flow. Like h, the heat transfer coefficient, h_m, depends on system geometry, fluid properties, flow situation, and the magnitude of the concentrations. Details of calculation of h_m are provided in Chapter 14.

10.3.2 *Example: Water Loss from a Reservoir*

Estimate the evaporative loss of water from a reservoir where the surface temperature, as well as the ambient air temperature, is 40°C. The air blowing over the reservoir leads to a surface mass transfer coefficient of 0.25 m/s. Assume the water surface is saturated with water vapor, and that the air is dry before contact with water.

Solution

Understanding and formulating the problem *1) What is the process?* Water evaporating from a water surface diffuses into the flowing air, which then gets convected away. *2) What are we solving for?* Rate of loss of water due to evaporation. *3) Schematic and given data:* A schematic is shown in Figure 10.18 with some of the given data superimposed on it. *4) Assumptions:* Air properties are constant and do not vary from the surface water into the bulk air. The mass transfer coefficient value provided is an average over the entire surface (more on this in Chapter 14).

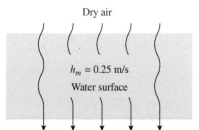

Figure 10.18: Schematic for Example 10.3.2.

Generating and selecting among alternate solutions This step is skipped here because it is a simple application of the convection–diffusion over a surface just presented (Eq. 10.33).

Implementing the solution The flux of water vapor is given by

$$n_{vapor} = h_m(c_{vapor,surface} - c_{vapor,bulk\ air})$$

The concentration of water vapor at the surface can be found from data on the partial pressure of water vapor at saturation at 40°C. Using the steam table (see Table C.12 on page 572), this value is 0.07318×10^5 Pa. The concentration corresponding to this partial pressure is calculated using the ideal gas law (see Section 9.1.1) as

$$\begin{aligned} c_{vapor,surface} &= \frac{p_{vapor,surface}}{RT} \\ &= \frac{0.07318 \times 10^5\ [\text{Pa}]}{8.314\ \left[\frac{\text{J}}{\text{mol K}}\right]\ 313\ [\text{K}]} \\ &= 2.812\ \frac{\text{mol}}{\text{m}^3} \end{aligned}$$

The flux of the evaporation water loss is given by

$$
\begin{aligned}
n_{water} &= h_m (c_{vapor,surface} - c_{vapor,bulk\ air}) \qquad (10.34)\\
&= \left(0.25\ \frac{\text{m}}{\text{s}}\right)\left(2.812\ \frac{\text{mol}}{\text{m}^3} - 0\right)\\
&= 0.703\ \frac{\text{mol}}{\text{m}^2\ \text{s}}\\
&= 1.265 \times 10^{-2}\ \frac{\text{kg}}{\text{m}^2 \cdot \text{s}}
\end{aligned}
$$

Evaluating and interpreting the solution *1) Does the flux value make sense?* Without experimental data, it is hard to conclude whether this is obviously wrong.

10.3.3 Convection–Dispersion Mass Transfer

As convection–diffusion mass transfer was the addition of bulk flow to the process of diffusion, convection–dispersion mass transfer is the added effect of bulk flow on the process of dispersion. Unlike the convection–diffusion mass transfer that we studied over a surface and focused on a thin layer called the boundary layer, we will study convection–dispersion mass transfer in the bulk fluid. Specifically, we will study in Chapter 13, convection–dispersion in a stream and in flow through a porous medium such as soil.

10.4 Comparison of the Modes of Mass Transfer

Molecular diffusion is the spontaneous movement of atoms and molecules of liquids, gases, and solids. Movement of mass occurs because of a concentration gradient, in contrast with bulk flow due to a pressure gradient.

Capillary "diffusion" is movement of mass due to capillary action in a porous medium. Such capillary transport can be treated mathematically similarly to molecular diffusion. In practice, capillary diffusion is often included under the general term diffusion. The expected range of capillary diffusivities in soils is illustrated in Figure 10.8.

Dispersion is the spreading of the fluid from its bulk flow path. Although its effects are quite similar to molecular diffusion, the mechanisms are entirely different. In fluids, for example, dispersion is primarily due to turbulence. In porous media, it is due to the random orientation of the shape, size, direction, and tortuosity of the pores. Generally, dispersion coefficients, as reported in the literature, include any molecular diffusion.

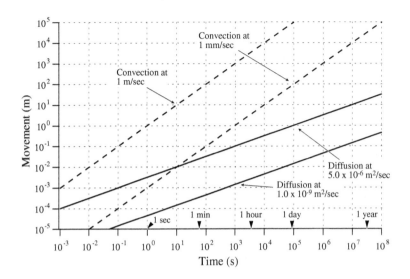

Figure 10.19: Relative movements of mass due to diffusion/dispersion and convection.

Convection refers to the addition of bulk flow to diffusion/dispersion. In practice, a convection–diffusion process is often referred to as simply convection. The surface convective mass transfer coefficient provides a measure of this combined convection–diffusion process for flow over a surface.

Advection is the transport of mass due to the bulk flow itself, without the diffusion or dispersion transport mechanism. It is due to a pressure gradient. Note, however, the terms convection and advection are sometimes used interchangably in practice and there is some ambiguity in their use. The term advection will not be used any further in this text.

It is useful to have a qualitative idea of the relative speeds of diffusion, dispersion, and advection. As shown in Figure 10.19, generally speaking, advective processes are relatively much faster and dominate if they are present simultaneously with diffusion.

10.5   Chapter Summary—Modes of Mass Transfer

- ## (Bulk) Flow through Porous Media (page 353)

    1. It is the bulk movement of a fluid in a porous media due to hydraulic forces

(pressure, matric attractive forces, and gravity).

2. It is described by Darcy's law (Eq. 10.1), which states that the volumetric flux of a fluid is proportional to the gradient of the hydraulic potential. The proportionality constant is termed the hydraulic conductivity.

3. Hydraulic conductivity depends on the fluid properties (density and viscosity) and matrix property called intrinsic permeability that takes into account the porosity, pore size distribution, shape of pores, and tortuosity.

4. When the porous medium is unsaturated, the flow is driven primarily by the matric attractive (capillary) forces. Flow in such a medium can be written in terms of Fick's law and an apparent (capillary) diffusivity.

• Molecular Diffusion (page 362)

1. It is the movement of mass from a higher to a lower concentration due to random molecular motion.

2. It is described by Fick's law (Eq. 10.18), which states that the flux of a mass species is proportional to the concentration gradient of that species. The proportionality constant is called the diffusivity.

3. Diffusivity measures speed of movement—it is one half of the mean-square displacement of a number of molecules per unit time through a given medium.

4. Diffusivity for gases is the highest due to their highest mobility. It is lower for liquids and the lowest for solids.

5. Diffusivity of a gas in a porous solid depends on the porous structure.

6. Diffusivity of a liquid in a porous medium is really a measure of capillary flow.

• Dispersion (page 373)

1. Dispersion is analogous to diffusion and is described by Eq. 10.31. However, dispersion is due to completely separate mechanisms— turbulent flow in a fluid or flow through uneven and tortuous paths in a porous medium.

2. Dispersion coefficients are generally much higher than diffusion coefficients.

• Convection (page 375)

1. It is the effect of adding bulk flow to diffusion or dispersion. Convection–diffusion mass transfer over a surface is described by Eq. 10.33 in complete analogy to heat transfer.

10.6   Problem Solving—Modes of Mass Transfer

▶ **Calculate flow rates and hydraulic conductivity through porous media**

- Darcy's law (Eq. 10.1) allows us to calculate the flow rate of a fluid through a porous medium. This is not to be confused with diffusion and other fluxes discussed later which describe transport of a component within the fluid. As an example, Darcy's law can describe flow of fluid through a tissue while diffusion formulas discussed later would describe how a drug will move through this fluid (which itself is moving through the tissue).

- For steady flow, Darcy's law, written as $n^v = \Delta \mathcal{H}/(\Delta x/K)$, is completely analogous to steady-state heat conduction (or mass diffusion later in this book), with $\Delta \mathcal{H}$, the pressure head, being the driving force and $\Delta x/K$ being the resistance. The same ideas as in steady-state heat conduction, such as combining resistances in series or parallel and relating heat flux and temperature between any two points are applicable here, relating mass flux and pressure head (see Problem 10.9).

- To calculate the average velocity through a porous medium, the flow is divided by the total cross-sectional area. This velocity represents bulk flow in the discussion on convection in Chapter 14.

- Flow in an unsaturated porous solid is calculated using the diffusion formula.

▶ **Calculate diffusion and dispersion coefficients and distances moved**

- Average distance moved can be calculated from the diffusivity or dispersion coefficient as $x = \sqrt{2Dt}$ or $x = \sqrt{2Et}$ in 1D.

- We can also get an average diffusion or dispersion velocity calculated from the above equation by getting $x/t = \sqrt{2D/t}$.

- Diffusivity in gases and liquids can be estimated from Eqs. 10.21 and 10.24, respectively, when needed. Change in diffusivity with temperature, size, and other parameters can be calculated using the same equations.

- When we need diffusivity for numerical problem solving, we typically use tabulated values in the appendix.

- The dispersion coefficient is mathematically treated the same way as diffusion, i.e., just replace the diffusion coefficient with the dispersion coefficient. Dispersion coefficients are harder to find. For example, one equation for flow in a stream is

$$E = \beta u + D$$

where u is the velocity of stream, D is the diffusion coefficient, and β is the dynamic dispersivity.

▶ **Calculate fluxes due to various modes.** This is done using the flux formulas in Table 10.1. Note that every flux has an implied direction, i.e., it is a vector quantity. These directions are also shown in the table.

▶ **Perform mass balance using mass fluxes.** The mass fluxes in Table 10.1 can be combined as part of mass balance for a species, using what is learned in Chapter 9:

$$\text{In} - \text{Out} + \text{Gen} = \text{Change in Storage}$$

10.7 Concept and Review Questions

1. What are the sources of matric forces in a porous medium?

2. Is the average fluid velocity inside the pores the same as the average velocity obtained by dividing the volumetric flow rate by the cross-sectional area of flow?

3. Does intrinsic permeability, k, of a porous medium depend on fluid properties?

4. In an unsaturated capillary porous material, why would water move or "diffuse" from higher to lower concentration?

5. What is molecular diffusion? How is it different from capillary flow? Dispersion? Convection?

6. What factors may cause molecular diffusion to occur?

7. In what ways are heat and mass transfer analogous and in what ways do they differ?

8. What are the units of mass diffusivity and thermal diffusivity? How do you physically interpret these diffusivity values?

9. Why are the diffusivities of gases larger than liquids and liquids larger than solids?

Table 10.1: Various fluxes and how to calculate them

Flux of a quantity, A	Expression	Direction	Comments
Diffusive or Dispersive	$-D\frac{\partial c_A}{\partial x}$	Positive x	Diffusion or dispersion alone (i.e., no contribution of flow)
Convective, at a surface	$h_m(c_{A,\text{surface}} - c_{A,\infty})$	From *surface* to ∞	Diffusion in a fluid from a surface, together with flow effect over the surface; h_m includes the effect of specific flow situation
Due to flow	uc_A	Along that of velocity u	Simply due to being carried with a velocity u
Total, due to diffusion/ dispersion and flow	$-D\frac{\partial c_A}{\partial x} + uc_A$	Combining terms assumes u is also along positive x direction	General equation for flux. Not to be confused with surface convective heat transfer equation (second item above) for a specific flow situation

10. Most convective processes also simultaneously involve molecular diffusion. Why does it make sense typically to ignore the molecular diffusion?

11. Why does the diffusivity typically increase with temperature?

12. Why does a larger molecule have a smaller diffusivity?

13. What bulk physical properties are related by the Stokes–Einstein equation? Verify that the equation is dimensionally consistent.

14. Do you expect gas diffusivity to increase as a material becomes more porous? What will be the limiting value of this diffusivity when the material is very very porous?

15. Why is Darcy's law needed, i.e., why can't we just use the normal equations for fluid flow that you learn in fluid mechanics class?

16. Explain how the difference in concentration of liquid in a porous medium drives capillary flow.

17. Consider a sandwich with two pieces of bread and a very wet (but not flowing) meat product in between. If you wrap this sandwich with plastic and leave it for a long time, would the bread and meat equilibrate in moisture content, i.e., will they have the same water content? Justify your answer. Hint: Consider capillary pressure by which water is held in the two different materials.

18. If the temperature of air in the room is raised from 25°C to 35°C, by what factor would the diffusivity of a gas in the air increase?

19. Explain the variation in diffusivity in Figure 10.8 in water between the two proteins myoglobin and tobacco mosaic virus using an appropriate equation.

20. During a lecture period of 50 minutes, would perfume molecules, having a diffusivity of $10^{-6} m^2/s$, reach from one end to the other end of a lecture room (*show calculations supporting your answer*)? We can typically smell it from the other end of the room, I believe. If this is true, what could be the reason?

21. Is dispersion just another name for diffusion? Can you have dispersion without any flow whatsoever? Can you have diffusion without any flow whatsoever?

22. Mention clearly what modes (can be more than one) of transport (molecular diffusion, capillary diffusion, dispersion, bulk flow) are present in a) smoke coming out of a chimney into blowing air; b) drug delivery through skin; c) drying of soil; d) movement of pollutants through a saturated soil; e) drying of skin surface; and f) water lost through evaporation from the surface of Cayuga Lake.

23. Consider two non-polar macromolecules, one being twice the effective radius as the other. What would be the ratio of the diffusivity of the larger macromolecule to that of the smaller macromolecule in a liquid? The temperature remains fixed.

24. *Be as specific as possible.* What are the mechanisms of dispersion in 1) radioactive water released into the ocean by the accident in Fukushima, Japan; 2) drug transport in a liver; and 3) pesticide movement in soil saturated with water?

Further Reading

Barrow, G. M. 1981. *Physical Chemistry for the Life Sciences*. McGraw-Hill, New York.

Campbell, G. S. 1985. *Soil Physics with Basic: Transport Models for Soil-Plant Systems*. Elsevier Science Publishing Company, Amsterdam.

Canny, M. J. 1990. Rates of apoplastic diffusion in wheat leaves. *New Phytologist* 116(2):263–268.

Cussler, E. L. 1997. *Diffusion Mass Transfer in Fluid Systems*. Cambridge University Press, New York.

Dainty, J. 1985. Water transport through the root. *Acta Horticulturae* 171:21–31.

Goldstick, T. K. and I. Fatt. 1970. Diffusion of oxygen in solutions of blood proteins. *Chem. Eng. Progress* 66(99):101–113.

Ingham, D. B. and I. Pop. (Editors) 1998. *Transport Phenomena in Porous Media*. Oxford Publishing Company, Danvers, MA.

Ishiguro M. 1991. Solute transport through hard pans of paddy fields. 1. Effect of vertical tubular pores made by rice roots on solute transport. *Soil Science* 152(6):432–439.

Kiranoudis, C. T., Z. B. Maroulis, D. Marinos-Kouris, and G. D. Saravacos. 1994. Estimation of the effective moisture diffusivity from drying data. Application to some vegetables, in *Developments in Food Engineering: Proceedings of the 6th International Congress on Engineering and Food* (T. Yano, R. Matsuno, and K. Nakamura, eds.), vol. 1, (New York), pp. 340–342, Blackie Academic & Professional.

McGrath, J. J. 1997. Quantitative measurement of cell membrane transport: Technology and applications. *Cryobiology* 34(4):315–334.

Monteith, J. L. and M. H. Unsworth. 1990. *Principles of Environmental Physics*. Edward Arnold, London.

Nobel, P. S. 1974. *Biophysical Plant Physiology*. W.H. Freeman and Company, San Francisco.

Rebolleda, G., F. J. Muñoz, and J. Murube. 1999. Audible pops during cyclodiode procedures. *Journal of Glaucoma* 8:177–183.

Siau, J. F. 1984. *Transport Processes in Wood*. Springer-Verlag, New York.

Stern, A. C. 1976. *Air Pollution*, 3rd Ed., Vol. I, Academic Press, New York.

Taura, T., Y. Iwaikawa, M. Furumoto, and K. Katou. 1988. A model for radial water transport across plant roots. *Protoplasma* 144:170–179.

Weiss, T. F. 1996. *Cellular Biophysics, Volume 1: Transport*. The MIT Press, Cambridge, MA.

10.8 Problems

10.1 Saturated Flow of Water in a Porous Medium

A layered soil column is positioned horizontally, as shown in Figure 10.20. It consists of 35 cm of a loam soil with a saturated hydraulic conductivity of 5 cm/day, followed by 100 cm of a sandy soil with a saturated hydraulic conductivity of 25 cm/day. The right end of the column is open to the atmosphere, as shown. Consider saturated flow at steady state. What is the volumetric flux through the soil column?

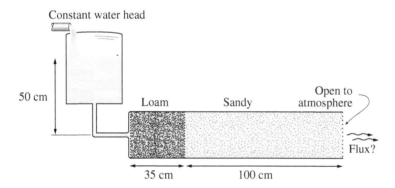

Figure 10.20: Schematic for Problem 10.1.

10.2 How Water Rises to the Top of a Tree

The numerous interstices in the cell wall of the xylem vessels of a tree form the capillaries by which water can rise to the top of tall trees. The needed properties of water can be found in the appendix. If a representative radius of these capillary channels is 5×10^{-9} m, show that the capillary suction in these channels can, in fact, raise the water to the top of any tree, at ambient temperature (25°C).

10.3   Molecular and Average Distances Moved during Diffusion

Consider gaseous diffusion with a diffusion coefficient of 10^{-6} m^2/s. 1) What is the average distance traveled by a group of molecules after 1 hour? 2) Is this the actual distance traveled by a single molecule during the same time? Explain.

10.4   Distance Moved by Diffusion and Dispersion

Calculate the average distance moved in one day by smoke particles in air for the following cases: 1) Smoke diffuses in air with a diffusion coefficient of 10^{-7} m^2/s. 2) Smoke disperses in air with a dispersion coefficient of 10^{-5} m^2/s.

10.5   Diffusivity Increase with Temperature

Starting at room temperature (25°C), approximately what temperature change is needed for diffusivity of a gas inside another gas (assuming gases are dilute and ideal) to go up by a factor of 10?

10.6   Estimating Diffusivity of Solutes in a Liquid

A protein macromolecule bovine serum albumin (BSA) and urea are separately dissolved in water at 23°C. Consider the effective radius of BSA to be 50 nm and that of urea is 0.16 nm. The viscosity of water at this temperature is 10^{-3} Pa·s. Calculate the diffusion coefficients of the 1) BSA, and 2) urea. 3) Qualitatively explain the difference between these two diffusivities using the fact that BSA has a molecular weight of 67,500 while urea has a molecular weight of 60.1.

10.7   Estimating Diameter from Diffusivity

The diffusivity of human immunoglobin (a protein) is 4×10^{-11} m^2/s in water at 20°C. The viscosity of water at this temperature is 1×10^{-3} Pa·s. Assuming the protein to be spherical, what is its effective radius?

10.8   Effect of Temperature on Liquid and Gas Diffusivity

Using appropriate equations and starting from 10°C, calculate how much the temperature of a species would have to be increased so that 1) its diffusivity in a *liquid* phase increases by a factor of two, and 2) its diffusivity in a *gas* phase increases by a factor of two. 3) Comment on the reason for the difference between the two diffusivities.

10.9   Flow through Porous Media

The schematic in Figure 10.21 shows two porous media (soils of different permeability) in parallel separated by an impermeable barrier. The right end is open to the atmosphere, and the left end is held at the same pressure for both media. Gravity is ignored. The hydraulic conductivity of the loam soil is 5 cm/hour while that of the sand is 10 cm/hour. The dimension perpendicular to the page is 10 cm. 1) Calculate the total volumetric flux of water through the two porous media. 2) Find the ratio of flux through sand to flux through loam and explain your answer in terms of possible difference in porosities of the two materials.

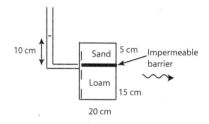

Figure 10.21: Flow through two soil columns in parallel.

10.10   Effective Diffusivity of an Inhomogeneous System

Diffusion of sucrose through the cell cytoplasm and the plasmodesmata that connects the cells is important in the root tips (e.g., rapid growth of corn root can depend on this diffusion). The arrangement of these two components is idealized as two slabs shown in the upper part of Figure 10.22. The fraction of cytoplasm, f, can be related to the sizes L_1 and L_2 of the cytoplasm and plasmodesmata by $f = L_1/(L_1 + L_2)$. Consider flux of sucrose between point A and point B.

1) Derive an expression from which you can solve for combined, or effective, diffusivity of the two components (D_{eff} in the bottom figure) in terms of their individual diffusivities, D_1 and D_2 (in the top figure), and the fraction f of one component. In other words, what combined diffusivity (as idealized in the bottom part of the figure) will provide the same flux between A and B for the same total concentration difference? 2) For diffusivities of sucrose in cytoplasm and plasmodesmata of 5.6×10^{-10} m^2/s and 7.8×10^{-12} m^2/s, respectively, what is the value of this effective diffusivity when $f = 0.5$?

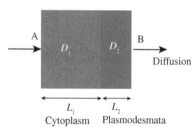

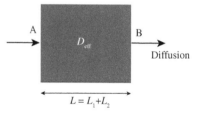

Figure 10.22: Schematic for effective diffusivity.

10.11   Flow Due to Pressures Generated from Evaporation

Moved to solved problems.

10.12   Permeability of a Tissue: Designing an Autosyringe

To estimate the pressure needed by an automated syringe to inject fluid medication into a tissue, it is required to measure saturated permeability through the tissue. In an experiment to measure permeability of a layer of muscle tissue 15 mm thick, a pressure of $0.6895 \times 10^5 \, Pa$ (10 psi) is applied using water over the tissue. The area of tissue over which the water flows is circular, having a diameter of 40 mm. The mass of water that flows through the tissue is collected on the other side and measured using a microbalance to be 0.1 g after 10 minutes. Assume the density of water at the temperature of measurement is 1 g/cm^3 and the viscosity of water is 0.001 Pa·s. Find the intrinsic permeability of the muscle tissue.

10.13   Effect of Temperature on Diffusion in Water

The viscosity of a liquid tends to decrease as temperature increases. Taking this change in viscosity into account, for a system that is initially at room temperature (25°C), approximately what temperature change is needed for the diffusivity of a molecule in water to go up by a factor of 2?

10.14   Permeability of a Microporous Tissue Scaffold

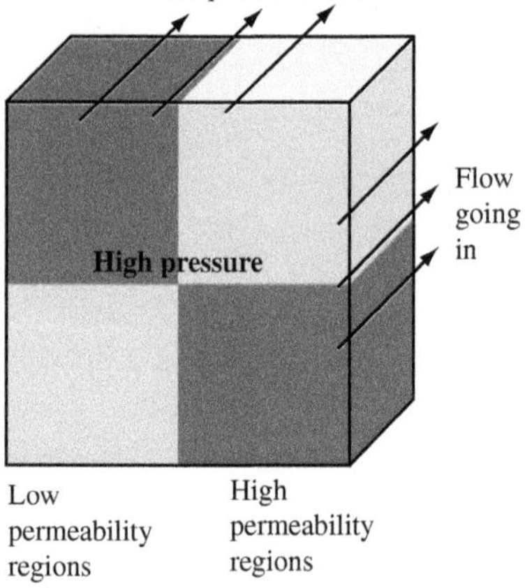

A certain tissue scaffold has regions of low intrinsic permeability (area A_1 with permeability k_1) and high intrinsic permeability (area A_2 with permeability k_2), as shown in Figure 10.23. Starting from Darcy's law, derive an expression for the effective permeability, k_{eff}, of the scaffold in terms of the two permeabilities and the volume fractions corresponding to those permeabilities, assuming volumes of regions with the same permeability are proportional to the areas. *You must show your work.*

Figure 10.23: Schematic of a scaffold surface with two different permeabilities.

Chapter 11

GOVERNING EQUATIONS AND BOUNDARY CONDITIONS OF MASS TRANSFER

CHAPTER OBJECTIVES

After you have studied this chapter, you should be able to

1. Identify the terms describing storage, convection, diffusion (or dispersion), and generation of mass species in the general governing equation for mass transfer.

2. Specify the three common types of mass transfer boundary condition.

KEY TERMS

- **diffusive mass flux**
- **convective mass flux**
- **total mass flux**
- **storage term**
- **convection term**

- **diffusion term**
- **generation term**
- **boundary concentration specified**
- **boundary flux specified**
- **convective boundary condition**

389

In this chapter we will develop a general equation that describes the concentration of a mass species as a function of time and position in a material. Later, in Chapters 12 through 14, we will solve this general equation for specific situations. The relationship of this chapter to other chapters in mass transfer is shown in Figure 11.1.

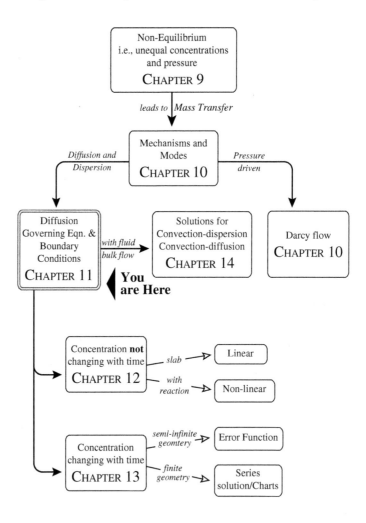

Figure 11.1: Concept map of mass transfer showing how the contents of this chapter relate to other chapters in the part on mass transfer.

11.1 Modified Fick's Law for Bulk Flow or Convection

Fick's law (Eq. 10.18), introduced in the previous chapter, provides the expression for diffusive flux. When bulk flow or convection is present, the expression for total flux needs the additional term due to convection, which is developed in this section. Since diffusional mass transfer involves at least two mass species, unlike diffusional heat transfer, certain additional quantities need to be defined for mass transfer.

11.1.1 Velocity and Mass Average Velocity

We know that individual molecules move randomly, producing an average velocity given by (Eq. 10.20)

$$u_{diff} = \sqrt{\frac{2D}{t}}$$

for 1D diffusion. Note that u_{diff} is the diffusion velocity of a species, for an aggregate of many molecules. It is not the velocity of one molecule; it is the average velocity of a large number of molecules contained in a small enough volume. The total velocity u_i with which a species i moves is this diffusion velocity superimposed on the bulk or mass average velocity, u. In other words, the velocity difference, $u_i - u$, of a species i is due to the diffusion velocity. The mass average velocity u of the fluid is found using the mass concentration of each species i:

$$u = \frac{\sum c_i u_i}{\sum c_i} \qquad (11.1)$$

where

$$c_i = \text{mass concentration of species } i$$
$$= \text{mass of species } i \text{ per unit volume}$$

The total mass concentration of all species, c, is given by

$$c = \sum_i c_i \qquad (11.2)$$

11.1.2 Flux Equation for a Convective Situation

If u_A is the velocity of mass species A and u is the mass average velocity of all the species, $u_A - u$ is the diffusion velocity of species A relative to mass average velocity and is due to diffusive movement relative to the bulk flow.

$$
\begin{aligned}
j_{A,x} &= \text{mass flux of } A \text{ in } x \text{ direction due to diffusion only} \\
&= c_A \left(u_{A,x} - u \right) \\
&= -D_{AB} \frac{\partial c_A}{\partial x} \text{ by Fick's law}
\end{aligned}
\tag{11.3}
$$

Alternatively:

Total velocity of species A in x direction		Velocity due to bulk flow in x direction		Movement relative to bulk flow in x direction
	=		+	

$$
u_{A,x} = u + u_{diffusion}
\tag{11.4}
$$

Multiplying both sides by c_A,

$$
c_A u_{A,x} = c_A u + \underbrace{c_A u_{diffusion}}_{\text{Diffusive flux}}
$$

$$
= c_A u - D_{AB} \frac{\partial c_A}{\partial x}
\tag{11.5}
$$

Equation for *total mass flux*:

$$
\begin{aligned}
n_{A,x} &= \text{total mass flux of } A \text{ in } x \text{ direction} \\
&= c_A u_{A,x} \\
&= \underbrace{-D_{AB} \frac{\partial c_A}{\partial x}}_{\substack{\text{Diffusive} \\ \text{flux}}} + \underbrace{c_A u}_{\substack{\text{convective} \\ \text{flux}}}
\end{aligned}
\tag{11.6}
$$

Equation 11.6 also shows how Fick's law is modified to include the effect of bulk flow or convection. Compare this with the *total energy flux* in heat transfer when convection is present:

$$
q''_x = \underbrace{-k \frac{\partial T}{\partial x}}_{\substack{\text{Conductive} \\ \text{flux}}} + \underbrace{\rho c_p (T - T_R) u}_{\substack{\text{Convective} \\ \text{flux}}}
\tag{11.7}
$$

Note that the quantity $\rho c_p (T - T_R)$ in the second term of Eq. 11.7, having the units of kJ/m^3, is the energy per unit volume and compares with c_A in Eq. 11.6, which is mass per unit volume. Note that, in terms of flux symbols, Eq. 11.6 can be written as

$$\underbrace{n_{A,x}}_{\text{Total}} = \underbrace{j_{A,x}}_{\text{Diffusive}} + \underbrace{c_A u}_{\text{Convective}} \tag{11.8}$$

When convective flux is zero or the flux is due to diffusion only, $n_{A,x} = j_{A,x}$. In such cases, either symbol $n_{A,x}$ or $j_{A,x}$ can be used to describe flux. In this text, $n_{A,x}$ will be used for flux in general and $j_{A,x}$ will be used only when there is a need to distinguish between diffusive and total flux.

Using the definition for the mass average velocity u for a binary system,

$$u = \frac{c_A u_{A,x} + c_B u_{B,x}}{c_A + c_B} = \frac{c_A u_{A,x} + c_B u_{B,x}}{c} \tag{11.9}$$

an alternative expression for total mass flux can be developed as

$$n_{A,x} = -D_{AB} \frac{\partial c_A}{\partial x} + \frac{c_A}{c}(c_A u_{A,x} + c_B u_{B,x}) \tag{11.10}$$

$$\underbrace{n_{A,x}}_{\substack{\text{Total flux} \\ \text{of } A \text{ in } x \\ \text{direction}}} = \underbrace{-D_{AB} \frac{\partial c_A}{\partial x}}_{\substack{\text{Diffusive} \\ \text{flux}}} + \underbrace{\underbrace{\omega_A}_{\substack{\text{mass fraction} \\ \text{of } A}} \underbrace{(n_{A,x} + n_{B,x})}_{\text{total flux}}}_{\substack{\text{convective flux due} \\ \text{to bulk motion}}} \tag{11.11}$$

where ω_A is the mass fraction of A in the system having the units of kg of species A per kg of total mass. Equation 11.11 is an alternative expression for mass flux, showing how the convective mass flux of species A is simply the mass fraction multiplied by the total flux. The units of fluxes $j_{A,x}$ and $n_{A,x}$ are kg of $A/\text{m}^2 \cdot \text{s}$. For the most part, we will use Eq. 11.6 as the expression for total mass flux.

11.2 Governing Equation for Mass Transfer Derived from Conservation of Mass Species and Fick's Law

In studying heat transfer, we said energy moves in the direction of decreasing temperature. The rate of energy movement was given by Fourier's law. We used energy conservation combined with Fourier's law to arrive at the differential equation of energy transfer that describes temperature as a function of position and time.

We will follow a completely analogous procedure using mass conservation and modified Fick's law (Eq. 11.6) for mass flux to describe the movement of mass in the direction of decreasing concentration. Consider an elemental control volume, as shown in Figure 11.2, fixed in space with mass flux into and out of the control volume. The total mass flux includes diffusive mass flux arising from the diffusion of the mass species A and convective mass flux due to the bulk movement of the fluid as a whole. The total mass flux $n_{A,x}$ is written from Eq. 11.6 as

$$n_{A,x} = \underbrace{-D_{AB}\frac{\partial c_A}{\partial x}}_{\substack{\text{Diffusive or} \\ \text{Dispersive flux}}} + \underbrace{c_A u}_{\substack{\text{Convective} \\ \text{flux}}} \tag{11.12}$$

where c_A is the concentration of mass species A in mass of A per unit volume. As the stored mass of species A increases in the control volume, it is reflected in an increased value of c_A. Conversely, if the stored amount decreases, the concentration decreases. Within the control volume, mass species A is generated at the rate of r_A in mass per unit volume. As in heat transfer, generation refers to transformation of other forms of mass into mass species A, as in chemical reactions. For example, O_2 in the fluid (mass species) can be used up due to biochemical reactions in the fluid, in which case r_{O_2} has a negative value. Using conservation of mass species A (Eq. 9.5).

$$\begin{array}{cccc} \text{Mass in} & - & \text{Mass out} & + & \text{Mass generated} & = & \text{Mass stored} \\ \text{(of species } A) & & \text{(of species } A) & & \text{(of species } A) & & \text{(of species } A) \end{array} \tag{11.13}$$

and referring to Figure 11.2, we can write the various quantities in Eq. 11.13 as

$$\begin{aligned} \text{Amount of } A \text{ in during time } \Delta t &= n_{A,x}\, \Delta y\, \Delta z\, \Delta t \\[2mm] \text{Amount of } A \text{ out during time } \Delta t &= n_{A,x+\Delta x}\, \Delta y\, \Delta z\, \Delta t \\[2mm] \text{Amount of } A \text{ generated during time } \Delta t &= r_A \Delta x \Delta y \Delta z \Delta t \\[2mm] \text{Amount of } A \text{ stored during time } \Delta t &= \Delta c_A \Delta x \Delta y \Delta z \end{aligned}$$

Substituting in Eq. 11.13 for mass balance on the control volume,

$$((n_{A,x} - n_{A,x+\Delta x})\, \Delta y\, \Delta z + r_A \Delta x\, \Delta y\, \Delta z)\, \Delta t = \Delta c_A\, \Delta x\, \Delta y\, \Delta z$$

$$\frac{n_{A,x} - n_{A,x+\Delta x}}{\Delta x} + r_A = \frac{\Delta c_A}{\Delta t}$$

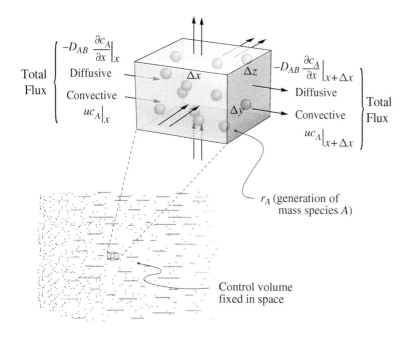

Figure 11.2: Control volume showing inflow and outflow of a mass species by diffusion and convection.

$$-\frac{\partial}{\partial x}\left(n_{A,x}\right) + r_A = \frac{\partial c_A}{\partial t} \tag{11.14}$$

which is the differential equation for mass balance for species A. Substituting Eq. 11.12 for $n_{A,x}$ in Eq. 11.14:

$$
\begin{aligned}
-\frac{\partial}{\partial x}\left(-D_{AB}\frac{\partial c_A}{\partial x} + c_A u\right) + r_A &= \frac{\partial c_A}{\partial t} \\
-\frac{\partial}{\partial x}\left(-D_{AB}\frac{\partial c_A}{\partial x}\right) - \frac{\partial}{\partial x}\left(c_A u\right) + r_A &= \frac{\partial c_A}{\partial t}
\end{aligned} \tag{11.15}
$$

If the diffusivity D_{AB} can be considered a constant,

$$D_{AB} \frac{\partial^2 c_A}{\partial x^2} - \frac{\partial}{\partial x}(c_A u) + r_A = \frac{\partial c_A}{\partial t}$$

Rearranging,

$$\frac{\partial c_A}{\partial t} + \frac{\partial}{\partial x}(c_A u) = D_{AB} \frac{\partial^2 c_A}{\partial x^2} + r_A \qquad (11.16)$$

Although Eq. 11.16 is for conservation of mass species A, the total mass of all species is also conserved. The conservation of total mass provides an additional relationship that can be used to rewrite Eq. 11.16. Writing conservation of total mass between the locations x and $x + \Delta x$,

$$(cuA)_x = (cuA)_{x+\Delta x}$$

where c is the total mass concentration. If the total mass concentration c does not change with x, we can rewrite this conservation equation for constant area A along the flow in the control volume of Figure 11.2 as

$$u_x = u_{x+\Delta x}$$

or

$$\frac{\partial u}{\partial x} = 0 \qquad (11.17)$$

Substituting Eq. 11.17 in Eq. 11.16, we get

$$\underbrace{\frac{\partial c_A}{\partial t}}_{\text{storage}} + \underbrace{u \frac{\partial c_A}{\partial x}}_{\substack{\text{flow or} \\ \text{convection}}} = \underbrace{D_{AB} \frac{\partial^2 c_A}{\partial x^2}}_{\text{diffusion}} + \underbrace{r_A}_{\text{generation}} \qquad (11.18)$$

Note that each term in Eq. 11.18 has the dimensions of mass per unit volume per unit time or $\text{kg/m}^3 \cdot \text{s}$. Compare this with Eq. 3.5, where, upon multiplying throughout by ρc_p, the terms have dimensions of energy per unit volume per unit time or $\text{J/m}^3 \cdot \text{s}$. Equation 11.18 is the general governing equation for mass transfer in a one-dimensional cartesian coordinate system with constant diffusivity. Compare this equation with Eq. 3.5 for heat transfer. As in Section 3.5, we can develop more general forms of Eq. 11.18 in other coordinate systems and in multiple dimensions, as presented later in this chapter. An excellent reference book for more general forms of the mass and heat transfer equations is Bird et al. (1960).

11.2.1 Meanings of Each Term in the Governing Equation

Note the similarity of Eq. 11.18 with Eq. 3.5, which is repeated here for convenience:

$$\underbrace{\frac{\partial T}{\partial t}}_{\text{storage}} + \underbrace{u \frac{\partial T}{\partial x}}_{\substack{\text{flow or} \\ \text{convection}}} = \underbrace{\frac{k}{\rho c_p} \frac{\partial^2 T}{\partial x^2}}_{\text{conduction}} + \underbrace{\frac{Q}{\rho c_p}}_{\text{generation}} \qquad (11.19)$$

Both equations represent the four basic terms of diffusion, convection, generation, and storage. Like the energy equation, the mass transfer equation is quite general and is useful for a variety of situations. Following the discussion in Section 3.1.1 on heat transfer, we only need to keep a few terms for the physical situations discussed in this text. The meanings of each term in Eq. 11.18 are summarized in Table 11.1.

Table 11.1: Various terms in the governing equation for mass transport (Eq. 11.18) and their interpretations

Term	What it represents	When we can ignore it
Storage	Rate of change of stored mass	Steady state (no variation of concentration with time)
Convection	Rate of net mass transport due to convection or bulk flow	Typically in a solid, with no bulk flow through it
Diffusion	Rate of net mass transport due to diffusion	Slow diffusion in relation to generation or convection. For example, bacterial species in sterilization of a solid food
Generation	Rate of generation or disappearance of mass	No chemical reaction leading to conversion from or to species A

11.3 General Boundary Conditions

This section parallels Section 3.2 on heat transfer. Like the description of a heat transfer problem is not complete without specifying the thermal conditions at the boundary, concentration conditions at the boundary are necessary for the full description of a

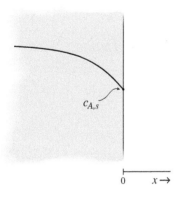

Figure 11.3: A surface concentration specified boundary condition.

mass transfer problem. The following are the three most common types of boundary conditions present in a mass transfer situation:

1. Surface concentration is specified A simple concentration condition that can occur on a surface is a specified concentration. For a one-dimensional mass transfer, as shown in Figure 11.3, this boundary condition can be expressed as

$$c_A\Big|_{x=0} = c_{A,s} \qquad (11.20)$$

where $c_{A,s}$, the concentration at the surface, can be specified as a constant or a function of time. At a solid–fluid interface, if we are setting up the mass transfer problem for the solid, we need $c_{A,s}$ for the solid. Sometimes, at such an interface, the boundary condition on the fluid side is provided instead. In such a case, $c_{A,s}$ for the solid is obtained from the boundary condition on the fluid side and equilibrium relations such as those in Figure 9.10. For example, suppose air at temperature 21°C and relative humidity 20% is blowing at a high velocity over the surface of a wet wood. From Figure 9.10, this maintains the surface of the wet wood at equilibrium with the air, at a moisture content value of 4.5%. Thus, the boundary condition at the wood surface will be given by

$$c_{water}\Big|_{x=0} = 0.045 \text{ kg/kg of total}$$

On the other hand, if we are setting up a mass transfer problem for a fluid (i.e., we need to calculate concentrations in the fluid), we need the boundary condition for the fluid. An example (see also Section 14.4) is the evaporation from the surface of free water or a very wet solid into air. Here the concentration of vapor in air at the boundary will be maintained at the vapor pressure of water. Thus, at a temperature of 30°C, this boundary condition for the fluid is written from Table C.10 on page 570 as

$$c_{vapor}\Big|_{x=0} = \frac{0.0424 \times 10^5 \text{ [N/m}^2\text{]}}{RT}$$
$$= 0.03 \frac{\text{kg}}{\text{m}^3}$$

Another example of a concentration specified boundary condition is the disappearance of a species at a surface, e.g., diffusing species O_2 is completely consumed at a surface, leading to $c_{O_2} = 0$ at the surface.

2. Surface mass flux is specified

Sometimes it is possible to know and specify the rate of mass transfer or mass flux on

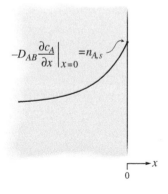

Figure 11.4: A mass flux specified boundary condition.

a surface. For a one-dimensional mass transfer, as shown in Figure 11.4, this boundary condition is expressed as

$$-D_{AB} \left. \frac{\partial c_A}{\partial x} \right|_{x=0} = n_{A,s} \qquad (11.21)$$

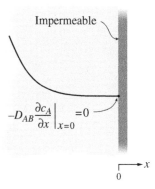

Here $n_{A,s}$, the surface mass flux, can be specified as a constant or a function of time. Two important special cases of surface mass flux occur frequently in practice. These are now discussed.

2a) Special case: Impermeable condition
Surfaces are sometimes made impermeable in an attempt to stop a particular type of mass movement. When the surface is highly impermeable to a particular species, the mass flux of that species is very small and can be approximated as zero, as illustrated in Figure 11.5. This boundary condition is given by

$$-D_{AB} \left. \frac{\partial c_A}{\partial x} \right|_{x=0} = 0 \qquad (11.22)$$

Figure 11.5: An impermeable (zero mass flux specified) boundary condition.

For example, when a biomaterial containing water is wrapped with aluminum foil to reduce moisture loss, the flux of water at the foil surface can be considered zero, leading to the above boundary condition.

2b) Special case: Symmetry condition
Another common situation arises in a mass transfer process when the geometry as well as the boundary conditions are symmetric, as shown in Figure 11.6 for the drying of a slab. Here the slab is of uniform thickness (making it symmetric about the centerline) and is being dried symmetrically (same boundary conditions on both faces). The resulting concentration profile in the slab will also be symmetric about the centerline, having a zero slope at the centerline. Thus, the symmetry boundary condition at the centerline can be written as

$$\left. \frac{\partial c_A}{\partial x} \right|_{x=L} = 0 \qquad (11.23)$$

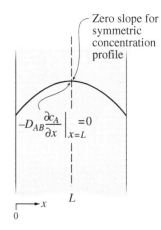

Note that this boundary condition resembles the impermeable condition mentioned above, since, to maintain symmetry, mass flux has to be zero at the line of symmetry.

Figure 11.6: A symmetry (zero mass flux) boundary condition at the centerline.

3. Convection at the surface
Another common mass transfer boundary condition occurs when a fluid is flowing over a surface, as shown schematically in Figure 11.7. At the surface ($x = 0$), the amount

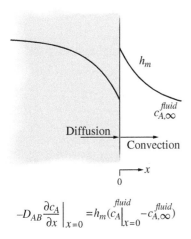

Figure 11.7: A convection mass transfer boundary condition.

$$-D_{AB}\frac{\partial c_A}{\partial x}\bigg|_{x=0} = h_m(c_A\big|_{x=0}^{fluid} - c_{A,\infty}^{fluid})$$

of mass diffused through solid or liquid is equal to the amount of mass convected away. This boundary condition is written simply as a mass balance at the surface, which is

$$\underbrace{-D_{AB}\frac{\partial c_A}{\partial x}\bigg|_{x=0}}_{\text{mass diffusion to the left}} = \underbrace{h_m(c_{A,x=0}^{fluid} - c_{A,\infty}^{fluid})}_{\text{mass convection to the right}} \qquad (11.24)$$

Here h_m is the mass transfer coefficient, defined analogously to the heat transfer coefficient and is explained in later chapters. In Eq. 11.24, it is important to note that the left hand side uses concentration of species A in the solid, c_A, while the right hand side uses concentration of species A in the fluid, c_A^{fluid}. A common example of this type of boundary condition is that at the surface of a wet solid with air flowing over it, after the solid has experienced some drying, i.e., when the surface is not very wet. In this case, the left hand side of Eq. 11.24 uses concentration of water in the solid, c_{water}, while the right hand side uses concentration of water vapor in air, c_{vapor}. Suppose the corresponding mass transfer coefficient for a situation is $h_m = 0.01$ m/s. Equation 11.24 is written for this specific case as

$$\underbrace{-D_{water,solid}\frac{\partial c_{water}}{\partial n}\bigg|_{surface}}_{\text{diffusion in wet solid}} = \underbrace{0.01(c_{vapor,surface} - c_{vapor,\infty})}_{\text{convection in air}}$$

At the surface, concentration of water in the solid, $c_{water,surface}$ and concentration of water vapor in air, $c_{vapor,surface}$ are not the same, but they are related by isotherms, as discussed in Section 9.3.4 on page 323.

Also, in analogy to the boundary conditions in Section 3.2 on page 47 of heat transfer, when $h_m \to \infty$ (such as in high fluid velocity), Eq. 11.24 leads to

$$c_{A,surface}^{fluid} = c_{A,\infty}^{fluid}$$

or the concentration in fluid at the surface becomes equal to the concentration in the bulk fluid. Thus, the boundary condition given by Eq. 11.24 reverts to the first kind of boundary condition given by Eq. 11.20 with the surface concentration in the solid being specified as that in equilibrium with the surface concentration in the fluid, $c_{A,surface}^{fluid}$.

11.4 Governing Equations for Mass Diffusion in Various Coordinate Systems

The governing equation for diffusion mass transfer, without the convection term, is shown here in some familiar coordinate systems. Control volumes for the various coordinate systems can be seen in Section 3.5. For a more general version of these governing equations, see Bird et al., 1960.

Cartesian

$$D_{AB}\left[\frac{\partial^2 c_A}{\partial x^2} + \frac{\partial^2 c_A}{\partial y^2} + \frac{\partial^2 c_A}{\partial z^2}\right] + r_A = \frac{\partial c_A}{\partial t} \tag{11.25}$$

Cylindrical

$$D_{AB}\left[\frac{1}{r}\frac{\partial}{\partial r}\left(r\frac{\partial c_A}{\partial r}\right) + \frac{1}{r^2}\left(\frac{\partial^2 c_A}{\partial \theta^2}\right) + \frac{\partial^2 c_A}{\partial z^2}\right] + r_A = \frac{\partial c_A}{\partial t} \tag{11.26}$$

Spherical

$$D_{AB}\left[\frac{1}{r^2}\frac{\partial}{\partial r}\left(r^2\frac{\partial c_A}{\partial r}\right) + \frac{1}{r^2 \sin^2\theta}\frac{\partial^2 c_A}{\partial \phi^2} + \frac{1}{r^2 \sin\theta}\frac{\partial}{\partial \theta}\left(\sin\theta\frac{\partial c_A}{\partial \theta}\right)\right]$$

$$+ r_A = \frac{\partial c_A}{\partial t} \tag{11.27}$$

Symbolically (any coordinate system)

$$D_{AB}\nabla^2 c_A + r_A = \frac{\partial c_A}{\partial t} \tag{11.28}$$

11.5 Chapter Summary—Governing Equations & Boundary Conditions of Mass Transfer

• Governing Equation of Mass Transfer (page 393)

1. The governing equation of mass transfer given by Eq. 11.18 comprises four terms representing storage, convection, diffusion (or dispersion), and generation.

2. Depending on the physical situation, some terms may be dropped.

3. Depending on the appropriate geometry of the physical problem, choose a governing equation in a particular coordinate system from the equations in Section 11.4.

• Boundary Conditions of Mass Transfer (page 397)

One of the following three types of boundary condition typically holds at a surface:

1. Concentration is specified (Eq. 11.20)—for example, when the convective mass transfer coefficient is high or when there is evaporation at the surface.

2. Mass flux is specified (Eq. 11.21)—for example, it is zero at an impermeable surface.

3. Convective mass transfer (Eq. 11.24)—neither concentration nor mass flux is specified. This is the most general situation.

11.6   Problem Solving—G.E. & B.C. of Mass Transfer

This section (which is really about problem formulation) is completely analogous to that for heat transfer, as described in detail in Section 3.7, and the reader is referred to that section for details.

Boundary conditions between two materials often require additional considerations. At an interface between two materials or phases such as that between a liquid and a gas, the equilibrium information (e.g., Henry's law) provides the boundary condition. In a problem, the boundary condition may not be given directly, but equilibrium with the surrounding phase could be mentioned, which then would be used to derive the boundary condition, using relations summarized in Table 9.3.

Table 11.2: Diffusive mass flux in terms of concentration in various coordinate systems

Coordinate system	Formulas for diffusive mass flux
Cartesian	$n_{A,x} = -D_{AB}\frac{\partial c_A}{\partial x}$; $n_{A,y} = -D_{AB}\frac{\partial c_A}{\partial y}$; $n_{A,z} = -D_{AB}\frac{\partial c_A}{\partial z}$
Cylindrical	$n_{A,r} = -D_{AB}\frac{\partial c_A}{\partial r}$; $n_{A,z} = -D_{AB}\frac{\partial c_A}{\partial z}$; $n_{A,\theta} = -D_{AB}\frac{1}{r}\frac{\partial c_A}{\partial \theta}$
Spherical	$n_{A,r} = -D_{AB}\frac{\partial c_A}{\partial r}$; $n_{A,\theta} = -D_{AB}\frac{1}{r}\frac{\partial c_A}{\partial \theta}$; $n_{A,\phi} = -D_{AB}\frac{1}{r\sin\theta}\frac{\partial c_A}{\partial \phi}$

11.7   Concept and Review Questions

1. How is mass generation different from storage of mass? Explain using an example.

Table 11.3: Various possibilities of the governing equation for mass transfer

Situation	What is ignored	Final equation
Steady diffusion	Transient (storage), flow, generation	$0 = \underbrace{D \dfrac{\partial^2 c}{\partial x^2}}_{\text{diffusion}}$
Steady diffusion with generation	Transient (storage), flow	$0 = \underbrace{D \dfrac{\partial^2 c}{\partial x^2}}_{\text{diffusion}} + \underbrace{r_A}_{\text{generation}}$
Transient diffusion	Flow, generation	$\underbrace{\dfrac{\partial c}{\partial t}}_{\text{transient}} = \underbrace{D \dfrac{\partial^2 c}{\partial x^2}}_{\text{diffusion}}$
Transient diffusion with generation	Flow	$\underbrace{\dfrac{\partial c}{\partial t}}_{\text{transient}} = \underbrace{D \dfrac{\partial^2 c}{\partial x^2}}_{\text{diffusion}} + \underbrace{r_A}_{\text{generation}}$
Transient convection without generation	Generation	$\underbrace{\dfrac{\partial c}{\partial t}}_{\text{transient}} + \underbrace{u \dfrac{\partial c}{\partial x}}_{\text{convection}} = \underbrace{D \dfrac{\partial^2 c}{\partial x^2}}_{\text{diffusion}}$
Transient convection with generation	None	$\underbrace{\dfrac{\partial c}{\partial t}}_{\text{transient}} + \underbrace{u \dfrac{\partial c}{\partial x}}_{\text{convection}} = \underbrace{D \dfrac{\partial^2 c}{\partial x^2}}_{\text{diffusion}} + \underbrace{r_A}_{\text{generation}}$

2. If we keep only the diffusion term in the governing equation, it becomes the Laplace equation. What other physical phenomena (besides mass or heat transfer) can be described by this equation?

3. Write Fick's law of mass diffusion using the cylindrical coordinate system.

4. What two basic physical laws are used to derive the diffusion equation?

5. What are the three types of boundary conditions we have discussed?

6. How is using a convective boundary condition for mass different from that used in heat? Hint: think of the two phases at the boundary.

7. Show that, when the mass transfer coefficient is high ($h_m \to \infty$), the boundary condition given by Eq. 11.24 approaches the boundary condition given by Eq. 11.20.

8. In what ways are heat and mass transfer analogous and in what ways do they differ?

Further Reading

Bird, R. B., W. E. Stewart, and E. N. Lightfoot. 1960. *Transport Phenomena*. John Wiley & Sons, New York. (Comprehensive source for governing equations.)

Crank, J. 1975. *The Mathematics of Diffusion*. Oxford University Press, Oxford, UK.

Shitzer, A. and R. C. Eberhart. 1985. *Heat Transfer in Medicine and Biology* Vol. 1, 2. Plenum Press, New York.

Welty, J. R., C. E. Wicks, R. E. Wilson, and G. Rorrer. 2001. *Fundamentals of Momentum, Heat, and Mass Transfer*. John Wiley & Sons, New York.

Zaritzky, N. E. and A. E. Bevilacqua. 1988. Oxygen diffusion in meat tissues. *International Journal of Heat and Mass Transfer* 31(5):923–930.

11.8   Problems

11.1   Governing Equation in Cylindrical Coordinate System

Starting from a mass balance, derive the governing equation for mass transfer in cylindrical coordinates, for the radial direction only (Eq. 11.26 with only the r dependence).

Problem Formulations

Problem formulation is one of the most important concepts that we should learn in preparing to solve practical problems. Often much of the problem is already formulated in an undergraduate class, as in typical homework. Here we will practice problem formulation. For each of the problems below, you need to provide 1) a schematic; 2) governing equation (do not solve); and 3) boundary and/or initial conditions. There is no right or wrong answer. The final governing equation and boundary conditions should only have terms relevant to this problem, i.e., you should not leave them as the general equations having all the terms. You will not be penalized for the degree to which you simplify. You are free to assume any parameter that you need, and you may need to make many approximations, but you should include a few words of justification for every assumption you make. To do these problems in the most realistic way, you need to know a lot more of the processes—this is not the goal here. You only need to be consistent so that your final set of governing equations and boundary conditions is solvable.

11.2   Cell Division with Diffusion of Microorganism in a Gel

Starting from the general governing equations in Section 11.4 on page 400, formulate the governing differential equation for mass transfer to describe the diffusion of a population of microorganisms placed in a stagnant fluid or gel, where the microorganism reproduces following a first-order reaction. Give two boundary conditions that could be used to solve the differential equation. (Adapted from Welty, et. al, 2001.)

11.3   Oxygen Diffusion in Meat Tissue

(See directions above) The color of meat (Figure 11.8) is determined by the relative proportions of myoglobin derivatives. For example, oxygen taken up by myoglobin converts the purple reduced pigment to the bright red oxygenated form, oxymyoglobin. Zaritzky et al. (1988) considered oxygen penetration in meat tissue by diffusion with simultaneous consumption of oxygen by the tissue. Consumption follows a zeroth-order reaction, with a rate constant that decays exponentially with (postmortem) time. (Contrast this with zeroth-order consumption of oxygen in living tissue in Problem 12.5.) At the maximum distance the gas penetrates, a zero flux boundary condition can be used. Formulate a one-dimensional problem with a constant dissolved oxygen concentration at the meat surface exposed to air.

Figure 11.8: Color of meat (beef) that changes with time. (iStock.com)

11.4   Drug Delivery Using a Patch

(See directions above) The Ortho Evra Birth Control Patch (Figure 11.9) is an effective alternative method of birth control. It releases two drugs, norelgestromin and ethinyl estradiol, and these two drugs diffuse through a person's skin and into their bloodstream. Assume the drugs are completely removed at the other end of the skin by the bloodstream. You would like to analyze the process of diffusion of norelgestromin from the patch and through the skin, i.e., find the rate at which drug reaches the bloodstream at any time.

11.5   Drug Delivery from a Stent

(See directions above) Balloon angioplasty and coronary stent deployment are powerful techniques in the treatment of individuals with advanced coronary artery disease. Recent advancements have led to the development of drug-eluting stents (Figure 11.10)

Figure 11.9: A patch for drug delivery. (Shutterstock.com–Tomasz Trojanowski)

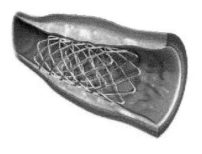

Figure 11.10: A stent that releases drugs.

Figure 11.11: Smoke coming from a chimney.

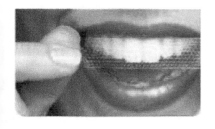

Figure 11.12: Whitening of teeth.

designed to release a drug polymer that inhibits restenosis, a bodily defense mechanism characterized by tissue ingrowth and reblockage of the artery. The stent material is impregnated with an anti-cancer or immunosuppressive drug that diffuses through the surrounding tissues over a critical period of time. In this study, you would like to model diffusion of the drug Rapamycin from the stent into the tissue. The stent geometry can be simplified to six concentric rings and it is sufficient to model diffusion from one of these rings.

11.6   Air Pollution

(See directions above) Consider air pollution from smoke coming out of a chimney (Figure 11.11). We would like to calculate how the smoke transports (moves and spreads) in the air due to convection and dispersion. The air flow and transport of the smoke are three dimensional. Let us assume that gravity can be ignored. Make any other assumptions as needed. For this problem only, provide only the governing equations, making sure to include terms representing all dimensions of the transport of the smoke.

11.7   Drug Delivery for Motion Sickness

(See directions above) Transdermal drug delivery systems are involved in the continuous administration of drug molecules from the surface of the skin into the circulatory system. Compared with oral administration, transdermal drug delivery offers better uniformity of drug concentrations in plasma throughout their duration of use. Scopolamine is the active ingredient in motion sickness medication that targets the nerve fibers in the inner ear. The scopolamine patch is effective for about three days, longer than if administered orally, which is effective for only several hours. We want to examine the rate of diffusion of transdermal scopolamine across the skin and into the systemic circulation.

11.8   Whitening of Teeth

(See directions above) Crest Whitestrips (Figure 11.12) are thin, liquid films adhered to a plastic exterior that can be applied directly to the tooth, enabling mass transfer of its active ingredient, hydrogen peroxide, to penetrate the tooth outer layer. The tooth outer layer consists of a layer of enamel followed by a layer of dentin. You want to model the concentration of hydrogen peroxide as it moves through the gel tooth outer layer for a period of 30 minutes.

Chapter 12

DIFFUSION MASS TRANSFER: STEADY STATE

CHAPTER OBJECTIVES

After you have studied this chapter, you should be able to

1. Formulate and solve for a diffusive mass transfer process in a slab geometry where concentrations do not change with time (steady state).

2. Extend the steady-state mass transfer concept for a slab to a composite slab using the overall mass transfer coefficient.

3. Extend the steady-state diffusive mass transfer in a slab to include a first-order decay of the mass species.

KEY TERMS

- **steady-state diffusion**
- **overall mass transfer coefficient**
- **mass transfer resistances in series**
- **diffusion with chemical reaction**

This chapter will deal with mass diffusion under steady-state conditions, closely following Chapter 4 for steady state heat conduction. We will solve the general governing equation derived in Chapter 11 for the particular situation of steady state. The relationship of this chapter to the other chapters in mass transfer is shown in Figure 12.1. As in Chapter 4 (page 71) for heat transfer, steady state is defined as the situation when mass concentrations are not changing with time. Note that steady state is

407

not the same as equilibrium, when the system is at the same concentration everywhere. In steady state, mass can continue to move if there are gradients in concentration or other driving forces. However, the concentration does not change with time.

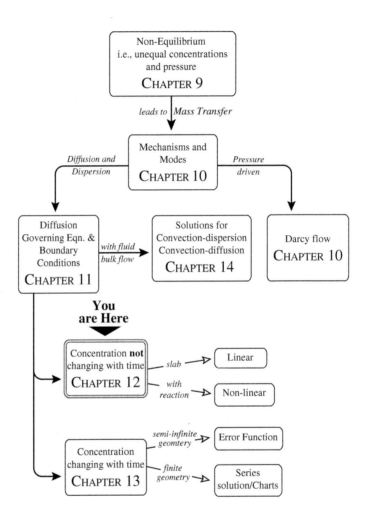

Figure 12.1: Concept map of energy transfer showing how the contents of this chapter relate to other chapters in the part on mass transfer.

12.1   Steady-State Mass Diffusion in a Slab

As discussed previously, there are many parallels between the systems of heat and mass transfer. This chapter deals with the solution to a one-dimensional, steady-state diffusion/dispersion problem. The topics closely parallel those of Chapter 4 for heat transfer, as the governing equations are almost identical. Thus, in a slab geometry as shown in Figure 12.2, if the concentrations are uniform over the surfaces at $x = 0$ and $x = L$, at locations not too close to the edges the concentration would vary only along the thickness of the slab. In other words, the mass transfer is one dimensional. At steady state with no mass generation and no bulk flow, the general governing equation is

$$\underbrace{\cancel{\frac{\partial c_A}{\partial t}}^{0}}_{\text{steady state}} + \underbrace{u\cancel{\frac{\partial c_A}{\partial x}}^{0}}_{\text{no convection}} = D_{AB}\frac{\partial^2 c_A}{\partial x^2} + \underbrace{\cancel{r_A}^{0}}_{\text{no reaction}} \tag{12.1}$$

which becomes

$$\frac{d^2 c_A}{dx^2} = 0 \tag{12.2}$$

after eliminating the terms. For the simplest kind of boundary condition where constant concentration c_{A_1} and c_{A_2} can be assumed at the two surfaces, we can write the boundary conditions as

$$c_A(x = 0) = c_{A_1}$$
$$c_A(x = L) = c_{A_2} \tag{12.3}$$

The solution to Eq. 12.2 is given by

$$c_A = \frac{c_{A_2} - c_{A_1}}{L}x + c_{A_1} \tag{12.4}$$

which shows a linear change in concentration from c_{A_1} to c_{A_2} at steady state. Note that this equation is identical to Eq. 4.4 on page 73, both equations showing a linear change (in concentration or temperature) in a slab under steady state. The diffusive mass flow rate at a position x is given by

$$\begin{aligned} N_{A,x} &= -D_{AB}A\frac{dc_A}{dx} \\ &= -D_{AB}A(c_{A_2} - c_{A_1})/L \quad \text{using Eq. 12.4} \\ &= D_{AB}A(c_{A_1} - c_{A_2})/L \end{aligned} \tag{12.5}$$

which is the same at all locations x along the thickness, signifying steady state (why?).

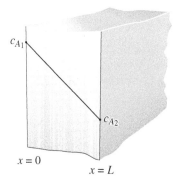

Figure 12.2: A linear concentration profile at steady state in a slab geometry.

12.1.1   *Example: Mass Transfer through a Biofilm*

A biofilm is a layer-like aggregation of microscopic organisms, plant or bacteria attached to a solid surface (Figure 12.3). In many natural aquatic systems, especially those having a high specific surface area and low nutrient concentrations, biofilms are 90% to 99.9% bacteria. Biofilms are found in or on stream-beds, ground-water aquifers, water pipes, ship hulls, piers, and aquatic plants and animals. Engineered processes that utilize films of bacteria include trickling filters, biologically activated carbon beds, and land treatment systems. Sometimes the biofilm can shield a surface undergoing a reaction, i.e., protection against corrosion. For simplicity, here we will consider such a biofilm where the the reactant is consumed only at the solid surface, maintaining a particular value of surface concentration. We want to calculate the reactant flux.

1) Calculate the concentration profile of the reactant in the biofilm at steady state. 2) Calculate the reactant flux. 3) How much of the reactant is contained in the biofilm per unit area at any time? Biofilm thickness is 0.01 cm and the diffusivity of the reactant in the biofilm is 0.8 cm^2/day. Assume the reactant concentration in the bulk fluid to be uniform at 3.2 mg/liter. At the surface, reactant concentration is 0.25 mg/liter.

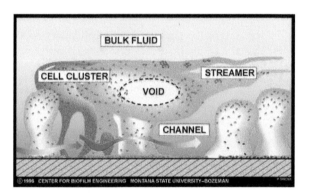

Figure 12.3: A biofilm. Picture courtesy of the Center for Biofilm Engineering, Montana State University.

Solution

Understanding and formulating the problem   *1) What is the process?* Reactant diffuses through a stagnant layer. *2) What are we solving for?* Flux of reactant through the biofilm at steady state. *3) Schematic and given data:* A schematic is shown in Figure 12.4 with some of the given data superimposed on it. Diffusivity of the reactant in the biofilm is 0.8 cm²/day. *4) Assumptions:* As already provided in the problem description, reactant being consumed at only the surface is an assumption that may not always be true. We have to treat this as a 1D problem since dimensions in the other directions are not provided.

Generating and selecting among alternate solutions *1) What solutions are possible?* For steady-state diffusive mass transfer, the solutions we have available are shown in the solution chart in Figure 12.19. *2) What approach is likely to work?* As we follow the solution chart in Figure 12.19, since there is no reaction present, we look on its left side. However, we are asked to find the concentration profile so the resistance formulation may not be the most appropriate. So we need to derive from scratch, except we should always check if our problem matches derivations already done in this text. Equation 12.4 is indeed for a slab with only diffusion and a concentration specified boundary condition on either surface, thus exactly matching the situation in this example.

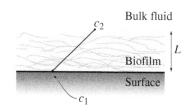

Figure 12.4: Schematic for Example 12.1.1.

Implementing the chosen solution 1) For concentration profile, we simply plug in Eq. 12.4 as

$$
\begin{aligned}
c_{reactant} &= \frac{c_{reactant_2} - c_{reactant_1}}{L} x + c_{reactant_1} \qquad (12.6)\\
&= \frac{3.2 - 0.25 \ [\text{mg/liter}]}{0.01 \ [\text{cm}]} x[\text{cm}] + 0.25 \ [\text{mg/liter}]\\
&= \frac{3.2 - 0.25 \ [\text{mg/liter}]}{0.01 \ [\text{cm}]} x[\text{cm}] + 0.25 \ [\text{mg/liter}]\\
&= 295x + 0.25 \ \text{mg/liter} \qquad (12.7)
\end{aligned}
$$

Note that, since the concentration profile in a slab is a straight line at steady state, we could also get a concentration profile simply by equating the slopes, i.e.,

$$
\frac{c - c_{reactant_1}}{x - 0} = \frac{c_{reactant_2} - c_{reactant_1}}{L - 0}
$$

which also leads to Eq. 12.6

2) The flux at steady state is given by Fick's law:

$$
n_{reactant} = -D_{reactant,biofilm} \frac{dc_{reactant}}{dx}
$$

Plugging in Eq. 12.7

$$= -D_{reactant,biofilm} \frac{d}{dx} (295x + 0.25)$$

$$= -0.8 \left[\frac{cm^2}{day}\right] 295 \left[\frac{mg}{liter \cdot cm}\right] \times 10^{-3} \left[\frac{liter}{cm^3}\right]$$

$$= -0.236 \text{ mg/cm}^2 \cdot \text{day}$$

Note that the flux comes out to be negative since it is in the negative x direction.

3) The amount of reactant in the film is given by

$$\int_0^L c\,dx = \int_0^L (295x + 0.25)dx$$

$$= 295\frac{L^2}{2} + 0.25L$$

$$= \left(295 \left[\frac{mg}{liter \cdot cm}\right] \frac{0.01^2 \text{ [cm}^2]}{2} + 0.25 \left[\frac{mg}{liter}\right] 0.01 \text{ [cm]}\right)$$

$$\times 10^{-3} \left[\frac{liter}{cm^3}\right]$$

$$= 17.25 \text{ } \mu\text{g/cm}^2$$

Evaluating and interpreting the solution   *1) Does the concentration profile make sense?* Yes. At steady state in a slab, flux is the same at any location, i.e., the gradient is the same or the profile is linear. *2) Does the flux value make sense?* This is hard to say without physical data. *3) What do we learn?* If the reactant is a substrate for the biofilm, such that it is consumed throughout the biofilm, a different formulation of the problem would be needed, as discussed in Section 12.2. This would lead to higher flux, as the consumption inside would provide larger gradients.

12.1.2   *Example: Diffusion in a Slab with Varying Diffusivity*

Sodium chloride has been traditionally used in meat curing processes, where it acts as a preservative and modifies the water holding capacity of the proteins. Consider diffusion of sodium chloride in a large slab of pig tissue with thickness L, with one side maintained at a concentration of sodium chloride of 0.1 g/cm^3 and the other side maintained at 0.03 g/cm^3. The diffusivity of sodium chloride in the tissue can be approximated as $D = (0.3 + 12c) \times 10^{-10}$ m^2/s, where c is the concentration of sodium chloride in g/cm^3.

1) Recall how we derived the governing equation for mass transfer using a control volume and In − Out + Gen = Change in Storage. For convenience, we are including below an intermediate step that ultimately leads to the governing equation:

$$-\frac{\partial}{\partial x}\left(-D_{AB}\frac{\partial c_A}{\partial x} + c_A u\right) + r_A = \frac{\partial c_A}{\partial t} \qquad (12.8)$$

Starting from this step, write the appropriate governing equation for steady-state diffusion of sodium chloride in the tissue when the diffusivity of sodium chloride in the tissue is *not* a constant. Include the boundary conditions. 2) Obtain the concentration profile of sodium chloride in the slab as a function of position x measured from the surface having the higher concentration. 3) What is the flux of sodium chloride at a location x? Explain whether or not the flux is constant over the thickness. 4) Plot approximately the concentration of sodium chloride vs. distance. Explain whether or not the slope is constant over the thickness.

Solution

Understanding and formulating the problem *1) What is the process?* Diffusion of sodium chloride through the water in a tissue. *2) What are we solving for?* Concentration profile and flux of sodium chloride. *3) Schematic and given data:* A schematic is shown in Figure 12.5 with some of the given data superimposed on it. *4) Assumptions:* None, other than what is already mentioned in the problem description.

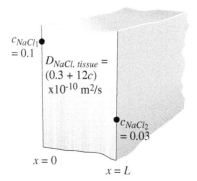

Figure 12.5: Schematic for Example 12.1.2.

Generating and selecting among alternate solutions *1) What solutions are possible?* For steady-state mass transfer, the solutions we have available are shown in the solution chart in Figure 12.19. *2) What approach is likely to work?* Using this chart, since there is no reaction mentioned, this is a case without reaction. A resistance formulation will not work since the derivation of resistance formula, $R = L/D_{AB}A$, assumes that D_{AB} is constant over the thickness L. Thus, we have to derive the concentration profile from scratch, using the governing equation and boundary conditions.

Implementing the chosen solution 1) Since this is a slab geometry, we start from the governing equation in cartesian coordinates but in this case the diffusivity is not constant so we cannot use Eq. 11.18, which assumed D_{AB} to be constant. Instead, we need to start from an earlier step in the derivation, where this assumption has not yet been made, Eq. 11.15:

$$\underbrace{\cancel{\frac{\partial c_A}{\partial t}}^{0}}_{\text{steady state}} = -\frac{\partial}{\partial x}\left(-D_{AB}\frac{\partial c_A}{\partial x}\right) - \underbrace{\cancel{\frac{\partial}{\partial x}(c_A u)}}_{\text{no flow}} + \underbrace{\cancel{r_A}^{0}}_{\text{no reaction}}$$

which simplifies to

$$\frac{\partial}{\partial x}\left(D_{AB}\frac{\partial c}{\partial x}\right) = 0 \tag{12.9}$$

where the subscript A, representing NaCl, has been dropped for simplicity. Here D_{AB} stands for $D_{NaCl,tissue}$. The boundary conditions provided are

$$c|_{x=0} = c_0 = 0.1 \tag{12.10}$$
$$c|_{x=L} = c_L = 0.03 \tag{12.11}$$

2) Integrating Eq. 12.9 once,

$$D_{AB}\frac{dc}{dx} = k_1 \tag{12.12}$$

where k_1 is the constant of integration. Substituting for D_{AB} in terms of concentration, $D_{AB} = a + bc$,

$$(a + bc)\frac{dc}{dx} = k_1$$

Integrating both sides,

$$\int (a + bc)dc = \int k_1 dx$$
$$ac + bc^2/2 = k_1 x + k_2$$

where k_2 is the second constant of integration. Applying the first boundary condition at $x = 0$, we get

$$ac_0 + bc_0^2/2 = k_1(0) + k_2$$

from which we get k_2 as

$$k_2 = ac_0 + bc_0^2/2$$

Applying the second boundary condition at $x = L$, we get

$$ac_L + bc_L^2/2 = k_1 L + k_2$$
$$= k_1 L + ac_0 + bc_0^2/2$$

From which we get k_1 as

$$k_1 = \frac{a(c_L - c_0)}{L} + \frac{b(c_L^2 - c_0^2)}{2L}$$

Substituting k_1 and k_2 into the equation for c, we get

$$ac + \frac{b}{2}c^2 = \left(\frac{a(c_L - c_0)}{L} + \frac{b(c_L^2 - c_0^2)}{2L} \right) x + ac_0 + bc_0^2/2$$

From the given expression for diffusivity, D_{AB}, above, $a = 0.3 \times 10^{-6}$ and $b = 12 \times 10^{-6}$. Plugging these and the concentrations $c_0 = 0.1$ g/cm^3 and $c_L = 0.03$ g/cm^3, we get

$$\left(0.3c + \frac{12}{2}c^2 \right) \times 10^{-6} = \left(0.3(0.03 - 0.1) + \frac{12}{2}(0.03^2 - 0.1^2) \right) \times 10^{-6} \frac{x}{L}$$
$$+ \left(0.3(0.1) + 12(0.1^2/2) \right) \times 10^{-6}$$

Simplifying,

$$0.3c + 6c^2 = -0.0756\frac{x}{L} + 0.09$$

$$c^2 + 0.05c - 0.015 + 0.0126\frac{x}{L} = 0$$

Solving for c,

$$c = \frac{-0.05 \pm \sqrt{0.05^2 - 4(1)(-0.015 + 0.0126\frac{x}{L})}}{2}$$
$$= \frac{-0.05 \pm \sqrt{0.0625 - 0.0504\frac{x}{L}}}{2}$$

Since c has to be between 0.1 g/cm^3 and 0.03 g/cm^3,

$$c = \frac{-0.05 + \sqrt{0.0625 - 0.0504(x/L)}}{2}$$

3) To find flux, we note that this problem is diffusion only, so the question must refer to only the diffusive flux, given by Fick's law, $-D\,\partial c/\partial x$. Using the expression we derived for concentration, we write

$$
\text{Flux} \quad = \quad -D_{AB}\frac{\partial c}{\partial x}
$$

$$
= \quad (0.3 + 12c) \times 10^{-10}[\text{m}^2/\text{s}]\frac{\partial}{\partial x}\left(\frac{-0.05 + \sqrt{0.0625 - 0.0504\frac{x}{L}}}{2}\right)
$$

which appears daunting but, from Eq. 12.12, the right hand side is also $-k_1$. Therefore,

$$
\text{Flux} \quad = \quad -\left(\frac{a(c_L - c_0)}{L} + \frac{b(c_L^2 - c_0^2)}{2L}\right)
$$

Notice that the flux is a constant (it has no x on the right hand side). This is due to the steady-state condition in a slab geometry. The slope of c vs. x, however, is not a constant, as can be written from the same equation:

$$
\frac{\partial c}{\partial x} \quad = \quad \frac{1}{D_{AB}}\left(\frac{a(c_L - c_0)}{L} + \frac{b(c_L^2 - c_0^2)}{2L}\right)
$$

$$
= \quad \frac{1}{a + bc}\left(\frac{a(c_L - c_0)}{L} + \frac{b(c_L^2 - c_0^2)}{2L}\right)
$$

which is not a constant since c varies with x.

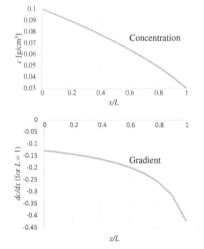

Figure 12.6: Solution for Example 12.1.2, showing concentration decreases with position while the magnitude of the concentration gradient increases with position to make up for the decreasing diffusivity.

Evaluating the solution *1) Is the computed concentration profile reasonable?* Since diffusivity decreases with concentration, diffusivity is higher at the left end where concentration is higher. Thus, the gradient on the left end will have to be lower so that gradient multiplied by diffusivity (= flux) is a constant. Thus, the concentration profile makes sense. *Is the computed concentration reasonable?* The concentration value everywhere in the slab is between the highest (left end) and the lowest (right end) concentrations. Since there is no generation or depletion of sodium chloride, the concentration values are reasonable given the boundary conditions. *2) What do we learn?* Concentration is not linear at steady state if the diffusivity is not a constant.

12.1.3 One-Dimensional Mass Diffusion through a Composite Slab— Overall Mass Transfer Coefficient

In practice, we often work with layers of two or more different materials, as was discussed under heat transfer (page 74). In fact, mass transfer through layers of skin, fat, and muscle, each having its own diffusivity, is an example of mass transfer through a composite material. For a slab material, with uniform surface concentrations, the concentration varies essentially in only one dimension. If there are no reactions (loss or generation of mass), for steady–state mass transfer, the same constant mass flow $N_{A,x}$ appears at any position x.

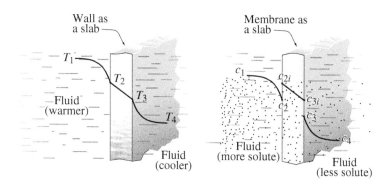

Figure 12.7: Distribution of concentrations around a slab with convection on both sides with analogy to distribution of temperatures around a slab with convection on both sides.

We can write this mass flow in terms of concentrations with analogy to heat transfer (see Figure 12.7) as shown in Table 12.1. In the flow expressions for mass transfer, note that the concentrations of the transporting species c_{2i} in the solid and c_2 in the liquid are not the same. These two concentrations at the interface are, however, related by the distribution coefficient discussed under solid–liquid equilibrium on page 326. For convenience, Figure 9.13 from page 326 is repeated here as Figure 12.8. Similarly for c_{3i} and c_3.

In the lower concentration region, the relationship is linear, as seen in Figure 12.8. In this region, the concentrations in the liquid and in the solid are related by

$$\frac{c_{2i}}{c_2} = K^* \quad \frac{c_{3i}}{c_3} = K^* \tag{12.13}$$

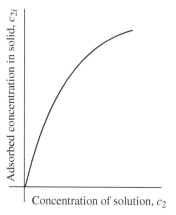

Figure 12.8: Equilibrium relationship between dissolved and surface adsorbed quantity.

Table 12.1: Analogy of heat and mass flows for a slab with convection on both sides

Flux	Mass Transfer	Heat Transfer Analog
Convection in fluid on left face	$N_{A,x} = \dfrac{c_1 - c_2}{\dfrac{1}{h_{m_1} A}}$	$q_x = \dfrac{T_1 - T_2}{\dfrac{1}{h_1 A}}$
Diffusion inside the slab	$N_{A,x} = \dfrac{c_{2i} - c_{3i}}{\dfrac{\Delta L}{D_{AB} A}} = \dfrac{c_2 - c_3}{\dfrac{\Delta L}{K^* D_{AB} A}}$	$q_x = \dfrac{T_2 - T_3}{\dfrac{\Delta L}{k_{solid} A}}$
Convection in fluid on right face	$N_{A,x} = \dfrac{c_3 - c_4}{\dfrac{1}{h_{m_2} A}}$	$q_x = \dfrac{T_3 - T_4}{\dfrac{1}{h_2 A}}$

The concentration equations are rewritten as

$$c_1 - c_2 = \frac{N_{A,x}}{h_{m_1} A}$$

$$c_2 - c_3 = \frac{N_{A,x}}{K^* D_{AB} A} \Delta L$$

$$c_3 - c_4 = \frac{N_{A,x}}{h_{m_2} A}$$

and their left and the right hand sides are added to obtain

$$c_1 - c_4 = N_{A,x} \left(\frac{1}{h_{m_1} A} + \frac{\Delta L}{K^* D_{AB} A} + \frac{1}{h_{m_2} A} \right)$$

Rewriting in terms of mass flow, we get

$$N_{A,x} = \frac{(c_1 - c_4)}{\underbrace{1/h_{m_1} A}_{\text{convective}} + \underbrace{\Delta L/(K^* D_{AB} A)}_{\text{diffusive}} + \underbrace{1/h_{m_2} A}_{\text{convective}}} \qquad (12.14)$$

$$= \frac{\text{concentration difference}}{\text{mass transfer resistances}}$$

As was done in Section 4.1.2, the terms in the denominator of Eq. 12.14 are identified as the mass transfer resistances. Thus, $1/h_m A$ is the convective mass transfer resistance and $\Delta L/K^* D_{AB} A$ is the diffusive mass transfer resistance. Since the inverse of resistance is termed conductance (by electrical analogy), h_m and $K^* D_{AB}/\Delta L$ are sometimes called conductances. An example of conductance values can be seen in Appendix D.4 for transport of water vapor during transpiration in a leaf.

Overall mass transfer coefficient   The diffusive and convective resistances just discussed are often combined and reported in terms of an overall mass transfer coefficient, U_m, such that

$$\frac{1}{U_m A} = \frac{1}{h_{m_1} A} + \frac{\Delta L}{K^* D_{AB} A} + \frac{1}{h_{m_2} A}$$

which simplifies to

$$\frac{1}{U_m} = \frac{1}{h_{m_1}} + \frac{\Delta L}{K^* D_{AB}} + \frac{1}{h_{m_2}} \tag{12.15}$$

In terms of this overall mass transfer coefficient, mass flow can be written as

$$N_{A,x} = U_m A (c_1 - c_4) \tag{12.16}$$

Thus, if U_m is known, mass flow can be calculated directly, without knowing the individual mass transfer resistances. An example of mass transfer through combined resistances is now shown.

12.1.4   Example: Dialysis to Remove Urea from Blood

The main function of the kidneys is to regulate the volume, composition, and the pH of body fluids. When kidneys fail to function, a process known as *hemodialysis* can be used. In this process, the person's blood is rerouted across an artificial membrane that "cleanses" it, removing substances that would normally be excreted in the urine. A patient must typically use this artificial kidney three times a week, for several hours each time. Figure 12.9 shows a capillary dialyzer. Such a dialyzer has a large number of tubes in a parallel arrangement. Figure 12.10 shows the cross-section of one these tubes inside the dialyzer. Blood flows through the inside of the tube and the dialyzing fluid flows on the outside. The tube wall acts as a semi-permeable membrane, letting some of the components (mostly the waste product) of blood to pass. The various mass fluxes are shown in Figure 12.10.

Calculate the flux and the rate of removal (for the total area, in g/hour) of urea from blood at steady state using a cellophane membrane dialyzer at 37°C. The membrane is 0.025 mm thick and has an area of 2.0 m^2. The mass transfer coefficient on the blood side is estimated as $h_{m_1} = 1.25 \times 10^{-5}$ m/s and that on the dialyzing fluid side is $h_{m_2} = 3.33 \times 10^{-5}$ m/s. The diffusivity of urea through the membrane is 1×10^{-10} m^2/s. The distribution coefficient is 2, i.e.,

$$2 = \frac{\text{concentration of urea in membrane at surface}}{\text{concentration of urea in solution at surface}}$$

The concentration of urea in the blood is 0.02 g/100 cc and that in the dialyzing fluid is assumed to be zero.

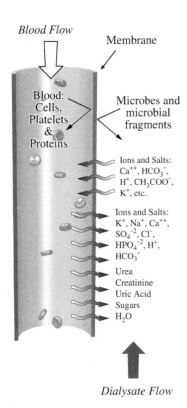

Blood Flow

Membrane

Blood:
Cells,
Platelets
&
Proteins

Microbes and
microbial
fragments

Ions and Salts:
Ca^{++}, HCO$_3^-$,
H$^+$, CH$_3$COO$^-$,
K$^+$, etc.

Ions and Salts:
K$^+$, Na$^+$, Ca^{++},
SO$_4^{-2}$, Cl$^-$,
HPO$_4^{-2}$, H$^+$,
HCO$_3^-$

Urea
Creatinine
Uric Acid
Sugars
H$_2$O

Dialysate Flow

Figure 12.10: Magnified representation of a single hollow fiber in a dialysis system. Picture courtesy of Prof. J. Hunter, Dept. of Biological and Environmental Engineering, Cornell University.

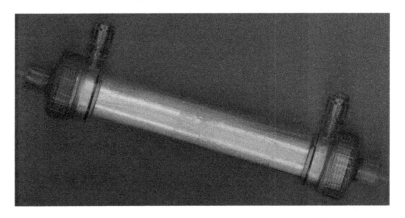

Figure 12.9: A capillary dialyzer.

Solution

Understanding and formulating the problem *1) What is the process?* Transport of urea occurs from higher concentration in the blood to lower concentration in the dialyzing fluid. *2) What are we solving for?* Flux of urea. *3) Schematic and given data:* A schematic is shown in Figure 12.11 with some of the given data superimposed on it. Additionally, $D_{urea,membrane} = 1 \times 10^{-10}$ m^2/s and distribution coefficient = 2. *4) Assumptions:* Since the tube wall is thin, it can be approximated as a slab, following discussion surrounding Eq. 4.48 on page 88.

Generating and selecting among alternate solutions *1) What solutions are possible?* For steady-state mass transfer, the solutions we have available are shown in the solution chart in Figure 12.19. *2) What approach is likely to work?* Using this chart, since there is no reaction mentioned, this is a case without reaction. A resistance formulation might work since the concentration difference, Δc, is given and other data are given from which resistances probably can be calculated. Urea has to transport through convection on the blood side, diffusion through the membrane, and convection on the dialyzate side, thus three resistances. The relevant resistance formula can be for a slab because of the thin wall assumption made above. These resistances are also in series since all of the urea would have to pass through every resistance (convective, diffusive, and convective again) in sequence.

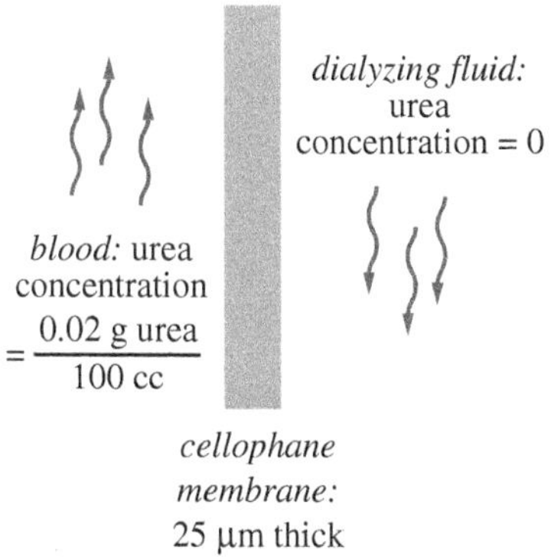

Implementing the chosen solution We could follow the directions for resistance formulation in Section 12.4. However, Section 12.1.3 covers mass transfer resistances in series, and, since this section matches exactly (three resistances in series), we can use Eq. 12.14 to calculate mass flux (the constant area terms on both sides have been cancelled) as

$n_{urea,x}$

$$
= \frac{(0.02 \times 10^4 - 0)\left[\frac{g}{m^3}\right]}{\underbrace{1/(1.25 \times 10^{-5})\left[\frac{s}{m}\right]}_{\text{convective}} + \underbrace{(0.025 \times 10^{-3})/(2)(1 \times 10^{-10})\left[\frac{s}{m}\right]}_{\text{diffusive}} + \underbrace{1/(3.33 \times 10^{-5})\left[\frac{s}{m}\right]}_{\text{convective}}}
$$

$$
= \frac{(0.02 \times 10^4 - 0)\left[\frac{g}{m^3}\right]}{\underbrace{80000\left[\frac{s}{m}\right]}_{\text{convective}} + \underbrace{125000\left[\frac{s}{m}\right]}_{\text{diffusive}} + \underbrace{30030\left[\frac{s}{m}\right]}_{\text{convective}}}
$$

$$
= 0.00085\left[\frac{g}{m^2 s}\right]
$$

Therefore, the rate of removal is

$$
= 0.00085\left[\frac{g}{m^2 s}\right] \times 2\left[m^2\right] \times 3600\left[\frac{s}{hour}\right]
$$

$$
= 6.127\left[\frac{g}{hour}\right]
$$

Evaluating and interpreting the solution *1) Does the computed value (rate of removal) make sense?* From the literature, an adult typically excretes about 25 g

dialyzing fluid: urea concentration = 0

blood: urea concentration $= \dfrac{0.02 \text{ g urea}}{100 \text{ cc}}$

cellophane membrane: 25 μm thick

Figure 12.11: Schematic for Example 12.1.4.

of urea per day; thus, the answer above ($\sim$6 g/hr) may not be completely wrong. *2) What do we learn?* Diffusive resistance through the membrane is the largest of the three resistances. To increase flux, resistances have to be decreased, and decreasing the largest resistance would make the greatest impact.

12.2 Steady-State Diffusion in a Slab with Chemical Reaction

Chemical reactions often occur simultaneously with diffusion. For example, as oxygen is diffusing in a tissue, it is also consumed by the tissue for metabolism. We will consider a slightly more complicated situation of steady-state mass transfer in a slab when the diffusing species decays with time, as illustrated in Figure 12.12. A simple but quite common situation of first-order decay, as explained under kinetics in Section 9.4.3 on page 330, will be considered. For steady-state diffusion in one dimension with first-order decay, the governing equation is

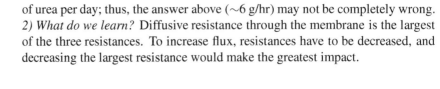

$$\underbrace{\frac{\partial c_A}{\partial t}}_{\text{at steady state}}^{0} + \underbrace{u \frac{\partial c_A}{\partial x}}_{\text{for no convection}}^{0} = D \frac{\partial^2 c_A}{\partial x^2} - k'' c_A$$

which simplifies to the governing equation

$$\frac{d^2 c_A}{dx^2} - \frac{k''}{D} c_A = 0 \tag{12.17}$$

We consider the boundary conditions:

$$c_A(x = 0) = c_{A0} \tag{12.18}$$
$$c_A(x = L) = 0 \tag{12.19}$$

Here L is the distance from the surface at which the concentration becomes zero. Note that the governing equation (Eq. 12.17) and the boundary conditions are almost identical to those for heat transfer in a fin, as described in Section 4.5 on page 99, with only a difference in the boundary condition given by Eq. 12.19. The analogy between the heat and mass transfer situations is quite straightforward. In heat transfer, energy is diffusing as well as being lost from the surface (decaying). In mass transfer, the species is diffusing as well as decaying as a first-order reaction. The solution to Eq. 12.17 is obtained following a process similar to the solution process for heat transfer (see Section 4.5 on page 99). Defining

$$m = \sqrt{\frac{k''}{D}} \tag{12.20}$$

Figure 12.12: Schematic of one-dimensional diffusion of a mass species in a slab with simultaneous consumption of the species.

Eq. 12.17 is written as

$$\frac{d^2 c_A}{dx^2} - m^2 c_A = 0$$

We assume a solution of the form

$$c_A = c_1 e^{-mx} + c_2 e^{mx} \tag{12.21}$$

where c_1 and c_2 are constants to be determined using the boundary conditions. Substituting the boundary conditions, we get

$$c_{A0} = c_1 + c_2$$
$$0 = c_1 e^{-mL} + c_2 e^{mL}$$

Solving for c_1 and c_2 (see Appendix G.5 on page 603 for details) and substituting into Eq. 12.21, we get the final solution for concentration c_A as a function of position as

$$\frac{c_A}{c_{A,0}} = \frac{-e^{-mL}}{e^{mL} - e^{-mL}} (e^{mx} - e^{-mx}) + e^{-mx} \tag{12.22}$$

The solution when a reaction is present (Eq. 12.22) is plotted in Figure 12.13 and compared with the situation when no reaction is present.

For a material that is quite thick (semi-infinite), the solution given by Eq. 12.22 can be simplified by taking its limits as $L \to \infty$. Note that, as $L \to \infty$,

$$\frac{e^{-mL}}{e^{mL} - e^{-mL}} = \frac{1}{e^{2mL} - 1} \to 0$$

so that Eq. 12.22 can be simplified to

$$\frac{c_A}{c_{A,0}} = e^{-mx} \tag{12.23}$$

which shows an exponential decay of concentration from the surface for a thick material.

The rate of mass transfer (the total quantity taken up) at the surface is often an important parameter. The flux at the surface can be compared to the situation with no reaction to find the effect of reaction. The flux at the surface is calculated from

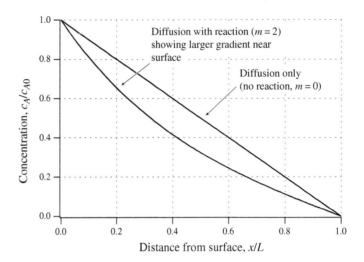

Figure 12.13: The presence of a chemical reaction (decay) produces a larger gradient of concentration near the surface and a higher flux. The figure is a plot of Eq. 12.22.

Eq. 12.22 for a finite thickness L as

$$
\begin{aligned}
n_{A,x} &= -D \frac{dc_A}{dx}\bigg|_{x=0} \\
&= -Dc_{A,0} \left(\frac{-e^{-mL}}{e^{mL} - e^{-mL}} (me^{mx} + me^{-mx}) - me^{-mx} \right)_{x=0} \\
&= -Dc_{A,0} \left(\frac{-e^{-mL}}{e^{mL} - e^{-mL}} (2m) - m \right) \\
&= -Dc_{A,0} \cdot \frac{m}{e^{mL} - e^{-mL}} \left(-2e^{-mL} - e^{mL} + e^{-mL} \right) \\
&= mDc_{A,0} \frac{e^{mL} + e^{-mL}}{e^{mL} - e^{-mL}} \\
&= \frac{Dc_{A,0}}{L} mL \frac{e^{mL} + e^{-mL}}{e^{mL} - e^{-mL}}
\end{aligned}
$$

When no reaction is present (diffusion only), the solution is a straight line given by

Eq. 12.4. From this equation, the flux can be calculated as

$$n_{A,x} = -D \frac{(0 - c_{A,0})}{L} = \frac{D c_{A,0}}{L} \qquad (12.24)$$

which is lower than when there is a reaction, since

$$mL \frac{e^{mL} + e^{-mL}}{e^{mL} - e^{-mL}} > 1 \qquad (12.25)$$

The two concentration profiles for reaction and no reaction are shown in Figure 12.13. In the presence of reaction, the gradient at the surface is higher, leading to higher flux.

12.2.1   *BioConnect: Photosynthesis and the Transport of Water Vapor and CO_2 in a Leaf*

The important process of *photosynthesis* can be modeled as diffusion with a chemical reaction. Photosynthesis is the manufacture of carbohydrates from carbon dioxide and water by chlorophyll-containing plant cells. It involves uptake of carbon dioxide as well as efflux of water vapor (*transpiration*) through the stomatal opening (Figure 12.14). Carbon dioxide from the atmosphere enters the leaf through diffusion in the air just outside the leaf surface and through the water vapor and air inside the leaf (in its substomatal cavity) and is ultimately absorbed by the mesophyll cell walls of the cavity. Water vapor evaporated from the mesophyll cell walls exits the leaf by diffusion through the air and CO_2 in the substomatal cavity and by convection outside the leaf surface.

The concentrations of CO_2 and water vapor along the substomatal cavity are shown in Figure 12.15. For more details, see Rand (1977) and Rand and Cooke (1996). The figure shows that CO_2 is absorbed into mesophyll cell walls generally throughout the deep interior of the leaf, while water vapor essentially evaporates from those cell walls near the substomatal cavity.

The flux of water vapor out of the leaf and the flux of CO_2 into the leaf has several resistances in series (see Section 12.1.3 on page 417). Intercellular air space is relatively unstirred and it offers primarily diffusive resistance. The stomatal opening through which gases must diffuse offers the second resistance. Finally, on the outside surface, there is the boundary layer resistance. When water vapor passes through the cuticle, which is a waxy layer, it has two resistances—the diffusive resistance of the waxy layer and the boundary layer resistance. The waxy cuticle layer greatly reduces the water loss (water in both liquid and vapor phases has difficulty diffusing through the wax). Also, the resistance of the stomata is less on sunny days when the stomata are fully opened.

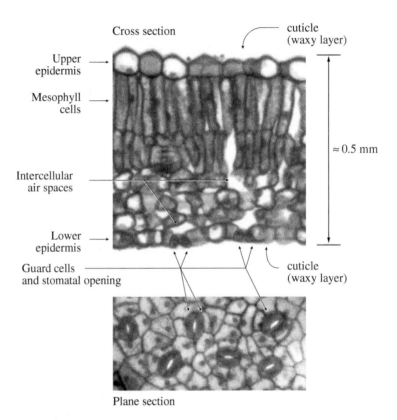

Figure 12.14: Cross section of a leaf showing opening and pathways for transport of CO_2 and water vapor.

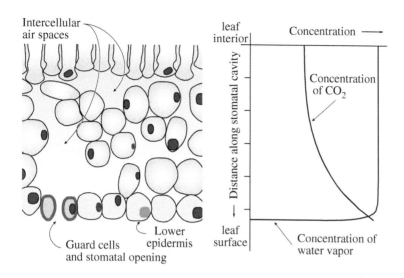

Figure 12.15: Schematic of a leaf cross-section and the concentrations of CO_2 and water vapor in the air space along the substomatal cavity.

12.2.2 Example: Spacing of Blood Vessels in Artificial Tissue

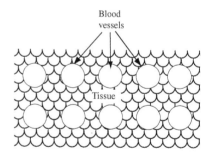

Figure 12.16: Schematic of a 2D geometry of artifical tissue showing blood vessels from which oxygen is transferred into the tissue.

Development of artificial tissues requires designing the vascular network. Consider capillaries as long cylindrical vessels of diameter 10 μm and oxygen transport to be one dimensional, in the radial direction, as shown in Figure 12.16. The spacing between capillaries needs to be small enough to make sure sufficient oxygen is available to all tissue areas. Due to symmetry with the neighboring capillaries, the concentration profile can be considered symmetric at a particular radius, 200 μm, resulting in a zero flux condition. On the capillary wall, a constant oxygen concentration of 130 μM is maintained by the flowing blood. The rate of oxygen consumption in the tissue is 30 μM/s. The diffusivity of oxygen in the tissue is 1.7×10^{-9} m^2/s. 1) Show the governing equation for steady-state oxygen transport with only the needed terms. 2) Show all the boundary conditions needed to solve the equation in step 1. 3) Solve the governing equation to obtain the concentration profile. 4) Assume a minimum concentration of 65 μM is needed in the tissue. Write the equation (with all appropriate values inserted, no need to solve) from which you can calculate the spacing between capillaries that ensures this minimum oxygen concentration in the tissue.

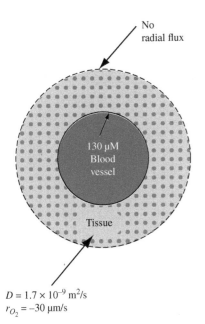

No radial flux

130 µM
Blood
vessel

Tissue

$D = 1.7 \times 10^{-9}$ m^2/s
$r_{O_2} = -30$ µm/s

Figure 12.17: Schematic for Example 12.2.2.

Solution

Understanding and formulating the problem *1) What is the process?* Oxygen is transported from the blood into the tissue with its simultaneous consumption into the tissue. *2) What are we solving for?* Concentration of oxygen as a function of position in the tissue. *3) Schematic and given data:* A schematic is shown in Figure 12.17 with some of the given data superimposed on it. *4) Assumptions:* Since the tube wall is thin, it can be approximated as a slab, following the discussion surrounding Eq. 4.48 on page 88.

Generating and selecting among alternate solutions *1) What solutions are possible?* For steady-state mass transfer, the solutions we have available are shown in the solution chart in Figure 12.19. *2) What approach is likely to work?* Using this chart, since there is a reaction (consumption of oxygen), we either derive from scratch using the governing equation and boundary conditions or choose a solution already derived in the text if conditions (governing equation, boundary conditions, and order of reaction) match. Since the derivation in Section 12.2 on page 422 is for a slab geometry and first-order reaction, it does not match the problem being solved here (which is a cylindrical geometry and has zeroth-order reaction). Thus, we have to derive the solution from scratch.

Implementing the chosen solution 1) Since the blood vessels are cylindrical, we have to start from the governing diffusion equation in cylindrical coordinates,

$$
\underbrace{\frac{\partial c}{\partial t}}_{\text{steady state}}^{0} = D_{AB}\left(\frac{1}{r}\frac{\partial}{\partial r}\left(r\frac{\partial c}{\partial r}\right) + \underbrace{\frac{1}{r^2}\left(\frac{\partial^2 c}{\partial \phi^2}\right)}_{\text{symmetry}}^{0} + \underbrace{\frac{\partial^2 c}{\partial z^2}}_{\text{no change with } z}^{0}\right) + r_A
$$

where the respective terms have been dropped out based on the reasons provided in the problem description. The final governing equation is

$$
D_{AB}\frac{1}{r}\frac{d}{dr}\left(r\frac{dc}{dr}\right) + r_A = 0 \qquad (12.26)
$$

This is a second-order equation that needs boundary conditions to solve. These are

$$
c = 130 \ \mu\text{m at } r = R_i \qquad (12.27)
$$

$$
\frac{dc}{dr} = 0 \text{ at } r = R_o \qquad (12.28)
$$

Integrating Eq. 12.26,

$$\frac{d}{dr}\left(r\frac{dc}{dr}\right) = -r\frac{r_A}{D_{AB}}$$

$$\int d\left(r\frac{dc}{dr}\right) = -\frac{r_A}{D_{AB}}\int r\,dr$$

$$r\frac{dc}{dr} = -\frac{r_A}{D_{AB}}\frac{r^2}{2} + k_1$$

Using the second boundary condition above (Eq. 12.28), we get

$$0 = -\frac{r_A}{D_{AB}}\frac{R_o^2}{2} + k_1$$

$$k_1 = \frac{r_A}{D_{AB}}\frac{R_o^2}{2}$$

Plugging in k_1 and integrating again,

$$r\frac{dc}{dr} = -\frac{r_A}{D_{AB}}\frac{r^2}{2} + \frac{r_A}{D_{AB}}\frac{R_o^2}{2}$$

$$\frac{dc}{dr} = -\frac{r_A}{D_{AB}}\frac{r}{2} + \frac{r_A}{D_{AB}}\frac{R_o^2}{2}\frac{1}{r}$$

$$\int dc = -\frac{r_A}{D_{AB}}\int \frac{r}{2}dr + \frac{r_A}{D_{AB}}\frac{R_o^2}{2}\int \frac{1}{r}dr$$

$$c = -\frac{r_A}{D_{AB}}\frac{r^2}{4} + \frac{r_A}{D_{AB}}\frac{R_o^2}{2}\ln r + k_2 \tag{12.29}$$

Using the first boundary condition (at $r = R_i$) above (Eq. 12.27),

$$130 = -\frac{r_A}{D_{AB}}\frac{R_i^2}{4} + \frac{r_A}{D_{AB}}\frac{R_o^2}{2}\ln R_i + k_2$$

which can be subtracted from 12.29 to eliminate k_2 and thus obtain the expression for concentration, c,

$$c - 130 = -\frac{r_A}{D_{AB}}\frac{r^2 - R_i^2}{4} + \frac{r_A}{D_{AB}}\frac{R_o^2}{2}\ln\frac{r}{R_i}$$

$$c = 130 - \frac{r_A}{D_{AB}}\frac{r^2 - R_i^2}{4} + \frac{r_A}{D_{AB}}\frac{R_o^2}{2}\ln\frac{r}{R_i} \tag{12.30}$$

2) Let r' be the distance at which we have the minimum oxygen. Thus we can write (note r_A is negative since oxygen is being consumed)

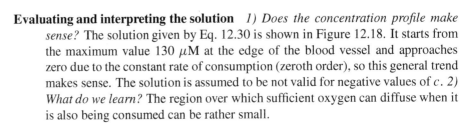

$$65 = 130 + \frac{30\left(r'^2 - \left(5 \times 10^{-6}\right)^2\right)}{4 \times 1.7 \times 10^{-9}} - \frac{30 \times \left(200 \times 10^{-6}\right)^2}{2 \times 1.7 \times 10^{-9}} \ln\left(\frac{r'}{5 \times 10^{-6}}\right)$$

$$65 = 130 + 4.41 \times 10^9 \left(r'^2 - 25 \times 10^{-12}\right) - 352.94 \ln\left(r'/(5 \times 10^{-6})\right)$$

from which r' is calculated as $r' = 6.01 \times 10^{-6}$m $= 6.01\,\mu$m.

Evaluating and interpreting the solution *1) Does the concentration profile make sense?* The solution given by Eq. 12.30 is shown in Figure 12.18. It starts from the maximum value 130 μM at the edge of the blood vessel and approaches zero due to the constant rate of consumption (zeroth order), so this general trend makes sense. The solution is assumed to be not valid for negative values of c. *2) What do we learn?* The region over which sufficient oxygen can diffuse when it is also being consumed can be rather small.

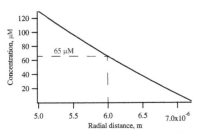

Figure 12.18: Plot of Eqn. 12.30

12.3 Chapter Summary—Steady-State Mass Diffusion with and without Chemical Reaction

- Steady-state diffusion in a slab produces a linear concentration profile (Eq. 12.4)

- Steady-state diffusion in a composite slab is described by overall mass transfer coefficient (Eq. 12.15).

- Presence of chemical reaction (decay) in the slab increases surface flux since a higher concentration gradient is maintained. The concentration in the slab is given by Eq. 12.23.

12.4 Problem Solving—Steady-State Diffusion

▶**Calculate concentration profiles and mass flows.** The steps in this solution process are completely analogous to those for heat transfer, described in Section 4.7 on page 105. Thus, the solution chart in Figure 12.19 shows an approach to solving steady-state diffusion problems. Steady-state diffusion problems can be divided into those with reaction and those without. One important complication is that the boundary condition in a solid, if not provided directly, needs to be obtained from concentration

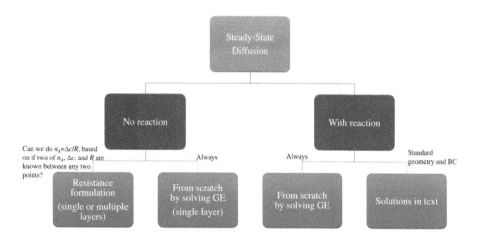

Figure 12.19: Problem solving in steady-state diffusion that is completely analogous to steady-state heat conduction in Chapter 4.

provided in the fluid and equilibrium relations. It is important to solve problems corresponding to each situation in the solution chart, from those provided at the end of this chapter.

▶▶**Resistance formulation** Calculation steps for resistance formulation can be as follows (Note: problems involving multiple materials are much easier to do using the resistance formulation):

1. How many resistances are there?

2. Resistances in what geometry? Diffusive resistances for each geometry, provided in Table 4.2, are to be used with the replacement of k by D. Convective resistance is $1/h_m A$.

3. Are the resistances in series, parallel, or mixed? For series, use a formula derived in Section 12.1.3 and, for parallel, use a formula analogous to that derived in Section 4.1.2.

4. Once mass flow, n_A, is known, using $n_A = \Delta c/R$, concentration can be found at any point when the resistance, R, is known.

▶▶ **Solution from scratch** Solution from scratch (i.e., starting from GE and BC) can be necessary when a resistance analysis won't work (for example, when diffusivity is not a constant, as in Example 12.1.2) or the problem does not match those in text, or the problem asks for a derivation from scratch.

1. Follow Figure 3.16, with the substitution of mass transfer quantities for heat transfer.

2. Use Fick's law to get mass flux from the concentration profile.

3. When there is generation, mass flow at the surface is equal to mass generated inside the domain.

12.5   Concept and Review Questions

1. By analogy to heat transfer, explain the concentration profile in a slab at steady state when no reaction is present.

2. Compare thermal resistance with mass transfer resistance. What parameter in mass transfer resistance replaces thermal conductivity?

3. Why is the overall mass transfer coefficient a useful parameter?

4. Provide a physical example of diffusion with reaction, outside of what has already been discussed.

5. What role does the distribution coefficient play in mass transfer involving a solid and a liquid phase (see Example 12.1.4).

Further Reading

Armstrong, W. 1989. Aeration in roots. *NATO ASI Series, Series G, Ecological Sciences* 19:197–206.

Armstrong, W., M. T. Healy, and T. Webb. 1982. Oxygen diffusion in pea. I: Pore space resistance in the primary root. *New Phytologist* 91(4):647–659.

Cooke, J. R. and R. H. Rand. 1980. Diffusion resistance models. In: *Predicting Photosynethesis for Ecosystem Models*, Vol. 1. J. D. Hesketh and J. W. Jones, Ed. CRC Press.

Dainty, J. 1985. Water transport through the root. *Acta Horticulturae* 171:21–31.

DeWilligen, P. and M. Van Noordwijk. 1984. Mathematical models on diffusion of oxygen to and within plant roots, with special emphasis on effects of soil-root contact. *Plant and Soil* 77(2/3):215–231.

Fournier, R. 2011. *Basic Transport Phenomena in Biomedical Engineering*. CRC Press, Boca Raton.

Heath, M. S., S. A. Wirtel, and B. E. Rittmann. Simplified design of biofilm processes using normalized loading curves. *Research Journal WPCF* 62(2):185–192.

Kim, B. R. and M. T. Suidan. 1989. Approximate algebraic solution for a biofilm model with the Monod kinetic expression. *Water Research* 23(12):1491–1498.

Nobel, P. S. 1996. Leaves and Fluxes. In: *Physiochemical and Environmental Plant Physiology*. Academic Press, San Diego.

Rand, R. H. 1977. Gaseous diffusion in the leaf interior *Transactions of the ASAE* 20(4):701–704.

Rand, R. H. and J. R. Cooke. 1996. Fluid mechanics in plant biology. In *Handbook of Fluid Dynamics and Fluid Machinery*, Edited by J. A. Schetz and A. E. Fuhs, John Wiley & Sons.

Van Bodegom, P., J. Goudriaan and P. Leffelaar. 2001. A mechanistic model on methane oxidation in a rice rhizosphere. *Biogeochemistry,* 55(2):145–177.

Welty, J. R., C. E. Wicks, and R. E. Wilson. 1984. *Fundamentals of Momentum, Heat, and Mass Transfer*. John Wiley & Sons, New York.

Young, T.-H., W.-Y. Chuang, N.-K. Yao, and L.-W. Chen. 1998. Use of a diffusion model for assessing the performance of poly(vinyl alcohol) bioartificial pancreases," CCC 0021-9304:385–391.

12.6 Problems

12.1 Diffusion of Oxygen through Soil

Consider the diffusion of O_2 gas through air in the pores of a column of dry sandy soil. The temperature of the system is 25°C and the total pressure of the system is 1 atmosphere. The soil bed depth is 1 m and the void fraction is 0.20. The partial pressure of O_2 at the top of the column is 0.23 atmosphere while at the bottom, the partial pressure is 0 atm. A tortuosity of 2 can be assumed. Calculate the rate of diffusion, in $g/m^2 \cdot$ hour, of O_2 in air at steady state. *(Hint: Due to the tortuous nature of the paths, the diffusivity in air will be reduced by a factor equal to the tortuosity.)*

12.2 Drug Delivery through Skin

For the simplified case of transport through skin, assume total diffusional resistance of skin is entirely due to the stratum corneum (see Figure 5.16) and none through the sweat ducts, etc. Consider application of some medication to the surface of skin, maintaining a concentration of 10 μg/cc of the medication at the skin surface. The inner surface of the stratum corneum is assumed to be maintained at essentially zero concentration since the molecules are removed as soon as they reach the microcirculation by a sufficiently high peripheral blood flow through skin. The thickness of the stratum corneum is 1 micron (10^{-6} m). The diffusivity through the stratum corneum for the medication (a low-molecular-weight non-electrolyte) is 10^{-10} cm^2/s. 1) Calculate the flux of medication through the skin at steady state. 2) Calculate the total amount of medication in μg/cm^2 of skin that resides in the stratum corneum at steady state.

12.3   Moisture Loss of Food through Packages

Consider food that is wrapped to keep it from drying out. The moisture in the small amount of air inside the wrapping reaches equilibrium with the moisture in the food. The moisture in the food is 0.04 kg of water per kg of dry solids, and the system follows the equilibrium moisture curve of Figure 9.18 on page 341. The outside air is quite dry, i.e., has zero moisture and is blowing such that the outside surface of the wrapping is also maintained at zero moisture. The system is at 60°C. Assume the diffusivity of water vapor in packaging material is 0.25×10^{-11} m^2/s. The density of dry air is 1.07 kg/m^3. Calculate the loss of water vapor in g/day at steady state for a wrapping of 0.08 mm thickness and a total surface area of 10 cm $\times$ 10 cm.

12.4   Diffusion of Gas through the Walls of a Tube

A gas diffuses at steady state through the walls of a cylindrical tube whose cross-section is shown in Figure 12.20. Assume the inside and outside walls are at gas concentrations c_i and c_0, respectively. 1) Starting from the governing equation, derive a relationship for the rate of diffusion of the gas through the tube in terms of the concentrations given, the diffusivity of the gas, and the tube dimensions. *(Hint: Follow the derivations in Section 4.2 for heat transfer.)* 2) Setting up a resistance formulation, show the mass transfer resistance of the tube.

12.5   Oxygen Transport in Human Tissue

Oxygen transport in human tissue has been considered a process involving diffusion from the blood vessels into tissue cylinders accompanied by a zero-order reaction, $-k''$ (Welty et al., 1984). This reaction represents the metabolic consumption of the oxygen to produce carbon dioxide. The tissue cylinders can be considered to be arranged in a bundle such that the boundary condition at the inner surface of the tissue cylinder is a constant c_i corresponding to the oxygen concentration in the arterial blood. Due to

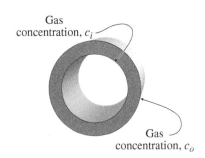

Gas concentration, c_i

Gas concentration, c_o

Figure 12.20: Tube cross-section.

symmetry at the outer surface of these tissue cylinders, a zero flux condition can be used. 1) Formulate the steady-state governing equation and the boundary conditions for oxygen concentration, c, in the tissue. 2) Determine the oxygen concentration as a function of radius, r. 3) Determine the rate of oxygen consumption in the inner surface of the tissue at steady state.

12.6 Oxygen Diffusion and Consumption in the Cornea

The cornea of the mammalian eye can be considered, as a first approximation, to be a membrane 0.05 cm thick. Assume the thickness of the cornea is much smaller than its surface area so it can be treated as a slab. The inside of the cornea is bathed by a dilute salt solution, the aqueous humor. The aqueous humor maintains a small oxygen concentration at the corneal inner surface which we assume for this problem to be zero. The cornea uses up oxygen at the rate of 8.045×10^{-10} mol O_2/cm^3 cornea per second. This oxygen consumption follows a zeroth-order reaction. The diffusivity of oxygen in the cornea is 1.08×10^{-9} m^2/s. 1) Assume the eye is open and the outer surface of the cornea is in equilibrium with the oxygen concentration (21%) in the air. The equilibrium Henry's law constant between oxygen at the corneal surface and that in the air is 47,000 atm/mole fraction of O_2. For conversion, use 1 mole fraction of oxygen in the cornea as approximately equal to 55.56×10^{-3} moles of O_2/cm^3 of cornea. Calculate the concentration of oxygen in the corneal outer surface in mol/cm^3. 2) Calculate the oxygen concentration profile as a function of depth in the cornea *under steady-state conditions* (plug in numbers and simplify). 3) Roughly sketch the oxygen concentration profile calculated in step 2. 4) Calculate the oxygen flux at the corneal outer surface.

12.7 Drug Delivery to the Brain

Since traditional methods often cannot be used for delivering drugs to the brain, researchers are working on implanting drug-containing polymers directly in the brain tissue. Consider steady one-dimensional diffusion in the x direction and first-order elimination of the drug dopamine near an implant, as shown in Figure 12.21. The boundary conditions of drug concentration are c_0 at the implant surface and zero far from the surface into the brain tissue. The reaction rate, k'', for elimination is 0.5/hour and the diffusivity of dopamine in brain tissue is 30×10^{-11} m^2/s. 1) Write the solution for drug concentration as a function of position x from the implant surface. 2) Find the distance in mm from the implant surface at which the drug concentration becomes 10% of its value at the implant surface. (The volume corresponding to this distance is termed the tissue treatment volume.) 3) If the *average* concentration of dopamine over the distance calculated in step 2 above needs to be 5 $\mu g/g$ of tissue for a certain

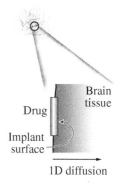

Figure 12.21: Schematic of drug delivery in a tissue.

treatment, what should the concentration c_0 at the implant surface for this treatment be?

12.8   Diffusion-Driven Aeration in Plants

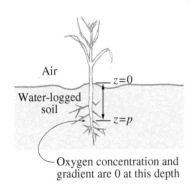

Air

Water-logged soil

$z=0$

$z=p$

Oxygen concentration and gradient are 0 at this depth

Figure 12.22:   Schematic for Problem 12.8.

Consider a plant growing in a water-logged plot, as shown in Figure 12.22. In such water-logged soils, most plants obtain little, if any, oxygen from the soil, so most of it comes from the atmosphere. It is desired to find the depth to which atmospheric oxygen concentration diffuses down the length of a submerged root so that we can know the portion of the root that is in need of oxygen. For simplicity, assume steady one-dimensional diffusion and neglect any generation of CO_2. The volumetric respiratory activity of the plants is constant at k''. 1) Write the governing equation for the problem. 2) There are three boundary conditions for this problem—the additional condition is needed since the submerged length, p, at which the concentration becomes zero is unknown. Two of them are at this submerged length— both concentration and fluxes are zero. The third boundary condition is a constant concentration c_0 at the soil surface. Write these three boundary conditions. 3) Find the submerged length (where concentration becomes zero) as a function of diffusivity, c_0 and k''.

12.9   Dialysis to Remove Urea from Blood

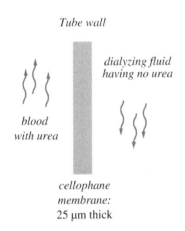

Tube wall

dialyzing fluid having no urea

blood with urea

cellophane membrane: 25 μm thick

Figure 12.23: Transport of urea through a dialysis membrane.

Consider the problem of designing a dialysis membrane (the individual tubes in a hollow fiber dialyzer, as discussed in Section 12.1.4). A schematic for transport of urea from the blood through a tube wall is shown in Figure 12.23 where the tube wall can be considered as a flat surface due to it being thin compared to the radius of the tube. Convection occurs on both sides of the wall. Through materials research, there is the potential to come up with a membrane material that has twice the diffusivity. With this new membrane material, the thickness can remain the same as the existing membrane material at 25 μm. The distribution coefficient of the existing membrane material for a urea solution is 2 and is also expected to remain unchanged. The concentrations of the dialyzing fluid in blood and urea will stay the same for both membranes as well. The diffusivity of the existing membrane material to urea is $1 \times 10^{-10} m^2/s$, and its flux value is 0.001 g of urea/m^{2}s. The concentration of urea in the blood is 0.0002 g/cm^3. 1) Through steady-state mass transfer analysis of transport through the membrane, show by what factor the flux would change (ratio of new flux to old flux) if the diffusivity were doubled. 2) Explain why the flux does not double even if the diffusivity is doubled. 3) If this new material were not available (diffusivity of the membrane could not be increased) and the membrane could not be made any thinner, how would you try to increase the rate of urea transport across the membrane?

12.10 Bioremediation

Some plants have been shown to be useful in the remediation of soils contaminated with volatile organic compounds, such as trichloroethylene (TCE). TCE removed from a contaminated soil by poplar trees is volatilized to the atmosphere, primarily through the stem. In this problem, we consider transport of TCE through radial diffusion in the outermost layers of the stem, which is composed of the phloem and the bark. The concentrations at these outermost layers are shown in Figure 12.24. The diffusivity of TCE in the phloem and bark is 5×10^{-7} m^2/s. 1) Calculate the steady-state concentration profile of TCE in the ring representing the phloem and the bark (plug in numbers and simplify). 2) Calculate the flux of TCE into the atmosphere.

12.11 Drug Transport through Skin with Metabolism Present

In addition to its function as a physical barrier, the skin also acts as a metabolic barrier capable of degrading a wide variety of compounds upon penetration. Consider a substrate diffusing through the skin and being metabolized (eliminated) as a first-order reaction, with a rate constant of $k'' = 10^3$/min. The substrate concentration at the surface is 300 nmol/ml, and the metabolizing skin tissue is thick enough such that this concentration reaches zero before the full thickness. The diffusivity of the substrate in the skin tissue is 2×10^{-10} m^2/s.

1) Write the governing equation for substrate diffusion and elimination at steady state. 2) Write the boundary conditions. 3) Write the solution, i.e., concentration as a function of position, and plug in the appropriate numerical values. 4) Find the depth of penetration, defined as the distance at which concentration is 10% of its surface value. 5) Find the rate in nmol/m$^2 \cdot$s at which the substrate penetrates the skin surface. 6) What is the depth of penetration when the substrate is consumed at twice the rate given by k''? 7) When the consumption rate doubles to $2k''$ and you still want to have the same concentration at the same depth as in step 4, what two parameters would you change? If you could change only one parameter at a time, what would you change them to?

12.12 Spacing of blood vessels in artificial tissue

Moved to solved example

12.13 Nutrient Transport in Designing Artificial Tissue

Tissue engineering utilizes synthetic scaffolds to serve as the extracellular matrix for culturing cells into tissue, which may be transplanted into a patient. Since the artifi-

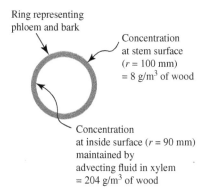

Ring representing phloem and bark

Concentration at stem surface (r = 100 mm) = 8 g/m^3 of wood

Concentration at inside surface (r = 90 mm) maintained by advecting fluid in xylem = 204 g/m^3 of wood

Figure 12.24: Transport of volatile organic compounds through a stem.

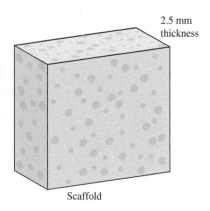

2.5 mm thickness

Scaffold

Figure 12.25: A scaffold for artificial tissue through which glucose and oxygen must diffuse.

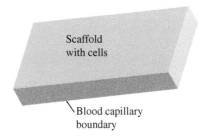

Scaffold with cells

Blood capillary boundary

Figure 12.26: Schematic of oxygen diffusion into a scaffold.

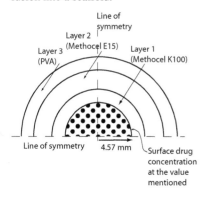

Line of symmetry

Layer 2 (Methocel E15)

Layer 3 (PVA)

Layer 1 (Methocel K100)

Line of symmetry   4.57 mm

Surface drug concentration at the value mentioned

Figure 12.27: Schematic of a tablet.

cial tissue is not vascularized, transport of nutrients and oxygen to cells is limited by diffusion through the matrix. Consider a collagen scaffold, approximated as a slab, as shown in Figure 12.25. Both sides of the slab are exposed in a symmetric manner to nutrients (including glucose) and oxygen that maintain a constant concentration at the surface, diffuse through the scaffold, and are depleted by the metabolism of the growing tissue. The rates of depletion are constant, at 54.4 $g/m^3 \cdot$ day for oxygen and 104 $g/m^3 \cdot$ day for glucose. The diffusivity of oxygen is 3.3×10^{-11} m^2/s and that of glucose is 6×10^{-11} m^2/s. 1) Write the governing equation from which you can find the concentration of glucose or oxygen at steady state. 2) Write the boundary conditions. 3) Solve for concentration as a function of position (no need to plug in numbers). 4) What is the minimum concentration for each of glucose and oxygen that must be maintained at the surface so that they reach across the entire thickness of the scaffold? *(Hint: At the minimum surface concentration, the concentration at the center of the slab is zero.)*

12.14   Oxygen Transport in Designing Artificial Tissue

Figure 12.26 shows a scaffold as a slab, used for generating tissue that has been seeded uniformly with cells. The blood capillary boundary at the bottom surface supplies oxygen that diffuses into the scaffold and is consumed by the cells in it. For the purpose of this problem, we have simplified the oxygen consumption to be a first-order process with a rate constant 4.5×10^{-6} s^{-1}. Oxygen is the cell growth limiting nutrient—the cells would die below a minimum oxygen concentration of 5 nmol/mL. The concentration of oxygen at the blood capillary boundary is 80 nmol/mL. The diffusivity of oxygen in the tissue is 1.75×10^{-9} m^2/s. Assuming steady state, what is the distance from the blood capillary boundary over which the cells are expected to survive?

12.15   Drug Diffusion through a Tablet

A schematic of a *spherical* tablet is shown in Figure 12.27. Here the main ingredient of the drug naproxen sodium (abbreviated as N from now on) diffuses through the three layers shown, which are designed to provide a more uniform release of the drug out of the tablet (among other things). The thicknesses of the layers 1, 2, and 3 and their diffusivities are shown in the table below. 1) Write the governing equation for

Layer	Thickness (mm)	Drug diffusivity (mm^2/s)
1 (Methocel K100)	0.89	9.7×10^{-2}
2 (Methocel E10)	0.76	0.15×10^{-2}
3 (PVA)	0.13	0.24×10^{-2}

the diffusion of N in *any* of the layers, assuming steady state. 2) For concentration specified boundary conditions $c(r = r_i) = c_i$ and $c(r = r_o) = c_o$, where r_i and r_o are the inner and the outer radius, respectively, solve the governing equation for c as a function of position. 3) Derive the expression for flow rate (of N) for a layer in terms of the boundary conditions. 4) Rewrite the expression for flow in terms of a concentration difference divided by resistance, and obtain the expression for resistance. 5) The concentration at the inner surface of layer 1 is 1.062 mg/mm^3, and the concentration at the outer surface of layer 3 is zero. Using this information along with the data in the table, calculate the instantaneous flow rate of N from the outer surface.

12.16 Diffusion in a slab with varying diffusivity

Moved to solved example.

12.17 Urea Transport in a Hollow Fiber Dialyzer

We would like to design a hollow fiber dialyzer, as shown in Figure 12.10. We will consider steady-state transport of urea through the membrane, which involves processes of diffusion across the wall and convection on both sides of the wall. The tubes are 0.5 mm in diameter and the thickness of the wall material is 25 μm. The diffusivity of urea is 1×10^{-10} m^2/s in the membrane material and 1.46×10^{-9} m^2/s in the blood. The distribution coefficient, K^*, of the membrane material (concentration in membrane divided by concentration in blood) for the urea solution is 2. The concentration of urea in the blood is 0.0002 g/cm^3 and that in the dialyzate fluid is zero. The mass transfer coefficient on the dialyzate side is 3.33×10^{-5} m/s. For the blood side, average velocity in the tube is 2.7 mm/s. The viscosity and density of blood are 3.2×10^{-3} Pa $\cdot$ s and 1025 kg/m^3, respectively. 1) Calculations for the mass transfer coefficient are analogous to those for the heat transfer coefficient. After choosing the equation for the appropriate flow situation, one simply replaces both the thermal conductivity and thermal diffusivity by mass diffusivity (more in Chapter 14). Using this analogy, calculate the mass transfer coefficient on the blood side. 2) What is the overall mass transfer coefficient for a tube? *(Hint: Like an overall heat transfer coefficient, an overall mass transfer coefficient relates flux to total concentration difference and includes both diffusive and convective resistances.)* 3) How many 24 cm long tubes do we need to obtain a total urea transport rate of 5 g/hour?

12.18 Oxygen Diffusion through a Contact Lens

In a hydrogel-based contact lens, oxygen transport from the air to the cornea (important for avoiding adverse clinical effects) occurs by the oxygen first getting absorbed into

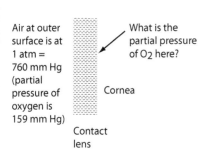

Air at outer surface is at 1 atm = 760 mm Hg (partial pressure of oxygen is 159 mm Hg)

What is the partial pressure of O$_2$ here?

Cornea

Contact lens

Figure 12.28: Schematic of oxygen transport through contact lens.

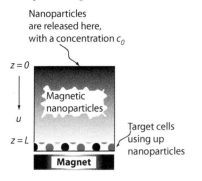

Nanoparticles are released here, with a concentration c_0

$z = 0$

u

Magnetic nanoparticles

$z = L$

Target cells using up nanoparticles

Magnet

Figure 12.29: Magnetic force pulls the magnetic nanoparticles toward the cells.

the water of the hydrogel and then diffusing through this gelled water. Oxygen flux through one particular contact lens (Figure 12.28), approximated as a slab that is 10 μm thick, is known to be $5\mu l/cm^2 \cdot$ hour at steady state (l stands for liter or 1000 cm^3). The density of oxygen is 1.308 g/l. For the purpose of this problem, we assume the contact lens to have the properties of water, so Henry's constant for oxygen–water equilibrium, 5×10^4 atm/mole fraction of O$_2$, can be used. The diffusivity of oxygen through the contact lens is 1.5×10^{-11} m^2/s. The corneal surface and the lens are at 35°C. The molecular weights of oxygen and water are 32 and 18, respectively. If the outer surface of the contact lens, open to the atmosphere, has a partial pressure of oxygen of 159 mm Hg, what is the steady-state concentration of oxygen (in terms of its partial pressure) at the inner surface of the contact lens that is in touch with the cornea?

12.19 Transport of Magnetic Nano-Carrier Particles in Gene Delivery

Transport using magnetic nano-carrier particles under an applied magnetic field can accelerate delivery of genes to target tissue. As shown in Figure 12.29, magnetic nanoparticles are released from the top of a fluidic chamber, where concentration is constant at c_0. They are then instantly taken up by target cells at the other end of the chamber upon reaching them. The magnet is configured in a way such that there is no radial movement, i.e., the only movements of the nanoparticles are along the z direction, with a steady velocity u. There is no effect of gravity.

1) Write the governing equation for steady-state transport of the nanoparticles, being sure to include magnet-driven "flow" in addition to diffusion. 2) Write the boundary conditions needed to solve the above problem. 3) Solve the equation to obtain concentration, c, as a function of position, z. (Hint: Rewrite your governing equation in the form $dc/dz + pc = q$ that you can solve—here p and q are constants). 4) Sketch the solution, i.e., c as a function of z, in step 3.

12.20 Oxygenation of Rice Rhizosphere

Rice paddies are suspected to contribute substantially to the increasing atmospheric methane concentration (van Bodegom et al., 2001). However, one way that methane can be used up is through oxidation of methane in the soil by oxygen. The rhizosphere is the region of soil in the vicinity of plant roots where methane in the soil is oxidized from the oxygen transported from the plant, as shown for a cylindrical region in Figure 12.30. The root releases oxygen and can be assumed to maintain a constant concentration of oxygen at the root surface. At the boundary of the aerobic soil (dashed line), the flux of oxygen is zero.

1) Write the governing equation for steady-state radial diffusion of oxygen and an assumed constant (zeroth-order) rate of consumption (through oxidation of methane and decomposition of organic matter) in the rhizosphere. Define any symbol that you use. 2) Write the boundary conditions needed to solve the above problem. 3) Solve the equation to obtain concentration as a function of radius, r. 4) For the following parameters, give the expression for the radius at which the concentration of oxygen becomes 1/100 of that at the root surface (plug in numbers, but no need to solve for a numerical value). The concentration of oxygen at the surface of the root is 6.8 mg O_2 per liter of total volume; the outer radius of the rhizosphere is 3 mm; the radius of the root is 0.1 mm; the diffusivity of oxygen in the rhizosphere is 2.41×10^{-9} m^2/s; and the consumption of oxygen is 5.1 mg O_2 per liter per hour. 5) For the system in step 4, calculate the rate of diffusive transport of oxygen into the soil from the entire surface of the root, if the root is 0.1 m long.

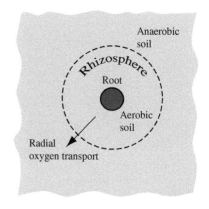

Figure 12.30: Schematic of a root and surrounding areas (rhizosphere).

12.21 Nutrient Supply in Engineered Cartilage

Since tissue-engineered cartilage is without blood vessels, nutrient (such as glucose) supply relies on diffusion. In a greatly simplified situation, shown in Figure 12.31, consider steady-state 1D transport of glucose into the cartilage construct where flow of glucose-containing liquid maintains a concentration of c_s mol/ml of glucose at the top surface. The bottom surface rests on an impermeable surface, i.e., glucose cannot go through this surface. The diffusivity of glucose is D, the constant rate of consumption of glucose is R mol of glucose per million cells/second, and the cell density is N (million cells/ml).

1) Write the governing equation for this problem, using the parameters given. 2) Write the boundary conditions for the problem. 3) Solve for concentration of glucose as a function of position into the construct. 4) Assuming cell density is positively correlated with the glucose concentration, at what location is the minimum cell density? 5) *(Be careful with units)* In a 2 mm construct, the surface concentration of glucose is $c_s = 5.56 \times 10^{-6}$ mol/ml, glucose diffusivity is 7.78×10^{-10} m^2/s, and glucose consumption rate is 3.61×10^{-11} mol of glucose per million cell/second. Assuming cells need a minimum glucose concentration of 0.2×10^{-6} mol/ml, what minimum cell density N can be maintained? 6) If we want to maintain a minimum cell density (million cell/ml) that is twice as large as what you calculated in step 5, for the same thickness of construct, same diffusivity, and same rate of consumption, what parameter would you change to maintain this minimum glucose concentration and what would its new value be?

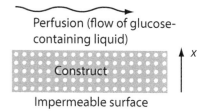

Figure 12.31: Cartilage construct with glucose-containing fluid flowing over the top surface.

Chapter 13

DIFFUSION MASS TRANSFER: UNSTEADY STATE

After you have studied this chapter, you should be able to formulate and solve for a time-varying (unsteady) diffusion mass transfer process for the following situations:

1. Where concentrations do not change with position.

2. In a simple slab geometry where concentrations vary with position.

3. Near the surface of a large body (semi-infinite region).

KEY TERMS

- **internal resistance**

- **external resistance**

- **mass transfer Biot number**

- **lumped parameter analysis**

- **1D diffusion in slab, cylinder, and sphere**

- **Heisler charts**

- **diffusion in a semi-infinite region**

In this chapter, like Chapter 5 on heat transfer, we will consider mass transfer situations where concentration is a function of both position and time. Such situations

443

are typically more complex as opposed to steady state studied in Chapter 12 where concentration did not vary with time. As mentioned in heat transfer, time-varying processes are called unsteady or transient. The relationship of this chapter to the other chapters in mass transfer is shown in Figure 13.1.

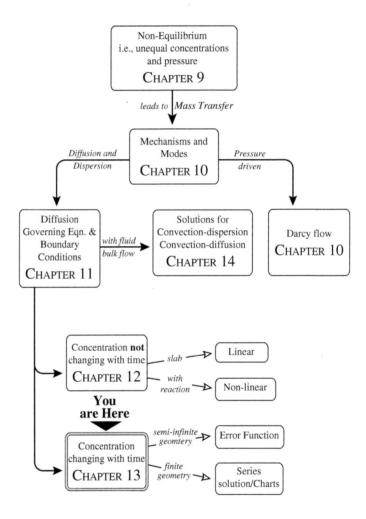

Figure 13.1: Concept map of mass transfer showing how the contents of this chapter relate to other chapters in mass transfer.

13.1 Transient Mass Transfer When Internal Diffusive Resistance Is Negligible: Lumped Parameter Analysis

In Chapter 11, we derived the general governing equation for mass transfer that was valid regardless of a small or large diffusivity. If the diffusivity in a particular solid is large, the governing equation for mass transfer in the solid can be described in a much simpler way. This approach is analogous to the lumped parameter analysis covered in heat transfer in Section 5.1 on page 129. As an example, consider the drying of a very wet solid such as a moist food due to dry air flowing over it, as shown in the schematic in Figure 13.2. Moisture evaporates from the surface of the solid and is removed by the flowing air. The convective resistance to mass (water vapor) transfer is the *external resistance*. Moisture removed from the surface of the solid creates a moisture gradient inside the solid that will cause capillary diffusion of water from the inside the solid to the surface. The (capillary) diffusional resistance of the water inside the solid is the *internal resistance*. In this example, when the solid is very wet (there is an abundance of water), water can diffuse very rapidly and equilibrate the concentration everywhere inside. Under such a condition, moisture concentration is *spatially uniform* at any instant. In other words, internal resistance to diffusive movement is negligible. The assumption of spatial uniformity or negligible concentration gradients is the essence of a *lumped parameter* approach. This is only an approximation, though, as, according to Fick's law, there cannot be moisture transport if the gradient is zero. But in the above example of a moist solid, we know intuitively that the solid will keep losing moisture, especially when it is very wet. Thus, there must be moisture gradients inside the solid for moisture transport to occur; it is just that the gradients are small.

We will now formulate a simple governing equation for mass transfer situations that can be described by the simple lumped parameter approach. If a solid of dry mass m_s and moisture content w (see Figure 13.2) is placed in air with moisture concentration c_∞, moisture loss over the surface will be described by convection. A mass balance on water content in the solid over time Δt during which its moisture content drops by Δw leads to

$$\underbrace{-m_s \Delta w}_{\substack{\text{moisture} \\ \text{lost}}} = \underbrace{h_m A (c_s - c_\infty) \Delta t}_{\substack{\text{moisture convected} \\ \text{away at surface}}} \tag{13.1}$$

where h_m is the convective mass transfer coefficient over the surface, A is the surface area of the solid, and c_s is the moisture (vapor) concentration in the air next to the solid surface. Note that c here denotes concentration of water vapor in air, $c_{vapor,air}$, where

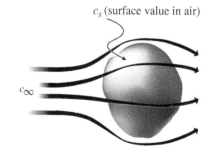

c_s (surface value in air)

c_∞

Figure 13.2: A moist solid with air flow over it.

the subscripts have been dropped for simplicity. Dividing both sides by $-m_s \Delta t$,

$$\frac{\Delta w}{\Delta t} = -\frac{h_m A}{m_s}(c_s - c_\infty)$$

Taking the limit as $\Delta t \rightarrow 0$,

$$\frac{dw}{dt} = -\frac{h_m A}{m_s}(c_s - c_\infty) \qquad (13.2)$$

Equation 13.2 is the governing equation describing concentration as a function of time for a lumped parameter mass transfer situation. Since the concentration w is only a function of time, we only need the initial condition

$$w(t = 0) = w_i \qquad (13.3)$$

to solve the equation. To solve Eq. 13.2, we note two different cases for which the surface concentration c_s will be obtained differently.

Case I: When solid surface has free water   This assumption remains valid for small amounts of drying within a very high moisture range. Since the surface has free water, concentration of water vapor at the surface will be given by the vapor pressure of water at the surface temperature, making the surface concentration a constant. Under such a condition, Eq. 13.2 can be integrated as

$$\int_{w_i}^{w} dw = \int_0^t \underbrace{-\frac{h_m A}{m_s}(c_s - c_\infty)}_{\text{constant}} dt \qquad (13.4)$$

with the final solution

$$w - w_i = -\frac{h_m A}{m_s}(c_s - c_\infty)t \qquad (13.5)$$

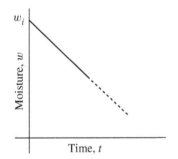

Figure 13.3: Linear decrease in concentration with time during drying of a very high moisture solid; illustration of Eq. 13.5.

Equation 13.5 shows that concentration w would change linearly with time, as illustrated in Figure 13.3. Thus, the rate of moisture change, dw/dt, is a constant, as illustrated in the first segment of the rate curve in Figure 13.4. The second segment of the rate curve will be discussed in the next section. When drying a solid, this situation is referred to as the constant rate of drying. Note that the analysis resulting in Eq. 13.5 corresponds to a high rate of evaporation, the same as evaporation from a free surface of water. This constant rate of moisture loss will not continue for too long. In the wet cloth example, as the moisture levels are reduced in the cloth, water will be held more tightly by the cloth and the diffusivity will decrease substantially. This decrease in diffusivity will increase the internal resistance to movement of water, and eventually the internal diffusive resistance will dominate over the external convective resistance, thus decreasing the rate of drying.

Case II: When solid surface has no free water   In this case, the vapor at the surface is in equilibrium with the moisture at the solid surface. Thus, c_s will be obtained from the equilibrium moisture relationships described on page 323 for air–wet solid equilibrium. If a linear relationship can be assumed between c_s and w for a small range of moisture, Eq. 13.2 becomes completely equivalent to Eq. 5.4 for heat transfer and the solution will be an exponential decay of moisture with time. However, this situation may not exist in reality for reasons mentioned in the previous paragraph, i.e., when the moisture level becomes lower, moisture diffusivity decreases (and therefore internal resistance increases) and the lumped parameter assumption is no longer valid.

Deciding When to Use Lumped Parameter Analysis   The lumped parameter analysis, being so simple, is preferred over other more complex analyses such as those described later. Obviously, the lumped parameter analysis would not be appropriate in every situation. Thus, it is important to know the conditions under which we can apply the lumped parameter approach. In an analogous to that way as described on page 131 for heat transfer, the parameter that compares the internal and the external resistance is defined as the mass transfer Biot number, Bi_m given by

$$\text{Bi}_m = \frac{h_m K^* L}{D_{AB}} = \frac{L/D_{AB}}{1/(h_m K^*)} = \frac{\text{diffusive resistance}}{\text{convective resistance}} \quad (13.6)$$

Here D_{AB} is the mass diffusivity and L is the characteristic length (same discussion on characteristic length as on page 132 holds). Note the additional variable K^*, the distribution coefficient, introduced here when the mass transfer is between two phases, one internal (i.e., solid) and one external (i.e., fluid flowing over the solid). The need for K^* in the Biot number equation for mass transfer can be seen by comparing Eqs. 13.22 and 13.25, described later. For a mass transfer Biot number $\text{Bi}_m < 0.1$, i.e.,

$$\frac{h_m K^* L}{D_{AB}} < 0.1 \quad (13.7)$$

the error in concentration calculation using lumped parameter analysis is less than 5%, and such analysis is often used. For diffusive mass transfer in wet solids, Biot numbers are typically high. For example, for a diffusivity value given by $D_{AB} = 10^{-8}$ m^2/s, a mass transfer coefficient given by $h_m = 0.01$ m/s, a K^* value of 0.05,[1] and a slab

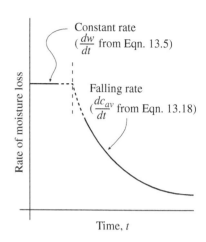

Figure 13.4: Initial rate of drying in a very high moisture solid is a constant corresponding to Figure 13.3, followed by a falling rate of drying when the moisture levels have reduced.

---

[1] Value corresponds approximately to equilibrium between wet wood and moist air at 20°C. Data from ICT, 1933.

half thickness of $L = 10$ cm, the mass transfer Biot number is

$$
\begin{aligned}
\text{Bi}_m &= \frac{0.01[\text{m/s}] \times 0.05 \times 0.1[\text{m}]}{10^{-8}[\text{m}^2/\text{s}]} \\
&\approx 5 \times 10^3
\end{aligned}
$$

which is quite large and the primary resistance to mass transfer is the internal diffusive resistance (see Eq. 13.6). As mentioned in heat transfer, the lumped parameter approach is typically suitable (Eq. 13.7 is satisfied) when surface to volume ratio is large or the diffusivity is large with respect to the convective mass transfer coefficient.

13.2   Transient Diffusion When Internal Resistance Is Not Negligible: Example of a Slab Geometry

As stated in the previous section, when the mass transfer Biot number is large, i.e., internal diffusional resistances are significant, a lumped parameter analysis is no longer applicable. This section parallels the analogous section on heat transfer (Section 5.3 on page 136). In this section, we will learn how to formulate and solve for the spatial distribution of concentration in the solid. Since concentration variation in the solid is significant, we will need to solve this problem by starting with the general governing equation on page 396. For simplicity, we will consider a one-dimensional (1D) slab where the concentration varies along the thickness, as shown in Figure 13.5. To keep things simple, we will consider the boundary condition at the surface to be a constant concentration, i.e., external fluid resistance is negligible. This corresponds to a very high convective mass transfer coefficient (see discussion on boundary conditions at the end of Section 11.3). The governing equation and boundary condition for symmetric drying or wetting of an infinite slab that has no chemical reaction are

$$
\frac{\partial c_A}{\partial t} + \underbrace{u \frac{\partial c_A}{\partial x}}_{\text{no bulk flow}}{}^{0} = D \frac{\partial^2 c_A}{\partial x^2} + \underbrace{r_A}_{\text{no reaction}}{}^{0} \tag{13.8}
$$

Thus, the simplified governing equation is

$$
\frac{\partial c}{\partial t} = D \frac{\partial^2 c}{\partial x^2} \tag{13.9}
$$

where we have dropped the subscript A in c_A for simplicity. Due to the second order in spatial (x) variation, two boundary conditions in space will be needed. For symmetric

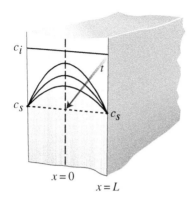

Figure 13.5: Schematic of a slab showing the line of symmetry at $x = 0$ and the two surfaces at $x = L$ and at $x = -L$ maintained at concentration c_s.

drying or wetting, the conditions are

$$\left.\frac{\partial c}{\partial x}\right|_{x=0,t} = 0 \quad \text{(due to symmetry)} \tag{13.10}$$

$$c(x = L, t) = c_s \quad \text{(surface concentration is specified)} \tag{13.11}$$

and the initial condition is given by

$$c(x, t = 0) = c_i \tag{13.12}$$

where c_i is the constant initial concentration and c_s is the constant concentration at the two slab surfaces at time $t > 0$. Note that, although c_s is at the surface, it is concentration in the solid, not in the surrounding fluid. Typically, concentration in the fluid is more readily available and is used to obtain c_s, concentration in the solid, using equilibrium relationships such as in Figure 9.9. Because the slab started at uniform initial concentration and was subjected to the same concentration on both sides, the concentration profile will remain symmetric during the transient process. Following the analogy from heat transfer on page 136 and the solution given in Appendix G.1 (page 593), the solution to the governing equation and the boundary and initial conditions mentioned above is

$$\frac{c - c_s}{c_i - c_s} = \sum_{n=0}^{\infty} \frac{4(-1)^n}{(2n+1)\pi} \cos \frac{(2n+1)\pi x}{2L} e^{-D\left(\frac{(2n+1)\pi}{2L}\right)^2 t} \tag{13.13}$$

The non-dimensional quantity Dt/L^2 in the exponential term of Eq. 13.13 is the Fourier number

$$Fo = \frac{Dt}{L^2}$$

where the diffusivity D replaces the thermal diffusivity, α, from heat transfer.

13.2.1   How Concentration Changes with Time

As discussed on page 137 for heat transfer, for large times the exponential terms in Eq. 13.13 drop off (decay) more rapidly and the equation can be simplified to

$$\frac{c - c_s}{c_i - c_s} = \frac{4}{\pi} \underbrace{\cos \frac{\pi x}{2L}}_{\text{spatial}} \underbrace{e^{-D\left(\frac{\pi}{2L}\right)^2 t}}_{\text{time}} \tag{13.14}$$

This simplified solution at large times shows that, for a given position, the concentration–time relationship is exponential. It can never reach the steady-state concentration. As the concentration difference decreases, the rate of mass flow decreases. If we take the natural logarithm of both sides of Eq. 13.14, we get

$$\ln \frac{c - c_s}{c_i - c_s} = \ln\left(\frac{4}{\pi} \cos \frac{\pi x}{2L}\right) - D\left(\frac{\pi}{2L}\right)^2 t \tag{13.15}$$

This linear relationship between $\ln(c - c_s)/(c_i - c_s)$ and time t can be seen in Figure B.1 on page 555 for various values of position x/L. Note that, at times close to $t = 0$, they are not linear. This is expected since by dropping terms in the series we fail to satisfy the initial condition.

13.2.2   Concentration Change with Position and Spatial Average

It can be easily seen from Eq. 13.14 that eventually, at a given time, the concentration varies as a cosine function. Note that it stays as a cosine function for all large values of time, except that the amplitude of the cosine wave drops exponentially with time. Often, in practice, a spatial average concentration is of interest. A spatial average concentration would be defined by

$$c_{av} = \frac{1}{L} \int_0^L c \, dx \tag{13.16}$$

Applying this definition of average to Eq. 13.15, we obtain an equation for the average concentration as

$$\ln \frac{c_{av} - c_s}{c_i - c_s} = \ln \frac{8}{\pi^2} - D\left(\frac{\pi}{2L}\right)^2 t \tag{13.17}$$

which shows that the average concentration changes exponentially with time, as illustrated in Figure 13.6. The time rate of change of the average concentration is obtained by differentiating Eq. 13.17 as

$$\frac{dc_{av}}{dt} = -\frac{2D}{L^2}(c_i - c_s)e^{-D(\pi/2L)^2 t} \tag{13.18}$$

which shows that the rate of change of average concentration drops exponentially with time. This is illustrated in Figure 13.4 for a typical drying process, along with the initial constant rate of drying discussed under lumped parameter analysis (Eq. 13.5). As mentioned earlier, for a typical biomaterial, constant rate of drying occurs for only a small time and the drying rate goes down for the rest of the time.

13.2.3   Concentration Change with Size

Consider Eq. 13.17 for average concentration as a function of time for large values of time. It can be rewritten as

$$\frac{Dt}{L^2} = -\frac{4}{\pi^2}\ln\left[\frac{\pi^2}{8}\left(\frac{c_{av} - c_s}{c_i - c_s}\right)\right] \tag{13.19}$$

This implies that, for a given change of concentration (measured in terms of fractional change of the total possible change $c_i - c_s$), the time required increases with the square of the thickness. This observation can be generalized for other systems by saying the time required is proportional to the square of the characteristic dimension, i.e.,

$$t \propto L^2 \tag{13.20}$$

where L is the characteristic dimension— it is half thickness for a slab and radius for a long cylinder or a sphere.

13.2.4   Charts Developed from the Solutions: Their Uses and Limitations

The solution given by Eq. 13.13 is identical to the solution given by Eq. 5.15 and therefore the same Heisler charts as shown in pages 555–557 can be used to calculate the concentration instead of temperatures. The vertical axis would now stand for $(c - c_\infty)/(c_i - c_\infty)$. Note that, as in heat transfer, these charts only provide the approximate solution at longer times given by Eq. 13.14.

Since the charts were developed from the analytical solution, the assumptions that went into the analytical solution are implicit in the charts. It is important to remind ourselves of these conditions that have to be satisfied in order to be able use the charts.

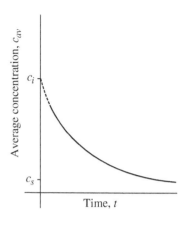

Figure 13.6: Plot of Eq. 13.17 showing average concentration decays exponentially with time. Note that the equation is not valid for times close to initial time.

These are 1) uniform initial concentration, 2) constant boundary fluid concentration, 3) perfect slab, cylinder, or sphere, 4) far from edges, 5) no chemical reaction ($r'_A = 0$), 6) constant diffusivity D. The discussion in Section 5.3.10 on using numerical methods to overcome the limitations of charts is also applicable for mass transfer.

13.2.5   When Both Internal and External Resistances Are Present: Convective Boundary Condition

So far in Section 13.2 we have considered a negligible external fluid resistance to mass transfer (resistance $1/h_m \to 0$ for $h_m \to \infty$) represented by the boundary condition of specified surface concentration. For a convective boundary condition on the slab surface given by

$$-D_{AB} \left.\frac{\partial c_A}{\partial n}\right|_{\text{surface}} = h_m(c_{A,\text{surface}}^{\text{fluid}} - c_{A,\infty}^{\text{fluid}}) \tag{13.21}$$

where h_m would be finite (instead of $h_m \to \infty$), external resistance of the fluid needs to be considered. This boundary condition appears similar to the convective heat transfer boundary condition

$$-k \left.\frac{dT}{dx}\right|_{x=0} = h(T\big|_{x=0} - T_\infty) \tag{13.22}$$

However, there is one important difference. Unlike temperature T, the variables c_A and c_A^{fluid} on either side of the equation are not the same, as illustrated in Figure 13.7. This complexity arises in mass transfer calculations since concentration on the surface in the solid, c_A, is different from concentration on the surface in the fluid, c_A^{fluid}, as discussed under boundary conditions in Section 11.3. This is unlike the case of heat transfer, where there is only one value of temperature at the surface. When c_A and c_A^{fluid} can be related in a simple manner as[2]

$$\frac{c_A^{\text{fluid}}}{c_A} = K^* \tag{13.23}$$

Equation 13.21 can be rewritten as

$$-D_{AB} \left.\frac{\partial c_A}{\partial n}\right|_{\text{surface}} = h_m(K^* c_{A,\text{surface}} - c_{A,\infty}^{\text{fluid}}) \tag{13.24}$$

$$-D_{AB} \left.\frac{\partial c_A}{\partial n}\right|_{\text{surface}} = h_m K^*(c_{A,\text{surface}} - c_{A,\infty}^{\text{fluid}}/K^*) \tag{13.25}$$

[2]Note that this definition of K^* is different from its earlier definition (Eqn. 9.10). See important discussion on page 326 on why it has to be so.

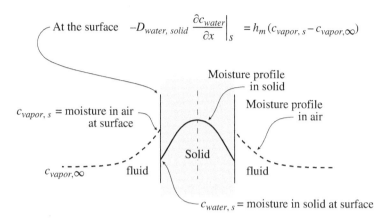

At the surface $-D_{water,\,solid} \left. \dfrac{\partial c_{water}}{\partial x} \right|_s = h_m (c_{vapor,\,s} - c_{vapor,\infty})$

Moisture profile
in solid

Moisture profile
in air

$c_{vapor,\,s}$ = moisture in air
at surface

Solid

$c_{vapor,\infty}$ fluid   fluid

$c_{water,\,s}$ = moisture in solid at surface

Figure 13.7: In a convective boundary condition, surface concentration is not the same as bulk fluid concentration, signifying additional resistance of the fluid to mass transfer. Also, surface concentrations in the fluid, c_{vapor}, and surface concentration in the solid, c_{water}, are not the same, but related through an equilibrium relationship.

Equation 13.25 is now completely analogous to the convective heat transfer boundary condition (Eq. 13.22). Thus, the solution to the diffusion mass transfer problem with convective boundary condition will follow that in heat transfer (see page 142), and the same charts on pages 555–557 can be used with appropriate replacement of parameters. As in the case of convective heat transfer, the concentration–time lines for a finite h_m value will become a function of h_m as well, which is why we have additional sets of lines on these charts. The parameters in the charts need to be replaced as follows: The y-axis will now be $(c - c_\infty/K^*)/(c_i - c_\infty/K^*)$ and $m = D_{AB}/h_m K^* L$ for the general situation of $K^* \neq 1$. For the special case of $K^* = 1$, the y-axis is $(c - c_\infty)/(c_i - c_\infty)$ and $m = D_{AB}/h_m L$. Note that m is the inverse of the mass transfer Biot number, Bi_m, analogous to how it was defined in heat transfer.

13.3   Transient Diffusion in a Finite Geometry— Multi-Dimensional Problems

There are mass transfer situations when we cannot approximate the geometry as one dimensional. Two- and three-dimensional effects need to be considered, as described for heat transfer problems in Section 5.4 on page 147. In a similar way as mentioned under heat transfer, to solve for multi-dimensional diffusion mass transfer problems, a finite geometry is considered as the intersection of two or three infinite geometries. A rectangular box, for example, is considered as the intersection of three infinite slabs and the concentration $c_{x,y,z,t}$ can be calculated as

$$\frac{c_{xyz,t} - c_s}{c_i - c_s} = \left(\frac{c_{x,t} - c_s}{c_i - c_s} \right)_{\substack{\text{infinite} \\ x \text{ slab}}} \left(\frac{c_{y,t} - c_s}{c_i - c_s} \right)_{\substack{\text{infinite} \\ y \text{ slab}}} \left(\frac{c_{z,t} - c_s}{c_i - c_s} \right)_{\substack{\text{infinite} \\ z \text{ slab}}} \quad (13.26)$$

Similarly, for a finite cylinder, the concentration $c_{r,z,t}$ can be calculated as

$$\frac{c_{r,z,t} - c_s}{c_i - c_s} = \left(\frac{c_{r,t} - c_s}{c_i - c_s} \right)_{\substack{\text{infinite} \\ \text{cylinder}}} \left(\frac{c_{z,t} - c_s}{c_i - c_s} \right)_{\substack{\text{infinite} \\ \text{slab}}} \quad (13.27)$$

Numerical computer-based solutions, mentioned in Section 5.3.10, can also handle multi-dimensional problems very efficiently.

13.3.1   Example: Drying of a Compost Pile

Consider drying of a layer of compost 30 cm thick, spread on a surface that is impermeable to water vapor, i.e., no vapor can pass through the surface. Air at 25°C and 20% relative humidity is blowing over the top of the compost, leading to an average mass transfer coefficient of 0.1 m/s. The equilibrium moisture content of the compost corresponding to the moisture in the air is approximated, for the purpose of this problem, to be that of paper, which is shown in Figure 13.9. From this figure, the partition coefficient, K^*, or the ratio of moisture content in the air to moisture content in the compost, is calculated to be 2.238×10^{-4}. The compost has an initial uniform moisture content of 1 kg of water/kg of dry matter. The diffusivity of moisture in compost is approximately 10^{-6} m^2/s. 1) How long does it take for the most moist location in the compost layer to reach 0.3 kg of water/kg of dry matter? 2) What is the moisture content at the surface exposed to the air (in the compost) at this time?

Figure 13.8: A compost pile.

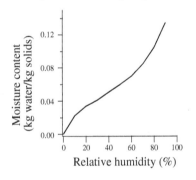

Figure 13.9: Equilibrium moisture content of paper. Moisture content is given in kg of water per kg of solids. Data from ICT, 1933.

Understanding and formulating the problem *1) What is the process?* Diffusion of moisture (water vapor through the air in the pores of the compost material and eventually into the outside air that gets convected away). *2) What are we solving for?* How long does it take to dry to a certain level of moisture and the value of the surface moisture at that time. *3) Schematic and given data:* A schematic is shown in Figure 13.10 with some of the given data superimposed on it. *4) Assumptions:* Since a constant value of diffusivity is provided, change in diffusivity with moisture, as the material dries, is ignored.

Generating and selecting among alternate solutions *1) What solutions are possible?* For unsteady-state (moisture is changing with time) mass transfer, the solutions we have available are shown in the solution chart in Figure 13.17. *2) What approach is likely to work?* Using this chart, since the most moist location will have a change of moisture (i.e., moisture changes everywhere), this is not a semi-infinite geometry. So, we are on the left side of the chart. To decide whether we can use a lumped parameter analysis, we check the Biot number,

$$\text{Bi}_m = \frac{K^* h_m L}{D_{AB}} = \frac{(2.238 \times 10^{-4})(0.1 \text{ m/s})(0.3 \text{ m})}{10^{-6} \text{m}^2/\text{s}} \approx 6.67, \text{ which is much higher}$$

than 0.1 and therefore we cannot use the lumped parameter analysis. We are left with the series solution. Although we cannot check the $D_{AB}t/L^2$ value since t is not known, this problem is unlikely to be for early times ($Dt/L^2 < 0.2$ in solution chart) since moisture is changing even at the center. Thus, a one-term series solution or Heisler chart is likely to be our appropriate solution technique. Note that, if a one-term solution turns out not to be the solution, we would not be able to read from the Heisler chart, anyway, since the chart only has plots for $Dt/L^2 > 0.2$.

Implementing the chosen solution 1) To read the Heisler chart, we calculate the various quantities needed for it:

$$m = \frac{D_{AB}}{K^* h_m L} = \frac{10^{-6} \text{ m}^2/\text{s}}{\left(2.238 \times 10^{-4}\right) 0.1 \frac{\text{m}}{\text{s}} \times 0.3\text{m}} \approx 0.15$$

The most moist region must be at the ground surface, i.e., $x = 0$,

$$n = \frac{x}{L} = 0$$

To calculate $(M - M_\infty)/(M_i - M_\infty)$, we need M_∞. From Figure 13.9, the equilibrium concentration in the compost, M_∞, for 20% RH is 0.03 kg H_2O/kg dry matter. Thus, for $M = 0.3$ kg H_2O per kg dry matter at $x = 0$,

$$\frac{M - M_\infty}{M_i - M_\infty} = \frac{0.3 - 0.03}{1 - 0.03} = 0.278$$

From the Heisler chart, interpolating between $m = 0$ and $m = 0.5$, we get $Fo = 0.8$, from which we calculate t as

$$t = Fo \frac{L^2}{D_{AB}} = 0.8 \times \frac{0.3^2}{10^{-6}} = 72000 \text{ s}$$

$$= 20 \text{ hours}$$

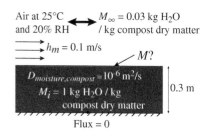

Figure 13.10: Schematic for Example 13.3.1.

2) For the top surface also we need to interpolate between m values, for $Fo =$ 0.8. To interpolate between the two m values, note that, for $m = 0$, at the surface ($n = 1$), we cannot read the chart. Instead, we know the surface moisture is equal to the equilibrium moisture, i.e.,

$$\frac{M - M_\infty}{M_i - M_\infty} = 0$$

For $m = 0.5$, $n = 1$, $Fo = 0.8$, from the chart,

$$\frac{M - M_\infty}{M_i - M_\infty} = 0.21$$

Interpolating for $m = 0.15$,

$$\frac{M - M_\infty}{M_i - M_\infty} = 0 + \frac{0.21 - 0}{0.5 - 0}(0.15 - 0) = 0.063$$

from which, M at the surface can be calculated as

$$\begin{aligned}
M &= M_\infty + 0.063(M_i - M_\infty) \\
&= 0.03 + 0.063(1 - 0.03) \\
&= 0.0911 \text{ kg water/kg dry solids}
\end{aligned}$$

Evaluating the solution *1) Is the calculated value (top moisture) reasonable?* The moisture at the top surface (0.0911) is between the most moist center (0.3) and the driest point, which is the bulk air that corresponds to the equilibrium moisture value (0.03), which makes sense. *2) What do we learn?* Mass transfer resistances (per unit area) can be calculated for inside and outside the compost as

$$(R_m)_{\text{internal}} = \frac{L}{D_{AB}} = \frac{0.3 \text{ m}}{10^{-6} \text{ m}^2/\text{s}} = 3 \times 10^5 \text{ s/m}$$

$$(R_m)_{\text{external}} = \frac{1}{K^* h_m} = \frac{1}{(2.238 \times 10^{-4})(0.1\text{m/s})} = 0.4468 \times 10^5 \text{ s/m}$$

Since $(R_m)_{\text{in}}$ and $(R_m)_{\text{out}}$ are comparable in this example, neither can be ignored.

13.4 Transient Diffusion in a Semi-Infinite Region

This section parallels the section on page 148 for heat conduction in a semi-infinite region. A semi-infinite region extends to infinity in two directions and has a single identifiable surface in the other direction, as shown in Figure 13.11. The semi-infinite region provides a useful idealization for many practical situations where we are interested in mass transfer for a relatively short period of time and/or in a relatively thick material. Examples can be short time exposure of a solid or liquid to a gas.

The governing equation for diffusion with no chemical reaction is

$$\underbrace{\frac{\partial c_A}{\partial t}}_{\text{storage}} + \underbrace{u\frac{\overset{0}{\cancel{\partial c_A}}}{\partial x}}_{\text{no bulk flow}} = \underbrace{D\frac{\partial^2 c_A}{\partial x^2}}_{\text{diffusion}} + \underbrace{\overset{0}{\cancel{r_A}}}_{\text{no reaction}} \qquad (13.28)$$

which leads to the simplified governing equation as

$$\frac{\partial c}{\partial t} = D\frac{\partial^2 c}{\partial x^2} \qquad (13.29)$$

where the subscript A from c_A has been dropped for simplicity. The boundary conditions at the surface and far from the surface (at infinity) are given by

$$c(x = 0) = c_s \qquad (13.30)$$

$$c(x \to \infty) = c_i \qquad (13.31)$$

The initial condition is given by

$$c(t = 0) = c_i \qquad (13.32)$$

Following the discussion on page 149, the solution for concentration c is given by

$$\frac{c - c_i}{c_s - c_i} = 1 - \text{erf}\left[\frac{x}{2\sqrt{Dt}}\right] \qquad (13.33)$$

The error function is tabulated on page 553. The solution given by Eq. 13.33 is plotted in Figure 13.12 for various times. The diffusive mass flux at the surface can be calculated from the concentration given by Eq. 13.33 as (see details on page 149)

$$n_s = -D\left.\frac{dc}{dx}\right|_{x=0}$$

$$= \sqrt{\frac{D}{\pi t}}(c_s - c_i) \qquad (13.34)$$

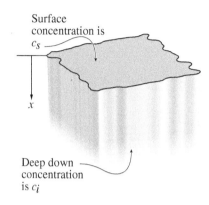

Figure 13.11: Schematic of a semi-infinite region showing only one identi-

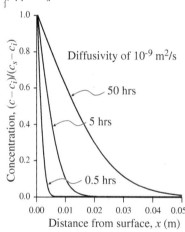

Figure 13.12: Plot of Eq. 13.33 showing how concentration profile changes with time.

We can see that the diffusive mass flux at the surface reduces with time. This is expected since the concentration gradient at the surface drops with time, as shown in Figure 13.12.

The amount of mass penetrated (diffused) through the surface into the semi-infinite region can also be of practical interest. For example, we may want to know how much drug has penetrated through skin or how much water has infiltrated (soaked) into a tissue. The total or cumulative amount of mass diffused into the material over time t per unit area can be obtained by integrating the flux n_s above over the time period t as

$$
\begin{aligned}
I &= \int_0^t n_s dt \\
&= \int_0^t \sqrt{\frac{D}{\pi t}} (c_s - c_i) dt \\
&= \sqrt{\frac{D}{\pi}} (c_s - c_i) \int_0^t t^{-1/2} dt \\
&= 2\sqrt{\frac{Dt}{\pi}} (c_s - c_i)
\end{aligned}
\tag{13.35}
$$

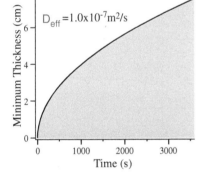

Figure 13.13: Illustration of minimum thickness of a material for which an error function solution can be used for diffusive mass transfer.

When can we use semi-infinite approximation? Semi-infinite approximation is true for thin materials over shorter time or thick materials over longer time. As described in Eq. 5.31 on page 150, at distances greater than $4\sqrt{Dt}$, i.e., for

$$
x \geq 4\sqrt{Dt}
\tag{13.36}
$$

the concentration change is less than 0.5%. Thus, any material whose thickness L is larger than $4\sqrt{Dt}$ can be approximated as semi-infinite for this purpose. This is illustrated in Figure 13.13 for a diffusivity of 10^{-8} m^2/s, a value chosen in the range of gas diffusivity in liquids. The material needs to be at least as thick as the minimum thickness shown to be able to use Eq. 13.33.

13.4.1 Summary of Solutions for Mass Diffusion in a Semi-Infinite Region for the Three Different Boundary Conditions

Concentration and surface mass flux for a semi-infinite region for the three common boundary conditions are summarized in Table 13.1. They are completely analogous to the corresponding equations derived for heat, in Section 5.5.2. Note that the surface mass flux approaches zero with time for the concentration specified boundary condition and convective boundary condition since the interior concentration approaches surface concentration. For a convective boundary condition, with time, the surface concentration also approaches the concentration in the surrounding fluid. The total amount of transport over time is the surface mass flux integrated over time.

Table 13.1: Summary of solutions for mass diffusion in a semi-infinite region for the three different boundary conditions

Boundary condition	Concentration	Flux at surface, $-D \left. \frac{\partial c}{\partial x} \right|_{x=0}$	Total amount over time per unit area, $\int_0^t \left(-D \left. \frac{\partial c}{\partial x} \right|_{x=0} \right) dt$
Concentration given	$\frac{c - c_i}{c_s - c_i}$ $= 1 - \mathrm{erf} \left(\frac{x}{2\sqrt{Dt}} \right)$	$\sqrt{\frac{D}{\pi t}}(c_s - c_i)$; drops with time	$2\sqrt{\frac{Dt}{\pi}}(c_s - c_i)$; increases as $\sqrt{t}$
Flux given	Substitute concentration for temperature, D for α or k, and surface mass flux for q_s'' in Eq. 5.33	Given value; would be a constant	Would be increasing linearly
Convective condition	Substitute concentration for temperature, D for α or k, and h_m for h in Eq. 5.34	Calculate $-D \left. \frac{\partial c}{\partial x} \right|_{x=0}$ from the concentration equation in the left column; would drop with time	Would be qualitatively similar to that for concentration given boundary condition

13.4.2 *Example: Oxygen Concentration and Bacterial Growth in a Silage*

Many of the problems associated with the storage of silage in a bunker silo arise during the unloading phase, through the exposed face (see Figure 13.14). Upon exposure to air, yeasts, molds, and aerobic bacteria begin to grow rapidly, causing a loss of nutrients, a rise in pH, and heating in particularly warm weather. Associated effects may include a decrease in animal production and in some cases toxicity and death. The mechanism of oxygen penetration into the silage through the exposed face has sometimes been assumed to be primarily diffusion (Pitt and Muck, 1993).

Assuming the oxygen concentration at the surface to be constant and equal to $0.21 m^3$ of O_2/m^3 of air, calculate the O_2 concentration within the silage at a depth of 0.15 m after 1 hour. The diffusivity of O_2 in air is 1.781×10^{-5} m^2/s at 0°C. Assume the silo to be long so that a 1D model can be used. Neglect microbial growth and CO_2 generated during respiration.

Figure 13.14: Schematic of a bunker silo showing the exposed face.

Solution

Understanding and formulating the problem *1) What is the process?* Oxygen diffuses from the surface through the air in the pores of a thick layer of biomaterial. *2) What are we solving for?* The concentration of O_2 at 0.15 m after 1 hour. *3) Schematic and given data:* A schematic is shown in Figure 13.15 with some of the given data superimposed on it. Additionally, initial O_2 concentration, c_i, is 0 and diffusivity of O_2 in air is $D_{O_2,\text{air}} = 1.781 \times 10^{-5}$ m^2/s. *4) Assumptions:* Diffusivity is a constant.

Generating and selecting among alternate solutions *1) What solutions are possible?* This is a transient problem. For transient diffusive mass transfer, the solutions we have available are shown in the solution chart in Figure 13.17. *2) What approach is likely to work?* As we follow the solution chart in Figure 13.17, to check whether we can use a simple semi-infinite solution for oxygen penetration over 1 hour, we would need to calculate Dt/L^2. However, L, the thickness of the silage, is not provided, although silage is mentioned as thick. If we traverse on the left side of the solution tree, it is not a lumped parameter situation (concentration changes with position) and we cannot use a series solution either since L is not known. We have no choice but to assume at this stage that the semi-infinite solution is valid. We can later come back to check this validity.

Implementing the chosen solution For a transient diffusion problem in a semi-infinite region, the solution is given by Eq. 13.33.

$$\frac{c - c_i}{c_s - c_i} = 1 - \text{erf}\left[\frac{x}{2\sqrt{Dt}}\right]$$

$$\frac{c - 0}{0.21 - 0} = 1 - \text{erf}\left(\frac{0.15}{2\sqrt{(1.781 \times 10^{-5})(3600)}}\right)$$

$$= 1 - \text{erf}(.296)$$

$$= 1 - .328 = .672$$

which leads to $c = .14$ m^{3}O$_2$/m^3 of air

Evaluating and interpreting the solution *1) Does the calculated value make sense?* This is hard to say, although the value is not obviously wrong since it is between the initial value of zero and the maximum value at the surface of 0.21. Also, was it correct to use the semi-infinite solution? The depth to which the oxygen will penetrate is given by Eq. 13.36 as $4\sqrt{Dt} = 1.01$ m, leading to our solution being correct as long as the silage is more than about 1 m thick. Judging from Figure 13.14, this is probably well satisfied. *2) What do we learn?* Within a short

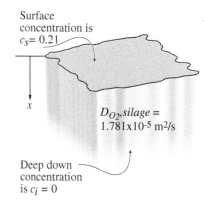

Surface concentration is $c_s = 0.21$

x

D_{O_2}silage = 1.781x10^{-5} m^2/s

Deep down concentration is $c_i = 0$

Figure 13.15: Schematic for Example 13.4.2.

period of 1 hour, the top layer of 15 cm has a very high oxygen concentration of 0.14 m³ O_2/m^3 of air. This layer is very much susceptible to bacterial and mold growth. Note, however, we have used the diffusivity value of oxygen in air whereas the diffusivity value of oxygen in the air contained in the silage can be significantly smaller depending on the tortuous paths that the oxygen would have to take in this porous medium.

13.4.3 *Example: Drug Release from a Matrix*

In developing immunotherapy, release characteristics of human immunoglobulin G (IgG) from a polymer matrix (in the shape of a slab that is 1 mm thick) are being studied. The slab is weighed and submerged in a beaker containing a phosphate solution that maintains the surface at zero concentration of the IgG. 1) For early times, the slab can be treated as very thick, with volume V and thickness $2L$. The total amount of the drug released over time t can be obtained from the corresponding formula for flux. From this formula, obtain the expression for total amount of drug released from *both surfaces* of the slab over time t. 2) Assuming the formula for early times is valid, if 30% of the drug is released from both surfaces in 108 hours, what is the estimated diffusivity of the drug? 3) For later times, diffusion through the slab is modeled using the one-term series solution. Assuming this solution is valid, obtain the expression for the total amount of drug released from *both surfaces* over time t. 4) Assuming the series solution is valid for these later times, if 80% of the drug is released in 208 hours, what is the estimated diffusivity of the drug? 5) Do you expect the two estimated diffusivities to be close? Explain.

Understanding and formulating the problem *1) What is the process?* Diffusion of drug from the inside of a polymer matrix to its surface where it is convected away. *2) What are we solving for?* Diffusivity of the drug at two different (early and later) times when the respective amounts of drug released are given. *3) Schematic and given data:* A schematic is shown in Figure 13.16 with some of the given data superimposed on it. *4) Assumptions:* Since a constant value of diffusivity is being asked for each time period in the problem description, diffusivity is assumed not to vary with concentration, i.e., amount of drug.

Generating and selecting among alternate solutions *1) What solutions are possible?* For unsteady-state (amount of drug is changing with time) mass transfer, the solutions we have available are shown in the solution chart in Figure 13.17. *2) What approach is likely to work?* The problem asks us to calculate diffusivity for two different periods: early and late. Early time is analogous to the material being thick, i.e., semi-infinite, so semi-infinite geometry solution should be appropriate for early times. For later times, the drug will be released from even

Early times

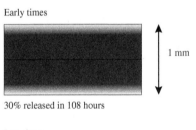

30% released in 108 hours

Late times

80% released in 208 hours

1 mm

1 mm

Figure 13.16: Schematic for Example 13.4.3.

the farthest location (center) so the geometry cannot be treated as semi-infinite. Looking to the left of Figure 13.17, a lumped parameter solution is not a possibility since the mass transfer Biot number is infinity ($h_m \to \infty$) corresponding to the concentration being given at the surface. We are left with the series solution. Although we cannot check the $D_{AB}t/L^2$ value (D_{AB} is not known), for later times, the one-term series solution or Heisler chart is very likely to be our appropriate solution technique.

Implementing the chosen solution Amount released from both surfaces

$$= 2 \int_0^t A\,(c_i - c_s) \sqrt{\frac{D}{\pi t}}\, dt \tag{13.37}$$

The integral, without the area term, has been evaluated in Eq. 13.35. Using it, the amount released over time t

$$= 2\,(c_i - c_s) \left(\frac{V}{L}\right) \sqrt{\frac{Dt}{\pi}} \tag{13.38}$$

Since 30% of the drug is released in 108 hours, plugging in the values, we get

$$0.3\, c_i V = 2\,(c_i - 0) \left(\frac{V}{0.0005\,[\mathrm{m}]}\right) \sqrt{\frac{D \times 108 \times 60 \times 60\ [\mathrm{s}]}{\pi}} \tag{13.39}$$

Canceling $c_i V$ and solving for D,

$$D = \left(\frac{0.3 \times 0.0005\,[\mathrm{m}]}{2}\right)^2 \times \frac{\pi}{108 \times 60 \times 60\ [\mathrm{s}]} \tag{13.40}$$

$$= 4.55 \times 10^{-14}\ \mathrm{m^2/s} \tag{13.41}$$

3) For later times, using the series solution, the average concentration of drug remaining in the slab at time t, c_{av}, is given by

$$\frac{c_{av} - c_s}{c_i - c_s} = \frac{8}{\pi^2} e^{-D\left(\frac{\pi}{2L}\right)^2 t}$$

$$c_{av} = (c_i - c_s)\frac{8}{\pi^2} e^{-D\left(\frac{\pi}{2L}\right)^2 t} + c_s$$

The amount of drug released over time t

$$= (c_i - c_{av}) V$$

$$= \left(c_i - c_s - (c_i - c_s) \frac{8}{\pi^2} e^{-D\left(\frac{\pi}{2L}\right)^2 t} \right) V$$

$$= (c_i - c_s) \left(1 - \frac{8}{\pi^2} e^{-D\left(\frac{\pi}{2L}\right)^2 t} \right) V$$

Since 80% is released in 208 hours,

$$0.8 \, c_i V = c_i V \left[1 - \frac{8}{\pi^2} e^{-D\left(\frac{\pi}{2 \times 0.0005m}\right)^2 \times 208 \times 60 \times 60 \, s} \right]$$

$$e^{-D\left(\frac{\pi}{0.001m}\right)^2 \times 748800 \, s} = \frac{\pi^2}{8} \times 0.2$$

$$-D\pi^2 \times \frac{748800 \, s}{(0.001m)^2} = \ln\left(\frac{\pi^2}{40}\right)$$

$$D = \frac{-\ln\left(\frac{\pi^2}{40}\right) \times (0.001)^2}{748800 \times \pi^2} \text{m}^2/\text{s}$$

$$= 1.89 \times 10^{-13} \text{ m}^2/\text{s}$$

Evaluating the solution *1) Do the values make sense?* Notice the diffusivity values are low; this makes sense since it is diffusion through a solid phase (polymer), which is much slower than diffusion through gas or liquid (see typical range of diffusivities in Figure 10.8 on page 364). Also, the two diffusivity values, estimated using two different solutions, should be the same ideally (same system), but the data provided here do not come from real experiments; even if they did come from experiments, the two solutions (one term series and semi-infinite) are both approximations when applied to a real system.

13.5 Chapter Summary—Transient Diffusive Mass Transfer

- **No Internal Resistance, Lumped Parameter (page 445)**

  1. For $Bi_m < 0.1$ in a finite size material, the diffusive mass transfer resistance in the solid can be ignored (the solid is considered lumped) in comparison with the convective mass transfer resistance at the solid surface.

2. As diffusive mass transfer resistance in the solid is ignored, concentration is not a function of position, is a function of time only, and is given by Eq. 13.5 on page 446.

- ## Internal Resistance Is Significant (pages 448, 453, 457)

  1. When internal diffusive resistance is significant ($Bi_m > 0.1$), concentration is a function of both position and time.

  2. For $Bi_m > 0.1$, $Fo > 0.2$ and for an infinite slab, infinite cylinder, and a sphere, solutions are given by the same Heisler charts as for conduction heat transfer (pages 555–557). The concentration parameter for the charts needs to be modified, as discussed on page 452.

  3. For a finite slab and a finite cylinder (page 454), the solutions are given by Eq. 13.26 and Eq. 13.27, respectively.

  4. Materials with thicknesses $L \geq 4\sqrt{Dt}$ are considered effectively semi-infinite (Eq. 13.36) and the solution is given by the error function (Eq. 13.33 on page 457).

13.6 Problem Solving—Unsteady-State Diffusion

▶**Calculate concentration profiles and mass flows.** The steps in this solution process are completely analogous to those for heat transfer. One important complication is that the boundary condition in solid, if not provided directly, needs to be obtained from the concentration provided in the fluid and equilibrium relations. Figure 13.17 shows the solution chart for transient diffusion. Some additional comments for problem solving are

- If L is not provided (or otherwise adequate information is not available), Dt/L^2 cannot be calculated—a semi-infinite approximation may have to be made to obtain a solution using what we know in this book.

- A semi-infinite solution may work for curved geometries as well, for small times.

- If surface concentration is provided, it amounts to $h_m \rightarrow \infty$ (discussed in Chapter 11), as would be needed in the Heisler chart.

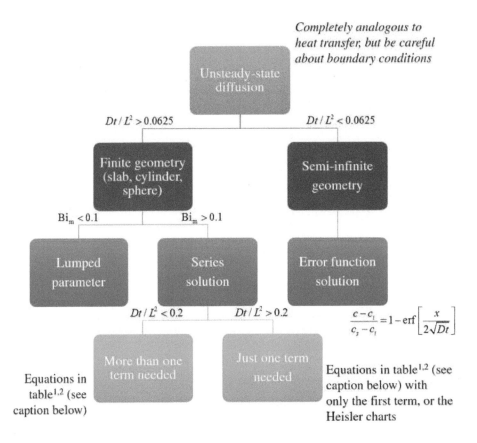

Figure 13.17: Problem solving in unsteady-state diffusion. [1]Table 5.3 on page 144 provides the series solution for slab for three different boundary conditions, whereas [2]Table 5.4 on page 145 provides the series solutions for slab, cylinder and sphere, for the temperature specified boundary condition. Appropriate parameters for mass transfer are substituted (c for T, D for α, D for k, and h_m for h) in these solutions to obtain the corresponding solution for mass transfer. When only one term is needed, we take the first term of the series or use the corresponding Heisler chart.

- Concentrations for Heisler charts are for the solid and therefore all fluid concentrations need to be converted to corresponding solid concentrations that are in equilibrium. We need to be careful about the calculation and use of parameter K^*. The parameter m in the Heisler chart is given by $m = D_{AB}/h_m L K^*$. For the same reason, c_∞ in the chart needs to be converted to the equivalent solid concentration.

- The non-dimensional quantity Dt/L^2 implies $t \propto L^2$.

13.7 Concept and Review Questions

1. In mass transfer, what parameter compares internal diffusive resistance to external convective resistance?

2. In lumped parameter analysis, we ignored the internal diffusional resistance. Is this physically possible? Explain.

3. Under what conditions can you approximate a slab as one dimensional?

4. What dimension would you choose for analyzing a long cylinder?

5. When can you approximate a geometry as semi-infinite?

6. Can you use Heisler charts for any duration of the diffusion process?

7. Soon after a process starts, the concentration inside can be obtained from (choose one or more) a) Heisler chart, b) solution with error function in it, c) series solution to heat equation.

8. Approximation to a semi-infinite region is possible for (choose one or more) a) thick material and long time, b) thin material and long time, c) thin material and short time, d) thick material and short time.

9. In the absence of external resistance, time to diffuse in a solid (infinite slab, long cylinder, sphere) varies as (choose one) a) square of the characteristic length, b) square root of the characteristic length, c) proportional to (i.e, linear with) the characteristic length.

Further Reading

Bartholomew, C. 1965. *Soil Nitrogen*. American Society of Agronomy, Madison, Wisconsin.

Bertola, N., A. Chaves, and N. E. Zaritzky. 1990. Diffusion of carbon dioxide in tomato fruits during cold storage in modified atmosphere. *International Journal of Food Science and Technology* 25:318–327.

Bjorklund, S., J. Engblom, K. Thuresson, and E. Sparr. 2010. A water gradient can be used to regulate drug transport across skin. *Journal of Controlled Release* 143(2):191–200.

Gorna-Binkul, A., K. Kaczmarski and B. Buszewski. 2001. Modeling of the sorption and diffusion processes of volatile organic air pollutants in grape fruits. *Journal of Agricultural and Food Chemistry,* 49(6): 2889-2893.

Hsieh, D. S. T., W. D. Rhine, and R. Langer. 1983. Zero-order controlled-release polymer matrices for micromolecules and macromolecules. *Journal of Pharmaceutical Sciences,* 72(1): 17-22.

ICT. 1933. International critical tables of numerical data, physics, chemistry and technology. Volume II. Published for the National Research Council of the United States of America by McGraw-Hill, New York.

Kohn, M. J. 2008. Models of diffusion-limited uptake of trace elements in fossils and rates of fossilization. *Geochimica Et Cosmochimica Acta,* 72(15): 3758-3770.

Lissik, E. A. 1988. Mechanisms of pathogen transport through the teat canal of a dairy cow. Ph.D. Thesis, Cornell University.

Mannapperuma, J. D., R. P. Singh, and M. E. Montero. 1991. Simultaneous gas diffusion and chemical reaction in foods stored in modified atmospheres. *Journal of Food Engineering* 14:167–183.

Pitt, R. E. and R. E. Muck. 1993. A diffusion model of aerobic deterioration at the exposed face of bunker silos. *J. Agric. Engr. Res.* 55(1):11–26.

Rhee, J. and L. E. Bode. 1990. Transport model of spray droplets above and within a plant canopy. Presented at the 1990 International Meeting of ASAE, Paper No. 90-1580. ASAE, St. Joseph, Michigan.

Saltzman, W. M. 2001. *Drug Delivery: Engineering Principles for Drug Therapy.* Oxford University Press, Oxford, UK.

13.8 Problems

13.1 Average Moisture Content in a Slab

Starting from the appropriate solution to the governing equation, show that, for a slab of material of length $2L$ dried symmetrically, with a constant moisture content at the surface, the average moisture of the slab as a function of time at long times is given by

$$\frac{c_{av} - c_s}{c_i - c_s} = \frac{8}{\pi^2} \exp(-D \, (\pi/2L)^2 \, t)$$

where c_{av} is the volume averaged moisture content.

13.2 Experimental Determination of Moisture Diffusivity

As you may recall, parameters such as diffusivity are typically found from experimentation. Consider a simple experiment where a slab 2 cm thick was initially weighed at 250 g. It is dried for 1 hour and weighed again to be 200 g. The weight of dry solids is known to be 70 g. The slab is thin enough in one direction compared to the other two directions so that the moisture transfer is along the thin direction only. Also, it is being dried from both faces equally, i.e., drying is symmetric. The dry air maintains a moisture content of zero at both faces of the slab. 1) Using this weight loss data, calculate the diffusivity of moisture in the slab. *(Hint: Use the expression for average moisture content as a function of time, with c being the moisture content on a dry weight basis.)* 2) How would you expect the diffusivity to change as the material dries?

13.3 Diffusion in Agar Gel

Agar gel is a gel matrix formed from agarose, a purified hydrocolloid isolated from agar or agar-bearing marine algae. This gel matrix is nearly ideal for diffusion and electrokinetic movement of biopolymers. Therefore, it is useful for studying processes such as electrophoresis and immunodiffusion. Consider an agar gel cylinder 20 mm in diameter, with a much larger length. The cylinder initially contains a uniform concentration of urea of $0.1 \ \mathrm{kmol/m^3}$ and is suddenly immersed in water containing no urea. The water flow is rapid enough such that the surface convective mass transfer resistance can be assumed to be negligible. The diffusivity of urea in the agar is $4.5 \times 10^{-10} \ \mathrm{m^2/s}$. 1) Calculate the concentration at the center of the cylinder *and* at 5 mm from the center after 20 hours. 2) If the diameter of the cylinder is doubled, what would the concentration at the center be after 20 hours?

13.4   Drying of a Cement Slab

Problem removed

13.5   Drying of a Compost Pile

Moved to solved example

13.6   Effective Diffusivity in Drying

During a drying process for raisins, the average moisture content of raisins measured at 30 minutes and 1 hour is 60 wt% and 50 wt% dry basis, respectively. The air being circulated over the raisins maintains its surface at its equilibrium moisture content of 5%, for its particular temperature and humidity. Consider an average raisin to be a slab of thickness 1 cm, with the air circulating on both sides for symmetric drying. Assume that shrinkage can be neglected during this period. Estimate the effective diffusivity of water in the raisins during this drying process. *(Hint: Although this is a transient problem, the moisture contents are average and therefore the Heisler charts on pages 555–557 cannot be used for this problem.)*

13.7   Diffusion of Oxygen through Alveoli

We would like to estimate how quickly the non-uniformities in gas composition in the alveoli are damped out. Consider an alveolus to be spherical with a diameter of 0.1 mm. Let the sphere have an initial uniform concentration of oxygen, c_i, and at a certain instant, the walls of the alveolus are raised to an oxygen concentration of c_∞ and maintained at this value. If the oxygen diffusivity in the alveolus is approximated as that in water, 2.4×10^{-9} m^2/s, how long does it take for the *concentration change* $(c - c_i)$ at the center to be 90% of the final concentration change?

Figure 13.18: Alveoli in lungs. (iStock.com)

13.8   Ripening of Tomatoes

As shown in the text example, ripening of tomatoes can be retarded by storing them in an atmosphere rich in CO_2 so that the CO_2 replaces the O_2 in the tissues. A tissue sample of a spherical tomato stored in such a CO_2-rich environment was found to have a CO_2 concentration of 0.1736 at a non-dimensional radius of 0.80 after a certain time. The radius of the tomato is 4 cm. The initial concentration of CO_2 in the tomato is 0.02. The tomato surface is maintained at a CO_2 concentration of 0.18. The diffusivity of CO_2 in the tomato is 2.3×10^{-8} m^2/s. 1) Calculate the concentration at the center of the tomato at this time. 2) If a concentration of 0.16 is required at the center for the completion of the diffusion process, what *additional* time is needed?

13.9   Size Effect in Diffusion

In Problem 13.8 of diffusion of CO_2 through a tomato, how would the time of exposure to CO_2 to reach the same concentration of CO_2 at the center change for tomatoes that are half the diameter? *(Note: No other numerical data are necessary.)*

13.10   Transport of Medication through Skin

For a simplified case of transport through skin, assume total diffusional resistance of skin is entirely due to the stratum corneum (see Figure 5.16), thus ignoring penetration in sweat ducts, etc. This would be sufficiently accurate for relatively rapidly penetrating molecules of small molecular weight. Consider application of a solute to the outer surface of a 1 micron thick slab of stratum corneum. The partition coefficient relating the membrane surface concentration to the concentration of applied material is 1. The inner (bottom) surface of the stratum corneum is assumed to be maintained at essentially zero concentration since the molecules are removed as soon as they reach the microcirculation by a sufficiently high peripheral blood flow through skin. Also, the viable layers below the stratum corneum have high diffusivity. Assume diffusivity for the stratum corneum for a low-molecular-weight non-electrolyte is 10^{-6} m^2/s. 1) How much total material enters the skin up to a given time? Simplify for large times and sketch a plot of your answer (concentration vs. time). 2) How much total material leaves the stratum corneum at the bottom surface into the viable tissue? Simplify for large times and sketch. 3) Using your results from steps 2 and 3, how much material remains in the stratum corneum? Simplify for large times and sketch.

13.11   Nitrogen Movement in Soils

Nitrogen is an extremely important element for proper growth and maintenance of plants since it is an essential part of the nucleic acids that plants synthesize. For this reason, a large part of any fertilizer is composed of nitrogen-based compounds. However, nitrogen compounds found in drinking water can be extremely toxic to humans, especially when the ground-water concentration of nitrogen is > 10 mg/l. Plants, however, generally need a concentration of at least 20 mg/l to thrive, depending on the crop species. For the analysis, assume that the soil region is a semi-infinite plane and assume the yearly average concentration of nitrogen in the soil directly beneath the crop layer is 20 mg/l. This is the first year that the field has been fertilized. The initial concentration of nitrogen in unfertilized soil is 6 mg/l. The diffusivity of nitrogen in soil is 1 cm^2/day. 1) Find the depth of soil beneath the crop layer where the nitrogen concentration is at least 10 mg/l by the end of this first year. 2) Do you think that the fertilizer practices on this field are adversely affecting the quality of the drinking water in that area? If not after 1 year, what if those practices continue for 10 years?

13.12 Unsteady-State Diffusion in a Semi-Infinite Region

Consider a semi-infinite region to have an initial concentration of a certain solute A of 0.01 kmol of A/m^3 of solid. It is suddenly exposed to a fluid that has a concentration of A of 0.10 kmol/m^3 of fluid. Assume a very high mass transfer coefficient at the surface. The distribution coefficient, K^*, for equilibrium at the surface (ratio of concentration in solid to concentration in fluid) is 2. The diffusivity of solute A through the solid is 5×10^{-9} m^2/s. 1) Calculate the concentration of A at the surface. 2) Calculate the concentration at 1 cm below the surface after 10 hours.

13.13 Evaporation from Wet Ground

Consider evaporation from wet ground (silt loam soil) at night. The ground was initially at a moisture content of 0.8 kg of water/kg of dry soil. Air is flowing over the ground at a relative humidity of 10%. The mass transfer coefficient for air flow over the surface is 0.01 m/s and the average diffusivity of moisture in the ground is 10^{-6} m^2/s. 1) For a distance of 100 m along the wind direction, show that the diffusive resistance to moisture movement (inside the ground) is a lot higher than the convective resistance to moisture movement (in air). 2) Due to our result from step 1, we can neglect convection and consider the moisture level at the surface of the ground to be in equilibrium with that in the flowing air. A graph relating the equilibrium moisture content of the soil (in kg of water/kg of solids) to the relative humidity in the air can be found in Figure 9.10. What is this equilibrium moisture in kg of water/m^3 at the surface in the soil, assuming the density of dry soil is 900 kg/m^3? 3) Continuing from step 2 in considering only diffusion to be significant, find the expression for rate of moisture loss per unit area and show that this moisture loss varies as the inverse of the square root of time. *(Hint: For small values of the argument ϕ, the error function, erf(ϕ), can be approximated as ϕ.)* 4) Using the expression you obtained in step 3, find the moisture flux in g/m^2·s after 6 hours, assuming a dry soil density of 900 kg/m^3.

13.14 Drying of Top Soil

Soil, being very wet at the end of spring (20% wet basis), begins to dry from the lack of rain during the summer. For purposes of calculation, moisture content values need to be converted to dry basis using the formula below. Consider the temperature of the top layers of the soil to be constant, i.e., assume the surface mass transfer coefficient is infinity. Consider average moisture diffusivity in the soil to be 10^{-7} m^2/s. After 1 month (30 days) of dry weather with the average relative humidity of air at 40% (see

Figure 9.10 for corresponding equilibrium moisture data), calculate the depth (*in cm*) at which the moisture levels would be down to 1/5th of the value it was at the end of spring.

$$\text{dry basis (fraction)} = \frac{\text{wet basis(fraction)}}{1 - \text{wet basis(fraction)}}$$

13.15 Diffusive Transport of Pathogens in a Teat Canal

Bacterial infection in cows leading to mastitis is a costly problem for the dairy industry. Mastitis results from infections due to the penetration of pathogens through the teat canal into the teat cistern (see Figure 13.19). One of the many possible mechanisms that have been considered is the diffusion of pathogens through the column of milk fluid (in the teat canal) into the teat cistern (Lissik, 1988).

1) Assume a one-dimensional diffusion model of a 1 cm long teat canal. The ratio of the length of the teat canal/diameter of canal is 125. The diffusivity of bacteria through milk is 3.8×10^{-13} m^2/s. Calculate the time required for the bacterial concentration to reach detectable amounts (8 nanogram/ml) at the entrance to the teat cistern from a high concentration of 250000 nanogram/ml held constant at the tip of the teat canal. (*Hint: You may need to use outside references for error function values other than the ones provided in your textbook.*) 2) Is this time to infect the entire length of the teat canal more than the intermilking period (12 hours)? 3) Which part of the anatomy is likely to be infected the most during these 12 hours?

13.16 Drug Release from Delivery Systems

Consider a drug delivery system (a tablet) consisting of a polymeric matrix containing the drug molecules. To keep things simple, consider a slab geometry for the tablet. The drug molecules diffuse to the surface of the slab symmetrically on both sides and are released to the surroundings. The concentration profile of drug in the polymer is given by

$$\frac{c - c_s}{c_i - c_s} = \sum_{n=0}^{\infty} \frac{4(-1)^n}{(2n+1)\pi} \cos \frac{(2n+1)\pi}{2} \frac{x}{L} \exp\left(-D\frac{(2n+1)^2\pi^2}{4L^2}\right)t \qquad (13.42)$$

1) Using a schematic to represent the tablet in Figure 13.20, write the governing equation (include only terms needed), boundary conditions, and initial condition that lead to this solution. 2) If we average the concentration over the slab, the average concentration is given by

$$\frac{c_{av} - c_s}{c_i - c_s} = \sum_{n=0}^{\infty} \frac{8}{(2n+1)^2\pi^2} \exp\left(-D\frac{(2n+1)^2\pi^2}{4L^2}t\right) \qquad (13.43)$$

Figure 13.20: A tablet with drug being released from the surfaces, to be treated as a slab geometry. (iStock.com/Anton_Sokolov)

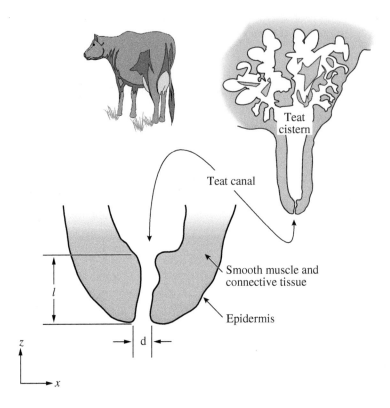

Figure 13.19: Schematic of an udder showing the length of the canal for possible infection.

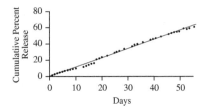

Figure 13.21: Constant rate of drug release achieved from the geometry shown in Figure 13.22.

Using Eq. 13.42, write the expression (no need to calculate) from which you can obtain the above expression (Eq. 13.43). 3) Consider the drug to be released only by diffusion. Write an expression for the total mass amount of the drug, M_t, released from the slab into the surroundings, in terms of c_{av}, that is, the average concentration at time t, and other parameters as needed. 4) For early stages of release (<u>small</u> time t), this expression for M_t can be written using Eq. 13.45 for c_{av} as (Saltzman, 2001)

$$\frac{M_t}{c_i A L} = 4\sqrt{\frac{Dt}{L^2 \pi}} \tag{13.44}$$

Provide the expression for the time rate of drug release as a function of time during the early stages of release. 5) Plot this rate of release as a function of time. 6) In a

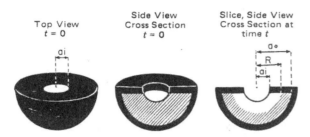

Figure 13.22: Geometry to achieve constant rate of drug release. Reprinted from *Journal of Pharmaceutical Sciences,* Vol. 72, No. 1, Hsieh, D. S. T., W. D. Rhine, and R. Langer, Zero-order controlled-release polymer matrices for micromolecules and macromolecules, pp. 17–22, 1983, with permission from Elsevier.

celebrated paper, Robert Langer of MIT published a different geometry instead of a slab where the release rate is constant, as shown by the plot of cumulative release in Figure 13.21. As shown in the schematic in Figure 13.22, a constant release rate was achieved by making the geometry a hemisphere and allowing release only through a small cavity cut into the center of the flat surface of the hemisphere. The rest of the surface area is covered with an impermeable coating through which no drug comes out. Describe qualitatively the likely reasons for getting a constant release from this geometry.

13.17 Tissue Shrinkage during an Operation

During the Lasik operation to change the focusing power of the eye, a thin protective layer of corneal tissue is folded back, as shown in Figure 13.23. As the operation proceeds, this layer could possibly lose moisture and thus shrink. The shrinkage in this high moisture region is linear with *average* moisture content, i.e., percent moisture loss is the same as percent shrinkage. Consider the moisture transport to be along the 0.1 mm thickness of the layer. The initial moisture content is 1.5 kg water/kg of dry matter and the surface is kept at 1.2 kg water/kg of dry matter. The diffusivity of water in the tissue is 10^{-9} m^2/s and the thickness of the layer is 0.1 mm. For a 20 minute long operation, calculate the shrinkage (in percentage) of the layer.

13.18 Permeability of Skin in Drug Delivery

Transdermal drug delivery systems have to overcome the major bottleneck of the low permeability of skin to drugs. To overcome this barrier, transdermal patches often con-

Slice of corneal tissue folded back

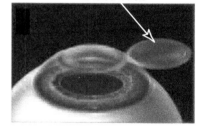

Figure 13.23: Lasik operation on a cornea. Adapted from the website of Eye Clinic, PC, Jackson, Tennessee.

tain penetration enhancers in addition to the drug. Some penetration enhancers, such as Azone, increase the permeability of skin toward protein-based drugs. Consider the application of a patch containing growth hormone on a region of skin. Assume none of the drug is present in the skin before applying the patch. The diffusion coefficient of growth hormone in the skin without the penetration enhancer is 2.9×10^{-11} m^2/s and with the penetration enhancer Azone is 7.7×10^{-11} m^2/s. Calculate the percentage increase in concentration of the growth hormone by using Azone as an enhancer (compared to using no enhancer) at a location 2 mm beneath the skin surface after 150 minutes.

13.19   Moisture Loss from Skin Surface

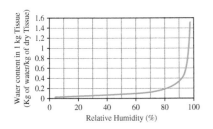

Figure 13.24: Equilibrium water content of skin as a function of water content (relative humidity) in air.

As you go out on a wintry day from a humid room (you have been cooking), the relative humidity (RH) of air drops, going from inside the room to the outside. Consider the initial moisture content of your skin tissue to be that corresponding to equilibrium with the air inside the room at 95% RH (see Figure 13.24). The outside air has an RH of 70%, but the wind is blowing hard (as it often does in Ithaca), which results in a surface moisture content of skin that is in equilibrium with air at 50% RH. The diffusivity of water in the skin and surrounding tissue is 10^{-10} m^2/s. 1) What is the initial moisture content in the tissue (in the humid room)? 2) What is the moisture content in the surface of the skin tissue when you are in the outside air? 3) At what depth in the skin, in mm, is the moisture content at 90% of the initial moisture content after 5 minutes? 4) What is the moisture content in the tissue at a depth of 1 mm after 5 minutes? 5) The density of dry tissue is 200 kg/m^3. What is the rate at which the skin is losing moisture per unit area at 5 minutes?

13.20   Drug Release Rates from a Slab at Early and after Long Times

Individuals falling short of immune response, i.e., production of antibodies in response to foreign bodies or antigens, can be treated with replacement therapy. One way to do this is to introduce a drug-incorporated polymer matrix into the body. Consider drug release from a slab of thickness $2L$, surface area for one side A, and volume V. The thickness is small enough compared to the other two dimensions so that transport of a drug out of the slab can be considered one dimensional. The released drug is the amount that is no longer in the slab. For steps 1 and 2 below, it may be helpful to consider the appropriate solutions of the diffusion equation at early times and late times.

1) Show that the total drug released over time t for short times can be approximated as

$$(c_s - c_i) \sqrt{\left(\frac{Dt}{\pi} \right)} \frac{2V}{L} \qquad (13.45)$$

2) Show that the total drug released at longer times can be approximated as

$$(c_i - c_s) \left(1 - \frac{8}{\pi^2} \exp(-\pi^2 Dt/4L^2) \right) V \qquad (13.46)$$

3) Plot approximately the two equations on the same graph, according to their corresponding time domains. *(Note that there is a region where neither one would be accurate because they are approximations.)*

13.21 Scaling of Drug Diffusion Time with Diffusivity

A certain drug is diffusing into the skin from a patch placed on it. This patch maintains a constant concentration of drug on the skin surface. There is no degradation of the drug in the skin and the skin is thick enough so that the other end (capillaries, etc.) is not reached for the time period we are interested in. If we change the drug to another one that diffuses only half as fast, how much longer would this drug take to reach the same concentration level at the same distance from the surface? Show your calculation.

13.22 Diffusion of Drug through DNA Gel

In the laboratory of Professor Luo at the BEE Dept. of Cornell Univ., researchers are looking at DNA hydrogels as a potential protein drug delivery system. Such DNA gels are novel and little information is known about how a drug can diffuse through them. An experiment was performed where a slab of hydrogel containing a uniform distribution of the protein BSA is washed using pure water (zero BSA) and the amounts of BSA that diffuses out into the pure water is measured. The thickness of the slab is 6 mm and its other two dimensions are considered large in comparison to the thickness. If the amounts of the BSA released (lost) after 1 day and 3 days are 4.15 g and 6.225 g, respectively, provide an equation from which you can calculate the diffusivity of BSA in the DNA gel. *(Important– you do not need to solve for the numerical value of diffusivity, D, since it requires iteration, but your final equation should have the only unknown, D.)* *(Hint: The amount of BSA __in__ the slab at any time is equal to its average concentration multiplied by the volume of the slab.)*

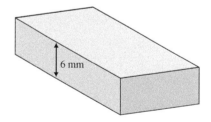

Figure 13.25: Schematic of a DNA slab where diffusion is considered primarily along the 6 mm thickness. Dimensions in the other directions are much larger than 6 mm.

13.23   Diffusion through Sediment at Lake Bottom

PCB (polychlorinated biphenyl), a synthetic organic chemical once widely used in electrical equipment, is highly toxic. Consider PCB dumped in a lake where it diffuses through sediment in the lake bottom. Before the dumping, the lake or the sediment contained no PCB. Samples collected at 1 cm from the top surface of the sediment after the dumping showed a PCB concentration of 30×10^{-9} g per gram of sediment. The diffusivity of PCB in the sediment is 1×10^{-12} m^2/s. Assuming the PCB content of the lake water stayed constant such that the top of the sediment layer always has 150×10^{-9} g per gram of sediment, how long has PCB been in the lake?

13.24   Drying Banana Slices

Banana slices of 4 mm thickness and 30 mm diameter are to be dried via exposure to air on both sides. Initial moisture content is 4 kg water/kg dry matter, and the final moisture content after drying is 0.25 kg water/kg dry matter. The moisture content at the surface of the slices is at equilibrium with the air throughout the duration of drying. This equilibrium moisture content corresponding to the temperature and humidity of the air is 0.2 kg water/kg dry matter. Diffusivity of moisture in the banana is estimated to be 10^{-10} m^2/s. 1) Provide reasons why the moisture transport of this particular geometry can be considered one dimensional. 2) What is the time needed to dry the slices? 3) As the banana loses moisture, it shrinks and the final thickness is 60% of the initial thickness. In our analytical solution, it is hard to consider the shrinkage effect, while in a numerical method it is readily considered. At one extreme, if you consider the final thickness as the thickness throughout the drying process, what would the drying time be? 4) Explain why the drying time from question 3 is lower/higher than what you calculated in question 2. 5) Would the times calculated in questions 1 and 3 provide an estimate of the range of drying time for the real situation? Explain.

Figure 13.26: Banana slices after drying. (iStock.com)

13.25   Drug release from a matrix

Moved to solved example

13.26   Random Walk of Insects

Insect movement has been successfully modeled as a diffusion process. Grasshoppers of quantity $M = 5000$ released on a golf course were found to spread radially (ignore any vertical movements) in an axisymmetric manner. The concentration (number per unit area) of grasshoppers at a radius, r, at time, t, is given by

$$c = \frac{M}{\pi \sigma^2} e^{-(r^2/\sigma^2)} \tag{13.47}$$

where $\sigma^2 = 2Dt$, D being the diffusivity of the grasshoppers, which is approximately 2.57×10^{-5} m^2/s. 1) Write the governing equation for which Eq. 13.47 is a solution. 2) What are the expected concentrations at 0, 5, and 10 m from the center after 1 day? 3) Plot approximately the data in question 2. 4) What is the radius within which 95% of the grasshoppers are contained after 1 hour? 5) What is the radius within which 95% of the grasshoppers are contained after 1 day?

13.27 Oxygen Transport in a Bacterial Aggregate

A homogeneous spherical aggregate of cells is receiving oxygen at its outer surface. The oxygen diffuses into the sphere and is consumed uniformly throughout the sphere. For simplicity, consumption of oxygen is considered at zeroth order with respect to oxygen concentration. The outer surface of the sphere is maintained at a constant concentration. We would like to obtain the steady-state oxygen concentration profile as a function of radius of this aggregate.

1) Write the governing equation, keeping only the terms needed. Assume all the variables and parameters needed are given. 2) Write all the boundary conditions needed. 3) Solve the governing equation to obtain concentration as a function of radius. 4) What is the radius of the oxygen-free core within the aggregate? 5) Write your answer to step 4 as an inequality that illustrates the condition under which an oxygen-free core radius will exist. 6) Provide a physical interpretation of the relationship you found in step 5 with respect to each of the parameters.

13.28 Drug Release Properties of a Sphere

We want to study release kinetics of drugs from uncoated resin spheres where all of the mass transfer resistance is due to diffusion in the sphere. Consider a sphere of diameter 100 μm. The initial amount of drug is 65 mg per 100 mg of resin. The diffusivity of the drug in the resin is 3.4×10^{-10} m^2/s. The surface of the sphere is maintained at zero concentration of the drug.

1) If 90% of the drug is released in 3 hours, what percentage of drug would be released in the same time when the diffusivity is halved while the diameter is reduced by a factor of $\sqrt{2}$? (Hint: Solution for a sphere is analogous to that for a slab, with appropriate substitutions.) 2) Explain how you got your answer in question 1.

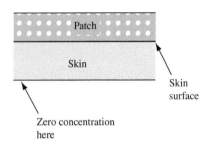

Figure 13.27: Schematic of skin with a polymeric drug patch over it.

13.29   Increasing Drug Transport from a Patch

Metronidazole is a drug with anti-bacterial properties and is used in topical formulations for the treatment of the chronic skin disease rosacea. An experiment was performed over 24 hours to study the diffusivity of metronidazole in the skin using two polymeric drug patches, PEG 1500 and PEG 4000 (Bjorklund et al., 2010). Assume diffusive resistance inside the patch itself is negligible. There is no drug in the skin at the start of the experiment, and the concentration at the surface of the skin is constant at 6.11 mg/liter (1 liter = 10^{-3} m^3) for the duration we are interested in.

1) The steady-state flux across the skin membrane is measured to be 8.2 μg/cm$^2 \cdot$ hr. If the drug is fully absorbed (drug concentration is zero) 0.1 mm below the skin surface, what is the effective diffusivity (in m^2/s) of metronidazole in the skin when using the PEG 1500 patch? 2) When the patch material is changed to PEG 4000, the flux as in question 1 was measured to be 7.0 μg/cm$^2 \cdot$ hr. What is the diffusivity when using this patch? If it is known that wetter skin has higher diffusivity and that one of the patches has more water content and thus hydrates the skin more, which patch has the higher water content? 3) Calculate the drug concentration at 25 μm from the surface at 1 second after the drug is applied, for the PEG 1500 patch. 4) Calculate the surface drug flux at the same instant as in question 3. 5) Compare the flux in question 4 with the steady-state flux in question 1 and explain why one is higher/lower than the other. 6) (Unrelated to above questions) If two patches have the same water content but are made up of different materials, how is it possible for one patch to lead to a higher surface skin moisture than the other?

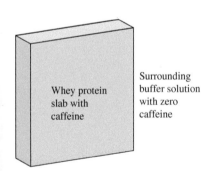

Figure 13.28: Slab containing caffeine being washed by a buffer solution containing zero caffeine.

13.30   Controlled Release of Components

Hydrogels are popular controlled drug release carriers. In an experiment to characterize diffusivity of caffeine in whey protein gels, slabs of these gels containing caffeine were put in a phosphate buffer solution, into which the caffeine would be released. Assume agitation of the buffer maintains a concentration of zero at the slab surface. The initial concentration of caffeine in the gel is c_i.

1) Treating the gel as a symmetrical 1D slab, find the equation for the total amount of caffeine released from its two surfaces as a function of time *during long times*. Define any other symbol that you use. 2) For a slab that is 1 cm thick, if 75% is released after 60 minutes, what is the diffusivity of caffeine in the hydrogel? 3) For what duration would your calculations in question 1 be valid?

13.31 Dating of Fossils

Many fossils are assumed to take up trace elements by diffusion, and the extent of this diffusion can be used to predict the age of the fossil (Kohn, 2008). Measurement of a trace element on a certain deer-like animal tooth (see Figure 13.29) showed that the distance from the surface at which the concentration is one-hundredth that of the surface concentration is 70 μm. Assume there is no trace element initially in the fossil. We suspect the trace element enters only a small distance from the surface. Diffusivity (in both enamel and dentin) of the trace element can be assumed to be the same, at 10^{-20} m^2/s.

1) What is the age of the fossil in years? 2) Hypothetically, if the trace element were also involved in reactions where it would be used up, is your estimate of age likely to be lower or higher than the true age? Provide reasoning. 3) What is the **total amount** of trace element uptake per unit area from one side of the tooth over a time period of t for a surface concentration of c_s (plug in numbers and simplify)? 4) By what factor would this total amount of uptake in question 3 change for a trace element whose diffusivity in the tooth is twice as much?

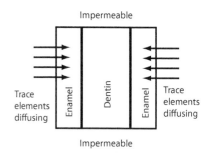

Figure 13.29: Schematic of a fossil tooth with trace elements diffusing.

13.32 Contaminated Wine?

Aerial contamination of vegetation is potentially an important source of human exposure to some toxic chemicals. We are interested in estimating toluene absorption from the air in grapes on a vine (Gorna-Binkul et al., 2001). A grape can be assumed to be spherical, with a typical radius of 0.0125 m, and containing no toluene initially. The toluene concentration in the air is 2.18×10^{-3} g/m^3 (which is well below the legal limit, just so that you know). The partition coefficient is $K^* = 3.39 \times 10^{-3}$; here K^* is used exactly as defined in Chapter 13 for convective boundary condition, i.e., $K^* = c_{\text{toluene}}^{\text{air}}/c_{\text{toluene}}^{\text{grape}}$ at the surface of the grape. Diffusivity of dissolved toluene in a grape can be approximated for all layers to be that of the pulp, at 1.15×10^{-7} m^2/s, and the diffusivity of toluene in air is 8.11×10^{-6} m^2/s. The average windspeed is 0.254 m/s. The viscosity and density of air are 1.98×10^{-5} kg/m·s and 1.24 kg/m^3, respectively.

1) Calculate the mass transfer coefficient at the grape's surface. *(Hint: Calculations are analogous to those for heat transfer coefficient, with the substitutions of mass diffusivity for both thermal conductivity and thermal diffusivity; See Chapter 14 if needed.)* 2) What is the exposure time at which the concentration of toluene in the center of the grape would reach 80% of the final concentration? 3) For a different air flow situation, when the mass transfer coefficient at the surface is effectively infinite, what is the *surface* concentration of toluene *in the grape*? 4) As you know, for a very short time after exposure, the grape can be considered very thick. With this

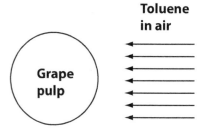

Figure 13.30: A spherical grape surrounded by flowing air with toluene in it.

assumption, and for a constant concentration condition given at the surface, how does the *rate at which toluene enters the grape* depend on the diffusivity of the grape's tissue? Discuss using the appropriate formula and symbols; no numerical calculations necessary.

Chapter 14

CONVECTION MASS TRANSFER

CHAPTER OBJECTIVES

After you have studied this chapter, you should be able to

1. Formulate and solve for convective–dispersive mass transfer in a flowing liquid, through a porous media, and in a gas.

2. Explain the process of convective–diffusive mass transfer in a fluid over a solid surface as a function of the relatively stagnant fluid layer over the surface called the boundary layer.

3. Calculate the convective mass transfer coefficient h_m knowing the flow situation.

KEY TERMS

- **convective mass transfer coefficient**

- **velocity boundary layer**

- **mass transfer boundary layer**

- **Sherwood number (mass transfer**

  Nusselt number)

- **Schmidt number**

- **mass transfer Grashof number**

- **natural and forced convection mass transfer**

In this chapter we will study how the mass transfer in a fluid is modified by the effects of bulk flow. We will do this by developing a more general version of the governing equation that we derived in Chapter 11, by keeping the term representing the bulk flow in addition to transport due to diffusion or dispersion. The relationship of this chapter to other chapters in mass transfer is shown in Figure 14.1.

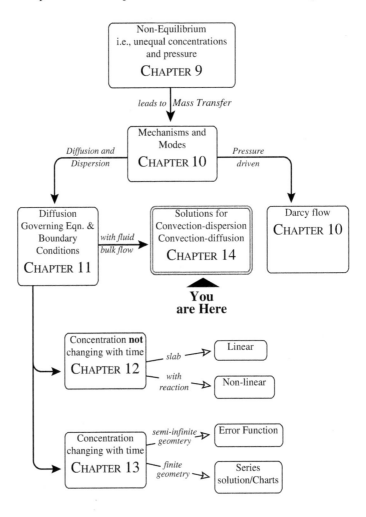

Figure 14.1: Concept map of mass transfer showing how the contents of this chapter relate to other chapters in mass transfer.

When bulk flow or convection is present, the governing equation for the transport of species A is given by Eq. 11.18 from page 396, which is repeated here.

$$\underbrace{\frac{\partial c_A}{\partial t}}_{\text{storage}} + \underbrace{u\frac{\partial c_A}{\partial x}}_{\substack{\text{bulk flow or} \\ \text{convection}}} = \underbrace{D_{AB}\frac{\partial^2 c_A}{\partial x^2}}_{\substack{\text{diffusion or} \\ \text{dispersion}}} + \underbrace{r_A}_{\text{generation}} \tag{14.1}$$

In Eq. 14.1, u is the velocity contributing to bulk flow. The term representing convection or bulk flow is the additional term retained in the governing equation in this chapter. We will consider four different scenarios of convective mass transfer using this equation that have important practical applications in environmental and biological transport processes. These four scenarios are 1) in an infinite fluid, 2) in a porous solid, 3) in a stagnant gas, and 4) over a surface. Remember that Eq. 14.1 can be used for both diffusion and dispersion. To differentiate between the two, we will use the symbol D for diffusion and E for dispersion, as introduced in Chapter 10. We will consider cases 1 and 2 in the context of dispersion, whereas cases 3 and 4 will be considered in the context of diffusion. To contrast this chapter on convective mass transfer with Chapter 6 on convective heat transfer, note that all of Chapter 6 was for flow over a surface, i.e., analogous to case 4 of the four scenarios considered here. The reason the first two cases considered here were not included in Chapter 6 is none other than that their practical applications are more abundant in mass transfer than the corresponding heat transfer situation.

14.1 Convection–Dispersion in an Infinite Fluid

One-dimensional convection–dispersion can be useful idealization for many practical transport situations, for example, pollutants being introduced in a flowing stream, as shown in Figure 14.2. Due to turbulence in the stream, pollutants will spread (disperse) in addition to being carried by the stream (bulk flow). If we can assume instantaneous mixing in the vertical and lateral directions (two directions perpendicular to the flow), the convective–dispersive transport can be approximated as one dimensional, often referred to as a plug-flow system. In this example, the pollutants can also degrade; thus the term containing mass depletion (negative of generation) would also need to be present. Thus, for a 1D dispersion with chemical reaction in a flowing medium that is of infinite extent, the governing equation can be written as

$$\frac{\partial c_A}{\partial t} + \underbrace{u\frac{\partial c_A}{\partial z}}_{\text{convection}} = \underbrace{E\frac{\partial^2 c_A}{\partial z^2}}_{\text{dispersion}} - k'' c_A \tag{14.2}$$

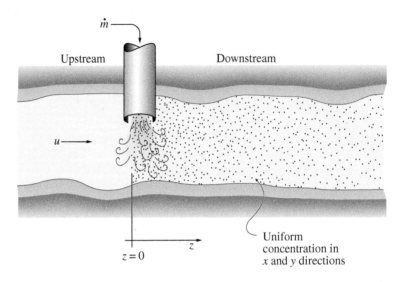

Figure 14.2: Schematic of a point source pollutant being introduced in a stream.

where k'' is the rate of reaction, E is the dispersion coefficient, and u is the bulk velocity along the flow (z direction). For steady state, Eq. 14.2 can be simplified to

$$u\frac{dc}{dz} = E\frac{d^2c}{dz^2} - k''c \tag{14.3}$$

where the subscript A in c_A has been dropped for simplicity. If mass is being introduced at the rate of $\dot{m}$ only at $z = 0$ (e.g., at a discharge point), boundary conditions can be written for upstream and downstream points far from $z = 0$ as

$$c(z \to \infty) = 0 \tag{14.4}$$
$$c(z \to -\infty) = 0 \tag{14.5}$$

The solution to Eq. 14.3, which is a second-order, ordinary, homogeneous differential equation, has the form

$$c = c^* e^{\lambda z} \tag{14.6}$$

where c^* and λ are to be determined. Substituting Eq. 14.6 in Eq. 14.3 to find λ,

$$c^*(u\lambda - E\lambda^2 + k'')e^{\lambda z} = 0$$

which would require that

$$u\lambda - E\lambda^2 + k'' = 0$$

Solving for λ,

$$\lambda = \frac{u \pm \sqrt{u^2 + 4Ek''}}{2E}$$
$$= \frac{u}{2E}(1 \pm \psi)$$

where

$$\psi = \sqrt{1 + \frac{4Ek''}{u^2}} \geq 1$$

Therefore, the two roots λ_1 and λ_2 satisfy the conditions

$$\lambda_1 = \frac{u}{2E}(1 + \psi) > 0 \tag{14.7}$$
$$\lambda_2 = \frac{u}{2E}(1 - \psi) < 0 \tag{14.8}$$

The general solution is

$$c(z) = c_1 e^{\lambda_1 z} + c_2 e^{\lambda_2 z}$$

Using the two boundary conditions (Eqs. 14.4 and 14.5), we get

$$c(z \to \infty) = 0 \quad \text{which leads to } c_1 = 0$$
$$c(z \to -\infty) = 0 \quad \text{which leads to } c_2 = 0$$

Thus, the general solution is

$$c(z) = c_1 e^{\lambda_1 z} \quad z < 0 \tag{14.9}$$
$$c(z) = c_2 e^{\lambda_2 z} \quad z > 0 \tag{14.10}$$

To evaluate c_1 and c_2, note that, in order for both Eqs. 14.9 and 14.10 to be satisfied at $z = 0$ (discharge point in Figure 14.2), c_1 has to be equal to c_2.

To find c_1, we need additional information. Note that we have not yet made use of the information provided on the rate at which pollutants are being introduced, $\dot{m}$. Our solution of course would depend on this information. We will utilize this information by performing a mass balance at $z = 0$, as shown in Figure 14.3, with the introduction of pollutants as another stream coming in, $\dot{m}$. Performing a mass balance, i.e., In – Out + Generation = Change in Storage, on a point control volume (thus change in storage and generation are zero), we can write

$$\underbrace{\left[-EA\frac{dc}{dz} + Auc\right]_{z=0^-}}_{\text{Pollutant transport from the left of } z=0} + \underbrace{\dot{m}}_{\substack{\text{pollutant source} \\ \text{at } z=0}} - \underbrace{\left[-EA\frac{dc}{dz} + Auc\right]_{z=0^+}}_{\text{Pollutant transport to the right of } z=0} = 0$$

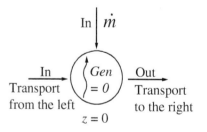

Figure 14.3: Schematic of mass balance at a point (shown enlarged) where the pollutants are being introduced.

where A is the cross-sectional area of the stream and thus Au is its flow rate. Substituting for c from Eqs. 14.9 and 14.10 and using $c_2 = c_1$,

$$-EA\,c_1\lambda_1 e^{\lambda_1 0} + Auc_1 e^{\lambda_1 0} + \dot{m} + EA\,c_1\lambda_2 e^{\lambda_2 0} - Auc_1 e^{\lambda_2 0} = 0$$
$$-EA\,c_1\lambda_1 + Auc_1 + \dot{m} + EA\,c_1\lambda_2 - Auc_1 = 0$$
$$EAc_1(\lambda_1 - \lambda_2) = \dot{m}$$

$$c_1 = \frac{\dot{m}}{EA(\lambda_1 - \lambda_2)}$$

Using Eqs. 14.7 and 14.8 to substitute for λ_1 and λ_2,

$$c_1 = \frac{\dot{m}}{A\,u\,\psi}$$

Solution for the general case of convection and dispersion

Substituting for c_1, we get the solution when both advection and dispersion effects are present:

$$c(z) = \begin{cases} \frac{\dot{m}}{Au\psi} e^{\frac{u}{2E}(1+\psi)z} & z < 0 \\ \frac{\dot{m}}{Au\psi} e^{\frac{u}{2E}(1-\psi)z} & z > 0 \end{cases}$$

This is illustrated graphically in Figure 14.4, showing greater concentration along the positive z direction where dispersion and bulk flow contribute to the transport. The small concentration along the negative z direction is due to dispersion only and has to work against the bulk flow.

When convection is the major mechanism

For flow in a freshwater river, when it is not under tidal influence, the dispersion coefficient E is often small. For small E, $\psi \approx 1$ and

$$c(0) = \frac{\dot{m}}{Au}$$

By setting $E = 0$ in Eq. 14.2 and solving,

$$c(z) = \begin{cases} 0 & z < 0 \\ \frac{\dot{m}}{Au} e^{-\frac{k''z}{u}} & z > 0 \end{cases}$$

This is illustrated graphically in Figure 14.5, showing the effect of only bulk flow (with decay) in the positive z direction. In the absence of dispersion, there is no concentration in the negative z direction.

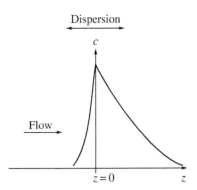

Figure 14.4: Concentration distribution due to a steady source when both convection and dispersion are present.

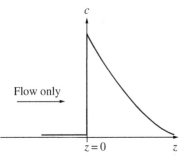

Figure 14.5: Concentration distribution due to a steady source in the presence of convection only.

Dispersion only

As the river approaches the sea, the dispersion coefficient E increases and the net downstream velocity u decreases. For $u \approx 0$,

$$u\psi = u\sqrt{\frac{u^2 + 4\,k''\,E}{u^2}}$$
$$\approx \sqrt{4\,k''\,E}$$

Substituting in Eq. 14.11 for $u\psi$ and $u \approx 0$, we get the solution when only dispersion is present:

$$c(z) = \begin{cases} \dfrac{\dot{m}}{A\sqrt{4k''E}}e^{+\sqrt{k''/E}\,z} & z < 0 \\[2mm] \dfrac{\dot{m}}{A\sqrt{4k''E}}e^{-\sqrt{k''/E}\,z} & z > 0 \end{cases}$$

This is illustrated graphically in Figure 14.6, where the profiles are symmetric due to the absence of bulk flow.

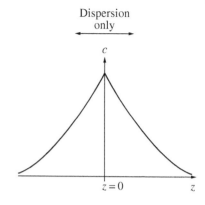

Figure 14.6: Concentration distribution due to a steady source in the presence of dispersion only.

14.1.1 Example: Radioactive Water Leak in Fukushima

The effect of radioactive water leakage from the Fukushima plant in Japan (Figure 14.7) on fish, and eventually humans, became an ongoing concern after the accident in March of 2011. This problem is not at all 1D, but we will do a somewhat related problem hoping to get some qualitative understanding of the problem. Consider radioactive water coming into a flowing stream that can be treated as 1D. Also, the process reaches a steady state, unlike the real one where the leak has since stopped. Assume Iodine-131 is the primary radioactive component that is of concern—it has a half-life of 8 days, following a first-order reaction (half-life for a first-order reaction is given by $0.5c_0 = c_0 e^{-k''t_{1/2}}$). Seven tons (7000 kg) of radioactive water containing 300000 becquerels per cm^3 continuously flows into the river every hour. The average velocity of water in the river can be approximated as 0.833 m/s (3 km/hr) and the volumetric flow rate of the stream is 280 m^3/s (one fourth of the Susquehanna river in New York). The turbulence in the current leads to a dispersion coefficient of 10 m^2/s. The density of water is 1000 kg/m^3. *Caution: Maintain several significant digits in your calculations until the end.*

1) Up to what distance downstream would the water be considered contaminated, given that the legal limit for radiation in water is 0.04 becquerels per cm^3? 2) Repeat question 1 to find the distance of contaminated water upstream. 3) What is the concentration in the river in becquerels per cm^3 at the point where the radioactive water is being dumped? 4) What transport mode contributes to contamination upstream?

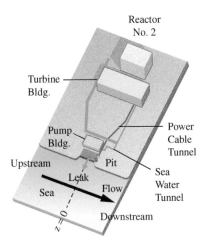

Figure 14.7: Radioactive water leaked into the ocean from the Fukushima power plant. Diagram based on illustration in the New York Times.

5) On the downstream side of the point where water is being dumped, what are the dispersive and convective fluxes of the radioactive water? 6) How do you expect the distances calculated here with a 1D model to change for the real 3D situation, where the Fukushima plant let out water into the Pacific Ocean? Provide reasoning for your answer (no calculations needed).

Solution

Understanding and formulating the problem *1) What is the process?* Contaminated radioactive water is being dumped into a stream at a continuous rate that moves and disperses with the streamflow while its radioactivity is decaying simultaneously. *2) What are we solving for?* Concentration profile and fluxes. *3) Schematic and given data:* A schematic is shown in Figure 14.8 with some of the given data superimposed on it. *4) Assumptions:* Dispersion coefficient remains constant.

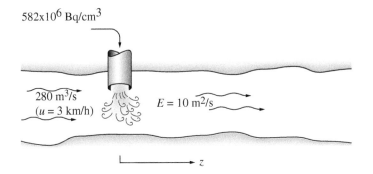

Figure 14.8: Schematic for Example 14.1.1.

Generating and selecting among alternate solutions *1) What solutions are possible?* For convective mass transfer, the solutions we have available are shown in the solution chart in Figure 14.24. *2) What approach is likely to work?* As we follow the solution chart in Figure 14.24, this has been given to us as a steady-state problem. There is no surface mentioned in this problem so it is not about fluid flowing over a surface. Transport (flow and dispersion) of contaminated water is occurring within the clean water in the flowing stream, i.e., it is a "flowing stream" problem.

Implementing the chosen solution   The governing equation and boundary conditions match those in Section 14.1. Thus, the solution can be used directly as

$$
c_I(z) = \begin{cases} \frac{\dot{m}_I}{Au\psi} \exp\left[\frac{u}{2E}(1+\psi)z\right] & z < 0 \\ \frac{\dot{m}_I}{Au\psi} \exp\left[\frac{u}{2E}(1-\psi)z\right] & z > 0 \end{cases} \tag{14.11}
$$

where

$$
\psi = \sqrt{1 + \frac{4Ek''}{u^2}} \tag{14.12}
$$

We can find k'' from the half life given (see Eq. 9.19) as

$$
k'' = \frac{0.693}{t_{1/2}} = 1.003 \times 10^{-6} \ [1/s] \tag{14.13}
$$

Plugging in,

$$
\psi = \sqrt{1 + \frac{4Ek''}{u^2}} = \sqrt{1 + \frac{4(10\ [\text{m}^2/\text{s}])(1.003 \times 10^{-6}\ [1/\text{s}])}{(0.833\ [\text{m/s}])^2}} \tag{14.14}
$$

$$
= 1.00003 \tag{14.15}
$$

from which we can calculate c_I as

$$
c_I(z) = \begin{cases} \dfrac{\dot{m}_I}{280 \times 1.00003} \exp\left[\frac{0.833}{2(10)}(1+1.00003)z\right] & z < 0 \\[4mm] \dfrac{\dot{m}_I}{280 \times 1.00003} \exp\left[\frac{0.833}{2(10)}(1-1.00003)z\right] & z > 0 \end{cases} \tag{14.16}
$$

$$
= \begin{cases} 0.00357\dot{m}_I \exp(0.083z) & z < 0 \\ 0.00357\dot{m}_I \exp\left(-1.25 \times 10^{-6}z\right) & z > 0 \end{cases} \tag{14.17}
$$

where m_I is the dumping rate of radioactivity, given by

$$
\dot{m}_I = 1.94\frac{\text{kg}}{\text{s}} \times \frac{\text{m}^3}{1000\ \text{kg}} \times 300000\frac{\text{Bq}}{\text{cm}^3} = 582\frac{\text{Bq} \cdot \text{m}^3}{\text{cm}^3 \cdot \text{s}} \tag{14.18}
$$

where the units have been left as $[\text{Bq/cm}^3][\text{m}^3/\text{s}]$ for ease of computation at the following step. Plugging in the legal limit and the dumping rate, we get

$$
0.04\frac{\text{Bq}}{\text{cm}^3} = \begin{cases} 0.00357\left[\frac{\text{s}}{\text{m}^3}\right]582\left[\frac{\text{Bq}\cdot\text{m}^3}{\text{cm}^3\cdot\text{s}}\right]\exp(0.083z) & z < 0 \\[4mm] 0.00357\left[\frac{\text{s}}{\text{m}^3}\right]582\left[\frac{\text{Bq}\cdot\text{m}^3}{\text{cm}^3\cdot\text{s}}\right]\exp\left(-1.25 \times 10^{-6}z\right) & z > 0 \end{cases}
$$

Simplifying,

$$-3.95 = \begin{cases} 0.083z & z < 0 \\ -1.25 \times 10^{-6}z & z > 0 \end{cases}$$

from which we calculate the upstream ($z < 0$) distance over which the legal limit would be exceeded as 0.048 km while the downstream ($z > 0$) distance over which the legal limit would be exceeded is 3160 km.

3) Concentration where the radioactive water is being dumped is

$$c \, (z = 0) \, \frac{\text{Bq}}{\text{cm}^3} = 0.00357 \left[\frac{\text{s}}{\text{m}^3} \right] 582 \left[\frac{\text{Bq} \cdot \text{m}^3}{\text{cm}^3 \cdot \text{s}} \right] = 2.08 \frac{\text{Bq}}{\text{cm}^3}$$

4) Dispersion/eddies/turbulence (not bulk flow).

5) The total flux is the sum of dispersive and convective fluxes, given by

$$n_{Iz}|_{z=0} = \underbrace{-E \frac{\partial c}{\partial z}\bigg|_{z=0}}_{\text{dispersive}} + \underbrace{(uc)|_{z=0}}_{\text{convective}}$$

Using the concentration on the downstream side ($z > 0$),

$$c_{Iz} = \underbrace{0.00357 \dot{m}_I}_{c(z=0)} \exp\left(-1.21 \times 10^{-6}z\right)$$

$$= 2.08 \left[\frac{\text{Bq}}{\text{cm}^3} \right] \exp\left(-1.21 \times 10^{-6}z\right)$$

and taking the derivative

$$\frac{\partial c}{\partial z}\bigg|_{z=0} = 2.08 \left[\frac{\text{Bq}}{\text{cm}^3} \right] \left(-1.21 \times 10^{-6} \left[\frac{1}{\text{m}} \right] \right)$$

$$= 2.08 \left[\frac{\text{Bq}}{\text{cm}^3} \right] \times -1.21 \times 10^{-6} \left[\frac{1}{\text{m}} \right] \times \frac{1 \text{ m}}{100 \text{ cm}}$$

$$= -2.51 \times 10^{-8} \frac{\text{Bq}}{\text{cm}^4}$$

Thus, the two flux components are

$$
n_{Iz}|_{z=0} = \underbrace{-10^5 \left[\frac{cm^2}{s} \right] \left(-2.51 \times 10^{-8} \left[\frac{Bq}{cm^4} \right] \right)}_{\text{dispersive}} + \underbrace{83.3 \left[\frac{cm}{s} \right] 2.08 \left[\frac{Bq}{cm^3} \right]}_{\text{convective}}
$$

$$
= \underbrace{2.51 \times 10^{-3} \left[\frac{Bq}{cm^2 \cdot s} \right]}_{\text{dispersive}} + \underbrace{173.3 \left[\frac{Bq}{cm^2 \cdot s} \right]}_{\text{convective}}
$$

6) The actual contaminated distance will be a lot less since that would be a 3D situation instead of the approximate 1D here. Fish/plants will also consume some of the radiation, changing the concentration.

Evaluating the solution *1) Does the concentration profile make sense?* This problem had to be highly idealized to obtain a simple solution, making it difficult to compare with possible available measured data. *2) What do we learn?* With significant flow present, convective flux dominates dispersive flux.

14.2 Convection–Dispersion in a Semi-Infinite Porous Solid

In the previous section, we discussed convection–dispersion in a flowing stream where the source of dispersion was the turbulence of the stream. In this section, we consider convection–dispersion in a saturated porous solid, that is, a porous solid where the carrier fluid completely fills the interstitial region between the porous media grains. The porous media of interest in this section is soil. Compared with dispersion in a stream, dispersion in soil is due to completely different mechanisms, such as the presence of pores of varying diameter and tortuosity, as explained in Section 10.2. We will study one-dimensional convection–dispersion in a semi-infinite region, as illustrated in Figure 14.9. Note that, although this figure may include an initial unsaturated zone, it is assumed that with rainfall, etc., the top region will first get saturated and the wetting front will move downwards, eventually reaching groundwater. Thus, the saturated condition under which we will study transport will be satisfied. One intended application of this section is the study of pollutant transport through soil.

The governing equation for convection–dispersion without any reaction is

$$
\underbrace{\frac{\partial c_A}{\partial t}}_{\text{storage}} + \underbrace{u \frac{\partial c_A}{\partial z}}_{\text{convection}} = \underbrace{E \frac{\partial^2 c_A}{\partial z^2}}_{\text{dispersion}}
$$

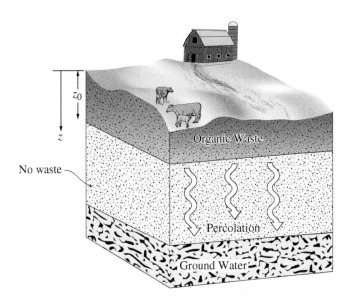

Figure 14.9: Schematic of pollutant transport through soil.

Here c_A is the concentration of dissolved solutes A in the liquid that is flowing through the porous medium with an average velocity u. Note that u is not the average velocity in the pores (see Example 10.1.1); rather, it is the volumetric flux, i.e., when u is multiplied by the total area of flow, we get the volumetric flow rate through the porous medium. Making a change of variable $\eta = z - ut$, as shown on page 601 in the appendix, this equation can be transformed to

$$\frac{\partial c}{\partial t} = E \frac{\partial^2 c}{\partial \eta^2} \tag{14.19}$$

where the subscript A from c_A has been dropped for simplicity. Note this is the familiar diffusion equation where the diffusion coefficient D has been replaced by E, the dispersion coefficient. For a semi-infinite region, this transformed equation was solved earlier (Eq. 13.29 on page 457), which we plan to use here. However, the initial condition corresponding to Figure 14.9 is different from those in Figure 13.11. Here the initial condition is a surface layer of thickness z_0 that will have the solute concentration raised to a value of c_i as a pulse, thus representing a finite amount of dispersing species near the surface at the initial time. For example, it could be a chemical that got spilled and mixed with the soil or it could be some pesticides applied to the soil. This initial condition can be written symbolically as

$$c(z,t) = c_i \; (z \le z_0, \; t = 0)$$
$$= 0 \; (z > z_0, \; t = 0) \tag{14.20}$$

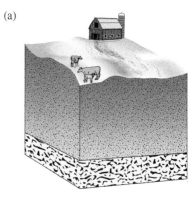

Due to this important difference in boundary condition from that studied in Section 13.4, the solution is not exactly the same as in that section but, as we will see, the solution to our problem here can be developed with the help of the solution process in Section 13.4.

The problem in Figure 14.9 can be treated as a combination of two problems, each having a concentration profile of c^* and c^{**}, respectively, of the dispersing species, as shown in Figure 14.10, such that

$$c = c^* - c^{**} \tag{14.21}$$

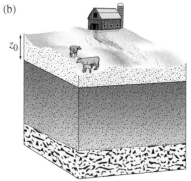

Note that subtraction of Figure 14.10b from Figure 14.10a does indeed lead to Figure 14.9. The first problem involving concentration c^* represents an initial concentration of material c_i extending from 0 to ∞. The governing equation is

$$\frac{\partial c^*}{\partial t} + u \frac{\partial c^*}{\partial z} = E \frac{\partial^2 c^*}{\partial z^2} \tag{14.22}$$

and the boundary conditions are

$$c^*(z \to \infty) = c_i$$
$$c^*(z = 0) = 0$$
$$c^*(t = 0) = c_i \tag{14.23}$$

The second problem involving concentration c^{**} also represents an initial concentration c_i of material, but extending from z_0 to ∞. The governing equation is

$$\frac{\partial c^{**}}{\partial t} + u \frac{\partial c^{**}}{\partial z} = E \frac{\partial^2 c^{**}}{\partial z^2} \tag{14.24}$$

and the boundary conditions are

$$c^{**}(z \to \infty) = c_i$$
$$c^{**}(z = z_0) = 0$$
$$c^{**}(t = 0) = c_i \tag{14.25}$$

Figure 14.10: Splitting up the problem shown in Figure 14.9 into two problems (a) and (b).

Using the transformation illustrated by Eq. 14.19, the first problem (Eq. 14.22) can be written as

$$\frac{\partial c^*}{\partial t} = E \frac{\partial^2 c^*}{\partial \eta^2} \tag{14.26}$$

which, together with the following boundary conditions for η (derived from Eq. 14.23),

$$c^*(\eta \to \infty) = c_i \tag{14.27}$$

$$c^*(\eta = 0) = 0 \tag{14.28}$$

is exactly equivalent to the problem in Section 13.4 with its solution given by Eq. 13.29. Thus, the solution to the first problem is

$$
\frac{0 - c^*}{0 - c_i} = \operatorname{erf}\left(\frac{\eta}{2\sqrt{Et}}\right)
$$

$$
= \operatorname{erf}\left(\frac{z - ut}{2\sqrt{Et}}\right) \tag{14.29}
$$

The solution to the second problem defined by Eq. 14.24 and Eq. 14.25 is almost identical to the first one, except for the distance z_0. If we first transform the distance variable as

$$z^* = z - z_0 \tag{14.30}$$

the second problem becomes identical to the first one and its solution is

$$
\frac{0 - c^{**}}{0 - c_i} = \operatorname{erf}\left(\frac{z^* - ut}{2\sqrt{Et}}\right) \tag{14.31}
$$

$$
= \operatorname{erf}\left(\frac{z - z_0 - ut}{2\sqrt{Et}}\right) \tag{14.32}
$$

Hence, the solution to the complete problem is

$$
c = c^* - c^{**}
$$

$$
= c_i\left[\operatorname{erf}\left(\frac{z - ut}{2\sqrt{Et}}\right) - \operatorname{erf}\left(\frac{z - z_0 - ut}{2\sqrt{Et}}\right)\right] \tag{14.33}
$$

Equation 14.33 is illustrated in Figure 14.11 at two different times, together with the initial uniform concentration distribution. Notice how the initial amount of solute flows (convects) down with time with a simultaneous dispersion (spreading) that increases with time. Since the solution assumes no reaction (such as decay), the total amount (area under the curve) remains constant with time. This solution does not include sorption whereby some of the solute "sticks" to the soil surfaces, which is described in the next section. However, this solution can represent, for example, transport of soluble salts through soil, such as nitrate ion that stays mostly dissolved in the water.

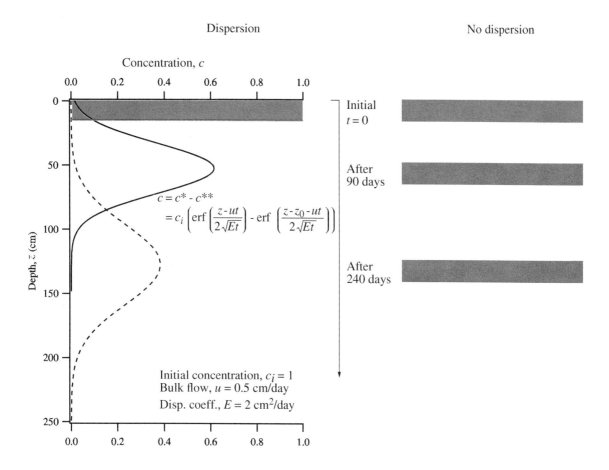

Figure 14.11: Movement and spreading of a layer of pollutant over time due to dispersion and convection (bulk flow) as predicted by Eq. 14.33. The concentration profiles shown on the left are at times 0, 90, and 240 days, respectively. On the right, for contrast, are how the profiles would have looked if there were no dispersion, i.e., all of the material moved uniformly with the bulk flow at 0.5 cm/day.

14.2.1 *Example: Transport of a Pesticide Spill into the Soil*

A pesticide spill occurred where the pesticide got incorporated in the soil up to a depth of 15 cm. This led to the initial concentration of pesticide in the soil to be 1 μg/cc. An average velocity of 0.5 cm/day due to percolation carries this pesticide into the soil.

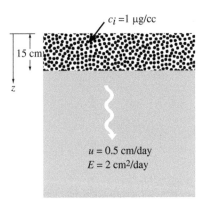

$c_i = 1$ µg/cc

15 cm

z

$u = 0.5$ cm/day
$E = 2$ cm²/day

Figure 14.12: Schematic for Example 14.2.1.

The dispersion coefficient is available at 2 cm²/day. Calculate 1) the concentration distribution in the soil after 90 and 240 days, respectively, 2) concentration at 2 m after 240 days, and 3) dispersive, convective, and total fluxes at any location. Assume no sorption or decay of the pesticide.

Solution

Understanding and formulating the problem *1) What is the process?* The pesticide dissolved in the water in soil flows with the water through the soil. However, this flow is not uniform (e.g., pores are not uniform in diameter) such that there is dispersion in addition to the flow. *2) What are we solving for?* Concentration and flux of pesticide in the soil as a function of position and time. *3) Schematic and given data:* A schematic is shown in Figure 14.12 with some of the given data superimposed on it. *4) Assumptions:* Dispersion coefficient remains constant.

Generating and selecting among alternate solutions *1) What solutions are possible?* For convective mass transfer, the solutions we have available are shown in the solution chart in Figure 14.24. *2) What approach is likely to work?* As we follow the solution chart in Figure 14.24, our problem has been given to us as unsteady, in a porous medium. It matches what we did in Section 14.2 in domain, governing equation, boundary conditions, and initial condition. Therefore, we can use this solution as is.

Implementing the chosen solution This problem is an application of Eq. 14.33. Thus, the concentration after 90 days is given by

$$c = (1)\left[\operatorname{erf}\left(\frac{z - (0.5)(90)}{2\sqrt{(2)(90)}}\right) - \operatorname{erf}\left(\frac{z - 15 - (0.5)(90)}{2\sqrt{(2)(90)}}\right)\right]$$

$$= \left[\operatorname{erf}\left(\frac{z - 45}{26.833}\right) - \operatorname{erf}\left(\frac{z - 60}{26.833}\right)\right]\ \mu g/cm^3$$

and that after 240 days is given by

$$c = (1)\left[\operatorname{erf}\left(\frac{z - (0.5)(240)}{2\sqrt{(2)(240)}}\right) - \operatorname{erf}\left(\frac{z - 15 - (0.5)(240)}{2\sqrt{(2)(240)}}\right)\right]$$

$$= \left[\operatorname{erf}\left(\frac{z - 120}{43.82}\right) - \operatorname{erf}\left(\frac{z - 135}{43.82}\right)\right]\ \mu g/cm^3 \qquad (14.34)$$

These are plotted in Figure 14.11. This calculation can be easily done on a spreadsheet where an error function is available.

2) The concentration value at $z = 200$ cm after 240 days can be calculated using Eq. 14.34 (from the same spreadsheet) as approximately 0.025 $\mu g/cm^3$.

3) From the c vs. z in Eq. 14.34, we can calculate the derivative, dc/dz numerically, i.e., as $\Delta c/\Delta z$, and thus the dispersive flux ($= -E\,dc/dz$). We can also calculate the convective flux ($= uc$), and the total flux ($= -E\,dc/dz + uc$). Here $E = 2$ cm^2/day and $u = 0.5$ cm/day. The calculated fluxes are shown in Figure 14.13.

Evaluating and interpreting the solution *1) Do the values (concentration profiles) make sense?* They do since they show a spread that is the characteristic of dispersion, move down with time (percolation or bulk flow), and their peaks are near 45 cm (= 0.5 cm/day × 90 days) and 120 cm (= 0.5 cm/day × 240 days), respectively. *2) What do we learn?* In this example, convective flux dominates dispersive flux. Also, interestingly, dispersive flux is positive (downward) in the lower regions while it is negative (upward!) in the upper regions. Dispersion here is in relation to the average movement.

14.2.2 Example: Drug Transport in a Liver

Elimination of drugs in the liver can be seen as a convection–dispersion process where the drug dissolved in the blood flows through the liver tissue, which can be modeled as a porous medium (Roberts and Rowland, 1985). Figure 14.14a shows blood vessels in the liver. As the blood travels through the liver, the dissolved drug reaches different lengths, i.e., disperses. Dispersion of the drug is due to variations in velocity and path lengths travelled by the elements of the blood and by branching and interconnection of the blood vessels in the tissue. A mass transfer schematic of the portion of the liver that is to the right of the dashed line in the illustration in Figure 14.14a is shown in Figure 14.14b. The removal of the drug in the tissue follows a first-order reaction,[1] with a reaction rate constant of 5×10^{-4} s^{-1}. For simplicity, we consider the boundary at $x = 0$ to be maintained at a drug concentration of $c = c_s$ from a continuous supply of drug. The bulk flow of the blood has a constant velocity of 0.27 mm/s and the dispersion coefficient of the drug in the direction of flow is 1.28×10^{-4} m^2/s.

1) For steady-state convection–dispersion of drug in the liver, the governing equation is given by

$$u\frac{dc}{dx} = E\frac{d^2c}{dx^2} - k''c \qquad (14.35)$$

Explain how we arrive at this equation starting from the general mass transfer governing equation, mentioning any missing or altered term(s) and their reason(s). 2)

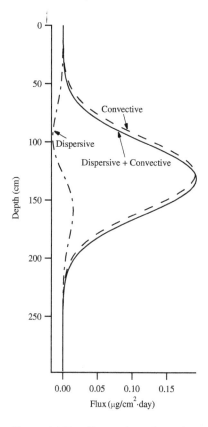

Figure 14.13: Convective, dispersive, and total fluxes calculated from the concentrations.

---

[1] Input data approximated from Roberts and Rowland (1985) and Izadifar and Alcorn (2012).

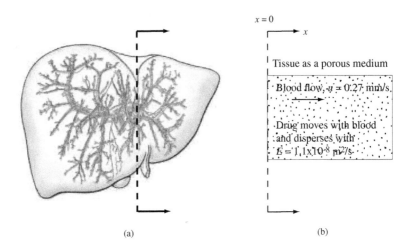

(a)   (b)

Figure 14.14: Blood vessels in a liver and a schematic of a one-dimensional convection–dispersion process in the liver.

For the situation when dispersion is negligible, solve the above equation to find the concentration profile of drug in the liver. 3) After plugging in the appropriate values, plot the concentration profile over the schematic in Figure 14.14b. 4) For the situation when dispersion is large, so that the effect of bulk flow (of blood) can be ignored, the problem reduces to one that is at a steady-state dispersion with reaction. Write the governing equation for this situation as a special case of Eq. 14.35. Then, write the solution to this equation (no need to derive the solution from scratch, but you must solve for constants in the solution), i.e., concentration profile as a function of position. *(Hint: At the far end of the liver, you can use the boundary condition $c(x \to \infty) = 0$. We have done this problem in the context of diffusion with reaction.)* 5) After plugging in the appropriate values, plot this concentration profile over the plot in step 3, labeling the solutions appropriately.

Solution

Understanding and formulating the problem *1) What is the process?* Drug dissolved in blood moves through the liver tissue but not uniformly, i.e., there is dispersion in addition to the flow. *2) What are we solving for?* Concentration profile under various conditions. *3) Schematic and given data:* A schematic is shown in Figure 14.14b with some of the given data superimposed on it. *4) Assumptions:* Dispersion coefficient remains constant.

Generating and selecting among alternate solutions *1) What solutions are possible?* For convective mass transfer, the solutions we have available are shown in the solution chart in Figure 14.24. *2) What approach is likely to work?* As we follow the solution chart in Figure 14.24, this has been given to us as a steady-state problem. Transport (flow and dispersion) of drug is occurring inside the porous medium, as opposed to over any particular surface. Of the two choices left, evaporation and "flowing stream," only "flowing stream" has flow with dispersion; thus, it is likely to be the solution of choice. As we will see, the flow and dispersion here are actually analogous to those in the flowing stream problem discussed in Section 14.1.

Implementing the chosen solution 2) For the situation when dispersion is negligible, the governing equation is

$$\underbrace{\frac{\partial c}{\partial t}^{0}}_{\text{steady state}} + \underbrace{u\frac{\partial c}{\partial x}}_{\text{convection}} = \underbrace{E\frac{\partial^2 c}{\partial x^2}^{0}}_{\text{no dispersion}} - kc$$

with the boundary condition

$$c(x = 0) = c_s$$

Integrating the governing equation once,

$$\ln c = -\frac{k}{u}x + C_1$$

Using the boundary condition

$$\ln c_s = C_1$$

Plugging in C_1,

$$\ln\left(\frac{c}{c_s}\right) = -\frac{k}{u}x$$

$$\frac{c}{c_s} = e^{-(k/u)x}$$

Plugging in values,

$$\frac{c}{c_s} = e^{(-5\times10^{-4}\ [1/\text{s}]/0.27\times10^{-3}\ [\text{m/s}])x}$$

$$= e^{-1.85x} \text{ where } x \text{ is in m}$$

3) This profile is plotted in Figure 14.15.

4) When dispersion is large such that bulk flow can be ignored,

$$\underbrace{\cancel{\frac{\partial c}{\partial t}}^{0}}_{\text{steady state}} + \underbrace{\cancel{u\frac{\partial c}{\partial x}}^{0}}_{\text{no convection}} = \underbrace{E\frac{\partial^2 c}{\partial x^2}}_{\text{dispersion}} -k''c$$

with two boundary conditions

$$c|_{x=0} = c_s$$
$$c|_{x\to\infty} = 0$$

A general solution to the above equation is given by

$$c = C_1 e^{\sqrt{k/E}\,x} + C_2 e^{-\sqrt{k/E}\,x}$$

Using the boundary condition at $x \to \infty$, C_1 has to be zero (otherwise the solution will blow up at $x \to \infty$); now using the boundary condition at $x = 0$, we get $C_2 = c_s$. Thus, the solution is

$$\frac{c}{c_s} = e^{-\sqrt{k/E}\,x}$$

Plugging in values,

$$\frac{c}{c_s} = e^{-\sqrt{(5\times10^{-4}\ [1/s])/(1.28\times10^{-4}\ [m^2/s])}\ x} = e^{-1.98x}$$

where x is in m. This profile is also plotted on Figure 14.15.

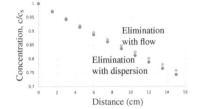

Figure 14.15: Concentration profiles for elimination with flow (top circles) and elimination with dispersion (bottom circles).

Evaluating and interpreting the solution   *1) Do the concentration profiles make sense?* The concentration profiles are reasonable, showing a decay. For the data used (velocity data is for rat liver whereas the other data are likely for human liver), the two profiles for flow and dispersion are close, signifying their effects would be similar for this set of data. *2) What do we learn?* Although a completely different physical system, the problem is analogous to convection–dispersion of pollutants in a stream (Section 14.1) by having the same three terms in the transport equation: bulk flow, dispersion, and reaction or elimination (boundary conditions, however, are different). Dispersion here is due to the blood path being tortuous through the tissue.

14.3 Convection–Dispersion in a Semi-Infinite Porous Solid: Inclusion of Sorption

In the previous section, we discussed convection–dispersion in a porous solid. We included convection and dispersion, but no reactions. Frequently, as applied to transport of contaminants through soil, contaminants can undergo sorption, where pollutants are removed from the aqueous phase (carrier fluid) and immobilized onto particles in the porous medium. Mechanisms of sorption of solutes to solid particles can be hydrophobic partitioning, electrostatic adsorption onto particle surfaces, and ion exchanges (Schnoor, 1996). In this section, we will include the sorption process during transport. In general, the effect of the sorption process will be to slow down the transport of the solutes (since they are retained by the solid particles).

As in the last section, we consider convection–dispersion in a semi-infinite region. As before, we will assume constant velocity of liquid flowing through the porous media, constant dispersion, and no decay. Our governing equation will now be only slightly modified to include a term due to sorption:

$$\underbrace{\frac{\partial c}{\partial t}}_{\text{Storage}} + \underbrace{u\frac{\partial c}{\partial z}}_{\text{Convection}} = \underbrace{E\frac{\partial^2 c}{\partial z^2}}_{\text{Dispersion}} - \underbrace{\frac{\partial c^{ad}}{\partial t}}_{\text{Adsorption}} \qquad (14.36)$$

where c is the concentration of dissolved solute in the liquid portion of the porous medium, but measured as per unit volume of the total porous medium. It is related to the dissolved concentration of solute, c^*, that is measured as per unit volume of the liquid only. For transport through soil, this relationship can be written as

$$c\left[\frac{\mu\text{g}}{\text{cm}^3 \text{ of soil}}\right] = c^*\left[\frac{\mu\text{g}}{\text{cm}^3 \text{ of water}}\right] * \theta_f\left[\frac{\text{cm}^3 \text{ of water}}{\text{cm}^3 \text{ of soil}}\right] \qquad (14.37)$$

where θ_f is the maximum water content of the soil. The quantity c^{ad} is the concentration of solutes adsorbed to the solid surfaces in the porous medium and is related to the dissolved solute concentration, c^*, as

$$c^{ad}\left[\frac{\mu\text{g}}{\text{cm}^3 \text{ of soil}}\right] = c^*\left[\frac{\mu\text{g}}{\text{cm}^3 \text{ of water}}\right] K^*\left[\frac{\text{cm}^3 \text{ of water}}{\text{g of soil}}\right] \rho_s\left[\frac{\text{g of soil}}{\text{cm}^3 \text{ of soil}}\right] \qquad (14.38)$$

where K^* is the distribution coefficient between the solute chemical and soil particles and ρ_s is the dry bulk density of soil. The total concentration of the solute chemical in the soil will be composed of the dissolved solute in the liquid and adsorbed solute on the solid surfaces, i.e., $c + c^{ad}$. Substituting c and c_{ad} into the governing equation,

$$\theta_f \frac{\partial c^*}{\partial t} + u\theta_f \frac{\partial c^*}{\partial t} = E\theta_f \frac{\partial^2 c^*}{\partial z^2} - \rho_s K^* \frac{\partial c^*}{\partial t} \tag{14.39}$$

and rearranging

$$\underbrace{\left(1 + \frac{\rho_s K^*}{\theta_f}\right)}_{R} \frac{\partial c^*}{\partial t} + u\frac{\partial c^*}{\partial z} = E\frac{\partial^2 c^*}{\partial z^2} \tag{14.40}$$

Defining R as the retardation factor, the governing equation becomes

$$\frac{\partial c^*}{\partial t} + \frac{u}{R}\frac{\partial c^*}{\partial z} = \frac{E}{R}\frac{\partial^2 c^*}{\partial z^2} \tag{14.41}$$

and the boundary conditions are

$$\begin{aligned} c^*(z,t) &= c_i^* \ (z \le z_0, \ t = 0) \\ &= 0 \ (z > z_0, \ t = 0) \end{aligned} \tag{14.42}$$

These governing equation and boundary conditions are identical in form to Eqs. 14.19 and 14.20, respectively. Therefore, the solution can be written from Eq. 14.33 as

$$c^*(z,t) = c_i^* \left[\mathrm{erf}\left(\frac{z - (u/R)t}{2\sqrt{(E/R)t}}\right) - \mathrm{erf}\left(\frac{z - z_0 - (u/R)t}{2\sqrt{(E/R)t}}\right) \right] \tag{14.43}$$

The total concentration of solute chemical in the porous medium is obtained by adding the concentration in the water to the concentration that is adsorbed on the solid surface:

$$\begin{aligned} c + c^{ad} &= (\theta_f + K^*\rho_s)c^* \\ &= \underbrace{(\theta_f + K^*\rho_s)c_i^*}_{c_i} \left[\mathrm{erf}\left(\frac{z - (u/R)t}{2\sqrt{(E/R)t}}\right) - \mathrm{erf}\left(\frac{z - z_0 - (u/R)t}{2\sqrt{(E/R)t}}\right) \right] \end{aligned}$$

$$\tag{14.44}$$

where c_i is the initial total concentration of solute chemical in the porous medium. The solution given by Eq. 14.44 will carry some features of Figure 14.11, in the sense that it also will have convection and dispersion. In contrast the solution from the previous section, Eq. 14.44 can represent transport of ions that do "stick" to the solid surface, for example, transport of ammonia and phosphorous ions that bind with aluminum and calcium in the soil.

14.4 Convection–Diffusion in a Stagnant Gas

In the previous two sections, we discussed convection–dispersion. We now pick up one important situation of convection–diffusion. Our goal is the same as before— to calculate concentration profiles and mass fluxes. This example will also serve to illustrate how molecular diffusion and convection in a gas are interrelated.[2] As an example, consider evaporation from a water surface and diffusion of the water vapor through the stagnant air column and eventually removed by the surrounding air from the top of the tube. At very low temperatures of water, the vapor pressure of water is low and the rate of evaporation is low. Water vapor moves slowly up the tube because of their thermal energy, that is, the molecular diffusion process. If, however, the water is at the boiling temperature, the situation is completely different. The constantly generated high volume of water vapor will rush up the tube, which is clearly a pressure-driven flow process or convection, in contrast with a slow diffusion process. At intermediate temperatures, both diffusion and convection are important, which is the subject of this section. Examples of such processes are many. A bioenvironmental example of diffusion through a stagnant gas is the evaporation and loss of water vapor from the soil surface when it is covered by mulch. The mulch maintains somewhat of a stagnant layer of air through which the water vapor moves from the soil surface into the surrounding air.

Consider steady-state diffusion of vapor A through the stagnant or motionless gas B, as shown schematically in Figure 14.16. The gaseous species B is not soluble in A in any significant quantity; thus there is no net flux of B into the liquid A. At the top of the tube, a large volume of gas (air) flows, maintaining the concentration of vapor A to be zero at the top. The governing equation for this convection–diffusion problem can be obtained starting from the general equation

$$\underbrace{\overset{0}{\cancel{\frac{\partial c_A}{\partial t}}}}_{\text{steady state}} + \underbrace{u\frac{\partial c_A}{\partial z}}_{\substack{\text{bulk flow or} \\ \text{convection}}} = \underbrace{D_{AB}\frac{\partial^2 c_A}{\partial z^2}}_{\text{diffusion}} + \underbrace{\overset{0}{\cancel{r_A}}}_{\text{no generation}} \tag{14.45}$$

and dropping the appropriate terms as shown. The source of the velocity u in the bulk flow term is due to continuous generation of vapor A, since the system is open at the top where vapor A is continually removed. This continuous generation of vapor "pushes" the bulk fluid upward. Also, for the gaseous species B, its bulk flow upward (due to the "push" from vapor A) and its diffusion downward (due to higher concentration of B near the top) balances exactly since there can be no net flux of B in liquid A (it is not soluble). This balancing of the fluxes causes B to stay stagnant. The final governing

---

[2] See the excellent illustration in Cussler (1997).

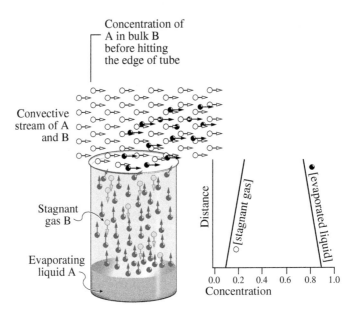

Figure 14.16: Schematic of a diffusive and convective process through a stagnant air column resulting from evaporation of a liquid.

equation is

$$u \frac{dc_A}{dz} \;=\; D_{AB} \frac{d^2 c_A}{dz^2} \tag{14.46}$$

The boundary conditions are given by

$$c_A \;=\; c_{A_1} \text{ at } z = z_1 \tag{14.47}$$

$$c_A \;=\; c_{A_2} \text{ at } z = z_2 \tag{14.48}$$

To solve Eq. 14.46, note that it is an ordinary, second-order differential equation with constant coefficients (see discussion later on u). The equation is rewritten as

$$\frac{d}{dz}\left(-D_{AB} \frac{dc_A}{dz} + u c_A\right) \;=\; 0 \tag{14.49}$$

Integrating,

$$-D_{AB}\frac{dc_A}{dz} + uc_A = n_A \tag{14.50}$$

where n_A is a constant. This constant, n_A, in Eq. 14.50 can also be physically interpreted as the total flux, since the two terms on the left side represent diffusive and convective flux, respectively. The coefficient u of the second term, the average velocity, is constant, as can be seen by approximating u under no net flux of B ($n_B = 0$), i.e., gas B is impermeable to liquid A, as

$$u = \frac{n_A + n_B}{c_A + c_B} = \frac{n_A}{c} \tag{14.51}$$

Substituting for u in Eq. 14.50, we can write

$$-D_{AB}\frac{dc_A}{dz} + \frac{n_A}{c}c_A = n_A$$

or

$$-D_{AB}\frac{dc_A}{dz} = n_A\left(1 - \frac{c_A}{c}\right)$$

Rearranging and integrating between the bottom and the top of the tube,

$$\int_{c_{A_1}}^{c_{A_2}} \frac{d(c_A/c)}{1 - c_A/c} = -\frac{n_A}{D_{AB}c}\int_{z_1}^{z_2} dz \tag{14.52}$$

$$\ln(1 - c_A/c)\Big|_{c_{A_1}}^{c_{A_2}} = -\frac{n_A}{D_{AB}c}(z_2 - z_1)$$

$$\ln\frac{(1 - c_{A_2}/c)}{(1 - c_{A_1}/c)} = -\frac{n_A}{D_{AB}c}(z_2 - z_1) \tag{14.53}$$

Since c_{A_1} and c_{A_2} are known, Eq. 14.53 can be used to find n_A, but we will postpone this for later. Integrating Eq. 14.52 between z_1 and any height z (where concentration is c_A), we can write an expression similar to Eq. 14.53 as

$$\ln\frac{(1 - c_A/c)}{(1 - c_{A_1}/c)} = -\frac{n_A}{D_{AB}c}(z - z_1) \tag{14.54}$$

Dividing the two sides of Eq. 14.54 by the two sides of Eq. 14.53,

$$\frac{\ln \frac{(1-c_A/c)}{(1-c_{A_1}/c)}}{\ln \frac{(1-c_{A_2}/c)}{(1-c_{A_1}/c)}} = \frac{z - z_1}{z_2 - z_1}$$

$$\ln \frac{(1 - c_A/c)}{(1 - c_{A_1}/c)} = \frac{z - z_1}{z_2 - z_1} \ln \frac{(1 - c_{A_2}/c)}{(1 - c_{A_1}/c)}$$

$$\frac{(1 - c_A/c)}{(1 - c_{A_1}/c)} = \left(\frac{(1 - c_{A_2}/c)}{(1 - c_{A_1}/c)} \right)^{\frac{z - z_1}{z_2 - z_1}} \tag{14.55}$$

Equation 14.55 provides concentration c_A of the evaporating species A as a function of z, the distance from the bottom. In terms of partial pressures, $c_A = p_A/RT$ and $c = P/RT$, we can write

$$\frac{(1 - p_A/P)}{(1 - p_{A_1}/P)} = \left(\frac{(1 - p_{A_2}/P)}{(1 - p_{A_1}/P)} \right)^{\frac{z - z_1}{z_2 - z_1}} \tag{14.56}$$

which provides partial pressure p_A as a function of z, the distance from the bottom. We can now use the concentration equations to calculate n_A, the flux of A, which is often of more practical interest. From Eq. 14.53, we can write the flux as

$$n_A = \frac{D_{AB}c}{z_2 - z_1} \ln \frac{(1 - c_{A_2}/c)}{(1 - c_{A_1}/c)} \tag{14.57}$$

In terms of partial pressures,

$$n_A = \frac{D_{AB}P}{RT(z_2 - z_1)} \ln \frac{(1 - p_{A_2}/P)}{(1 - p_{A_1}/P)} \tag{14.58}$$

These two equations for the flux n_A provide the means to calculate the net loss of A through the top of the stagnant gas column.

14.5   Convection–Diffusion over a Surface

In contrast with Sections 14.2 and 14.3 that dealt with convection–dispersion and Section 14.4 that dealt with convection–diffusion, all involving bulk fluid, this section will deal with convection–diffusion over a surface, i.e., convective transport of mass from a surface into the fluid. Recall that convective mass transfer over a surface is described by the convective mass transfer coefficient, h_m, using the equation

$$N_{A_{1-2}} = h_m A(c_1 - c_2) \tag{14.59}$$

where $N_{A_{1-2}}$ is the mass flow rate of species A from the surface (designated as 1) to the bulk fluid (designated as 2) and c is the corresponding concentrations of species A. Most of this section will deal with the details of what makes up h_m and how we can calculate it. The section will be completely analogous to the contents of Chapter 6 on convective transport of heat from a surface. As in Chapter 6 for heat transfer, mass transfer can be analyzed in terms of thin layers formed over the surface called the boundary layer. Convective mass transfer coefficients will be defined inside this mass transfer boundary layer and various formulas to predict the convective mass transfer coefficient will be introduced.

14.5.1 Concentration Profiles and Boundary Layers over a Surface

Development of a boundary layer for flow over a solid was explained in detail in Chapter 6. The fluid velocity changes from zero on the solid surface to close to free stream velocity at the edge of the boundary layer. Diffusion of a mass species near the surface is completely analogous to diffusion of energy near the surface. For example, consider dry air being blown over a wet surface. Water vapor at the wet surface first diffuses into the air and is eventually removed by the bulk flow. Like the thermal boundary layer, a concentration boundary layer will therefore develop. This concentration boundary layer is superimposed in Figure 14.17 on top of the velocity and thermal boundary layers shown in Figure 6.2 for a flat plate. Since the fluid in contact with the surface is at rest, the species in the liquid will come into equilibrium with the species in the solid. The concentration will vary from the value $c_{A,s}$ at the surface to the value $c_{A,\infty}$ in the free stream. Like the velocity and temperature variation, the change in concentration through space is also asymptotic such that the concentration is never exactly equal to $c_{A,\infty}$. The thickness of the concentration boundary layer, δ_{conc}, is defined as the distance from the surface at which concentration c_A is given by

$$\frac{c_{A,s} - c_{A,\delta_{conc}}}{c_{A,s} - c_{A,\infty}} = 0.99 \qquad (14.60)$$

which is the distance over which most of the concentration change takes place. Like the thermal boundary layer, the fluid can be considered to have two distinct concentration regions: the concentration boundary layer and outside the concentration boundary layer. In the boundary layer, concentration gradients are large, whereas, outside the boundary layer, the concentration gradients are small as the concentration at any point is within 1% of the free stream concentration. In other words, the effect of the flat plate on the mass concentration in the fluid is essentially restricted to the concentration boundary layer. Just as a thermal boundary layer exists only if there is a temperature difference between the surface and the flowing fluid, a concentration boundary layer

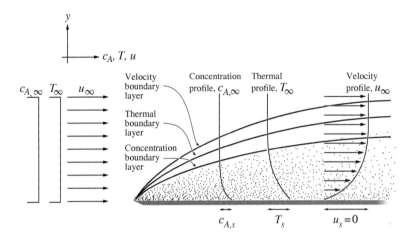

Figure 14.17: Schematic showing one example of a concentration profile and mass transfer boundary layer together with velocity and thermal boundary layers.

exists only if there is a concentration difference between the surface and the flowing fluid.

Like the thermal boundary layer, the thickness of the concentration boundary layer can be related to that of the velocity boundary layer using the characteristic number Sc or Schmidt number, that is analogous to the Prandtl number, Pr (see Section 6.2 on heat transfer). The Schmidt number is defined as

$$Sc = \frac{\nu}{D_{AB}} \qquad (14.61)$$

For a flat plate, the relationship between the two boundary layer thicknesses is

$$\frac{\delta_{velocity}}{\delta_{conc}} = Sc^{1/3} = \left(\frac{\text{Momentum diffusivity}}{\text{Mass diffusivity}} \right)^{1/3} \qquad (14.62)$$

or

$$\delta_{conc} = \frac{\delta_{velocity}}{Sc^{1/3}}$$

which implies that mass transfer boundary layer thickness, like other boundary layer thicknesses, decreases as velocity increases.

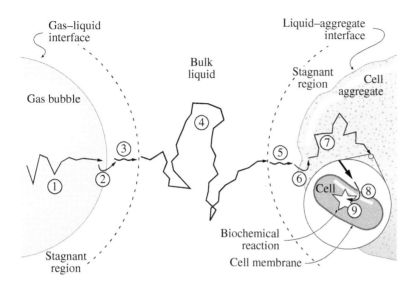

Figure 14.18: Boundary layer combined with other resistances in oxygen transfer from a bubble to a microorganism. Adapted from Bailey and Ollis. © 1986 McGraw-Hill, Inc. Reproduced with permission from McGraw-Hill, Inc.

14.5.2 *BioConnect: Boundary Layer in Oxygen Transport from a Gas Bubble to a Cell or an Organism*

As an example of a boundary layer, consider the situation where oxygen is transferred from a rising air bubble into a liquid phase containing cells (Bailey and Ollis, 1986). The oxygen must pass through a series of transport resistances, the relative magnitudes of which depend on bubble (droplet) hydrodynamics, temperature, cellular activity and density, solution composition, interfacial phenomena, and other factors. The total resistance arises from various combinations of the following resistances (Figure 14.18):

1. Diffusion from bulk gas to the gas–liquid interface.

2. Movement through the gas–liquid interface.

3. Diffusion of the solute through the relatively unmixed liquid region adjacent to the bubble into the well-mixed bulk liquid. *This is the boundary layer region just studied.*

4. Transport of the solute through the bulk liquid to a second relatively unmixed liquid region surrounding the cells.

5. Transport through the second unmixed liquid region associated with the cells. *This is the second boundary layer region.*

6. Diffusive transport into the cellular floc, mycelia, or soil particle.

7. Transport across cell envelope and to intra-cellular reaction site.

When the organisms take the form of individual cells, the sixth resistance disappears. Microbial cells themselves have some tendency to adsorb at interfaces. Thus, cells may preferentially gather at the vicinity of the gas-bubble–liquid interface. Then, the diffusing solute oxygen passes through only one unmixed liquid region and no bulk liquid before reaching the cell. In this situation, the bulk dissolved O_2 concentration does not represent the oxygen supply for the respiring microbes.

14.5.3   Convective Mass Transfer Coefficient Defined

The convective mass transfer coefficient over a surface is defined quite similarly to the convective heat transfer coefficient. For fluid flowing over a surface, the fluid layer in contact with the surface is at rest, as illustrated in Figure 14.17. Thus, at the surface ($y = 0$), mass transfer *in the fluid* is by diffusion only and the flux is given by

$$\begin{array}{c}\text{Diffusive mass}\\\text{flux in the fluid}\end{array} = \left. -D_{AB}\frac{\partial c_A}{\partial y}\right|_{y=0,\text{ in fluid}} \tag{14.63}$$

where c_A is the concentration of species A *in the fluid* at a distance y from the surface and D_{AB} is the diffusivity of species A *in the fluid*. However, the same mass flux can be written in terms of a surface convective mass transfer coefficient, h_m, as

$$\begin{aligned}\text{Convective mass flux} &= h_m(c_{A,y=0,\text{in fluid}} - c_{A,\infty})\\ &= h_m(c_{A,s} - c_{A,\infty})\end{aligned} \tag{14.64}$$

where $c_{A,s}$ is the concentration at the surface *in the fluid* and $c_{A,\infty}$ is the concentration in the fluid far from the surface. Since Eqs. 14.63 and 14.64 represent the same mass flux,

$$\left. -D_{AB}\frac{\partial c_A}{\partial y}\right|_{y=0,\text{in fluid}} = h_m(c_{A,s} - c_{A,\infty})$$

which leads to the defining equation for h_m as

$$h_m = \frac{\left. -D_{AB}\frac{\partial c_A}{\partial y}\right|_{y=0,\text{in fluid}}}{c_{A,s} - c_{A,\infty}} \tag{14.65}$$

The concentration profile $c_A(y)$ can be obtained by solving the governing equation for convection (Eq. 14.1). Knowing $c_A(y)$, we can calculate $\partial c_A / \partial y$ in Eq. 14.65 and therefore can get h_m, the mass transfer coefficient. Since $c_A(y)$ is obtained by solving diffusion and flow equations, the *convective mass transfer coefficient includes contributions from both bulk flow and diffusion.*

14.5.4 Significant Parameters in Convective Mass Transfer

To obtain the convective mass transfer coefficient h_m for many practical situations, one would like to obtain a solution to the convection–diffusion equation or simply the convection equation (Eq. 14.1). The governing equation for convective mass transfer with its appropriate boundary conditions can be solved in much the same way as the governing equation for convective heat transfer. In order to solve this equation, we need the fluid flow equations which describe the velocity u as it influences the rate of mass transfer. This solution process for convective mass transfer is completely analogous to the process described for convective heat transfer, so that we will not describe the process in the same detail. The convection equation (Eq. 14.1) for steady two-dimensional flow over a flat plate can be written in non-dimensional form following Section 6.5 for heat transfer as

$$u^* \frac{\partial c_A^*}{\partial x^*} + v^* \frac{\partial c_A^*}{\partial y^*} = \frac{D_{AB}}{u_\infty L} \frac{\partial^2 c_A^*}{\partial y^{*2}} \qquad (14.66)$$

By defining a non-dimensional quantity, the Schmidt number Sc as

$$\text{Sc} = \frac{\mu}{\rho D_{AB}} = \frac{\nu}{D_{AB}} = \frac{\text{momentum diffusivity}}{\text{mass diffusivity}} \qquad (14.67)$$

eq. 14.66 can be rewritten (like Eq. 6.22) as

$$u^* \frac{\partial c_A^*}{\partial x^*} + v^* \frac{\partial c_A^*}{\partial y^*} = \frac{1}{\text{Re}_L \text{Sc}} \frac{\partial^2 c_A^*}{\partial y^{*2}} \qquad (14.68)$$

The solution to this equation, together with the momentum equation that provides for the velocities, u^* and v^*, can be written in functional form as

$$c_A^* = f(x^*, y^*, \text{Re}_L, \text{Sc}) \qquad (14.69)$$

Using the definition of the convective mass transfer coefficient (Eq. 14.65), h_m can be written in terms of a non-dimensional concentration gradient as

$$
\begin{aligned}
h_m &= \frac{-D_{AB} \frac{\partial c_A}{\partial y}\Big|_{y=0,\text{in fluid}}}{c_{A,s} - c_{A,\infty}} \\
&= \frac{D_{AB}}{L} \frac{\partial \left(\frac{c_A - c_{A,s}}{c_{A,\infty} - c_{A,s}} \right)}{\partial \left(\frac{y}{L} \right)}\Bigg|_{y=0} \\
&= \frac{D_{AB}}{L} \frac{\partial c_A^*}{\partial y^*}\Big|_{y^*=0}
\end{aligned}
\tag{14.70}
$$

which can be rewritten using Eq. 14.69 as

$$
\begin{aligned}
\frac{h_m L}{D_{AB}} &= \frac{\partial c_A^*}{\partial y^*}\Big|_{y^*=0} \\
&= f(x^*, \text{Re}_L, \text{Sc})
\end{aligned}
\tag{14.71}
$$

This shows that the mass transfer coefficient h_m is a function of position, x^*, in general. The average mass transfer coefficient would not depend on the position and would be given by

$$
\frac{h_{mL} L}{D_{AB}} = f(\text{Re}_L, \text{Sc})
\tag{14.72}
$$

The subscript L in h_{mL} is generally dropped, as averaging is typically assumed. This equation shows the functional relationship between the mass transfer coefficient h_m and the non-dimensional parameters Re_L and Sc. The non-dimensional quantity on the left of Eq. 14.72 is defined as the Sherwood number having the symbol Sh (sometimes referred to as the mass transfer Nusselt number):

$$
\begin{aligned}
\text{Sh} &= \frac{h_m L}{D_{AB}} \\
&= \frac{\partial c_A^*}{\partial y^*}\Big|_{y^*=0}
\end{aligned}
$$

using Eq. 14.70. Thus we see that the Sherwood number, like the Nusselt number, is a non-dimensional concentration gradient. The non-dimensional parameters defined in this section are summarized and compared with their heat transfer equivalents in Table 14.1.

14.5.5 Calculation and Physical Implications of Convective Mass Transfer Coefficient Values

Convective mass transfer coefficients over a surface can be obtained in an analogous manner to convective heat transfer coefficients by either solving the convection–diffusion mass transfer equation or through experiments. Since the solution to the convection–diffusion mass transfer equation would result in formulas identical in form to those of heat transfer, the formulas in Section 6.6 on page 193 can be used for convective mass transfer over a surface with the substitution of appropriate dimensionless parameters for mass transfer, as described in Table 14.1. Note that the The Schmidt number, Sc, in that table is the equivalent in mass transfer of the Prandtl number, Pr, from heat transfer. In the Schmidt number, mass diffusivity replaces thermal diffusivity. The Schmidt number compares the rates of momentum (or viscous) diffusion with mass diffusion. The Sherwood number, Sh, in mass transfer replaces the Nusselt number, Nu, in heat transfer. As mentioned earlier, the Sherwood number compares the diffusive and convective resistances *(both in fluid)* over a surface. The diffusive resistance in the Sherwood number is defined in terms of mass diffusivity, in contrast with the conductive resistance in Nusselt number defined in terms of thermal conductivity (not thermal diffusivity). The Grashof number for mass transfer is defined in terms of the density difference that drives the flow in contrast with the temperature difference (that drives the flow in heat transfer) used in the heat transfer Grashof number. The Rayleigh number for mass transfer is defined in analogy with the Rayleigh number for heat transfer.

For example, for flow over a sphere (Figure 14.19), the formulas for mass transfer coefficients can be written from Eqs. 6.51 and 6.52 on page 202 for the heat transfer coefficient. Thus, for natural convection mass transfer, we have

$$Sh_D = 2 + 0.43 Ra_{mD}^{\frac{1}{4}} \quad \text{for } 1 < Ra_{mD} < 10^5, \text{ Sc} \cong 1 \tag{14.73}$$

For forced convection mass transfer over a sphere, the correlation would be

$$Sh_D = 2 + \left(0.4 Re_D^{1/2} + 0.06 Re_D^{2/3}\right) Sc^{0.4} \tag{14.74}$$

$$\text{for } 3.5 < Re_D < 7.6 \times 10^4 \text{ and } 0.71 < Sc < 380$$

In these equations, Nu, the Nusselt number for heat transfer, has been replaced by Sh, the Sherwood number. Likewise, $Ra = Gr \times Pr$, the Rayleigh number for heat transfer, has been replaced by $Ra_m = Gr_{AB} \times Sc$, the Rayleigh number for mass transfer and the Prandtl number, Pr, has been replaced by Schmidt number, Sc. See Table 14.1 for the complete list of non-dimensional parameters in mass transfer and their heat transfer equivalent.

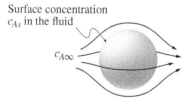

Surface concentration c_{As} in the fluid

$c_{A\infty}$

Figure 14.19: Mass transfer over a sphere.

To understand the implication of the correlations for mass transfer coefficients over a surface, we turn to refreshing our knowledge of the correlations for heat transfer coefficients over a surface, as discussed in Section 6.6 on page 193. Identical conclusions can be made about the implications of the correlations. For example, as Re increases, the thickness of the mass transfer boundary layer decreases. Since this boundary layer is the slower moving layer that provides the resistance to mass transfer, reduced thickness will mean a higher rate of mass transfer and a larger value of h_m, the mass transfer coefficient. Likewise, turbulence will increase the rate of mass transfer.

14.5.6 Example: Maximum Oxygen Uptake of a Microorganism

Calculate the maximum possible rate of oxygen uptake at 37°C of microorganisms having a diameter of 1 μm suspended in an unagitated aqueous solution. It is assumed that the surrounding liquid is saturated with O_2 from air at 1 atm abs pressure, leading to a saturation concentration of 2.26×10^{-4} kmol/m^3. It will be assumed that the oxygen is utilized by the microorganism much faster than it can diffuse to it. The microorganism has a density very close to that of water. The diffusivity of O_2 in water is 3.25×10^{-9} m^2/s. (Adapted from Geankoplis, 1993.)

Understanding and formulating the problem *1) What is the process?* Oxygen is transported from the bulk water to the surface of a microorganism that is suspended in water. *2) What are we solving for?* Flux of oxygen over the microorganism surface. *3) Schematic and given data:* A schematic is shown in Figure 14.20 with some of the given data superimposed on it. *4) Assumptions:* We assume there will be no relative motion between the microorganism surface and surrounding water. The concentration of O_2 is zero at the surface of the microorganism since oxygen is consumed faster than it can diffuse to the surface.

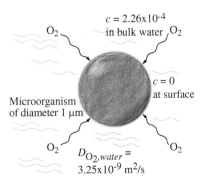

Figure 14.20: Schematic for Example 14.5.6.

Generating and selecting among alternate solutions *1) What solutions are possible?* For convective mass transfer, the solutions we have available are shown in the solution chart in Figure 14.24. *2) What approach is likely to work?* Going down this solution chart, we first note that it is a steady-state (there is no mention of variation with time) convection–diffusion problem, where the interest is over a fluid (microorganism) surface (as opposed to inside the bulk fluid, i.e., water). The solution is given by the convection formulas. We follow the steps outlined in Section 14.7 to calculate h_m. In implementing these steps, we see that the geometry is a sphere. The characteristic length for flow over a sphere is its diameter. It is neither natural nor forced. Since the density of the microorganism is close to that of water, it will not be moving by buoyant forces, so there is no velocity (Re = 0). For natural convection, the density difference across the

Table 14.1: Non-dimensional parameters in mass transfer and their heat transfer equivalents. To obtain the correlations for convective mass transfer coefficient, an appropriate non-dimensional parameter needs to be substituted in the equations in Section 6.6 on page 193. The quantity $\Delta\rho$ in the Grashof number for mass transfer is the density difference due to concentration gradient

Mass transfer	Heat transfer equivalent
Schmidt number	*Prandtl number*
$\text{Sc} = \dfrac{\mu/\rho}{D_{AB}}$ $= \dfrac{\text{momentum diffusivity}}{\text{mass diffusivity}}$	$\text{Pr} = \dfrac{\mu/\rho}{\alpha_{fluid}}$ $= \dfrac{\text{momentum diffusivity}}{\text{thermal diffusivity}}$
Sherwood number	*Nusselt number*
$\text{Sh} = \dfrac{h_m L}{D_{AB}} = \dfrac{\frac{L}{D_{AB}}}{\frac{1}{h_m}}$ $= \dfrac{\text{diffusive resistance}}{\text{convective resistance}}$	$\text{Nu} = \dfrac{h L}{k_{fluid}} = \dfrac{\frac{L}{k_{fluid}}}{\frac{1}{h}}$ $= \dfrac{\text{conductive resistance}}{\text{convective resistance}}$
Grashof number[1]	*Grashof number*
$\text{Gr}_{AB} = \dfrac{g\rho\Delta\rho L^3}{\mu^2}$	$\text{Gr} = \dfrac{\beta g\rho^2 L^3 \Delta T}{\mu^2}$
Rayleigh number	*Rayleigh number*
$\text{Ra}_m = \text{Gr}_{AB} \times \text{Sc}$	$\text{Ra} = \text{Gr} \times \text{Pr}$

[1]The quantity $\Delta\rho$ in the mass transfer Grashof number is the magnitude of the difference between the two densities across the boundary layer.

boundary layer (from water at the microorganism surface to bulk water) would be needed, but has not been provided. Perhaps we can treat this problem as forced convection with a limiting zero Reynolds number.

Implementing the chosen solution The calculation steps for the convection formulas are provided in Section 14.7. Thus, we start from the equation for forced convection over a sphere which, for heat transfer, is Eq. 6.52 and rewrite it for mass transfer by replacing the various parameters as (see Table 14.1)

$$\frac{h_m D}{D_{AB}} = 2 + \left(0.4 \text{Re}_D^{1/2} + 0.06 \text{Re}_D^{2/3}\right) \text{Sc}^{0.4} \tag{14.75}$$

where D is the diameter of the microorganism. For $\text{Re}_D \approx 0$, Eq. 14.75 reduces to

$$h_m = \frac{2 D_{AB}}{D}$$

which shows that the convective mass transfer coefficient is simply due to the contribution to transport from diffusion (limiting case of no bulk flow). Plugging in numbers, we get

$$\begin{aligned} h_m &= 2 \times \frac{3.25 \times 10^{-9} [\text{m}^2/\text{s}]}{1 \times 10^{-6} [\text{m}]} \\ &= 6.5 \times 10^{-3} \text{ m/s} \end{aligned}$$

The flux of O_2 to the microorganism is given by

$$\begin{aligned} n_{O_2} &= h_m \left(c_{O2,\infty} - c_{O2,\text{surface}}\right) \\ &= (6.5 \times 10^{-3} [\text{m/s}])(2.26 \times 10^{-4} [\text{kmol/m}^3] - 0) \\ &= 1.47 \times 10^{-6} \text{ kmol } O_2/\text{m}^2\text{s} \end{aligned}$$

Interpreting the solution *1) What do we learn?* As the microorganism grows bigger (radius increases), the surface area increases as the square of the radius, while the volume (thus likely its oxygen requirement) as the cube of the radius, i.e., faster than the surface area. Thus, the microorganism can grow only so big before oxygen becomes limiting in stagnant water. Agitation would increase the relative velocity, u_∞, between the fluid and the microorganism, thus increasing $\text{Re}_D = u_\infty D\rho/\mu$ in the right-hand side of Eq. 14.75, increasing h_m.

14.5.7 *Example: Convective Loss of Moisture from Soil Surface*

Estimation of evaporation from soil is normally very complicated. For simplicity, consider dry air blowing over a field of very wet silt loam soil having a flat and bare surface, i.e., no grass or other cover. Calculate the moisture loss per unit area per unit time (in $g/m^2 \cdot s$) from this soil surface under the following conditions: Length of the field along the wind direction is 100 m, soil surface and air temperature is 20°C, moisture content of saturated air at 20°C is 0.0173 kg water/m^3 saturated air, average wind speed is 20 km/hour, diffusivity of water vapor in air is 2.2×10^{-5} m^2/s, viscosity of air is 2×10^{-5} Pa·s, and density of air is 1.14 kg/m^3.

Solution

Understanding and formulating the problem *1) What is the process?* Water evaporates from the wet soil surface and the water vapor diffuses into the flowing air that convects it away. *2) What are we solving for?* Flux of water vapor from the wet surface into the air. *3) Schematic and given data:* Schematic is shown in Figure 14.21 with some of the given data superimposed on it. *4) Assumptions:* Fluid properties are constant and do not vary from the surface into the bulk fluid.

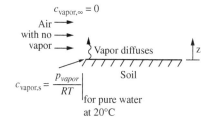

Figure 14.21: Moisture loss from a wet, bare surface due to dry air blowing over it.

Generating and selecting among alternate solutions *1) What solutions are possible?* For convective mass transfer, the solutions we have available are shown in the solution chart in Figure 14.24. *2) What approach is likely to work?* As we follow the solution chart in Figure 14.24, we see this is a steady-state problem (time is not mentioned in the problem). Moisture transport *over a (wet soil) surface* is our interest, as opposed to *inside the air*. Thus, we simply need to choose the appropriate formula for the mass transfer coefficient, h_m, following the steps in Section 14.7. Once h_m is known, the flux of moisture is $h_m(c_{vapor,s} - c_{vapor,\infty})$.

Implementing the chosen solution We simply follow the steps in Section 14.7. The geometry here is flat plate. The length L along the flow is 100 m. It is a forced convection problem as air velocity is provided. We are not interested in how the moisture loss varies along the flow, but we need the total moisture loss from the surface; thus, we need h_{mL}, not h_{mx}. For forced convection over a plate, the formula we choose depends on the type of flow, laminar or turbulent, to find which, we need to calculate the Reynolds number:

$$
\begin{aligned}
\mathrm{Re}_L &= \frac{uL\rho}{\mu} \\
&= \frac{20 \times 10^3\,\mathrm{m} \times 100\ \mathrm{m} \times 1.14\ \mathrm{kg/m^3}}{3600\ \mathrm{s} \times 2 \times 10^{-5}\ \mathrm{kg/m \cdot s}} = 3.167 \times 10^7
\end{aligned}
$$

Therefore, the flow is turbulent. The appropriate formula is chosen from Section 6.6 on page 195 with substitutions made for mass transfer as:

$$\text{Sh}_L = 0.0360 \text{Re}_L^{4/5} \text{Sc}^{1/3}$$

$$\frac{h_m L}{D_{AB}} = 0.0360 \text{Re}_L^{4/5} \left(\frac{\mu/\rho}{D_{AB}} \right)^{1/3}$$

Calculating the Schmidt number Sc as

$$\text{Sc} = \frac{\mu}{\rho D_{AB}} = \frac{2 \times 10^{-5} \text{ Pa} \cdot \text{s}}{(1.14 \text{ kg/m}^3)(2.2 \times 10^{-5} \text{ m}^2/\text{s})} = 0.80$$

Plugging in all the quantities,

$$\frac{h_{mL}(100 \text{ m})}{2.2 \times 10^{-5} \text{ m}^2/\text{s}} = 0.036 (3.16 \times 10^7)^{4/5} (0.80)^{1/3}$$

from which we calculate h_{mL} as

$$h_{mL} = 7.35 \times 10^{-3} \text{ m/s}$$

Moisture loss per unit area is $h_{mL}(c_{vapor,s} - c_{vapor,\infty})$. We calculate the vapor concentration at the surface, $c_{vapor,s}$, from the vapor pressure data on page 570 as

$$p_{vapor,s} = 0.02338 \times 10^5 \text{ N/m}^2$$

from which the vapor concentration (assuming perfect gas) is calculated as

$$\begin{aligned}
c_{vapor,s} &= \frac{p_{vapor} M}{RT} \\
&= \frac{0.02338 \times 10^5 [\text{N/m}^2] \ 18 \times 10^{-3} [\text{kg/mol}]}{8.314 [\text{J/mol} \cdot \text{K}] 293 [\text{K}]} \\
&= 0.0173 \text{ kg/m}^3
\end{aligned}$$

The bulk air is dry; thus

$$c_{vapor,\infty} = 0$$

leading to

$$\text{Moisture loss} = (7.35 \times 10^{-3} \text{ [m/s]})(0.0173 - 0)\frac{\text{kg}}{\text{m}^3}$$

$$= 0.127 \times 10^{-3}\frac{\text{kg}}{\text{m}^2\text{s}} = 0.127\frac{\text{g}}{\text{m}^2\text{s}}$$

Evaluating and interpreting the solution *1) Does the flux value make sense?* Whether
0.1 g/m$^2 \cdot$ s is a reasonable value is hard to determine as such numbers are not
easily available from experiment. An important assumption here is that the soil
surface continues to stay very wet, i.e., saturated.

14.5.8 *Example: Reduction of Moisture Loss Using a Cover*

Consider the same very wet soil surface at 20°C as in Example 14.5.7 but now with
mulch (loose straw) of thickness 2.5 cm on it. Because the mulch maintains a layer
of stagnant air over the soil surface, we cannot calculate evaporation from the soil
surface in the same manner as for Example 14.5.7. Instead, the water vapor from the
soil surface is now considered to diffuse through the air held stagnant by the mulch.
Assume flow of dry air over the mulch maintains a water vapor concentration of zero
at the top of the mulch. The system is at atmospheric pressure of 1 atm, and the vapor
pressure of water at 20°C is 0.0231 atm (1 atm = 1.013×10^5 N/m^2). The molecular
weight of water is 18 g/mol. 1) Calculate the loss of moisture per unit area of soil (in
g of water per m^2 per second) for this mulched condition. 2) Compare the evaporative
water loss for the mulched condition you calculated in question 1 with evaporative loss
from bare soil (no mulch) calculated in Example 14.5.7, and explain the difference.

Solution

Understanding and formulating the problem *1) What is the process?* Water evaporates from the wet soil surface and the water vapor diffuses through the stagnant
air column maintained by the mulch. *2) What are we solving for?* Flux of water vapor from the wet surface into the air. *3) Schematic and given data:* A
schematic is shown in Figure 14.22 with some of the given data superimposed
on it. *4) Assumptions:* Temperature and fluid properties remain constant from
soil surface to the top surface of mulch.

Generating and selecting among alternate solutions *1) What solutions are possible?* For convective mass transfer, the solutions we have available are shown
in the solution chart in Figure 14.24. *2) What approach is likely to work?* As

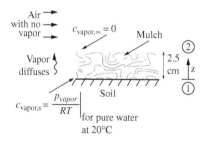

Figure 14.22: Moisture loss from a wet
surface with stagnant air maintained by
mulch due to dry air blowing over.

we follow the solution chart in Figure 14.24, we see this is a steady-state problem (time is not mentioned in the problem description). Water vapor transport is occurring inside the fluid, i.e., stagnant air, in contrast with Example 14.5.7, where transport was over a (soil) surface. Thus, this is a case of evaporation in stagnant air and we simply need to apply the formula derived for this situation.

Implementing the chosen solution The flux of water vapor through the stagnant air is given by

$$N_{vapor,z} = \frac{D_{vapor,air}\, P}{RT\,(z_2 - z_1)} \ln \frac{1 - p_{vapor,2}/P}{1 - p_{vapor,1}/P}$$

Here

$$P = \text{total pressure} = 1\ \text{atm} = 1.013 \times 10^5\ \text{N/m}^2$$
$$T = \text{temperature of the system} = 293\ \text{K}$$
$$p_{vapor,1} = \text{vapor pressure at } 20^\circ\text{C}$$
$$= 0.02338 \times 10^5\ \text{N/m}^2$$
$$= 0.02338 \times 10^5\ \text{N/m}^2 \frac{1\ \text{atm}}{0.101325 \times 10^6\ \text{N/m}^2} = 0.0231\ \text{atm}$$
$$p_{vapor,2} = 0$$
$$z_2 - z_1 = 2.5 \times 10^{-2}\ \text{m}$$

Plugging into the equation for $N_{vapor,z}$,

$$N_{vapor,z} = \frac{2.2 \times 10^{-5}\frac{\text{m}^2}{\text{s}} \times 1.013 \times 10^5\frac{\text{N}}{\text{m}^2}}{8.314\frac{\text{J}}{\text{mole K}}(293\ \text{K})\left(2.5 \times 10^{-2}\text{m}\right)} \ln \frac{1 - 0/1}{1 - 0.0231/1}$$
$$= 0.000855\frac{\text{mol}}{\text{m}^2 \cdot \text{s}}$$
$$= 0.0154\frac{\text{g}}{\text{m}^2\text{s}}$$

Evaluating and Interpreting the solution *1) Does the flux value make sense?* This is hard to say, as in the previous example, but it is consistent with the previous example in that the flux value is lower as there is the added resistance to moisture transfer of the stagnant air. An important assumption here is that the soil surface continues to stay very wet, i.e., saturated. Note that this flux value (0.0154

m^2/s) is higher than what would be obtained by considering pure diffusion, i.e., Fick's law:

$$N_{vapor,z} = -D_{vapor,z} \frac{c_{vapor,2} - c_{vapor,1}}{z_2 - z_1}$$

$$= -2.2 \times 10^{-5} m^2/s \frac{0 - 0.0173 \text{ kg/m}^3}{2.5 \times 10^{-2} m}$$

$$= 1.52 \times 10^{-5} \frac{\text{kg}}{\text{m}^2 \cdot \text{s}} = 0.0152 \frac{\text{g}}{\text{m}^2 \cdot \text{s}}$$

The stagnant air formula includes a small amount of bulk flow as vapor keeps getting generated. Thus, although the stagnant air formula leads only to a very slightly ($\approx$ 1%) higher flux than pure diffusion, it is, nevertheless, in the right direction. This provides us some additional confidence in what we calculated. By the way, the difference between the stagnant air formula and simple Fick's law will become more significant at a higher temperature when more vapor is present.

14.5.9 Convection–Diffusion of Heat and Mass (Simultaneous) over a Surface: Example of a Wet Bulb Thermometer

Heat and mass transfer are often simultaneous. In this section, we will illustrate simultaneous heat and mass transfer over a surface, using the wet bulb thermometer. A schematic of a wet bulb thermometer is shown in Figure 14.23. The mercury bulb of one of the thermometers is covered with a thin wet cloth or wick and the cloth is kept wet by a continuous supply of water. To keep a continuous flow of moving air, the wet bulb thermometer is swung about an axis. Evaporation cools the cloth and the temperature drops compared to another identical thermometer that is placed side-by-side but without the wet cloth. This second thermometer without the wet cloth measures the ambient air temperature and its temperature is referred to as the dry bulb temperature. The wet bulb temperature, together with the dry bulb temperature, can provide the humidity or the relative humidity of air, as we will see shortly. Psychrometric charts, such as Figure C.9 on page 569 in the appendix, can be used to obtain humidity or relative humidity from the dry and the wet bulb temperature.

We will now develop a convection heat and mass transfer analysis of the wet bulb thermometer. Heat transferred due to the temperature difference between the ambient air temperature (or the dry bulb temperature) and the wet bulb temperature causes evaporation of water in the wet bulb. At steady state, when the temperature of the wet bulb stays constant, the heat gained by the wet bulb balances the heat lost due to

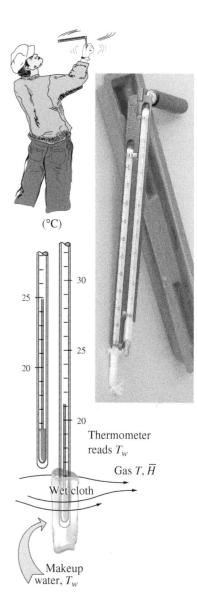

Figure 14.23: Schematic of a wet bulb thermometer (lower left figure). The picture on the right shows an actual thermometer and the upper left figure illustrates how the thermometer is rotated around the handle during operation.

evaporation:

$$\text{Convected heat gain} \quad = \quad \text{Evaporated heat loss}$$

$$= \quad \Delta H_{vap} \times \text{Convected mass loss}$$

where ΔH_{vap} is the latent heat of evaporation of water. Substituting formulas for convective heat and mass transfer:

$$\underbrace{h(T - T_s)A}_{\substack{\text{Convected} \\ \text{Heat gain}}} = \Delta H_{vap} \underbrace{h_m(c_s - c)A}_{\substack{\text{Convected} \\ \text{mass loss}}} \tag{14.76}$$

where subscript s stands for the surface of the wet wick. Concentration of water vapor in the air, c, is in kg of vapor per m^3 of dry air. Humidity of the air, $\bar{H}$, is defined as kg of water vapor per kg of dry air and is therefore related to c as

$$c = \bar{H}\rho_0$$

where ρ_0 is the density of dry air in kg of dry air per m^3 of dry air. Substituting in Eq. 14.76 and simplifying,

$$\frac{\bar{H}_s - \bar{H}}{T - T_s} = \frac{h}{h_m}\frac{1}{\Delta H_{vap}\rho_0} \tag{14.77}$$

where $\bar{H}_s$ is the humidity at the wet surface, or saturation humidity. Assuming the bulb is a cylindrical geometry and crossflow occurs over it, we can write the general equations for h and h_m as (see Section 6.6.5 on page 201)

$$\frac{hD}{k_{air}} = B\mathrm{Re}_D^n \mathrm{Pr}^{1/3}$$

$$\frac{h_m D}{D_{wa}} = B\mathrm{Re}_D^n \mathrm{Sc}^{1/3}$$

where D is the diameter of the wet bulb. From these two equations, we can calculate the ratio h/h_m as

$$\frac{h}{h_m} = \frac{k_{air}}{D_{wa}}\left(\frac{\mathrm{Pr}}{\mathrm{Sc}}\right)^{1/3} \tag{14.78}$$

$$= \frac{k_{air}}{D_{wa}}\left(\frac{D_{wa}}{\alpha_{air}}\right)^{1/3} \tag{14.79}$$

where k_{air} and α_{air} are the thermal conductivity and thermal diffusivity of air, respectively, and D_{wa} is the diffusivity of water vapor in air. Thus, substituting Eq. 14.79 in Eq. 14.77, we get

$$\frac{\bar{H}_s - \bar{H}}{T - T_s} = \frac{k_{air}}{D_{wa}} \left(\frac{D_{wa}}{\alpha_{air}} \right)^{1/3} \frac{1}{\Delta H_{vap}\rho_0} \qquad (14.80)$$

Note that the right hand side depends only on properties of air and diffusivity of vapor in air. It turns out that the right hand side has only a small variation over a large temperature range and can be treated approximately as a constant. Thus, Eq. 14.80 can be used to relate humidity $\bar{H}$ to the wet bulb temperature, T_s, all other variables being known.

14.6 Chapter Summary—Convective Mass Transfer

- **Convection–Dispersion in Infinite (page 485) and Semi-Infinite Region (page 493)**

    1. For an infinite region in a fluid, the solution is given by Eq. 14.11.

    2. For a semi-infinite region in a porous medium, the solution is given by Eq. 14.33.

- **Convection–Diffusion through a Stagnant Gas (page 505)**

    1. When a gas is diffusing through another stagnant gas, it is still a diffusion–convection situation. The flux is given by Eq. 14.58 and the concentration is given by Eq. 14.55.

- **Convection–Diffusion over a Surface or Convective Mass Transfer Coefficient (page 509)**

    1. The convective mass transfer resistance of fluid flowing over a solid surface is restricted to a thin layer on the surface where the fluid moves relatively slowly. This layer is called a mass transfer boundary layer.

    2. The convective mass transfer coefficient, h_m, represents the mass transfer resistance of the boundary layer.

    3. h_m includes the effect of diffusion in the fluid and flow. It therefore also depends on the flow parameter Re. Thus h_m should not be confused with a material property such as the mass diffusivity, D.

4. In complete analogy to the heat transfer coefficient, h, the many significant variables such as the flow velocity which h_m depends on are grouped into similar dimensionless parameters. The new dimensionless parameters are Sherwood number (replacing Nusselt number), Schmidt number (replacing Prandtl number), and Grashof number for mass transfer.

5. Since the heat and mass transfer coefficients are completely analogous, the values of h_m for a situation are given by the same formulas as for the heat transfer coefficient (Section 6.6, page 193) with appropriate substitution of parameters noted in Table 14.1.

14.7   Problem Solving—Convective Mass Transfer

The map in Figure 14.24 shows how the different scenarios covered in this chapter relate to each other. One wonders about the apparent loss of complete analogy to heat transfer. The analogy remains. The scenarios that are new here are also possible for heat transfer—just that the mass transfer applications of the same physics are more common. Boundary layer and convective mass transfer coefficient related computations are identical in form to those in Chapter 6, with appropriate substitutions from Table 14.1. The types of problems that can be solved here are

Convection–diffusion over a surface

▶ **Calculate flux due to diffusion/dispersion and flow**   This is done using Table 10.1.

▶ **Calculate boundary layer thicknesses.**   This is done using Eq. 14.62.

▶ **Calculate mass flow for a flat plate when h_m is changing with x.**   Obtain flux at x as $h_{mx}(c_{\text{surface}} - c_\infty)$. This is flux over a strip at x. Integrate this mass flux over the entire area, i.e., all strips.

▶ **Calculate h_m.**   Calculation steps for h_m can be as follows:

1. Choose geometry (flat plate, cylinder, sphere).

2. Find L based on dimensions provided (L for a flat plate is the total length along the flow; for a cylinder or sphere, L is the diameter).

3. Note natural or forced (velocity not provided may be natural).

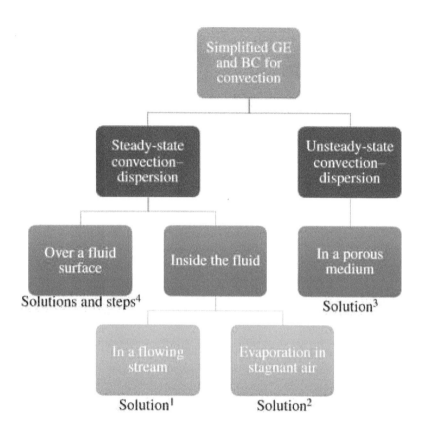

Figure 14.24: Problem solving in convection. Final solutions are on following pages:
[1]page 488, [2]page 496, and [3]page 508. [4]Steps for calculating h_m are on page 526.

4. We need h_{mx} if we desire h_m at a location x. We need h_{mL} if we need the average over the entire length L.

5. Calculate Reynolds number to choose between laminar/turbulent. Velocity u_∞ is velocity in the bulk flow away from the surface.

6. Choose the appropriate equation from Section 6.6 for heat transfer and substitute the quantities for mass transfer (see Table 14.1 on page 517), i.e., $h_m L / D_{AB}$ for hL/k, $\mu/(\rho D_{AB})$ for $\mu c_p/k$, and $\Delta\rho$ for $\rho\beta\Delta T$ in Grashoff number.

7. All properties in the equations for h_m are properties of fluid flowing over the surface of interest. Temperature and concentration variation of properties across the boundary layer are typically ignored in mass transfer.

▶**Calculate how h_m varies with different parameters.**   This is calculated from the respective equation.

▶**Perform mass balance using h_m.**   Once calculated, h_m can also be used to perform mass balance. Two examples are

- Steady state. As shown in Figure 14.25, by performing mass balance at steady state, we can write

$$\text{Rate of mass generated} \quad = \quad -\text{Rate of mass in by convection}$$
$$-r_A \times \text{Volume} \quad = \quad -h_m A(c_\infty - c_{\text{surface}})$$

where r_A is the magnitude of the rate of mass consumption; note the negative sign in front denoting consumption. Note also the concentration on the convective mass transfer formula is switched to make the mass flow from *surroundings* to *surface*, i.e., mass flow In. If r_A is known, h_m or other items can be calculated.

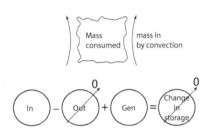

Figure 14.25: Steady state with generation.

- Unsteady state. As shown in Figure 14.26, by performing mass balance at unsteady state (without mass generation/depletion), we can write

$$-\text{Rate of mass out by convection} \quad = \quad \text{Rate of change in mass storage}$$
$$-h_m A(c_{\text{vapor,surface}} - c_{vapor,\infty}) \quad = \quad \frac{dw}{dt}$$

where w is the mass contained in the domain.

Convection–diffusion through a stagnant gas

▶**Calculate concentration profile or flux.**   Concentration profile is calculated using Eq. 14.55 and flux using 14.57.

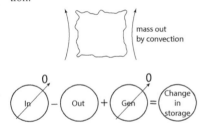

Figure 14.26: Unsteady state with generation.

Convection–dispersion in a fluid

▶**Calculate concentration profile or flux**   Concentration profile is calculated using Eq. 14.11 and flux using the general definition of dispersive or total flux provided in Table 10.1 on page 382.

Convection–dispersion in a porous medium

▶**Calculate concentration profile or flux** Concentration profile is calculated using Eq. 14.33 and flux using the general definition of dispersive or total flux provided in Table 10.1 on page 382. Flux will likely be easier to evaluate numerically, as discussed in Example 14.2.1.

14.8 Concept and Review Questions

1. Can you have dispersion without convection or bulk flow? Explain.

2. Can you think of a heat transfer analog of the convection–dispersion in an infinite fluid? Hint: Consider thermal pollution.

3. In Section 14.1 (page 488), plot the concentration profile for the general case of convection and dispersion, when there is no decay ($k'' = 0$).

4. In convection–dispersion in a porous solid, does the dispersion coefficient depend on the magnitude of the average velocity, u? How?

5. In pollutant transport through soil, what role do macropores play?

6. How does sorption in the porous matrix affect the transport of a species through the porous matrix? Explain qualitatively.

7. What is meant by a concentration boundary layer? How is its thickness defined?

8. Are the velocity and thermal boundary layers always thicker than the concentration boundary layer? Explain.

9. Give an example of natural convection mass transfer, identifying the two densities, the difference between which causes natural convection.

10. What dimensionless group in natural convection mass transfer substitutes for the Reynolds number in forced convection?

11. What are the mass transfer analogs of the Grashoff and Prandtl numbers? Describe their physical significances.

12. The analysis of the wet bulb thermometer has several assumptions. It is important to understand these assumptions in the context of our simple analysis.

    (a) Why can you treat the surface concentration of moisture as a constant? Is this a good assumption if you keep reading the thermometer for days without adding water?

(b) What is the reason the temperature reaches steady state?

(c) As we rotate the psychrometer, would the dry bulb temperature change? Why or why not?

(d) The faster we rotate, the higher is the relative velocity between the air and the thermometer. Thus, according to the formulas for the heat transfer coefficient, the rate of heat transfer should keep increasing. Would our temperature readings depend on how fast we rotate the psychrometer?

13. You can have a mass transfer coefficient over a surface when there is no flow (choose one): a) Yes; b) No; c) Sometimes.

14. If there were no velocity boundary layer (velocity on the solid surface is the same as bulk velocity, due to slip on the surface), the mass transfer coefficient would be (choose one): a) Very small; b) Infinite; c) Undefined.

15. On a flat plate, due to the presence of flow, the rate of transport is faster because of (choose one): a) Continuous removal of diffused material that maintains a higher gradient; b) Turbulence caused by the fluid; c) Not always true.

16. When the size of a material increases, the time required for the same amount of mass (or heat) transfer increases as (choose one): a) Square of the characteristic length; b) Square root of the characteristic length; c) Proportional to (i.e, linear with) the characteristic length.

17. Figure 14.27 shows three situations where a beaker is partially filled with water and is open to the atmosphere (1 atm). For each situation, pick the most appropriate approach to calculate the flux of water vapor from the beaker into the atmosphere from the following list: 1) Fick's law (FL); 2) using bulk flow with velocity u (BF); and 3) convection–diffusion equation derived for stagnant gas (CD).

18. Show that, although the units for the heat transfer coefficient (W/m^2K) look quite different from those for the mass transfer coefficient (m/s), they are still completely analogous to each other.

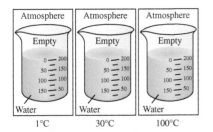

Figure 14.27: Beakers, partially filled with water, at three different temperatures.

Further Reading

Arogo, J., R. H. Zhang, G. L. Riskowski, L. L. Christianson, and D. L. Day. 1999. Mass transfer coefficient of ammonia in liquid swine manure and aqueous solutions. *Journal of Agricultural Engineering Research* 73(1): 77–86.

Bailey, J. E. and D. F. Ollis. 1986. *Biochemical Engineering Fundamentals.* McGraw-Hill, New York.

Clark, M. M. 1996. *Transport Modeling for Environmental Engineers and Scientists.* John Wiley & Sons, New York.

Cussler, E. L. 1997. *Diffusion Mass Transfer in Fluid Systems.* Cambridge University Press, Cambridge, UK.

Enfield, C. G. 1982. Approximating pollutant transport to groundwater. *Ground Water* 20(6):711–722.

Geankoplis, C. J. 1993. *Transport Processes and Unit Operations.* P T R Prentice Hall, Inc., Englewood Cliffs, NJ.

Gish, T. J. and W. A. Jury. 1983. Effect of plant roots and root channels on solute transport. *Trans. of the ASAE* 26(2):440–444, 451.

Izadifar, M. and J. Alcorn. 2012. Mass transfer modeling of hepatic drug elimination using local volume averaging approach. *Frontiers in Heat and Mass Transfer* 3(3):1–7.

Johnson, W. B., R. C. Sklarew, and D. B. Turner. 1976. *Urban Air Quality Simulation Modeling, in Air Pollution: Air Pollutants, Their Transformation, and Transport.* A.C. Stern, ed. Academic Press, New York.

Kasting, G. B. and N. D. Barai. 2003. Equilibrium water sorption in human stratum corneum. *Journal of Pharmaceutical Sciences.* 92(8): 1624–1631.

Knox, R. C., D. A. Sabatini, and L. W. Cantor. 1993. *Subsurface Transport and Fate Processes.* Lewis Publishers, Boca Raton, FL.

Leij, F. J., Jiri Simunek, D. L. Suarez, and M. Th. Van Genuchten. 1999. Nonequilibrium and multicomponent transport models. In *Agricultural Drainage* by R.W. Skaggs and J. van Schilfgaarde, American Society of Agronomy, Inc., Madison, Wisconsin, pp. 405–430.

Loucks, D. P., J. R. Stedinger, and D. A. Haith. 1980. *Water Resource Systems Planning and Analysis.* Prentice Hall, Englewood Cliffs, NJ.

Mongkholkhajornsilp, D., S. Douglas, P. L. Douglas, A. Elkamel, W. Teppaitoon, and S. Pongamphai. 2005. Supercritical CO_2 extraction of nimbin from neem seeds A modelling study. *Journal of Food Engineering* 71(4): 331–340.

Montes, F. J., Galan, M. A. and R. L. Cerro. 1999. Mass transfer from oscillating bubbles in bioreactors, *Chemical Engineering Science* 54:3127–3136.

Patterson, M. R. 1992. A mass transfer explanation of metabolic scaling relations in some aquatic invertebrates and algae. *Science* 255:1421–1423.

Pitt, R. E. and R. E. Muck. 1993. A diffusion model of aerobic deterioration at the exposed face of bunker silos. *Journal of Agricultural Engineering Research* 55(1):11–26.

Roberts, M. S. and M. Rowland. 1985. Hepatic elimination-dispersion model, *Journal of Pharmaceutical Sciences* 74(5):585–587.

Schnoor, J. L. 1996. *Environmental Modeling: Fate and Transport of Pollutants in Water, Air and Soil*. John Wiley & Sons, New York.

Schulin, R., P. J. Wierenga, H. Flühler, and J. Leuenberger. 1987. Solute transport through a stony soil. *Journal of the Soil Science Society of America* 51:36–42.

Skjervold, P. O., S. O. Fjaera and L. Snipen. 2002. Predicting live-chilling dynamics of Atlantic salmon (Salmo salar). *Aquaculture*, 209:185–195.

Waldschmidt, S. R. and W. P. Porter. 1987. A model and experimental test of the effect of body temperature and wind speed on ocular water loss in the lizard Uta stansburiana. *Physiological Zoology* 60(6):678–86.

Weber, E. 1982. *Air Pollution: Assessment Methodology and Modeling* Plenum Press, New York.

14.9   Problems

14.1   Convection–Dispersion in Soil

Consider a pollutant on a soil surface being dissolved in the rainfall and snowmelt and carried through the soil depth into the groundwater, adsorbing onto soil particles along the way. The initial concentration of pollutant is 333 μg/cm^3 of water, and its depth of incorporation in the soil is 15 cm. The pollutant is carried into the groundwater by percolation, with a mean velocity of 0.5 cm/day, and gets dispersed with a dispersion coefficient of 2 cm^2/day. The partition coefficient between the pollutant and soil (cm^3 of water per gram of soil) is 2. The saturated soil has 0.3 cm^3 of water per cm^3 of soil, the retardation factor is 9, and the dry bulk density of the soil is 1.2 g/cm^3 of soil. Calculate the concentration profiles of the pollutant in the soil after 2 and 5 years.

14.2 Transport of Material Injected in a Vein

Patients are sometimes administered saline mineral water that is injected into the veins. We would like to consider the transport of the saline solution in the vein. A simple model would be to consider the diffusion of the saline water with diffusivity D, as the entire liquid (blood + saline water) moves with some average velocity, u. Consider a continuous supply of the saline substance of known concentration to be maintained at a location $z = 0$ along the length of the vein. The vein has an initial concentration of saline water of c_i. 1) Draw a schematic for this problem. 2) Set up the one-dimensional governing equation for this problem, keeping only the relevant terms. 3) Set up the boundary and initial conditions for this problem. 4) Write the governing equation to provide concentration as a function of time and position along the vein *(Hint: See Section 14.2.)*

14.3 Diffusion through a Stagnant Gas

The water surface in an open cylindrical tank with a diameter of 2 m is 7.5 m below the top. Stagnant air fills this empty space in the tank above the water. Dry air is blown over the top of the tank so that no water vapor is at the top. The entire system is maintained at 18°C and 1 atm total pressure. The diffusivity of water vapor through air at this temperature is 2.77×10^{-5} m^2/s. Determine the rate of water loss per unit area through the top in g/m^2·s. *(Note: Some required data is in the appendix.)*

14.4 Diffusion without a Stagnant Gas

Consider dry air at 18°C flowing at 15 km/hr over a circular water surface with a 2 m diameter exposed to atmospheric pressure. The diffusivity of vapor in air is given by the same value as in Problem 14.3. 1) Calculate the rate of water loss per unit area in g/m^2·s. 2) Compare the water loss just calculated with that calculated in Problem 14.3. Comment on which is greater and explain why.

14.5 Convective Loss of Moisture from Soil Surface

Moved to solved examples

14.6 Reduction of Moisture Loss Using a Cover

Moved to solved examples

Figure 14.28: Schematic for Problem 14.7.

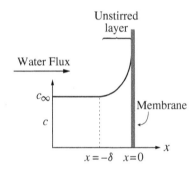

Figure 14.29: Concentration profile of solute near membrane.

14.7 Thickness of Various Boundary Layers

Consider dry air flowing at 10 cm/s over a wet sponge of size 10 cm × 5 cm, in a direction perpendicular to the longer side, as shown in Figure 14.28. Air is warmer than the sponge surface. The average or film temperature is 20°C. The properties at this temperature are given as $\mu_{air}/\rho_{air} = 15.89 \times 10^{-6}$ m^2/s, $\alpha_{air} = 2.08 \times 10^{-5}$ m^2/s, and $D_{vapor,air} = 2.6 \times 10^{-5}$ m^2/s. Calculate and compare the thicknesses of the velocity and the thermal and moisture boundary layers over this sponge at 5 cm (trailing edge) from where the flow hits the sponge.

14.8 Solute Concentration Profile in a Root Zone

Water and solute transport inside roots is a much-researched topic. Some researchers have indicated that the solute concentration very near the membrane of the root (i.e., in the unstirred surface layer shown in Figure 14.29) is much greater than the solute concentration in the bulk water. This increase occurs because the membrane has a very low solute permeability and a very high water permeability. This makes for a high flux of water out across the membrane, but the solute cannot pass through the membrane as quickly as the water can. As a result, there is a subsequent solute build-up very near the membrane.

1) The concentration profile of the solute in the unstirred layer very near the membrane, as shown in Figure 14.29, is steady state because, as concentration builds up in the unstirred layer, diffusion of solute occurs in the direction opposite to convection (water flux) and eventually balances it. Write the governing equation for this steady-state concentration profile of solute in the unstirred layer. 2) The boundary condition at the membrane surface ($x = 0$) is zero total flux (convective+diffusive). The other boundary condition at $x = -\delta$ (the thickness of the unstirred layer) is given by c_∞, the concentration of solute in bulk solution. Write the two boundary conditions. 3) From steps 1 and 2, solve for the concentration profile as a function of position x, plugging in the following parameters: for a concentration of solute in bulk solution of 2 mol/liter, bulk velocity of water at 30×10^{-9} m/s, diffusivity of solute in water of 2×10^{-10} m^2/s, and a thickness of the unstirred layer of 8×10^{-9} m.

14.9 Ocular Evaporative Water Loss in Lizards

Reptiles are often able to successfully colonize arid habitats because their bodies have a very low rate of evaporative water loss (EWL). Although a reptile's outer covering is virtually impermeable to water, its wet ocular surfaces freely evaporate water to the environment. While the eye area comprises less that 0.03% of the lizard's total surface area, the water loss from these moist surfaces can be very significant. Consider the data below for a lizard habitat.

Assume that the surrounding air is completely dry. The percentage of time the eyes are open is 60. The total surface area of both eyes combined is 8.5×10^{-7} m^2, vapor concentration in the air at the eye surface corresponds to the vapor pressure of water (see Table C.10 on page 570) at the lizard's surface temperature of 37°C, diffusivity of vapor in air is 2.6×10^{-5} m^2/s, thermal conductivity of air is 0.027 W/m·K, density of air is 1.14 kg/m^3, and specific heat of air is 1 kJ/kg·K.

1) Write the equation for the average mass transfer coefficient of forced air flow over the eye, assuming the situation can be modeled as laminar flow over a flat plate (no need to plug in numbers). 2) For a wind velocity that leads to an average heat transfer coefficient of 11 W/m^2·K, calculate the average mass transfer coefficient for air flow over the eye. 3) Find the rate of evaporative water loss (EWL) from the two eyes, in mg/hr. 4) If the total evaporative water loss (total EWL) from the two eyes and the body surface is 1.77 mg/hr, find the EWL from the two eyes as a percentage of total EWL.

14.10 Evaporation from a Falling Drop of Water

Consider evaporation from a falling drop of water (perfectly spherical) in dry air at an average temperature of 35°C. At a certain instant, the diameter of the drop is 0.001 m and its velocity is 3 m/s. At this temperature, the diffusivity of water vapor in air at 35°C is 0.273×10^{-4} m^2/s, and the vapor pressure of water at 35°C is 5.624×10^3 Pa. Additional air properties can be found in Appendix C.8. Calculate the instantaneous rate of evaporation per unit surface area from this falling drop of water.

14.11 Time to Completely Evaporate a Drop

Consider convective mass transfer from tiny spherical droplets of water suspended in dry air as evaporation occurs at the droplet surface. The droplets being very small, gravitational force in relation to viscous drag is not as dominating and there are also small updrafts leading to no relative velocity between the droplets and the air. The droplet temperature remains constant at 40°C. The diffusivity of water vapor in air at this temperature is 0.273×10^{-4} m^2/s, and the density of water is 1000 kg/m^3. Vapor pressure data can be found in Appendix C.10. 1) Equate the time rate of change of a droplet mass to the water loss through convective mass transfer at its surface so that you end up with a differential equation for radius, r. 2) Solve the equation in step 1 to find an expression for the time it takes for the droplet to shrink to a radius, r, starting from an initial radius, r_i. 3) Calculate the time it takes for a 0.1 mm radius drop to completely evaporate.

Figure 14.30: Egyptian desert agama lizard. (iStock.com/Paul Vinten)

14.12   Mass Transfer to a Rising Bubble

A bubble of pure chlorine gas 0.5 cm in diameter and at 1.0 atm is rising at a velocity of 20 cm/s in pure water at 16°C. Chlorine from the bubble surface is dissolved in water and diffuses away, making the bubble smaller. The maximum amount of chlorine that can dissolve in water (saturation solubility) at this temperature, which is also the concentration at the surface of the bubble, is 0.823 g of chlorine per 100 g of water. The diffusivity of chlorine gas in water at this temperature is 1.26×10^{-9} m^2/s. The density and viscosity of water are 1000 kg/m^3 and 1.155×10^{-3} kg/m·s, respectively. What is the rate (in grams per second) of absorption of the bubble by the water when the bubble is 0.5 cm in diameter?

14.13   Oxygen Diffusion to Sustain Microorganisms in Water

Consider a spherical colony of microorganisms having an effective diameter of 0.1 mm in stagnant water at 25°C that is neutrally buoyant, i.e., not floating up or down. Oxygen diffuses through water to the surface of the microorganisms where it is completely consumed, maintaining a zero concentration at the surface. Oxygen in the bulk water is in equilibrium with air (21% oxygen). Henry's constant for oxygen and water is 4.4×10^4 atm/mole fraction at 25°C. Diffusivity of oxygen in water is 3.25×10^{-9} m^2/s and density of water is 1000 kg/m^3. 1) Write the governing equation for oxygen diffusion in water at steady state, considering radial variation only. 2) Write the boundary conditions needed. 3) Solve for the oxygen concentration profile as a function of radius (leave in terms of symbols). 4) Calculate the oxygen concentration in bulk water in mol/m^3 assuming air and water are in equilibrium. 5) Calculate the rate of oxygen consumption by the microorganism colony in mol/m^2·s.

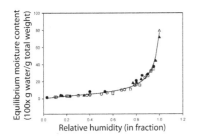

Figure 14.31: Equilibrium moisture content of human skin. Adapted from Kasting and Barai (2003).

14.14   Evaporation from Pure Liquid and Wet Solid Compared

1) A drop of water is falling in dry air at an average temperature of 35°C. At a certain instant, the diameter is 0.01 m and its velocity is 3 m/s. Diffusivity of vapor in air is 0.26×10^{-4} m^2/s. Air properties are available in the appendix (interpolate as necessary). Calculate the instantaneous rate of evaporation from the surface of the falling drop at this instant. 2) Consider replacing the drop of water in question 1 with a piece of skin tissue that is the same size as the drop (i.e., 0.01 m) and at the same temperature (i.e., 35°C). Air flows over the tissue at 3 m/s. The tissue water content is 50% wet basis and is in equilibrium with the air (see Figure 14.31 for conversion to relative humidity in air). For the purpose of this problem, you can obtain the concentration of water vapor in kg/m^3 from the relative humidity (RH) by multiplying the RH value in the fraction by 0.0396. Calculate the instantaneous rate of evaporation from the

surface of the tissue.

14.15 Pollutant Movement through Soil

Consider convection–dispersion of pollutants dissolved in water moving through a soil column (a porous media) for which the concentration profiles are as shown in Figure 14.11. The units for concentration, c, in the figure can be assumed as $\mu g/cm^3$. Answer briefly the following questions:

1) What are the governing equations, boundary conditions, and initial conditions that were solved to get the solution noted on the figure? 2) How do you interpret the variable $z - ut$ in the solution? 3) Show what the concentration profile would look like at 90 and 240 days, respectively, if there were no dispersion. 4) In Figure 14.11, why does the peak concentration get smaller over time (i.e., from 90 days to 240 days)? 5) Can we have dispersion in this problem without convection? Explain. 6) Calculate approximately the downward flux at a depth of 30 cm after 90 days using the concentration profile shown in the figure. (*Note that there is dispersion as well as bulk flow or convection in the figure.*)

14.16 Oxygen Supply to a Bioartificial Liver

Bioartificial livers can provide temporary support for patients with a liver failure. In a bioartificial liver, hepatocytes are embedded in a collagen gel matrix. Survivability of the embedded hepatocytes can depend on the extent of oxygen reaching them by diffusion through the matrix. Figure 14.33 illustrates such a matrix as a slab of 335 μm thickness with blood plasma containing oxygen flowing over it. This oxygen is supplied to one side of the slab by the flowing blood plasma (assume laminar flow), and the other side is considered impermeable to oxygen. Consider a steady-state situation with the rate of oxygen consumption inside the matrix as 5.76×10^{-1} g/m$^3 \cdot$ s, that is required to stay constant. Concentrations of O_2 in the bulk plasma and in the plasma on the matrix surface are 7 g/m^3 and 1.35×10^{-2} g/m^3, respectively. Temperature can be assumed to be constant at 37°C. For simplicity, diffusivity of oxygen in blood plasma can be assumed to be that of oxygen in water at 25°C. Other thermal and mass transport properties of blood plasma can also be treated as those for water. What should the minimum bulk velocity of plasma be to provide the required oxygen supply?

14.17 Effect of Air Humidity on Wind Chill Temperature

Your shirt got wet in the rain, and you are letting it dry by hanging it in flowing air. The shirt feels colder than the ambient air. Assume that the process is at steady state, and the rate of heat gain by your shirt is equal to the enthalpy equivalent of the rate

Figure 14.32: Cross section of a soil column. (J. Helgason—Shutterstock.com)

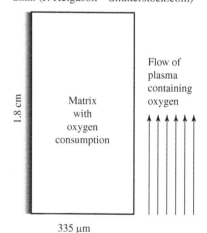

Figure 14.33: Schematic for plasma flow over a flat plate.

of water mass lost due to phase change (evaporation) from its water mass lost from convection, i.e., the energy equivalent of mass loss from evaporation. 1) Considering the shirt surface, write the expression for the rate of convective heat gain of the shirt surface from the air. 2) Write the expression for the rate of convective moisture loss from the shirt surface into the air. 3) At steady state, equating the heat loss with the energy equivalent of mass loss, show how the shirt will not feel as cold if the humidity of the air is higher.

14.18   Moisture Transfer Boundary Layer Thickness over Skin

A breeze at 5 mph (2.2 m/s) is moving over your cheek (length 0.05 m) and is causing the skin to lose water. The viscosity, μ, of air is 1.8465×10^{-5} kg/m · s, density of air, ρ, is 1.1769 kg/m^3, and diffusivity of water vapor in air, $D_{vapor,air}$, is 2.538×10^{-4} m^2/s. What is the thickness of the mass transfer boundary layer at half the distance (0.025 m) from the edge of the cheek surface over which this water loss is taking place?

14.19   Aeration and Deforming Bubbles

Aeration in bioreactors involves transfer of oxygen from air bubbles into the liquid where biological reactions occur. One way to achieve aeration is by stirring. Consider bubbles of air containing O_2 that are 0.5 cm in diameter and moving at 15 cm/s due to stirring by a sieve plate agitator. At the interface between bubble and water, water is saturated with oxygen at 8.24 mg/l (l stands for liter which is equal to 1000 cm^3). Far away from the bubble, there is no oxygen in the water. The system is maintained at 25°C. The density and viscosity of water at this temperature are 997 kg/m^3 and 0.894×10^{-3} kg/m·s, respectively, and the diffusivity of oxygen in water is 2.41×10^{-9} m^2/s. 1) What is the mass transfer coefficient for a single bubble? 2) What is the rate of oxygen transfer, in g/m^3 · s, from the bubbles into the reactor when there are 10^6 bubbles per m^3 of solution? 3) An oscillating agitator can make the bubbles oscillate. For oscillating bubbles, surface tension affects the frequency and the amplitude of the oscillations, and these in turn affect the mass transfer coefficient. The rigid sphere assumed in the previous steps also can no longer be assumed. This situation uses a new correlation for bubbles that deform (as in Figure 14.34):

$$Sh_D = \frac{2}{\sqrt{\pi}} Re^{1/2} Sc^{1/2} \left(1.1 + 0.027 We^{1/2} \right) \qquad (14.81)$$

where We is yet another dimensionless number called the Weber number that incorporates the surface tension effects. The Weber number is calculated as $We = D u_\infty^2 \rho / \sigma$, where σ is the surface tension, having a value of 0.072 N/m between air and water at

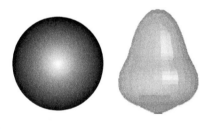

Undeformed   Deformed
spherical bubble   bubble

Figure 14.34: Schematic of a rigid (undeformed) and a deformed bubble.

this temperature. Also, D is the diameter of the oscillating bubble that, for the sake of comparison, can be assumed to be 0.5 cm, i.e., that of the undeformed bubble. Other symbols have their usual meanings as in step 1. Assuming all other parameters and transport properties are kept the same, what is the mass transfer coefficient for this deforming bubble situation, using Eq. 14.81? 4) For the non-oscillating bubbles, describe qualitatively how you expect the mass transfer coefficient to change if the bubbles are made smaller.

14.20 Drug transport in a liver

Moved to solved example

14.21 Air Pollution from Animal Waste

Release of ammonia from animal waste is a major air pollution problem. Consider ammonia release from liquid manure over an experimental flat surface, as shown in Figure 14.35a. The bulk air flow over the surface has a velocity of 0.1 m/s and its ammonia content can be assumed to be zero. Assume the temperature of both liquid surface and air is 20°C. Air density and viscosity at the temperature needed (293 K) are 1.205 kg/m^3 and 1.81×10^{-5} kg/m·s, respectively. Diffusivity of ammonia in the air is 0.22×10^{-4} m^2/s. 1) Calculate the average mass transfer coefficient for this flow situation. 2) The concentration of ammonia in the surface of the liquid manure is 25 mg/l and the Henry's law constant for NH$_3$ in the liquid–gas system is 6×10^{-6} MPa · m^3/mol. Molar mass of ammonia is 17 g/mol. Calculate the rate at which ammonia is lost over the surface. 3) If the air velocity is changed from 0.1 m/s to 0.5 m/s, calculate the ratio by which the mass transfer coefficient changes. 4) Now compare the ratio in question 3 with the corresponding ratio of experimentally measured mass transfer coefficients shown in Figure 14.35b (estimate the average of the two sets of data points shown) and comment on whether they are close.

14.22 Effects of Tablet Location on Dissolution Testing

Three tablets undergoing dissolution testing (water flowing over the tablets to test their disintegration), located on the bottom of the dissolution vessel as shown in the schematic in Figure 14.36, can stay fixed at different distances from the center. We would like to study the effect of this distance on the dissolution rate for tablets that are non-disintegrating (surface stays flat). We will consider only flow over the top flat surface of the tablet, as shown in Figure 14.36b. For a rotational speed of $N = 50$ revolutions per minute of the impeller, the bulk velocity of water over the tablets is given by $u_\infty = 2\pi N R$, where R is the radial location of the tablet from the center.

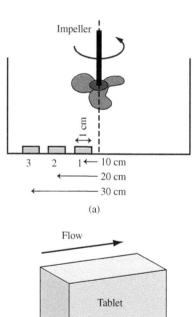

Figure 14.36: a) Dissolution apparatus with tablets at the bottom and b) enlargement of one tablet.

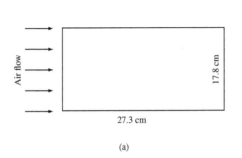
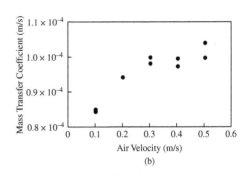

(a)   (b)

Figure 14.35: a) Dimensions of the flat surface of liquid manure. b) Measured mass transfer coefficient for different air velocities (Arogo et al., 1999). Two points for the same velocity represent scatter in the experimental data (you can take the average of them).

The length of tablet along the flow is 1 cm. The viscosity of water is 0.0007 Pa · s, density of water is 1000 kg/m^3, and diffusivity of the drug in the water is 1.47×10^{-9} m^2/s.

  1) Find the mass transfer coefficient for the tablet at location 1. 2) Find the mass transfer coefficients for the tablets at locations 2 and 3. 3) Would the ratio of mass transfer coefficients between two locations (say, 1 and 2) change with rotational speed, $\omega = 2\pi N$? Explain.

14.23   Extraction of Medicinal Components From a Seed

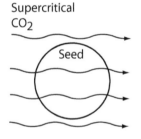

Figure 14.37: Schematic of a spherical seed with fluid flowing over it.

Nimbin, an important chemical component with medicinal value found in the seeds of the "Neem" tree in South Asia, is to be extracted using supercritical carbon dioxide (a fluid) (Mongkholkhajornsilp et al., 2005). Consider steady flow of supercritical CO_2 over a single spherical seed of diameter 1.5 mm at a velocity of 1 m/s. Supercritical CO_2 has a density of 800 kg/m^3 and a viscosity of 3×10^{-4} Pa·s, and the diffusivity of Nimbin in it is 4×10^{-9} m^2/s.

  1) Calculate the convective mass transfer coefficient of Nimbin with supercritical CO_2 flowing over the seed surface. 2) Using the mass transfer coefficient you just calculated, write an expression for the flux of Nimbin from the seed surface at any time, defining each of the variables you use. 3) Is the mass transfer coefficient zero when the supercritical CO_2 is not flowing with respect to the seed, i.e, the seed is in a stationary fluid? Why or why not?

Part III

APPENDIX

Appendix A

SUMMARY
AND ANALOGIES BETWEEN
HEAT AND MASS TRANSFER

A.1 Basic Heat and Mass Transfer Parameters and Fluxes

	Heat Transfer	Mass Transfer
Quantity of Interest	$T\,[°C]$	$c_A\,\left[\text{kg/m}^3\right]$
Equation for diffusive flux	$q'' = -k\dfrac{\partial T}{\partial z}\left[\frac{\text{W}}{\text{m}^2}\right]$ $= -\alpha\dfrac{\partial(\rho c_p T)}{\partial z}$	$j_A = -D_{AB}\dfrac{\partial c_A}{\partial z}\left[\frac{\text{kg}}{\text{m}^2\text{s}}\right]$
Diffusivity	$\alpha\,\left[\text{m}^2/\text{s}\right]$	$D_{AB}\,\left[\text{m}^2/\text{s}\right]$
Equation for total flux	$q'' = \underbrace{-k\dfrac{\partial T}{\partial z}}_{\text{conductive}} + \underbrace{\rho c_p T u}_{\text{convective}}$	$n_{A,z} = \underbrace{-D_{AB}\dfrac{\partial c_A}{\partial z}}_{\text{diffusive}} + \underbrace{c_A u}_{\text{convective}}$
Convective transfer coefficient	$h\,\left[\text{W/m}^2\cdot\text{K}\right] = \frac{q''}{\Delta T}$	$h_m\,[\text{m/s}] = \frac{j_A}{\Delta c}$
Overall coefficient	$\dfrac{1}{U} = \dfrac{1}{h_1} + \dfrac{\Delta L}{k} + \dfrac{1}{h_2}$	$\dfrac{1}{U_m} = \dfrac{1}{h_{m_1}} + \dfrac{\Delta L}{K^* D_{AB}} + \dfrac{1}{h_{m_2}}$

A.2   Heat and Mass Transfer Governing Equations and Boundary Conditions

	Heat Transfer	Mass Transfer
Governing Equation	$\frac{\partial T}{\partial t} + u\frac{\partial T}{\partial z}$	$\frac{\partial c_A}{\partial t} + u\frac{\partial c_A}{\partial z}$
	$= \alpha\frac{\partial^2 T}{\partial z^2} + \frac{Q}{\rho c_p}$	$= D_{AB}\frac{\partial^2 c_A}{\partial z^2} + r_A$
	Q is the rate of heat generation or absorption per unit volume	r_A is the rate of mass generation or disappearance per unit volume
Boundary Conditions		
	$T\Big|_{surface} = T_s$ surface temperature is specified, e.g., *condensing steam*	$c_A\Big|_{surface} = c_{A,s}$ surface concentration is specified, e.g., *later stages of drying*
	$-k\frac{\partial T}{\partial z}\Big|_{surface} = q_s''$ surface heat flux is specified, e.g., *insulated (zero heat flux)*	$-D_{AB}\frac{\partial c_A}{\partial z}\Big|_{surface} = n_{A,s}$ surface mass flux is specified, e.g., *impermeable (zero mass flux)*
	$-k\frac{\partial T}{\partial z}\Big|_{surface}$ $= h(T\Big|_{surface} - T_\infty)$ convection over surface	$-D_{AB}\frac{\partial c_A}{\partial z}\Big|_{surface}$ $= h_m(c_A\Big|_{surface} - c_{A,\infty})$ convection over surface

A.3 Generic Transport Equation and the Physical Meanings of the Constituent Terms

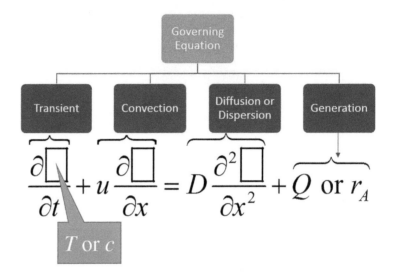

A.4 Generic Boundary Conditions and Initial Condition Needed to Solve the Transport Equation in A.3

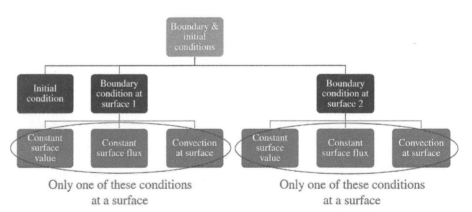

A.5   Solution Map for Heat Conduction and Mass Diffusion Problems (without Bulk Flow)

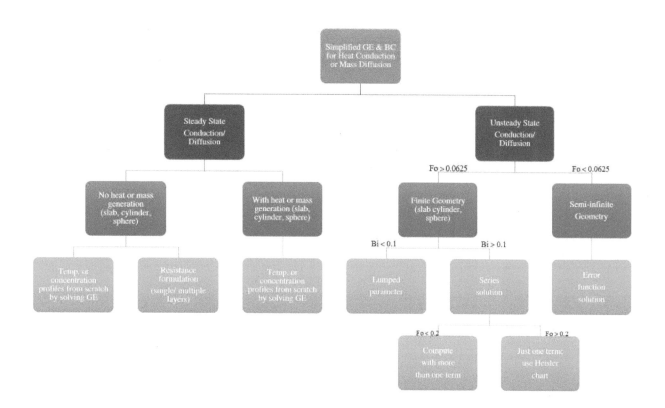

Appendix B

PHYSICAL CONSTANTS, UNIT CONVERSIONS, AND MATHEMATICAL FUNCTIONS

B.1 Physical Constants

Quantity	Symbol	Value
Universal Gas Constant	R_g	8.205×10^{-2} m^3· atm/kmol·K
		8.314×10^{-2} m^3· bar/kmol·K
		8.314 kJ/kmol·K
Avogadro's Number	N	6.024×10^{23} molecules/mol
Planck's Constant	h	6.625×10^{-34} J· s/molecule
Boltzmann's Constant	κ	1.380×10^{-23} J/K·molecule
Speed of Light in Vacuum	c_o	2.998×10^8 m/s
Stefan–Boltzmann Constant	σ	5.670×10^{-8} W/m^2· K^4
Gravitational Acceleration (Sea Level)	g	9.807 m/s^2
Normal Atmospheric Pressure	p	$101{,}325$ N/m^2

B.2   Some Useful Conversion Factors

To Convert From	to	Multiply by
Area		
acre	m^2	$4.046\,856 \times 10^3$
acre	ft^2	$4.356\,000 \times 10^4$
ft^2	m^2	$9.290\,304 \times 10^{-2}$
m^2	ft^2	$1.076\,391 \times 10^1$
cm^2	m^2	$1.000\,000 \times 10^{-4}$
ft^2	$in.^2$	$1.440\,000 \times 10^2$
$in.^2$	m^2	$6.451\,600 \times 10^{-4}$
Energy		
Btu (international)	kJ	$1.055\,056$
Btu	kcal	$2.519\,958 \times 10^{-1}$
kcal	kJ	$4.186\,800$
$kW \cdot h$	kJ	$3.600\,000 \times 10^3$
$kW \cdot h$	Btu	3.413×10^3
$hp \cdot h$	Btu	2.454×10^3
J	ergs	$1.000\,000 \times 10^7$
eV	J	$1.602\,19 \times 10^{-19}$
Btu	ft-lbf	$7.781\,693 \times 10^2$
$ft \cdot lbf$	kJ	$1.355\,818 \times 10^{-3}$
Force		
lbf	N	$4.448\,222$
dyne	N	$1.000\,000 \times 10^{-5}$
Length		
angstrom	m	$1.000\,000 \times 10^{-10}$
caliber	m	$2.540\,000 \times 10^{-4}$
fathom	m	$1.828\,800$
ft	m	$3.048\,000 \times 10^{-1}$
inch	m	$2.540\,000 \times 10^{-2}$
mile	m	$1.609\,344 \times 10^3$
mile	ft	5.280×10^3
yard	m	$9.144\,000 \times 10^{-1}$

To Convert From	to	Multiply by
Mass		
carat	kg	$2.000\,000 \times 10^{-4}$
grain	kg	$6.479\,891 \times 10^{-5}$
lb_m (pound mass)	kg	$4.535\,924 \times 10^{-1}$
kg	lb_m	$2.204\,622$
kg	slug	$6.852\,1 \times 10^{-2}$
slug	lb_m	$3.217\,405 \times 10^{1}$
ton	lb_m	2.000×10^{3}
ton	kg	$9.071\,847 \times 10^{2}$
Mass per Unit Volume (density)		
lb_m/ft^3	kg/m^3	$1.601\,846 \times 10^{1}$
$lb_m/in.^3$	kg/m^3	$2.767\,990 \times 10^{4}$
lb_m/gal	kg/m^3	$1.198\,264 \times 10^{2}$
Power		
Btu/h	W (watt)	$2.930\,711 \times 10^{-1}$
J/s	W	$1.000\,000$
erg/s	W	$1.000\,000 \times 10^{-7}$
ft · lbf/s	W	$1.355\,818$
hp	W	$7.456\,999 \times 10^{2}$
hp	ft · lbf/s	5.50×10^{2}
hp	ft · lbf/min	$3.300\,0 \times 10^{4}$
hp (boiler)	Btu/h	$3.347\,14 \times 10^{4}$
hp (boiler)	kW	$9.809\,50$
Pressure		
std atm	kPa	$1.013\,25 \times 10^{2}$
std atm	$lbf/in.^2$	$1.469\,6 \times 10^{1}$
cm Hg (0°C)	kPa	$1.333\,22$
in. Hg (32°F)	kPa	$3.386\,389$
in. H_2O (60°F)	kPa	$2.488\,4 \times 10^{-1}$
bar	kPa	$1.000\,000 \times 10^{2}$
bar	$lbf/in.^2$	$1.450\,377 \times 10^{1}$
$lbf/in.^2$	kPa	$6.894\,757$
in. Hg (32°F)	$lbf/in.^2$	$4.911\,542 \times 10^{-1}$
ft H_2O (60°F)	$lbf/in.^2$	$4.330\,943 \times 10^{-1}$
ft H_2O (60°F)	kPa	$2.986\,08$
$dyne/cm^2$	kPa	$1.000\,000 \times 10^{-4}$

To Convert From	to	Multiply by
Specific Heat		
Btu/lb$_m$ · °F	kJ/kg · K	4.186 800
cal/g · °C	Btu/lb$_m$ ·°F	1.000 000
Thermal Conductivity		
Btu/ft · h ·°F	W/m · K	1.730 6
279×10^{-1}		
cal/cm · h ·°C	Btu/ft · h ·°F	$6.719\ 69 \times 10^{-2}$
Thermal Conductance		
Btu/ft^2· h ·°F	W/m^2· K	5.674 466
Temperature		
°C	K	$T_K = T_{°C} + 273.15$
°F	°R	$T_{°R} = T_{°F} + 459.67$
K	°R	$T_{°R} = T_K \times 1.8$
°C	°F	$T_{°F} = (T_{°C} + 273.15) \times 1.8 - 459.67$
°F	°C	$T_{°C} = (T_{°F} + 459.67)/1.8 - 273.15$
Velocity		
ft/s	m/s	$3.048\ 888 \times 10^{-1}$
miles/h	km/h	1.609 344
knot	m/s	$5.144\ 444 \times 10^{-1}$
Viscosity		
centipoise	Pa · s	$1.000\ 000 \times 10^{-3}$
lb$_m$/ft · s	Pa · s	1.488 164
lbf · s/ft^2	Pa · s	$4.788\ 026 \times 10^{1}$
Volume		
acre · ft	m^3	$1.233\ 483 \times 10^{3}$
barrel (42 gal)	m^3	$1.589\ 873 \times 10^{-1}$
ft^3	m^3	$2.831\ 685 \times 10^{-2}$
in.^{3}	m^3	$1.638\ 706 \times 10^{-5}$
ft^3	gallons	7.480 52
liter	m^3	$1.000\ 000 \times 10^{-3}$
gallons	liter	3.785 412
gallons	m^3	$3.785\ 412 \times 10^{-3}$
gallons	ft^3	$1.336\ 81 \times 10^{-1}$
quart (U.S. liquid)	m^3	$9.463\ 529 \times 10^{-4}$
pint (U.S. liquid)	liter	$4.731\ 765 \times 10^{-1}$

B.3 Error Function Tabulated

ϕ	erf(ϕ)	ϕ	erf(ϕ)	ϕ	erf(ϕ)
0.000	0.0000	0.850	0.7707	1.700	0.9838
0.025	0.0282	0.875	0.7841	1.725	0.9853
0.050	0.0564	0.900	0.7969	1.750	0.9867
0.075	0.0845	0.925	0.8092	1.775	0.9879
0.100	0.1125	0.950	0.8209	1.800	0.9891
0.125	0.1403	0.975	0.8321	1.825	0.9901
0.150	0.1680	1.000	0.8427	1.850	0.9911
0.175	0.1955	1.025	0.8528	1.875	0.9920
0.200	0.2227	1.050	0.8624	1.900	0.9928
0.225	0.2497	1.075	0.8716	1.925	0.9935
0.250	0.2763	1.100	0.8802	1.950	0.9942
0.275	0.3027	1.125	0.8884	1.975	0.9948
0.300	0.3286	1.150	0.8961	2.000	0.9953
0.325	0.3542	1.175	0.9034	2.025	0.9958
0.350	0.3794	1.200	0.9103	2.050	0.9963
0.375	0.4041	1.225	0.9168	2.075	0.9967
0.400	0.4284	1.250	0.9229	2.100	0.9970
0.425	0.4522	1.275	0.9286	2.125	0.9973
0.450	0.4755	1.300	0.9340	2.150	0.9976
0.475	0.4983	1.325	0.9390	2.175	0.9979
0.500	0.5205	1.350	0.9438	2.200	0.9981
0.525	0.5422	1.375	0.9482	2.225	0.9983
0.550	0.5633	1.400	0.9523	2.250	0.9985
0.575	0.5839	1.425	0.9561	2.275	0.9987
0.600	0.6039	1.450	0.9597	2.300	0.9989
0.625	0.6232	1.475	0.9630	2.325	0.9990
0.650	0.6420	1.500	0.9661	2.350	0.9991
0.675	0.6602	1.525	0.9690	2.375	0.9992
0.700	0.6778	1.550	0.9716	2.400	0.9993
0.725	0.6948	1.575	0.9741	2.425	0.9994
0.750	0.7112	1.600	0.9763	2.450	0.9995
0.775	0.7269	1.625	0.9784	2.475	0.9995
0.800	0.7421	1.650	0.9804		
0.825	0.7567	1.675	0.9822		

Note: erf$(-\phi) = -$erf(ϕ) and erfc$(\phi) = 1 - erf(\phi)$

B.4   Charts for Unsteady Diffusion

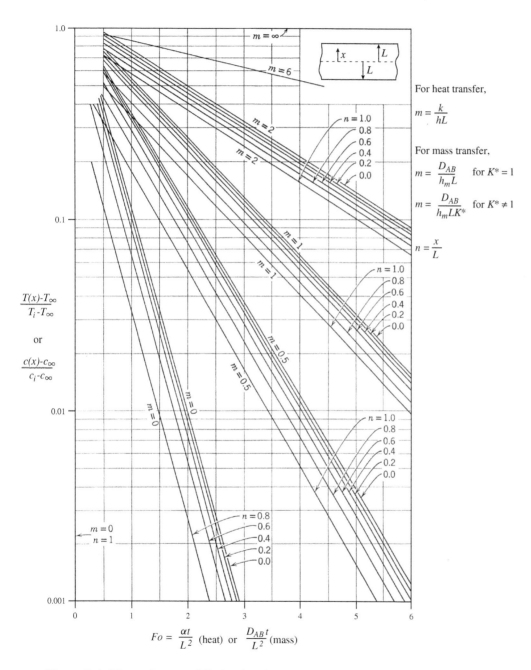

For heat transfer,

$$m = \frac{k}{hL}$$

For mass transfer,

$$m = \frac{D_{AB}}{h_m L} \quad \text{for } K^* = 1$$

$$m = \frac{D_{AB}}{h_m L K^*} \quad \text{for } K^* \neq 1$$

$$n = \frac{x}{L}$$

$$\frac{T(x)-T_\infty}{T_i - T_\infty}$$

or

$$\frac{c(x)-c_\infty}{c_i - c_\infty}$$

$$Fo = \frac{\alpha t}{L^2} \text{ (heat) or } \frac{D_{AB} t}{L^2} \text{ (mass)}$$

Figure B.1: Unsteady-state diffusion in a large slab. Before using in mass transfer for $m \neq 0$, refer to discussion in Section 13.2.5 on page 452. From *Principles of Unit Operations* by A. S. Foust et al., © 1960 by John Wiley & Sons, Inc. Reprinted by permission of John Wiley & Sons, Inc.

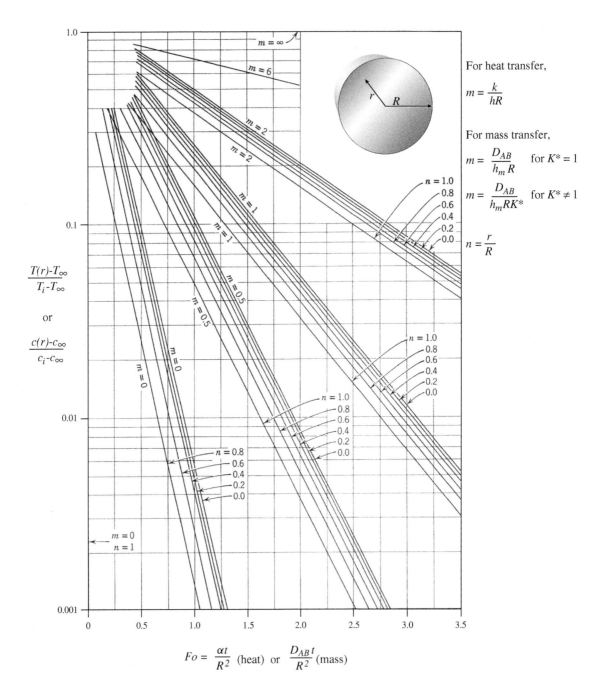

Figure B.2: Unsteady-state diffusion in a long cylinder. Before using in mass transfer for $m \neq 0$, refer to discussion in Section 13.2.5 on page 452. From *Principles of Unit Operations* by A. S. Foust et al., © 1960 by John Wiley & Sons, Inc. Reprinted by permission of John Wiley & Sons, Inc.

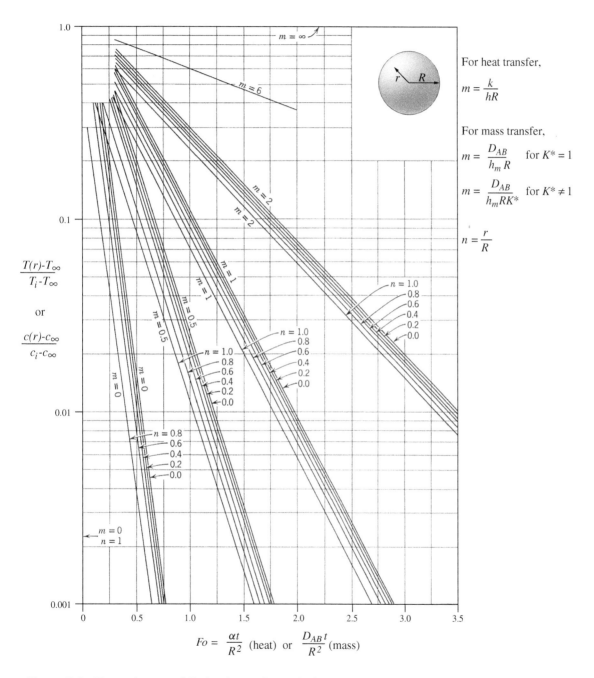

Figure B.3: Unsteady-state diffusion in a sphere. Before using in mass transfer for $m \neq 0$, refer to discussion in Section 13.2.5 on page 452. From *Principles of Unit Operations* by A. S. Foust et al., © 1960 by John Wiley & Sons, Inc. Reprinted by permission of John Wiley & Sons, Inc.

Appendix C

HEAT TRANSFER AND RELATED PROPERTIES

C.1 Basal Metabolic Rate for a Few Animals

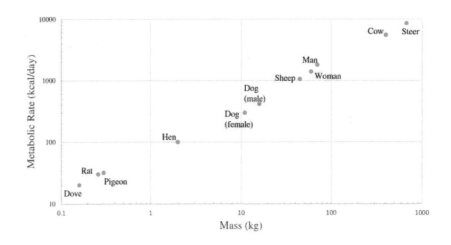

Data source: Greenberg, L. H. 1975. *Physics for Biology and Pre-med Students.* W. B. Saunders Company, Philadelphia.

C.2 Typical Metabolic Rate for Various Human Activities

Activity	Metabolic rate per unit body surface area W/m^2
Sleeping	41
Reclining	33
Sitting quietly	58
Standing relaxed	70
Dressing and undressing	74
Walking on the level at 3–6 km/hr	116–221
Driving	
Car	87
Motorcycle	116
Heavy vehicle	186
House cleaning	116–198
Shopping	81–105
Carpentry, metal working, industrial painting	150
Dancing	140–256
Swimming	314
Jogging (at about 9 km/hr)	357
Walking up stairs	690

Adapted from Shitzer, A. and R. C. Eberhart. 1985. Heat generation, storage, and transport processes. 1:137–151. In: *Heat Transfer in Medicine and Biology* edited by A. Shitzer and R. C. Eberhart. Plenum Press, New York. Reprinted with permission of Springer.

C.3 Blood Flow, Oxygen Consumption, and Metabolic Rate for Various Human Organs

Organ	Mass kg	Blood Flow ml/min	Oxygen Consumption ml/min	Metabolic Rate W
Heart muscle	0.3	250–1800	30–90	10–31
Skeletal muscle	31.0	1200–24000	50–1000	17–350
Skin	3.6	400–2800	12–85	4–30
Liver	2.6	1500	51	18
Kidney	0.3	1260	18	6
Brain	1.4	750	49	17

Adapted from Shitzer, A. and R. C. Eberhart. 1985. Heat generation, storage, and transport processes. 1:137–151. In: *Heat Transfer in Medicine and Biology* edited by A. Shitzer and R. C. Eberhart. Plenum Press, New York. Reprinted with permission of Springer.

C.4 Thermal Properties of Animal Materials

Description	Thermal Conductivity k W/m · K	Thermal Diffusivity $\alpha \times 10^7$ m^2/s
A. Materials in vivo		
(values depend on blood perfusion)		
Bone, bovine and caprine	0.33–3.1	
Brain, cat	0.56–0.66	1.1–1.2
Cartilage, scapula, bovine	1.8–2.8	
Cartilage, scapula, caprine	1.4–1.9	
Kidney, sheep	0.60–1.2	2.0–4.3
Liver, canine	0.60–0.90	1.5–2.4
Muscle, canine	0.70–1.0	0.7–1.3
Skin, animal and human	0.48–2.8	0.4–1.6
B. Materials in vitro		
(room to body temperatures)		
Bone, fresh to several months postmortem	0.41–0.63	
Dry bone	0.22	
Bone marrow, bovine	0.22	
Brain, bovine, cat, and human	0.16–0.57	0.44–1.4
Fat	0.094–0.37	
Heart	0.48–0.59	1.4–1.5
Kidney	0.49–0.63	1.3–1.8
Kidney, cortex, and medulla	$0.664w - 0.04$[a]	
Liver	0.42–0.57[a]	1.1–2.0
Liver, parenchyma	0.32	1.7–2.0
Lung		2.4–2.8
Muscle	0.34–0.68	1.8
Skin, animal and human	0.21–0.41	0.82–1.2
Spleen	0.45–0.60	1.3–1.6
Tumors, 37°C, General range	0.47–0.58	

(continued next page)

[a] w is the mass fraction of water.

Description	Thermal Conductivity k W/m $\cdot$ K	Thermal Diffusivity $\alpha \times 10^7$ m^2/s
C. Biological fluids **(at room to body temperatures)**		
Agar gel, 1–1.75%	0.60–0.70	1.6
Blood	0.48–0.60	
Blood, human	0.57–0.12H^b	
Blood, whole		
−10°C	1.6	8.7
−20°C	1.7	10.4
−40°C	1.9	13.6
−60°C	2.1	16.9
−80°C	2.4	20.4
−100°C	2.7	23.7
Blood plasma	0.57–0.60	
Humor, aqueous and vitreous	0.58–0.59	
Milk, regular and skimmed	0.53–0.59	
Cream (double Devon)	0.31	
Cod liver oil	0.17	
Egg white	0.56	
Egg yolk	0.34–0.42	
Gastric juice	0.44	
Urine	0.56	
Water	0.59–0.63	

[b] H is the hematocrit fraction.

Adapted from Chato, J. C. 1985. Selected thermophysical properties of biological materials. 2:413–418. In: *Heat Transfer in Medicine and Biology* edited by A. Shitzer and R. C. Eberhart. Plenum Press, New York. Reprinted with permission of Springer.

C.5   Thermal Conductivities of Some Animal Hair Coats

Material	Thermal conductivity, k W/m · K
Arctic mammals	0.036–0.106
Various wild animals	0.038–0.051
Merino sheep	0.037–0.048
Newborn Merino	0.065–0.107
Cattle	0.076–0.147
Rabbit	0.038–0.100
Kangaroo	0.043–0.064
Harp seal pups	0.047–0.065
Penguin	0.031–0.046
Gosling	0.036–0.046
Artifical fur	0.040–0.067
Woven fabric	0.040
Dry air (for comparison) at 10°C	0.025

Adapted from Cena, K. and J. A. Clark. 1979. Transfer of heat through animal coats and clothing. *International Review of Physiology, Environmental Physiology III*, Vol. 20, pp. 1–42, Edited by D. Robertshaw, University Park Press, Baltimore.

C.6 Thermal Properties of Some Agricultural Materials

Material	Thermal conductivity, k W/m · K
Wood, G = 0.45 M = 12% $\perp$	0.13
Wood, G = 0.45 M = 12%	0.31
Wood, G = 0.70 M = 12% $\perp$	0.18
Wood, G = 0.70 M = 12%	0.44
Cell-wall substance $\perp$	0.44
Cell-wall substance	0.88
Douglas-fir plywood	0.12
Concrete	0.93
Expanded polyurethane	0.02
Copper	386.74
Aluminum	201.96
Stainless steel	16.3
Sand	
Porosity - 0.4, Volumetric wetness = 0.0	0.29
Porosity - 0.4, Volumetric wetness = 0.2	1.76
Porosity - 0.4, Volumetric wetness = 0.4	2.18
Clay	
Porosity - 0.4, Volumetric wetness = 0.0	0.25
Porosity - 0.4, Volumetric wetness = 0.2	1.17
Porosity - 0.4, Volumetric wetness = 0.4	1.59
Peat	
Porosity - 0.8, Volumetric wetness = 0.0	0.06
Porosity - 0.8, Volumetric wetness = 0.4	0.29
Porosity - 0.8, Volumetric wetness = 0.8	0.50
Snow	
Porosity - 0.95, Volumetric wetness = 0.05	0.06
Porosity - 0.8, Volumetric wetness = 0.2	0.13
Porosity - 0.5, Volumetric wetness = 0.5	0.71

C.7 Thermal Properties of Food Materials (Representative Values)

Material	Thermal Conductivity W/m · K	Density kg/m^3	Specific Heat kJ/kg · K
High moisture, frozen			
Apple[a] at − 40°C	1.669	785	2.29
Beef, lean	1.42		1.68
Strawberries, tightly packed	1.1		1.14
High moisture, unfrozen			
Apple	0.418		3.60
Beef, lean	0.506		3.35
Egg white	0.558		3.88
Low moisture, non-porous			
Butter	0.197		1.38
Low moisture, porous			
Apple, dried	0.219		2.27
Apple, freeze dried	0.0405		
(at pressure 0.2880 × 10^4 Pa)			
Beef, freeze dried	0.0652		
(at pressure of 1 atm)			
Egg albumin gel, freeze dried	0.0393		
(at pressure of 1 atm)			

Data from 1998 ASHRAE Refrigeration Handbook, unless otherwise mentioned. Note that the thermal conductivity, density, and specific heat are not always from the same study. Thus they should be treated as representative values. Due to the large variability in composition, thermal properties of foods are sometimes estimated using empirical correlations based on composition, such as the one in Choi, Y. 1985. Food thermal property prediction as affected by temperature and composition. PhD Thesis, Purdue University.

[a]From Heldman, D. R. and R. P. Singh. 1986. Thermal properties of frozen foods. In: *Physical and Chemical Properties of Foods*, M. R. Okos, editor. ASAE, St. Joseph, MI.

C.8   Thermal Properties of Air at Atmospheric Pressure

Temp K	Density ρ kg/m^3	Specific Heat c_p kJ/kg $\cdot$ K	Viscosity* $\mu \times 10^5$ kg/m $\cdot$ s	Thermal Conductivity k W/m $\cdot$ K	Thermal Diffusivity* $\alpha \times 10^5$ m^2/s	Prandtl Number Pr
200	1.7690	1.0064	1.3286	0.01809	1.0161	0.739
250	1.4133	1.0054	1.5992	0.02227	1.5673	0.722
260	1.3587	1.0054	1.6504	0.02308	1.6896	0.719
270	1.3082	1.0055	1.7005	0.02388	1.8154	0.716
280	1.2614	1.0057	1.7504	0.02467	1.9447	0.714
290	1.2177	1.0060	1.7985	0.02547	2.0792	0.710
300	1.1769	1.0063	1.8465	0.02624	2.2156	0.708
310	1.1389	1.0068	1.8929	0.02701	2.3556	0.705
320	1.1032	1.0073	1.9392	0.02779	2.5008	0.703
330	1.0697	1.0079	1.9855	0.02853	2.6462	0.701
340	1.0382	1.0085	2.0302	0.02928	2.7965	0.699
350	1.0086	1.0092	2.0748	0.03003	2.9503	0.697
360	0.9805	1.0100	2.1177	0.03078	3.1081	0.695
370	0.9539	1.0109	2.1606	0.03150	3.2666	0.693
380	0.9288	1.0120	2.2018	0.03223	3.4289	0.691
390	0.9050	1.0130	2.2447	0.03295	3.5942	0.690
400	0.8822	1.0142	2.2859	0.03365	3.7609	0.689
450	0.7842	1.0212	2.4849	0.03710	4.6327	0.684
500	0.7057	1.0300	2.6703	0.04041	5.5594	0.681

*Read the data as follows: For example, at 200 K, thermal diffusivity is 1.0161 $\times$ 10^{-5} m^2/s. Adapted from *Tables of Thermal Properties of Gases*, National Bureau of Standards Circular 564, Washington, DC (1955).

C.9 Psychrometric Chart

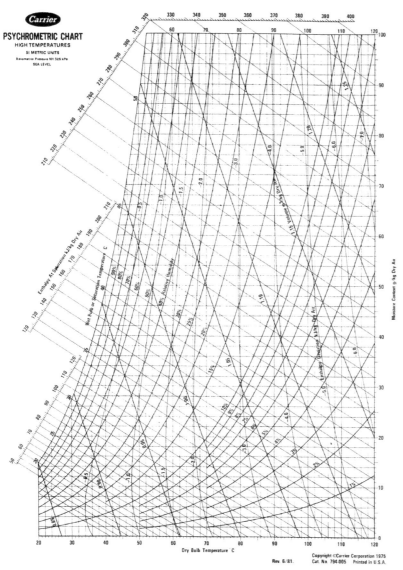

Courtsey of Carrier Corporation.

C.10 Vapor Pressure of Liquid Water from 0 to 100°C

Temp	Vapor Pressure	
°C	mm Hg	$\times 10^{-5}$ N/m^2
0	4.58	0.00611
5	6.54	0.00872
10	9.21	0.01228
15	12.79	0.01705
20	17.54	0.02338
25	23.76	0.03168
30	31.82	0.04242
35	42.18	0.05624
40	55.32	0.07375
45	71.88	0.09583
50	92.51	0.12334
60	149.4	0.19918
70	233.7	0.31157
80	355.1	0.47343
90	525.8	0.70101
95	633.9	0.84513
100	760.0	1.01325

C.11   Steam Properties at Saturation Temperatures

Temp	Pressure	Density Liquid	Density Vapor	Sp. enthalpy Liquid	Sp. enthalpy Vapor	Heat of vaporization
°C	bar	kg/m³	kg/m³	kJ/kg	kJ/kg	kJ/kg
0.01	0.006108	999.80	0.004847	0.00	2501	2501
1	0.006566	999.90	0.005192	4.22	2502	2498
2	0.007054	999.90	0.005559	8.42	2504	2496
3	0.007575	999.90	0.005945	12.63	2506	2493
4	0.008129	999.90	0.006357	16.84	2508	2491
5	0.008719	999.90	0.006793	21.05	2510	2489
6	0.009347	999.90	0.007257	25.25	2512	2489
7	0.010013	999.90	0.007746	29.45	2514	2485
8	0.010721	999.80	0.008264	33.55	2516	2482
9	0.011473	999.70	0.008818	37.85	2517	2479
10	0.012277	999.60	0.009398	42.04	2519	2477
15	0.017041	999.00	0.01282	62.97	2528	2465
20	0.02337	998.20	0.01729	83.90	2537	2454
25	0.03166	997.01	0.02304	104.81	2547	2442
30	0.04241	995.62	0.03037	125.71	2556	2430
35	0.05622	993.94	0.03962	146.60	2565	2418
40	0.07375	992.16	0.05115	167.50	2574	2406
45	0.09584	990.20	0.06544	188.40	2582	2394
50	0.12335	988.04	0.08306	209.3	2592	2383
55	0.15740	985.71	0.1044	230.2	2600	2370
60	0.19917	983.19	0.1302	251.1	2609	2358
65	0.2501	980.49	0.1613	272.1	2617	2345
70	0.3117	977.71	0.1982	293.0	2626	2333
75	0.3855	974.85	0.2420	314.0	2635	2321
80	0.4736	971.82	0.2934	334.9	2643	2308
85	0.5781	968.62	0.3536	355.9	2651	2295
90	0.7011	965.34	0.4235	377.0	2659	2282
95	0.8451	961.91	0.5045	398.0	2668	2270
100	1.0131	958.31	0.5977	419.1	2676	2257
110	1.4326	951.02	0.8264	461.3	2691	2230
120	1.9854	943.13	1.121	503.7	2706	2202
130	2.7011	934.84	1.496	546.3	2721	2174
140	3.614	926.10	1.966	589.0	2734	2145
150	4.760	916.93	2.547	632.2	2746	2114
200	15.551	864.68	7.862	852.4	2793	1941
250	39.776	799.23	19.28	1085.7	2801	1715
300	85.92	712.45	46.21	1344.9	2749	1404.3
374.15	221.297	306.75	306.75	2100	2100	0.0

Adapted from Irvine, T. F., Jr. and J. P. Hartnett. 1976. *Steam and Air Tables in SI Units*. Hemisphere Publishing Co., Washington, DC.

C.12　Thermophysical Properties of Saturated Water

Temp K	Pressure $P \times 10^{-5}$ Pa	Specific Heat c_p kJ/kg·K	Viscosity $N \cdot s/m^2$ $\mu \times 10^6$	Thermal Conduc. k W/m·K	Prandtl Number Pr	Expansion Coefficient $\beta \times 10^6$ K^{-1}
273.15	0.00611	4.217	1750	0.569	12.99	−68.05
275	0.00697	4.211	1652	0.574	12.22	−32.74
280	0.00990	4.198	1422	0.582	10.26	46.04
285	0.01387	4.189	1225	0.590	8.81	114.1
290	0.01917	4.184	1080	0.598	7.56	174.0
295	0.02617	4.181	959	0.606	6.62	227.5
300	0.03531	4.179	855	0.613	5.83	276.1
305	0.04712	4.178	769	0.620	5.20	320.6
310	0.06221	4.178	695	0.628	4.62	361.9
315	0.08132	4.179	631	0.634	4.16	400.4
320	0.1053	4.180	577	0.640	3.77	436.7
325	0.1351	4.182	528	0.645	3.42	471.2
330	0.1719	4.184	489	0.650	3.15	504.0
335	0.2167	4.186	453	0.656	2.88	535.5
340	0.2713	4.188	420	0.660	2.66	566.0
345	0.3372	4.191	389	0.668	2.45	595.4
350	0.4163	4.195	365	0.668	2.29	624.2
355	0.5100	4.199	343	0.671	2.14	652.3
360	0.6209	4.203	324	0.674	2.02	697.9
365	0.7514	4.209	306	0.677	1.91	707.1
370	0.9040	4.214	289	0.679	1.80	728.7
373.15	1.0133	4.217	279	0.680	1.76	750.1
375	1.0815	4.220	274	0.681	1.70	761
380	1.2869	4.226	260	0.683	1.61	788

Adapted from Mills, A. F. 1995. *Basic Heat and Mass Transfer*. Irwin, Chicago.

Appendix D

MASS TRANSFER PROPERTIES

D.1   Apparent Diffusivities in Solids

Solid	Diffusing material	Temp °C	Diffusivity $D \times 10^{11}$ m²/s	Ref
Agar gel (0.79% solids)	glucose	5	3.3	1
Cellulose acetate (5% moisture)	water	25	0.20	2
Polyethylene	oxygen	30	6	1
Rubber	oxygen	25	21	1
Agarose gel (2% solids)	NaCl	25	14	1
Meat muscle (fresh)	NaCl	2	22	3
Meat muscle (frozen-thawed)	NaCl	2	39	3
Starch gel (60% moisture dry basis)	water	30	15	4
Apple (air dried, 12% moisture db)	water	30	0.65	5
Apple (freeze dried, 12% moisture db)	water	30	12	5
soil, clay (water content 0.45)	chloride ion		57.87	
soil, clay (water content 0.25)	chloride ion		18.52	
soil, loam (water content 0.25)	chloride ion		40.51	
soil, silt (water content 0.248)	chloride ion		31.83	
soil, sand (water content 0.402)	chloride ion		138.89	
soil, sand (water content 0.168)	chloride ion		4.86	

[1]Geankoplis (1992). [2]Roussis(1981). [3]Dussap and Gros (1980). [4]Saravacos and Raouzeos (1984); Saravacos (1967). [5]Saravacos (1967).

D.2   Apparent Diffusivities in Liquids

Liquid	Diffusing material	Temperature °C	Diffusivity $D \times 10^9$ m^2/s	Reference
Water	acetic acid (1 kmol/m^3)	18	0.96	5
Water	chlorine (0.12 kmol/m^3)	16	1.26	5
Water	hydrogen chloride (0.5 kmol/m^3)	16	2.44	5
Water (dilute)	caffeine	25	0.63	3
Water (dilute)	carbon dioxide	25	2.00	1
Water (dilute)	ethanol	25	1.28	2
Water (dilute)	oxygen	25	2.41	1
Water (dilute)	sodium chloride	25	1.61	4
Water (dilute)	soybean protein	25	0.03	1
Water (dilute)	sucrose	25	0.56	3
Water (dilute)	urea	25	1.37	3

[1] Geankoplis (1992).
[2] Loncin and Merson (1979).
[3] Perry and Chilton (1973).
[4] Schwartzberg and Chao (1982).
[5] Welty et al. (1984).

D.3   Diffusivities in Air at 1 atm (1.013×10^5 Pa)

Temp K	Diffusivity, $D \times 10^4$ m^2/s				
	H_2O	O_2	CO_2	CO	SO_2
200	0.1095	0.095	0.074	0.098	0.058
300	0.2538	0.188	0.157	0.202	0.126
400	0.4606	0.325	0.263	0.332	0.214
500	0.6983	0.475	0.385	0.485	0.326
600	0.9403	0.646	0.537	0.659	0.440
700	1.2093	0.838	0.684	0.854	0.576
800	1.5037	1.05	0.857	1.06	0.724

Adapted from Mills (1995).

D.4 Representative Values of Resistances to Water Vapor Transport out of Leaves

Component Condition	Conductance $\times 10^3$ m/s
Boundary layer	
thin	80
thick	8
Stomata	
large area–open	19
small area–open	1.7
closed	0
mesophytes–open	4 – 20
xerophytes and trees–open	1 – 4
Cuticle	
crops	0.1 – 0.4
many trees	0.05 – 0.2
many xerophytes	0.01 – 0.1
Intercellular air spaces	
calculation	24 – 240
typical	40 – 100
waxy layer (typical)	50 – 200
Leaf (lower surface)	
crops–open	2 – 10

The values are provided in terms of conductance values, which are the inverse of resistances. See Section 12.1.3 for discussion on resistances and conductances. To calculate flux, concentration differences need to be in mol/m^3 and multiplied by the conductances. Adapted from *Physicochemical and Environmental Plant Physiology* by P. S. Nobel, Academic Press, San Diego, 1996.

D.5 Approximate Ranges of Dispersion Coefficients in Surface Water

Condition	Dispersion Coefficient, m^2/s
Compacted sediment	$10^{-11} - 10^{-9}$
Bioturbated sediment	$10^{-9} - 10^{-8}$
Lakes-vertical	$10^{-6} - 10^{-3}$
Large rivers-lateral	$10^{-2} - 10^{-1}$
Large rivers-longitudinal	$10^{0} - 10^{2}$
Estuaries-longitudinal	$10^{2} - 10^{3}$

Source: Adapted from *Environmental Modeling*. Republished by permission of John Wiley and Sons, Inc., from *Environmental Modeling: Fate and Transport of Pollutants in Water, Air and Soil*, Schnoor, J. L., 1996; Permission conveyed through Copyright Clearance Center, Inc.

D.6 Approximate Ranges of Dispersion Coefficients in Porous Media

Disperion coefficients used in contaminant transport in porous media such as groundwater are mostly empirical, often obtained from tracer studies. The table below can be used to obtain some order of magnitude estimates. The dispersion coefficients are a strong function of length scale involved and the velocities. The longitudinal dispersion coefficient can be calculated from

$$E = \alpha u + D \tag{D.1}$$

where E is the longitudinal dispersion coefficient, α is the factor depending on length scale, u is the longitudinal velocity in m/s, and D is the molecular diffusion coefficient. This formula also shows how the molecular diffusion is added on to the mechanical dispersion, although the molecular diffusion component is typically much smaller and can be ignored.

Condition	Scale, m	Average longitudinal α, m
Laboratory	< 1	0.001–0.01
Field, small scale	1–10	0.1–1.0
Field, large scale	10–100	25

Source: Adapted from *Environmental Modeling*. Adapted from *Environmental Modeling*. Republished by permission of John Wiley and Sons, Inc., from *Environmental Modeling: Fate and Transport of Pollutants in Water, Air and Soil*, Schnoor, J. L., 1996; Permission conveyed through Copyright Clearance Center, Inc.

D.7   Kinetic Parameters for Various First-Order Rate Processes in Biological and Environmental Systems

Application	k'' 1/min	Temp. for k'', °C	Activation energy, kJ/mole
Food pasteurization and sterilization			
E. coli O157:H8 in ground beef	4.9 – 8.86	62.8	401
L. monocytogenes in milk	3.97 – 10.47	63.3	386
S. aureus in milk	2.56	60	224
C. botulinum B in vegetable products	0.185 – 4.7	110	267 – 390
Thiamine (vitamin B_1) in pea puree	0.00933	121	111.5
Pesticide degradation in soil			
Aldicarb	1.604×10^{-5}		
Atrazine	0.802×10^{-5}		
Carboxin	6.875×10^{-5}		

Source for microorganisms in foods: Microorganisms in foods. *Microbiological specifications of Food Pathogens/ICMSF*. Book 5. Blackie Academic and Professional Publications, London, 1996 (cited in Kinetics of Microbial Inactivation for Alternative Processing Technologies. A report of the Institute of Food Technologists for the Food and Drug Administration of the U.S. Department of Health and Human Services). Thiamine data from Karel et al. (1975).

Source for pesticides: These values are to be used only as representative data since the actual rate constant depends on pH, temperature, soil type, and crop. For a comprehensive database of pesticide properties, see the document via the internet "The ARS Pesticide Properties Database" at http://wizard.arsusda.gov/acsl/ppdb.html (last accessed on May 3, 2001).

D.8   Surface Tension of Water

Temperature T °C	Surface tension γ N/m
0	0.07550
10	0.07440
20	0.07288
30	0.07120
40	0.06948
50	0.06777
60	0.06607
70	0.06436
80	0.06269
90	0.06079
100	0.05891
110	0.05697
120	0.05496
130	0.05290
140	0.05079
150	0.04868
160	0.04651
170	0.04438
180	0.04219
190	0.04000
200	0.03777

Adapted from Vargaftik (1983).

D.9   References to Data on Heat and Mass Transfer and Related Properties

Air Force Cambridge Research Laboratories (U.S.). 1961. *Handbook of Geophysics.* The Macmillan Company, New York.

ARS. 2001. Pesticide Properties Database. Document on the internet (last accessed on May 3, 2001) at http://wizard.arsusda.gov/acsl/ppdb.html.

ASHRAE. 1997. *ASHRAE Handbook: Fundamentals.* American Society of Heating, Refrigeration and Air-Conditioning Engineers, Inc., Atlanta, GA.

Calvert, S. and H. M. Englund. 1984. *Handbook of Air Pollution Technology.* John Wiley & Sons, New York.

Cena, K. and J. A. Clark. 1979. Transfer of heat through animal coats and clothing. *International Review of Physiology, Environmental Physiology III*, Vol. 20, pp. 1–42, Edited by D. Robertshaw, University Park Press, Baltimore.

Construction of a database of physical properties of foods. 2000. EU Project ERB FAIR CT96-1063. Document on the internet, http://www.nel.uk/fooddb/ (last accessed June 19).

Chato, J. C. 1985. Selected thermophysical properties of biological materials. 2:413–418 In: *Heat Transfer in Medicine and Biology* edited by A. Shitzer and R. C. Eberhart. Plenum Press, New York.

Choi, Y. 1985. Food thermal property prediction as affected by temperature and composition. PhD Thesis, Purdue University.

Dussap, G. and J. B. Gros. 1980. Diffusion-sorption model for the penetration of salt in pork and beef muscle. In: *Food Process Engineering* edited by P. Linko, Y. Malkki, J. Olkku, and J. Larinkari. Applied Science Publishers, London.

Foust, A. S., L. A. Wenzel, C. W. Clump, L. Maus, and L. B. Anderson. 1960. *Principles of Unit Operations.* John Wiley & Sons, Inc.

Geankoplis, C. J. 1992. *Transport Processes and Unit Operations.* Prentice Hall, New Jersey.

Greenberg, L. H. 1975. *Physics for Biology and Medical Students.* W. B. Saunders Company, Philadelphia.

Heldman, D. R. and R. P. Singh. 1986. Thermal properties of frozen foods. In: *Physical and Chemical Properties of Foods*, M. R. Okos, editor. ASAE, St. Joseph, MI.

Incropera, F. P. and D. P. Dewitt. 1990. *Introduction to Heat Transfer*. John Wiley & Sons, New York.

Irvine, T. F., Jr. and J. P. Hartnett. 1976. *Steam and Air Tables in SI Units*. Hemisphere Publishing Co., Washington, DC.

Karel, M., O. R. Fennema, and D. B. Lund. 1975. *Physical Principles of Food Preservation*. Marcel Dekker, Inc., New York.

Loncin, M. and R. L. Merson. 1979. *Food Engineering*. Academic Press, New York.

Mills, A. F. 1995. *Basic Heat and Mass Transfer*. Irwin, Chicago.

O'Brien, W. J. 2000. Biomaterials Properties Database. Document on the internet, http://www.lib.umich.edu/dentlib/Dental_tables/toc.html. (last accessed June), Univ. of Michigan, Ann Arbor.

Perry, R. H. and C. H. Chilton. 1973. *Chemical Engineers Handbook*. McGraw-Hill, New York.

Rao M. A. and S. S. H. Rizvi. 1986. *Engineering Properties of Foods*. Marcel Dekker, Inc., New York.

Roussis, P. P. 1981. Diffusion of water vapor in cellulose acetate: 2. Permeation and integral sorption kinetics. *Polymer*, 22:1058–1063.

Schnoor, J. L. 1996. *Environmental Modeling: Fate and Transport of Pollutants in Water, Air and Soil*. John Wiley & Sons, New York.

Schwartzberg, H. G. and R. Y. Chao. 1982. Solute diffusivities in the leaching processes. *Food Technology* 36(2):73–86.

Saravacos, G. D. 1967. Effect of drying method on the water sorption of dehydrated apple and potato. *Journal of Food Science* 32:81–84.

Saravacos, G. D. and G. S. Raouzeos. 1984. Diffusivity of moisture in air drying of starch gels. In: *Engineering and Food*, 1:499–507. Edited by B.M. McKenna, Elsevier Applied Science Publishers, London.

Saravacos, G. D. and Z. B. Maroulis. 2001. *Transport Properties of Foods*. Marcel Dekker, New York.

Shitzer, A. and R. C. Eberhart. 1985. Heat generation, storage, and transport processes. 1:137–151. In: *Heat Transfer in Medicine and Biology* edited by A. Shitzer and R. C. Eberhart. Plenum Press, New York.

Singh, R. P. 1995. Food Properties Database 2.0 Windows. CRC Press, Boca Raton, FL.

Vargaftik, N. B. 1983. *Handbook of Physical Properties of Liquids and Gases.* Hemisphere Publishing Corporation, Washington, DC.

Welty, J. R., C. E. Wicks, and R. E. Wilson. 1984. *Fundamentals of Momentum, Heat, and Mass Transfer.* John Wiley & Sons, New York.

Appendix E

MISCELLANEOUS ENVIRONMENTAL DATA

E.1   Atmospheric Temperature, Pressure, and Other Parameters as Function of Altitude

Altitude m	Temp K	Pressure mb	Gravity $m\,s^{-2}$	Density $kg\,m^{-3}$	Mean Free Path m
−4000	314.18	1.5960×10^3	9.8190	1.7698	4.5904×10^{-8}
−3000	307.67	1.4297×10^3	9.8159	1.6189	5.0181×10^{-8}
−2000	301.16	1.2778×10^3	9.8128	1.4782	5.4959×10^{-8}
−1000	294.66	1.1393×10^3	9.8097	1.3470	6.0310×10^{-8}
0	288.16	1.01325×10^3	9.8067	1.2250	6.6317×10^{-8}
1000	281.66	8.9876×10^2	9.8036	1.1117	7.3079×10^{-8}
2000	275.16	7.9501×10^2	9.8005	1.0066	8.0710×10^{-8}
3000	268.67	7.0121×10^2	9.7974	9.0926×10^{-1}	8.9347×10^{-8}
4000	262.18	6.1660×10^2	9.7943	8.1935×10^{-1}	9.9151×10^{-8}
5000	255.69	5.4048×10^2	9.7912	7.3643×10^{-1}	1.1032×10^{-7}
6000	249.20	4.7217×10^2	9.7882	6.6011×10^{-1}	1.2307×10^{-7}
7000	242.71	4.1105×10^2	9.7851	5.9002×10^{-1}	1.3769×10^{-7}
8000	236.23	3.5651×10^2	9.7820	5.2578×10^{-1}	1.5451×10^{-7}
9000	229.74	3.0800×10^2	9.7789	4.6706×10^{-1}	1.7394×10^{-7}
10,000	223.26	2.6500×10^2	9.7759	4.1351×10^{-1}	1.9646×10^{-7}
20,000	216.66	5.5293×10^1	9.7452	8.8909×10^{-2}	9.1374×10^{-7}
40,000	260.91	2.9977	9.6844	4.0028×10^{-3}	2.0296×10^{-5}
60,000	253.68	2.5657×10^{-1}	9.6241	3.5235×10^{-4}	2.3056×10^{-4}
80,000	165.7	1.008×10^{-2}	9.564	2.120×10^{-5}	3.831×10^{-3}
100,000	199.0	2.138×10^{-4}	9.505	3.743×10^{-7}	2.171×10^{-1}

Adapted from *Handbook of Geophysics*, The Macmillan Company, New York, 1961.

E.2 National (U.S.) Primary Ambient Air Quality Standards

Pollutant	Averaging Time	Primary Standards[a]
Sulfur oxides	Annual arithmetic mean	80 μg/m^3 (0.03 ppm)
	24 hr	365 μg/m^3 (0.14 ppm)
Particulate matter	Annual geometric mean	75 μg/m^3
	24 hr	260 μg/m^3
Carbon monoxide	8 hr	10 mg/m^3
	1 hr	40 mg/m^3 (35 ppm)
Ozone (corrected for NO_2 and SO_2)	1 hr	240 μg/m^3 (0.12 ppm)
Hydrocarbons (corrected for methane)	3 hr	160 μg/m^3 (0.24 ppm)
Nitrogen oxides	Annual arithmetic mean	100 μg/m^3 (0.05 ppm)
Lead	3 months	1.5 μg/m^3
Ozone	1 hr	235 μg/m^3 (0.12 ppm)

[a]Except for annual means, standards are not to be exceeded more than once a year.

Adapted from *Handbook of Air Pollution Technology*, Calvert, S. and H. M. Englund, John Wiley & Sons, New York, 1984.

Appendix F

EQUATIONS OF MOTION IN VARIOUS COORDINATE SYSTEMS

F.1 The Equation of Motion in Rectangular Coordinates (x, y, z)

x-component

$$\rho \left(\frac{\partial v_x}{\partial t} + v_x \frac{\partial v_x}{\partial x} + v_y \frac{\partial v_x}{\partial y} + v_z \frac{\partial v_x}{\partial z} \right) = -\frac{\partial p}{\partial x}$$
$$+ \mu \left(\frac{\partial^2 v_x}{\partial x^2} + \frac{\partial^2 v_x}{\partial y^2} + \frac{\partial^2 v_x}{\partial z^2} \right) + \rho g_x \tag{F.1}$$

y-component

$$\rho \left(\frac{\partial v_y}{\partial t} + v_x \frac{\partial v_y}{\partial x} + v_y \frac{\partial v_y}{\partial y} + v_z \frac{\partial v_y}{\partial z} \right) = -\frac{\partial p}{\partial y}$$
$$+ \mu \left(\frac{\partial^2 v_y}{\partial x^2} + \frac{\partial^2 v_y}{\partial y^2} + \frac{\partial^2 v_y}{\partial z^2} \right) + \rho g_y \tag{F.2}$$

z-component

$$\rho\left(\frac{\partial v_z}{\partial t} + v_x\frac{\partial v_z}{\partial x} + v_y\frac{\partial v_z}{\partial y} + v_z\frac{\partial v_z}{\partial z}\right) = -\frac{\partial p}{\partial z}$$

$$+\mu\left(\frac{\partial^2 v_z}{\partial x^2} + \frac{\partial^2 v_z}{\partial y^2} + \frac{\partial^2 v_z}{\partial z^2}\right) + \rho g_z \tag{F.3}$$

F.2   The Equation of Motion in Cylindrical Coordinates (r, θ, z)

r-component

$$\rho\left(\frac{\partial v_r}{\partial t} + v_r\frac{\partial v_r}{\partial r} + \frac{v_\theta}{r}\frac{\partial v_r}{\partial \theta} - \frac{v_\theta^2}{r} + v_z\frac{\partial v_r}{\partial z}\right) = -\frac{\partial p}{\partial r}$$

$$+\mu\left(\frac{\partial}{\partial r}\left(\frac{1}{r}\frac{\partial}{\partial r}(rv_r)\right) + \frac{1}{r^2}\frac{\partial^2 v_r}{\partial \theta^2} - \frac{2}{r^2}\frac{\partial v_\theta}{\partial \theta} + \frac{\partial^2 v_r}{\partial z^2}\right) + \rho g_r \tag{F.4}$$

θ-component

$$\rho\left(\frac{\partial v_\theta}{\partial t} + v_r\frac{\partial v_\theta}{\partial r} + \frac{v_\theta}{r}\frac{\partial v_\theta}{\partial \theta} + \frac{v_r v_\theta}{r} + v_z\frac{\partial v_\theta}{\partial z}\right) = -\frac{1}{r}\frac{\partial p}{\partial \theta}$$

$$+\mu\left(\frac{\partial}{\partial r}\left(\frac{1}{r}\frac{\partial}{\partial r}(rv_\theta)\right) + \frac{1}{r^2}\frac{\partial^2 v_\theta}{\partial \theta^2} + \frac{2}{r^2}\frac{\partial v_r}{\partial \theta} + \frac{\partial^2 v_\theta}{\partial z^2}\right) + \rho g_\theta \tag{F.5}$$

z-component

$$\rho\left(\frac{\partial v_z}{\partial t} + v_r\frac{\partial v_z}{\partial r} + \frac{v_\theta}{r}\frac{\partial v_z}{\partial \theta} + v_z\frac{\partial v_z}{\partial z}\right) = -\frac{\partial p}{\partial z}$$

$$+\mu\left(\frac{1}{r}\frac{\partial}{\partial r}\left(r\frac{\partial v_z}{\partial r}\right) + \frac{1}{r^2}\frac{\partial^2 v_z}{\partial \theta^2} + \frac{\partial^2 v_z}{\partial z^2}\right) + \rho g_z \tag{F.6}$$

F.3   The Equation of Motion in Spherical Coordinates (r, θ, ϕ)

In the following equations,

$$\nabla^2 = \frac{1}{r^2}\frac{\partial}{\partial r}\left(r^2\frac{\partial}{\partial r}\right) + \frac{1}{r^2\sin\theta}\frac{\partial}{\partial \theta}\left(\sin\theta\frac{\partial}{\partial \theta}\right) + \frac{1}{r^2\sin^2\theta}\left(\frac{\partial^2}{\partial \phi^2}\right) \tag{F.7}$$

the three equations of motion are

r-component

$$\rho\left(\frac{\partial v_r}{\partial t} + v_r\frac{\partial v_r}{\partial r} + \frac{v_\theta}{r}\frac{\partial v_r}{\partial\theta} + \frac{v_\phi}{r\sin\theta}\frac{\partial v_r}{\partial\phi} - \frac{v_\theta^2 + v_\phi^2}{r}\right)$$

$$= -\frac{\partial p}{\partial r} + \mu\left(\nabla^2 v_r - \frac{2}{r^2}v_r - \frac{2}{r^2}\frac{\partial v_\theta}{\partial\theta} - \frac{2}{r^2}v_\theta\cot\theta\right.$$

$$\left. - \frac{2}{r^2\sin\theta}\frac{\partial v_\phi}{\partial\phi}\right) + \rho g_r \tag{F.8}$$

θ-component

$$\rho\left(\frac{\partial v_\theta}{\partial t} + v_r\frac{\partial v_\theta}{\partial r} + \frac{v_\theta}{r}\frac{\partial v_\theta}{\partial\theta} + \frac{v_\phi}{r\sin\theta}\frac{\partial v_\theta}{\partial\phi} + \frac{v_r v_\theta}{r} - \frac{v_\phi^2\cot\theta}{r}\right)$$

$$= -\frac{1}{r}\frac{\partial p}{\partial\theta} + \mu\left(\nabla^2 v_\theta + \frac{2}{r^2}\frac{\partial v_r}{\partial\theta} - \frac{v_\theta}{r^2\sin^2\theta} - \frac{2\cos\theta}{r^2\sin^2\theta}\frac{\partial v_\phi}{\partial\phi}\right) + \rho g_\theta \tag{F.9}$$

ϕ-component

$$\rho\left(\frac{\partial v_\phi}{\partial t} + v_r\frac{\partial v_\phi}{\partial r} + \frac{v_\theta}{r}\frac{\partial v_\phi}{\partial\theta} + \frac{v_\phi}{r\sin\theta}\frac{\partial v_\phi}{\partial\phi} + \frac{v_\phi v_r}{r} + \frac{v_\theta v_\phi}{r}\cot\theta\right)$$

$$= -\frac{1}{r\sin\theta}\frac{\partial p}{\partial\phi} + \mu\left(\nabla^2 v_\phi - \frac{v_\phi}{r^2 sin^2\theta} + \frac{2}{r^2 sin\theta}\frac{\partial v_r}{\partial\phi}\right.$$

$$\left. + \frac{2\cos\theta}{r^2\sin^2\theta}\frac{\partial v_\theta}{\partial\phi}\right) + \rho g_\phi \tag{F.10}$$

Appendix G

SOME USEFUL MATHEMATICAL BACKGROUND

G.1 Series Solution to the One-Dimensional Heat Equation

In this section, the detailed steps in obtaining a series solution to the diffusion equation (Eq. 5.11 on page 136 for heat diffusion and Eq. 13.9 on page 448 for mass diffusion) are described. To simplify solution, temperature T in Eq. 5.11 is transformed using a non-dimensional temperature θ defined as

$$\theta = \frac{T - T_1}{T_0 - T_1} \tag{G.1}$$

Using this non-dimensional temperature θ, the derivatives in Eq. 4.1 are transformed as

$$\frac{\partial \theta}{\partial t} = \frac{1}{T_0 - T_1} \frac{\partial T}{\partial t}$$

$$\frac{\partial^2 \theta}{\partial x^2} = \frac{1}{T_0 - T_1} \frac{\partial^2 T}{\partial x^2}$$

593

Substituting these derivatives back into Eq. 5.11, the transformed governing equation and boundary conditions are

$$\text{G.E.} \quad \frac{\partial \theta}{\partial t} = \alpha \frac{\partial^2 \theta}{\partial x^2} \tag{G.2}$$

$$\text{B.C.} \quad \left. \frac{\partial \theta}{\partial x} \right|_{x=0, t>0} = 0 \tag{G.3}$$

$$\theta(L, t > 0) = 0 \tag{G.4}$$

$$\text{I.C.} \quad \theta(x, t = 0) = 1 \tag{G.5}$$

To solve by separation of variables, first assume that the solution $\theta(x, t)$ can be written as product of a function in x and a function in t as

$$\theta(x, t) = F(x)G(t)$$

Substituting into Eq. G.2,

$$F(x)G'(t) = \alpha F''(x)G(t)$$

Separating functions in x and t as below, we note functions in two unrelated variables x and t can be equal only if each of them is equal to a constant. Let $-p^2$ be this negative constant. We will later find out that this constant has to be negative; therefore, we start with a negative value to avoid difficulty later.

$$\frac{F''(x)}{F(x)} = \frac{G'(t)}{\alpha G(t)} = -p^2 \tag{G.6}$$

The two equations from Eq. G.6 are

$$F''(x) + p^2 F(x) = 0 \tag{G.7}$$
$$G'(t) + \alpha p^2 G(t) = 0 \tag{G.8}$$

A general solution to Eq. G.7 for F is

$$F(x) = A \sin(px) + B \cos(px)$$

To calculate the coefficients A and B, we need boundary conditions for F. The first boundary condition is derived from Eq. G.3 as

$$\theta'(0, t) = 0$$
$$F'(0)G(t) = 0$$

This leads to $F'(0) = 0$ at the centerline ($x = 0$). At the surface,

$$\theta(L,t) = 0$$
$$F(L)G(t) = 0$$

which leads to $F(L) = 0$ on the surface. Using the first condition $F'(0) = 0$,

$$Ap\cos(px)|_{x=0} - Bp\sin(px)|_{x=0} = 0$$
$$-Ap = 0$$
$$A = 0$$

Using the second condition $F(L) = 0$,

$$B\cos(px)|_{x=L} = 0$$
$$B\cos pL = 0$$

Now B cannot be equal to zero simply because if it is so, we get zero as the trivial solution. Thus,

$$\begin{aligned}
\cos(pL) &= 0 \\
pL &= (2n+1)\frac{\pi}{2} \qquad \text{where } n = 0, 1, 2, \dots \\
p &= (2n+1)\frac{\pi}{2L}
\end{aligned}$$

Therefore,

$$F_n(x) = B_n \cos\frac{(2n+1)\pi}{2L}x, \qquad n = 0, 1, 2, \dots \tag{G.9}$$

Now let us solve Eq. G.8. Noting that p can take multiple values, we put a subscript to p to make it p_n and rewrite Eq. G.8 as

$$G'(t) + \alpha p_n^2 G(t) = 0$$

The general solution to this equation is given by

$$G(t) = Ce^{-\alpha p_n^2 t}, \text{ where } p_n = (2n+1)\frac{\pi}{2L}$$

Denoting each solution of G (for each p_n) by G_n,

$$G_n(t) = C_n e^{-\alpha t \left(\frac{(2n+1)\pi}{2L} \right)^2}, \quad n = 0, 1, 2, \dots \tag{G.10}$$

Combining Eqs. G.9 and G.10, we get the differential equation for θ as

$$
\begin{aligned}
\theta_n(x, t) &= F_n(x) G_n(t) \\
&= D_n \cos \frac{(2n+1)\pi x}{2L} e^{-\alpha t \left(\frac{(2n+1)\pi}{2L} \right)^2}
\end{aligned}
\tag{G.11}
$$

This satisfies the governing equation and the boundary conditions for θ. However, it does not satisfy the initial condition given by Eq. G.5. Also, $D_n = B_n C_n$ is an unknown in the above equation. Thus, we have an unknown and we have not used one of the conditions. We now use the initial condition to find D_n. Let us consider a solution of the form

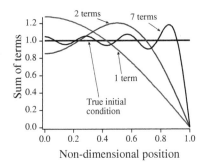

$$
\begin{aligned}
\theta(x, t) &= \sum_{n=0}^{\infty} \theta_n(x, t) \\
&= \sum_{n=0}^{\infty} D_n \cos \frac{(2n+1)\pi x}{2L} e^{-\alpha t \left(\frac{(2n+1)\pi}{2L} \right)^2}
\end{aligned}
\tag{G.12}
$$

Figure G.1: As more terms are added to the series in Eq. G.12, the summation can better approximate the initial condition of constant temperature.

Why this step? The initial condition looks like a square wave. Our initial solution (Eq. G.12) gave us a cosine wave! Thus, we will add an infinite number of cosine waves to get the square wave, as shown in Figure G.1. Note that, although we add an infinite number of terms, the magnitude of each term drops quickly, as shown in Fig. G.2

Using the initial condition $\theta(x, 0) = 1$ in Eq. G.12, we get

$$\sum_{n=0}^{\infty} D_n \cos \frac{(2n+1)\pi}{2L} x = 1$$

To solve for D_n, both sides of the above equation are multiplied by $\cos(2n+1)\pi \frac{x}{2L}$

and integrated to obtain

$$
\begin{aligned}
D_n &= \frac{2}{L} \int_0^L 1 \cdot \cos \frac{(2n+1)\pi}{2L} x \, dx \quad \text{by orthogonality} \\
&= \frac{2}{L} \frac{2L}{(2n+1)\pi} \sin \frac{(2n+1)\pi x}{2L} \Big|_0^L \\
&= \frac{4}{(2n+1)\pi} \sin \frac{(2n+1)\pi}{2} \\
&= \frac{4}{(2n+1)\pi} (-1)^n, \quad n = 0, 1, 2, \ldots
\end{aligned}
$$

Substituting D_n into Eq. G.12 and transforming θ back to temperature T, we get the final solution of temperature T as a function of position x and time t:

$$
\frac{T - T_1}{T_0 - T_1} = \sum_{n=0}^{\infty} \frac{4(-1)^n}{(2n+1)\pi} \cos \frac{(2n+1)\pi x}{2L} e^{-\alpha \left(\frac{(2n+1)\pi}{2L} \right)^2 t} \tag{G.13}
$$

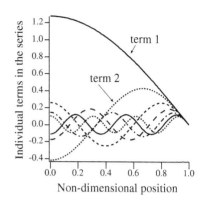

Figure G.2: Relative magnitudes of individual terms in the series solution.

G.2   Similarity Transformation of the Heat Equation

For certain situations, the heat equation, which is normally a partial differential equation in x and t, can be transformed to an ordinary differential equation. The final form of the equation developed here is used in Section 5.5.

$$\eta = \frac{x}{2\sqrt{\alpha t}}$$

$$t : \frac{\partial \eta}{\partial t} = \frac{x}{2\sqrt{\alpha}} \cdot t^{-3/2}(-1/2) = -\frac{1}{2t}\frac{x}{2\sqrt{\alpha t}}$$

$$\frac{\partial T}{\partial t} = \frac{\partial T}{\partial \eta} \cdot \frac{\partial \eta}{\partial t} = \frac{\partial T}{\partial \eta}\left(-\frac{1}{2t}\frac{x}{2\sqrt{\alpha t}}\right)$$

$$x : \frac{\partial \eta}{\partial x} = \frac{1}{2\sqrt{\alpha t}}$$

Therefore,

$$\frac{\partial T}{\partial x} = \frac{\partial T}{\partial \eta} \cdot \frac{\partial \eta}{\partial x} = \frac{\partial T}{\partial \eta} \cdot \frac{1}{2\sqrt{\alpha t}}$$

$$\frac{\partial^2 T}{\partial x^2} = \frac{\partial}{\partial x}\left(\frac{\partial T}{\partial \eta} \cdot \frac{1}{2\sqrt{\alpha t}}\right)$$

$$= \frac{\partial}{\partial \eta}\left(\frac{\partial T}{\partial \eta} \cdot \frac{1}{2\sqrt{\alpha t}}\right)\frac{\partial \eta}{\partial x}$$

$$= \frac{\partial^2 T}{\partial \eta^2} \cdot \frac{1}{2\sqrt{\alpha t}}\frac{1}{2\sqrt{\alpha t}} = \frac{1}{4\alpha t}\frac{\partial^2 T}{\partial \eta^2}$$

Substituting into the governing equation,

$$\frac{\partial T}{\partial t} = \alpha \frac{\partial^2 T}{\partial x^2}$$

$$-\frac{1}{2t} \cdot \frac{x}{2\sqrt{\alpha t}} \cdot \frac{dT}{d\eta} = \alpha \frac{d^2 T}{d\eta^2} \cdot \frac{1}{4\alpha t}$$

$$\frac{d^2 T}{d\eta^2} + \frac{x}{\sqrt{\alpha t}}\frac{dT}{d\eta} = 0$$

$$\frac{d^2 T}{d\eta^2} + 2\eta\frac{dT}{d\eta} = 0$$

Let

$$\theta = \frac{T - T_i}{T_s - T_i}$$

In terms of the new variable θ, the governing equation can be written as

$$\frac{d^2\theta}{d\eta^2} + 2\eta\frac{d\theta}{d\eta} = 0$$

The new boundary conditions (now only in η) can be derived from the old boundary conditions. Using $T(x \to \infty) = T_i$ and noting $\eta \to \infty$ when $x \to \infty$,

$$\theta(\eta \to \infty) = 0$$

Note that this condition also incorporates the initial condition $T(t = 0) = T_i$ since $\eta \to \infty$ when $t = 0$. Likewise, using $T(x = 0) = T_s$ and noting $\eta = 0$ when $x = 0$,

$$\theta(\eta = 0) = 1$$

Solution:

Let

$$\frac{d\theta}{d\eta} = Z$$

Substituting into the equation for θ, we get

$$\frac{dZ}{d\eta} + 2\eta Z = 0$$

Integrating,

$$\int \frac{dZ}{Z} = \int -2\eta \, d\eta$$

$$\ln Z = -\eta^2 + c_1$$

$$Z = e^{-\eta^2 + c_1} = e^{c_1} \cdot e^{-\eta^2} = c_2 e^{-\eta^2}$$

Substituting for Z and integrating again,

$$\frac{d\theta}{d\eta} = c_2 e^{-\eta^2}$$

$$\int_1^\theta d\theta = \int_o^\eta c_2 e^{-\eta^2} d\eta$$

$$\theta - 1 = c_2 \int_o^\eta e^{-\eta^2} d\eta$$

where $\theta(\eta = 0) = 1$ has been used. Using initial condition $\theta(\eta \to \infty) = 0$,

$$-1 = c_2 \cdot \frac{\sqrt{\pi}}{2} \quad \text{since} \quad \int_o^\infty e^{-\eta^2} d\eta = \frac{\sqrt{\pi}}{2}$$

$$c_2 = -\frac{2}{\sqrt{\pi}}$$

Substituting back,

$$\theta = 1 - \frac{2}{\sqrt{\pi}} \int_o^\eta e^{-\eta^2} d\eta$$

$$= 1 - \text{erf}(\eta)$$

$$= 1 - \text{erf}\left[\frac{x}{2\sqrt{\alpha t}}\right]$$

where

$$\text{erf}(\eta) = \frac{2}{\sqrt{\pi}} \int_o^\infty e^{-\eta^2} d\eta$$

Therefore, the solution is

$$\frac{T - T_i}{T_s - T_i} = 1 - \text{erf}\left[\frac{x}{2\sqrt{\alpha t}}\right]$$

G.3   Lumped Parameter Analysis as Related to the Energy Equation

Here we show how the lumped parameter equation for energy (Eq. 5.4 on page 130) is is in fact a variation of the general energy equation

$$\frac{\partial T}{\partial t} = \frac{k}{\rho c_p} \nabla^2 T$$

that we had derived (Eq. 3.24 on page 56, in the absence of heat generation and convection). Integrating over the volume of the material,

$$\iiint \frac{\partial T}{\partial t} dV \;=\; \frac{k}{\rho c_p} \iiint \nabla^2 T \, dV$$

Using Green's theorem, with n as the direction normal to surface

$$=\; \frac{k}{\rho c_p} \iint \frac{\partial T}{\partial n} \, dA$$

using the boundary condition $-k\frac{\partial T}{\partial n} = h(T - T_\infty)$

$$=\; \frac{-h}{k} \frac{k}{\rho c_p} \iint (T - T_\infty) dA$$

$$=\; -\frac{h}{\rho c_p} \iint (T - T_\infty) dA$$

When the spatial variation of temperature is insignificant,

$$\frac{\partial T}{\partial t} \iiint dV \;=\; -\frac{h(T - T_\infty)}{\rho c_p} \iint dA$$

$$\frac{\partial T}{\partial t} V \;=\; -\frac{h(T - T_\infty)}{\rho c_p} A$$

$$\frac{\partial T}{\partial t} \;=\; \frac{hA}{\rho V c_p}(T - T_\infty)$$

$$=\; \frac{hA}{m c_p}(T - T_\infty)$$

which is the lumped parameter equation.

G.4 Transformation of the Convection–Diffusion Equation to the Diffusion Equation

The convection–diffusion equation

$$\frac{\partial c}{\partial t} + u\frac{\partial c}{\partial z} = D\frac{\partial^2 c}{\partial z^2} \tag{G.14}$$

was transformed to a diffusion equation and used in Section 14.2. To see how this transformation can be done, we define two new variables as

$$x = z - ut$$
$$\eta = t$$

From these definitions, we can find the partial derivatives as

$$\frac{\partial x}{\partial t} = -u \qquad \frac{\partial \eta}{\partial t} = 1$$
$$\frac{\partial x}{\partial z} = 1 \qquad \frac{\partial \eta}{\partial z} = 0$$

Using the chain rule,

$$\frac{\partial c}{\partial t} = \frac{\partial c}{\partial x}\frac{\partial x}{\partial t} + \frac{\partial c}{\partial \eta}\frac{\partial \eta}{\partial t} = -u\frac{\partial c}{\partial x} + \frac{\partial c}{\partial \eta}$$

$$\frac{\partial c}{\partial z} = \frac{\partial c}{\partial x}\frac{\partial x}{\partial z} + \frac{\partial c}{\partial \eta}\frac{\partial \eta}{\partial z} = \frac{\partial c}{\partial x},$$

$$\frac{\partial^2 c}{\partial z^2} = \frac{\partial}{\partial z}\left(\frac{\partial c}{\partial z}\right) = \frac{\partial}{\partial x}\left(\frac{\partial c}{\partial z}\right)\frac{\partial x}{\partial z} + \frac{\partial}{\partial \eta}\left(\frac{\partial c}{\partial z}\right)\cdot\frac{\partial \eta}{\partial z}$$

$$= \frac{\partial}{\partial x}\left(\frac{\partial c}{\partial x}\right)\cdot 1 + \frac{\partial}{\partial \eta}\left(\frac{\partial c}{\partial x}\right)\cdot 0 = \frac{\partial^2 c}{\partial x^2}$$

Substituting in Eq. G.14,

$$\frac{\partial c}{\partial t} + u\frac{\partial c}{\partial z} = -u\frac{\partial c}{\partial x} + \frac{\partial c}{\partial \eta} + u\frac{\partial c}{\partial x} = \frac{\partial c}{\partial \eta}$$

and using $\eta = t$, we get the diffusion equation

$$\frac{\partial c}{\partial t} = D\frac{\partial^2 c}{\partial x^2}$$

G.5 Solution to Steady One-Dimensional Diffusion in a Slab with Chemical Reaction

The governing equation for steady one-dimensional diffusion with a first-order reaction (Eq. 12.17 on page 422) is

$$\frac{d^2 c_A}{dx^2} - m^2 c_A = 0$$

The transformed boundary conditions are

$$c_A(x = 0) = c_{A0} \tag{G.15}$$
$$c_A(x = L) = 0 \tag{G.16}$$

We assume a solution of the form

$$c_A = c_1 e^{-mx} + c_2 e^{mx} \tag{G.17}$$

where c_1 and c_2 are constants to be determined using the boundary conditions. Substituting the boundary conditions, we get

$$c_{A0} = c_1 + c_2$$
$$0 = c_1 e^{-mL} + c_2 e^{mL}$$

Using the second equation,

$$c_1 = -c_2 \, e^{2mL}$$

Plugging in the first one,

$$-c_2 \, e^{2mL} + c_2 = c_{A,0}$$
$$c_2 (1 - e^{2mL}) = c_{A,0}$$
$$c_2 = \frac{c_{A,0}}{1 - e^{2mL}}$$

and

$$c_1 = -\frac{c_{A,0} \, e^{2mL}}{1 - e^{2mL}}$$

Substituting into Eq. G.17,

$$c_A = c_{A,0} \left[\frac{-e^{2mL}}{1-e^{2mL}} \cdot e^{-mx} + \frac{1}{1-e^{2mL}} \cdot e^{mx} \right]$$

$$= \frac{c_{A,0}}{e^{-mL}-e^{mL}} \left[-e^{mL} \cdot e^{-mx} + e^{-mL} \cdot e^{mx} \right] \tag{G.18}$$

Rearranging,

$$\frac{c_A}{c_{A,0}} = \frac{e^{-mx} \left(-e^{mL} + e^{-mL} \right) + e^{-mL} \left(e^{mx} - e^{-mx} \right)}{e^{-mL} - e^{mL}}$$

$$\frac{c_A}{c_{A,0}} = e^{-mx} + \frac{e^{-mL}}{e^{-mL} - e^{mL}} \left(e^{mx} - e^{-mx} \right)$$

$$= e^{-mx} + \frac{-e^{-mL}}{e^{mL} - e^{-mL}} \left(e^{mx} - e^{-mx} \right) \tag{G.19}$$

Appendix H

ANSWERS TO PROBLEMS

Chap 1

1.1	$39.19°C$
1.2	**1)** $Q\rho C_p T_{in}\Delta t - Q\rho C_p T\Delta t + 4800\Delta t = \rho C_p\Delta T$
	2) $T = 41.22 - 16.22e^{-0.00556t}$
	3) $41.22\ °C$
1.3	$h_{eff} = 33.7\ W/m^2\cdot K$

Chap 2

2.2	$q'' = 106.46\ W/m^2$
2.3	$q'' = 650\ W/m^2$
2.4	$q'' = 6.416\times 10^7\ W/m^2$
2.5	$q'' = 497.13\ W/m^2$, not net radiative heat loss
2.6	**1)** 2.2 W; **2)** 9.01 W; **3)** 3.72 mW; **4)** 11.21 W

Chap 3

3.1	$k\frac{1}{r^2}\frac{\partial}{\partial r}\left(r^2\frac{\partial T}{\partial r}\right) + Q = \rho c_p\frac{\partial T}{\partial t}$
3.2	See derivation of Eq. 3.17. Finite h means additional convective resistance to heat transfer.

Chap 4

4.1	q (single pane) = 374.4 W, q (thermopane) = 93.7 W
4.2	q (summer) = 487.63 W, q (winter) = 124.93 W
4.3	$L = 6.49$ mm
4.4	$T = h\frac{(T_\infty - T_2)}{hL + k}x + T_2$
4.5	29 mm
4.6	**1)** $q = \dfrac{T_0 - T_i}{\frac{1}{2\pi Lhr_0} + \frac{\ln(r_0/r_i)}{2\pi kL}}$
	2) $r_0 = k/h$
	4) $r_0 = 4$ mm
4.7	**1)** $q/L = 131.9$ W/m
	2) $q/L = 134.6$ W/m
	3) Resistance is increased by insulation but decreased by larger surface area available for convection.

4.8	$T = T_s + (Q_m/2k)(L^2 - x^2)$
4.9	$h = 23.33\ W/m^2\cdot K$
4.10	**1)** $\frac{\partial^2 T}{\partial x^2} = -\frac{Q}{k}$
	BC: $T_{x=0} = T_1; -k\left.\frac{\partial T}{\partial x}\right|_{x=L} = h(T_L - T_\infty)$
	2) $T = -35x^2 + 80.06x + 20°C$
	3) $T = 65.8°C$ at 1.14 m from the bottom
	4) $T = 40.1°C$
4.11	$T = T_s + (Q_m/4k)(R^2 - r^2)$
4.13	$\frac{qr|_{r=R}}{L} = 1.82$ W/m
	2) $T_{max} - T_s = 0.35°C$
4.14	**1)** 83.5°C; **2)** 107.67°C; **3)** Yes. Difference of 24.17°C
4.15	**1)** $T_{max} - T_\infty = 7.96°C$; **2)** Yes, $T_{max} - T_\infty$ is directly proportional to Q
4.16	**1)** $Q = 624\ W/m^3$; **2)** $q_L = 4.9$ W/m
4.17	0.043 mm
4.18	$0.0535\ W/m\cdot K$
4.19	$800\ W/m^3$
4.20	$0.197°C$
4.21	**3)** $T = Q/(4\pi kr) + T_\infty$; **4)** $R = 1.26$ nm
4.22	**1)** $\frac{\partial^2 T}{\partial x^2} = -\frac{Q}{k}$ **3)** 2.26 m above ground
4.23	$T = 180.28 - 35x^2\ °C$
4.24	$20.78°C$
4.25	3912.57 W; $317.7\ W/m^3$
4.26	**2)** $T = -\frac{Q}{2k}x^2 + (T_2 - T_1)\frac{x}{L} + \frac{QL}{2k}x + T_1$
	3) $x_{max} = \frac{k}{Q}\frac{(T_2 - T_1)}{L} + \frac{L}{2}$
4.27	$T_s = \dfrac{T + (hL/k)T_\infty}{1 + (hL/k)}$
4.28	**1)** $T_{avg} = 315.42$ K
	2) $T_{max} = 334.05$ K
4.29	**1)** 0.11 mm; **2)** 0.33 mm
4.30	**4)** $R_{total} = \left(\dfrac{\ln(r_p/r_s)}{2\pi kL} + \dfrac{1}{hA + 4\sigma T_\infty^3 A}\right)$; **5)** 13.2 W; **7)** The outside B.C. is $-kA\left.\frac{dT}{dr}\right|_s = hA\left(T|_s - T_\infty\right) + 4\sigma T_\infty^3 A\left(T|_s - T_\infty\right)$

4.31   2) $T = -\frac{R(T_\infty - T_i)}{r} + T_\infty$; 4) 344.67 K; 5) 6.27×10^{-7} W

4.32   1) 82.9%; 2) 30.8%; 3) 31595 W/m³; 4) 4.07

4.33   3) $\frac{R_o - R_i}{4\pi k R_o R_i}$; 4) 6.64 W; 5) 0.8%

4.34   2) $R = \frac{1}{4\sigma T_\infty^3 (2\pi r_o L) + h(2\pi r_o L)} + \frac{\ln \frac{r_o}{r_i}}{2\pi k L}$; 3) 0.62 W; 4) 1.14 W

4.35   2) $T = -\frac{Q}{2k}x^2 + \left(\frac{QL - h\left(-\frac{Q}{2k}L^2 + T_g - T_\infty\right)}{(hL+k)} \right)x + T_g$; 3) 30.45°C (at 3.22 m) and 20.5°C (at 1.91 m); 4) $T_{max} = -\frac{h(T_g - T_\infty)}{(hL+k)}L + T_g$

4.36   3) $T = -\frac{Q}{2k}x^2 + \left[\frac{h\left(\frac{Q}{2k}L^2 - (T_b - T_\infty)\right) + \frac{Q}{k}L}{(k+hL)} \right]x + T_b$; 4) 0.008°C

Chap 5

5.1   Bi $= 1.42 \times 10^{-4} \ll 0.1$ so can we use lumped parameter; 2) 166 s; 3) Based on equation, the bead never exactly reaches the air temperature of 200°C.

5.3   1) Thickness is very small compared to surface area so lumped parameter approach can be used. 2) Initial temperature of leaf, air temperature, mass of leaf, specific heat of leaf, leaf area

5.4 t (without agitation) = 54.8 s; t (with agitation) = 23.1 s

5.5   1) For $n = 100$, $R = 5.5$ μm, $h = 2.31$ W/m²·K; For $n = 10^5$, $R = 55$ μm, $h = 0.23$ W/m²·K; For $n = 10^6$, $R = 119$ μm, $h = 0.107$ W/m²·K; 2) For $n = 100$, Bi $= 2.59 \times 10^{-5}$; for $n = 10^5$, Bi $= 2.59 \times 10^{-5}$; for $n = 10^6$, Bi $= 2.59 \times 10^{-5}$; 3) For $n = 100$, $t = 14.8$ s; for $n = 10^5$, $t = 0.40$ h; for $n = 10^6$, $t = 1.84$ h

5.6   1) $-91 \exp(-t/488.3) + 121$ in °C with t in s; 22 min

5.7   1) 5.75 h; 2) 8.86 h; 3) 1.5 times as long

5.8 $\frac{\Delta T}{\Delta t} = 0.48$°C/hr

5.9   1) $T_{b_{out}} = T_t - (T_t - T_{b_{in}})e^{-\frac{L}{3.345}}$

2) $T_{b_{out}} = 36.88$°C

3) Artery length is not enough to reach equilibrium

5.10 $\alpha = 2.51 \times 10^{-7}$ m²/s

5.11 $(T_{av} - T_s)/(T_i - T_s) = \frac{8}{\pi^2} \exp\left[-\alpha \left(\frac{\pi}{2L}\right)^2 t \right]$

5.12 $t_2 = t_1/4$

5.13   1) $q'' = \frac{2k}{L}(T_i - T_s) \exp\left(-\alpha(\pi/2L)^2 t\right)$ 2) Heat flux drops exponentially with time since the temperatures become closer to surface temperature

5.14 $t = 3200$ s

5.15   1) $T = 0.137$°C; 2) $T = 0.02738$°C

5.16   5 s

5.17   1) 3200 s; 2) 1280 s; 3) New method reduces time required for potatoes to cook

5.18   Analytical $x = 1.36$ mm, numerical $x = 1.37$ mm

5.19 $x = 5.08$ mm

5.20   1) Heisler charts $T = 0.99$°C; 2) 3.374°C

5.21   1) $x \approx 0.033$ m; 2) $T_s = -6.98$°C; 3) $q'' = 195.3$ W/m²

5.22   1) GE: $\frac{\partial T}{\partial t} = \alpha \frac{\partial^2 T}{\partial z^2}$ BC: $T|_{z\to\infty} = T_{avg}$; $T|_{z=0} = T_{avg} + (T_{max} - T_{avg}) \cos \omega t$; IC: $T|_{t=0} = T_{avg}$

5.23   1) $T = \frac{Q_0}{\rho c_p} e^{-\frac{x}{\delta}} t + T_i$

5.24   766.76 s

5.25   1) 165.8 s

5.26 -174.3°C, -187.32°C

5.29   323.84 s

5.30 $T = -2.06$°C. The arm experiences frostbite

5.31 $t_2 = 1.59 t_1$

5.32 0.7 μm, 23.6 μm, 748 μm

5.33 **1)** 23 min; **2)** 363 min; **3)** $\approx$ 16. This is because time is proportional to the square of the distance.

5.35 $T = 278.26 + 19.74e^{-0.00105t}$

5.36 35.73 hr

5.37 0.14 K, 42 s

5.38 0.168°C/min

5.39 $1.538 \times 10^{-7} \text{m}^2/\text{s}$, 9.49 mm

5.40 324 s

5.41 **1)** 522 s; **2)** 144 s; **3)** 80.9 s; **4)** No

5.42 **2)** $T = \frac{B+CT_i}{C} \exp(Ct) - \frac{B}{C}$; $B = \frac{Q_m V - \dot{m}\lambda V(1000) + (\rho_a c_{pa} V_a V + hA)T_a}{\rho V c_p}$, $C = \frac{-\rho_a c_{pa} V_a V - hA}{\rho V c_p}$

5.44 **1)** 255,000 W/m²; **2)** 39.9°C; **3)** Not beneficial since temperature is higher

5.45 $N = N_0 e^{\left[-\int_0^t k_0 \exp\left(-\frac{E}{R} \frac{1}{T_s + (T_i - T_s)\frac{4}{\pi} \exp\left(-\alpha(\pi/2L)^2 t \right)} \right) \right] dt}$

5.46 **2)** $7 = 0.434 \int_0^t k_0 \exp\left[-\frac{E}{R} \frac{1}{T_s + (T_i - T_s)\frac{4}{\pi} \exp\left(-\alpha(\pi/2L)^2 t \right)} \right] dt$

5.47

5.48 **2)** $T = 17.8e^{-4.40 \times 10^{-4}t} + 2$ in °C; **4)** $\sim 7\%$

5.49 0.105 ns

5.50 **2)** 52.5°C; **3)** 45°C

5.51 **2)** $T = T_b + (T_i - T_b)\exp\left(-\frac{hA}{V\rho C_p}t \right)$; **4)** Plot $\ln((T-T_b)/(T_i-T_b))$ vs. time and relate the slope.

Chap 6

6.1 $\delta_{vel} = 0.00416$ m ; $\delta_{ther} = 0.00466$ m

6.3 **1)** $h = (\text{const})x^{(-1/2)}$; **2)** As cold air comes in contact with hot surface initially, the heat transfer is very high. **3)** $h_w = 395.32 (u_\infty/x)^{1/2}$, $h_a = 1.962 (u_\infty/x)^{1/2}$ so water cools 201 times faster; **4)** $\sqrt{2}$ times as high

6.4 $h = \frac{2k}{D}$

6.5 **1)** 245 times faster; **2)** 1.414 times faster

6.6 **1)** QL; **2)** $\pi DL(T_s - T_\infty)\frac{k}{D} B \left(\frac{u_\infty D}{\nu} \right)^n \text{Pr}^{\frac{1}{3}}$; **3)** $u_\infty^n = Q/\left(\pi(T_s - T_\infty)kB \left(\frac{D}{\nu} \right)^n \text{Pr}^{\frac{1}{3}} \right)$; **4)** 13.12 m/s

6.7 4 times more heat is lost on treadmill. While seated (46.41 W) and on treadmill = 190.3 W

6.8 **1)** $Q = 4.664(T_s - 20)$; **2)** 18%

6.9 **1)** $h_x = 817.8x^{-1/5}$; **2)** Rate of melting = $5.74 \times 10^{-5}x^{-1/5}$ m/s; **4)** 1.558 m/day. **5)** Not feasible. The ice will melt by the time it is towed to Southern California. Slower velocity will reduce h but it would take longer to tow.

6.10 **1)** $\frac{h_{byrd}}{h_{plate}} = 8 \times 10^6$; **2)** $\frac{q''_{u=0}}{q''_{u=40}} = 0.320$; **3)** $-80°C$

6.11 **1)** Steady state is reached when the heater surface temperature increases to a value where the natural convection heat loss balances the heat production. **2)** For the first iteration, the assumed temperature is 400 K and the final calculated temperature is 4585 K. To get the correct answer, additional iterations are needed.

6.12 $t = 115$ minutes

6.13 Power = 30.2 W

6.14 Surface area in the problem should be 0.04 m² instead of 0.017 m². **3)** $\ln \frac{T_{av}-T_\infty}{T_i-T_\infty} = \frac{A}{mc_p}(120000)\left(e^{-t/600} - 1 \right)$ **4)** $t = 386$ s

6.15 1.353 times

6.16 200 W/m² · K

6.17 6.71

6.18 **1)** No. $h = (\text{constant})x^{-1/2}$ **3)** $h_a = 1.95 (u_\infty/x)^{1/2}$; $h_w/h_a = 202$ times; $h_w = 393.7 (u_\infty/x)^{1/2}$; **4)** h becomes $\sqrt{2}$ times its original value

6.19 1) 73.75 W/m²·K; 2) 430.31 W/m²·K

6.20 $h = 44.42$ W/m² · K; Average heat flux = 444.2 W/m²; $T_\infty = -24.42°$C

6.21 $h = 2.69$ W/m² · K

6.22 1) $h = 575.73$ W/m² · K; 2) $h = 47.85$ W/m² · K; 3) $h = 617.63$ W/m² · K–it increases and thus would not. 4) No; 5) shivering increases by local turbulence; 6) Yes.

6.23 1) $h = 3.81$ W/m² · K; 2) $h = B(0.167)(104s)^n$ W/m² · K; 3) Yes.

Chap 7

7.1 1) % frozen = 71.8; 2) % frozen = 89.6; 3) The answer will be different for a more concentrated solution, because depression in freezing point is directly proportional to concentration.

7.2 $-3.6°$C is the lowest temperature

7.3 $t = \frac{\Delta H_f \rho}{(T_m - T_s)} \left(\frac{R^2}{4k} + \frac{R}{2h} \right)$

7.4 $t = \frac{\Delta H_f \rho}{k(T_m - T_s)} \frac{R^2}{6}$

7.5 $t_{slab} = \frac{\Delta H_f \rho}{k(T_m - T_s)} \frac{L^2}{2}$; With all conditions equal, thicker slab takes 4 times as long.

7.6 Thawing time is 2.5 times as long.

7.7 1) h = 9.81 W/m²·K; 2) $t_f = 1.5$ h

7.8 $t_f = 5.18$ h

7.9 For long cylinder, $t = 2955$ s; for small cylinder, $t = 2984$ s.

7.10 1) Depth = 0.737 m; 2) $t = 186.7$ days; 3) Thawing is slower since layer through which heat transfer takes place is water instead of ice and water has a lower thermal conductivity.

7.11 2 m

7.12 $t = \frac{\Delta H_f \rho}{k(T_m - T_p)} \left[\frac{r^2}{2} \left(\ln \frac{r}{r_i} - \frac{1}{2} \right) + \frac{r_i^2}{4} \right]$

7.13 $t = \frac{\rho \Delta H_f}{(T_m - T_p)k r_i} \left(\frac{r^3}{3} + \frac{r_i^3}{6} - \frac{r_i r^2}{2} \right)$

714 1) Freezing time for the thicker slab is four times that for thinner

7.15 1) 2.37

7.16 1) 10.37 hr; 2) 4.37 hr

7.17 1) 0.023 s; 2) 0.575 s

7.18 $t_f = \frac{\Delta H_s \rho}{(T_\infty - T_m)} \left(\frac{L^2}{2k} + \frac{L}{h} \right)$

7.19 4) $t = \frac{\rho \lambda}{T_m - T_a} \left[\frac{1}{3} \left(\frac{1}{hr_0^2} - \frac{1}{k r_0} \right) (r_0^3 - r_i^3) + \frac{1}{2k} (r_0^2 - r_i^2) \right]$

7.20 1) $t_F = 5.96$ s; 2) 0.8; 3) 283,968 J/m²; 4) Sensible heat not included; 5) 494,000 J/m²

7.21 1) 97.2 W/m² · K; 2) 221 K;); 4) $t_F = \left(\frac{\lambda \rho}{T_m - T_\infty} \right) \left(\frac{r_o}{3h} + \frac{r_o^2}{6k} \right)$

7.22 1) $t_F = \frac{\Delta H_f \rho}{T_m - T_\infty} \left(\frac{L_{plastic}}{k_{plastic}} L_{bio} + \frac{L_{bio}^2}{2k_{frozen}} + \frac{L_{bio}}{h} \right)$

7.23 9 when $h \to \infty$

Chap 8

8.1 $q'' = \sigma T_{sun}^4 = 6.42 \times 10^7 \frac{W}{m^2}$; This is much higher than the flux on earth's surface since only a small fraction of this energy in intercepted by earth.

8.2 1) $\lambda_1 = 0.378$ μm; 2) $\lambda_1 = 1.616$ μm; 3) $\lambda_{max} = 0.5$ μm; $E_{b,\lambda_{max}} = 8.41 \times 10^7$ W/m²· μm

8.3 1) $F_{0-(5800)(0.7)} - F_{0-(5800)(0.35)} = 0.42$; 2) Outside the atmosphere, it is the same fraction. However, on the earth's surface, it will not be the same fraction since radiation at different wavelengths is absorbed differently by the atmosphere.

8.4   Energy flux incident outside the earth's surface = 1377 W/m²

8.5   Total radiation received = 1.211×10^{17} W; Radiative loss into space = 1.935×10^{17} W; 1.6 times

8.6   **1)** $q_{human \rightarrow room} = 128$ W; **2)** $\lambda_{max} = 9.47$ μm; **3)** Infrared region; **4)** $F_{\lambda_1 - \lambda_2} = 0.5464$; **5)** Heat loss = 72.9 W; **6)** No, radiative loss is double the convective loss

8.7   **1)** 468.70 W/m²; **2)** 445.26 W/m²

8.8   **1)** $q = 309$ W; **2)** Thermal radiation emitted = 918 W; Difference = 609 W; This example explains the importance of thermal radiation environment of an object, and explains why people feel cold when sitting near a large window in cold weather, even if indoor air temperature is adequate. The glass is cold and net thermal radiation exchange with it is more.

8.9   **1)** $\sigma A T_{sky}^4 + \sigma A T_{grd}^4 - 2\sigma A T^4 + 2hA(T_{air} - T) = 0$; **2)** $-0.17°C$; **3)** Higher heat transfer coefficient brings the leaf closer to air temperature.

8.10   **1)** Energy coming into the glass = 980 W/m²; **2)** Energy going out = $2\varepsilon\sigma T^4 A + \underbrace{hA(T - T_{air})}_{inside} + \underbrace{hA(T - T_{air})}_{outside}$ **3)** 23 °C

8.11   **1)** 661.80 W/m²; **2)** $661.80 + \sigma T_{gr}^4 = 2\sigma T^4 + 2h(T - T_{air})$; **3)** $T \approx 22.2°C$

8.12   27.64°C

8.13   281.46 W/m²

8.14   265 W

8.15   **1)** 510.25 W/m²;
2) $([0.45^2 + 4L^2]^{1/2} - [0.15^2 + 4L^2]^{1/2})/0.3 \times 5.67 \times 10^{-8} \times (773^4 - 308^4) = 200$;
3) 60 W/m²; **4)** 140 W/m²; **5)** Evaporation

8.16   **1)** 2.904×10^4 W; **2)** $E_{solar} + E_{radiation\ from\ plants} = E_{radiation\ to\ plants} + E_{convective\ outside} + E_{convective\ inside}$; **3)** $T_{air\ inside} = 61.7°C$

8.17 $500 = 0.36(T_s - 298) + 1.78 \times 10^{-9}(T_s^4 - 298^4)$; $T_s = 679$ K

8.18   **3)** $T_0 = \left(\frac{F_s(1-\rho)}{4\sigma(1-\alpha/2)}\right)^{1/4}$; $T_0 = 287.9$ K; As α increases, T_0 increases

8.19   0.127

8.20 $h_r = 3.66$ W/m² · K; Ratio = 1.36

8.21 $T_{air} = 472K - 0.565T_{wall}$ in K

8.22   **1)** $E = 7.46 \times 10^{-14}$ W/m² · μm; **2)** 3.69×10^{-4} W/m²

8.23 $\phi = 64.7°$

8.24   **1)** $a\left[\sqrt{\left(\frac{50}{a} + \frac{30}{a}\right)^2 + 4} - \sqrt{\left(\frac{30}{a} - \frac{50}{a}\right)^2 + 4}\right] = 39.3$; **2)** Increasing wind will lower temperature, may or may not be below ignition temperature; **3)** View factor increases, chance of fire increases.

8.25   **1)** $h = 10.06$ W/m² · K; **3)** 34.9°C; **4)** 54.4°C

8.26   10.9 cm

8.27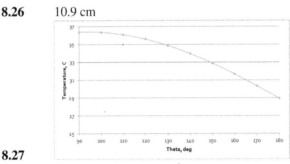

8.28   **1)** $h = 2.91$ W/m² · K; **2)** Calculate $F_{0-2.8\times303} - F_{0-0.28\times303}$; **4)** 76.4°C

Chap 9

9.1   **1)** $c_{N_2} = 32.83$ mol/m³; $c_{O_2} = 8.73$ mol/m³

2) $\rho_{N_2} = 0.919 \text{ kg/m}^3$; $\rho_{O_2} = 0.279 \text{ kg/m}^3$
3) Mole fraction $X_{N_2} = 0.79$; $X_{O_2} = 0.21$
4) Mass fraction $X_{N_2} = 0.767$; $X_{O_2} = 0.233$

9.2 **1)** Average change in soil concentration = 3320 kg/ha; **2)** Salt concentration after first year is 4320 kg/ha (dangerously high)

9.3 **1)** 0.064 ppm; **2)** $\dot{V} = 1090 \text{ m}^3/\text{h}$

9.4 Fraction of O_2 in blood = 0.25

9.5 $c_{O_2} = 4.95 \times 10^{-4}$ %

9.6 **1)** Three times more N_2 is dissolved. **2)** At lower pressure, the N_2 comes out of the blood rapidly, leading to a discomfort known as the "bends."

9.8 Saturation concentration = 9.3 mg/L

9.9 Partial pressure of CO_2 = 3879 Pa

9.10 Final moisture content = 0.0732

9.11 **1)** 34 g; **2)** 4.5 g

9.12 Moisture gained = $0.163 \frac{\text{kg H}_2\text{O}}{\text{kg dry wood}}$

9.13 **1)** $T = 76.5°\text{C}$; **2)** $T = -3.4°\text{C}$

9.14 $t_{v2,\text{aldicarb}} = 29.6$ days

9.15 **1)** $(c/c_0)|_{\text{bacteria}} = 1.54 \times 10^{-38}$; $(c/c_0)|_{\text{nutrients}} = 0.919$
2) $t = 7.327$ seconds
3) $(c/c_0)|_{\text{nutrients}} = 0.994$
4) 140°C

9.16 Time = 63.74 minutes

9.17 $c = 0.079 \left(1 - e^{-1.79t}\right) \text{ mg/m}^3$, t in hr

9.18 **1)** 0.275 kg of O_2/m^3; **2)** 0.5×10^{-3} kg of O_2/min; **5)** $c = c_0 - (nr/Q)(1 - \exp(-(Q/V)t))$; **6)** 0.2432 kg of O_2/m^3 of air

9.19 27.62 s

9.20 1.25

9.22 3.8 kg H_2O/kg dry solids

9.23 **5)** $c_{ss} = \dfrac{(F/V)c_i - k_1 + k_2 + k_3 c_b}{(F/V) + k_3}$

9.24 55.87°C

9.25 0.625

9.26 0.2 g H_2O/ g total weight

9.27 0.873 m³/s

9.28 **3)** $c = \dfrac{m_f a}{V(k-a)}\left(e^{-at} - e^{-kt}\right)$; **4)** $c = 0.00218$ ml alcohol/ml blood for 1 drink and $c = 0.00168$ ml alcohol/ml blood for 3 drinks; **5)** 0.514 h

9.29 **2)** $c = \dfrac{-Nk + k_2 c_0}{k_2} - \dfrac{(-Nk + k_2 c_0 - k_2 c_i)\exp(-k_2 t)}{k_2}$; **3)** $c = c_0 - \dfrac{Nk}{k_2}$

9.30 **3)** $c_p = \dfrac{\dot{V}_{in} c_{p,in} - B\exp(A)}{\dot{V}_{out} + kM}$ where $B = \dot{V}_{in} c_{p,in} - c_{pi}(\dot{V}_{out} + kM)$ and $A = (-t(\dot{V}_{out} + kM))/M$; **4)** $\dot{V}_{in} C_{p,in} - \dot{V}_{out} C_p - k C_p M(t) = \dfrac{d(C_p M(t))}{dt}$

9.31 **1)** 56529 K; **2)** 11.1 h

9.32 **2)** $M_i = \dfrac{k_s M_0}{(k_i + k_a - k_s)}\left[\exp(-k_s t) - \exp(-(k_i + k_a)t)\right]$; **3)** $F k_a M_i - k_{el} M_b + 0 = \dfrac{dM_b}{dt}$

Chap 10	
10.1	$n^v = 4.55$ cm/day
10.2	2937 m
10.3	0.085 m
10.4	13.14 cm; 1.314 m
10.5	4.64 times
10.6	**1)** $D_1 = 4.334 \times 10^{-12} \text{m}^2/\text{s}$; **2)** $D_2 = 1.354 \times 10^{-9} \text{m}^2/\text{s}$
10.7	5.363 nm
10.8	**1)** 566 K final temperature; **2)** 449.23 K final temperature; **3)** Diffusion in gases is easier than in liquids and increases more rapidly
10.9	**1)** 3.125 cm/h; **2)** 2 times
10.10	$1.54 \times 10^{-11} \text{ m}^2/\text{s}$
10.11	$P = P_a + 81489.68$ Pa

10.12 2.89×10^{-17} m^2

10.13 $\sim 56°$C

10.14 1) $k_{eff} = k_1\phi_1 + k_2\phi_2$ where ϕ is volume fraction

---

Chap 11

11.3 $\frac{\partial c}{\partial t} = D\frac{\partial^2 c}{\partial x^2} + r_A$; BC: $\frac{\partial c}{\partial x}\Big|_{x=L} = 0, c(x = 0) = c_s$; IC: $c(t = 0) = c_i$

11.4 Assuming no consumption of drug in tissue and no partition coefficient, δ as thickness of patch, and x measured from patch-skin interface. In tissue: $\frac{\partial c}{\partial t} = D\frac{\partial^2 c}{\partial x^2}$; BC: $c(x = L) = 0$; IC: $c(t = 0) = 0$; In patch: $\frac{\partial c}{\partial t} = D_1\frac{\partial^2 c}{\partial x^2}$; BC: $\frac{\partial c}{\partial x}\Big|_{x=-\delta} = 0$; IC: $c(t = 0) = c_i$

11.8 Assuming no partition coefficient between gel and enamel, δ as thickness of gel, and x measured from top surface of gel. In enamel: $\frac{\partial c}{\partial t} = D\frac{\partial^2 c}{\partial x^2}$; BC: $c(x = L) = 0$; IC: $c(t = 0) = 0$; In gel (over thickness δ): $\frac{\partial c}{\partial t} = D_1\frac{\partial^2 c}{\partial x^2}$; BC: $\frac{\partial c}{\partial x}\Big|_{x=0} = 0$; IC: $c(t = 0) = c_i$

Chap 12

12.1 2.0367 g/m$^2 \cdot$ hour

12.2 1) Flux $= 10^{-5}$ μg/cm^{2}s; 2) Medication residue at steady state $= 5 \times 10^{-4}$ μg/cm^2

12.3 3.756×10^{-4} g/day

12.4 Resistance $= \ln\frac{r_o}{r_i}/2\pi DL$

12.5 1) GE: $\frac{1}{r}\frac{d}{dr}\left(r\frac{dc}{dr}\right) = \frac{k''}{D}$; BC: $c = c_i$ at $r = r_i$; $\frac{dc}{dr} = 0$ at $r = r_0$

2) $c - c_i = \frac{k''}{4D}(r^2 - r_i^2) - \frac{k''}{2D}r_0^2 \ln(r/r_i)$

3) Flux $= -\frac{k''}{2}\left(r_i - \frac{r_0^2}{r_i}\right)$;

Flow/length $= k''\pi\left(r_o^2 - r_i^2\right)$

12.6 1) At outer surface, concentration $= 2.482 \times 10^{-7}$mol/cm^{3}cornea; 2) $c(x) = 3.725 \times 10^{-5}x^2 + 3.103 \times 10^{-6}x$ mol/cm^3

4) $N_{out} = -7.373$ mol/cm$^2 \cdot$ s

12.7 1) $\frac{c_A}{c_0} = \frac{-e^{-mL}}{e^{mL}-e^{-mL}}(e^{mx} - e^{-mx}) + e^{-mx} \approx e^{-mx}$

Assuming $\frac{1}{e^{2mL}-1} \to 0$ as $L \to \infty$

2) $x = 3.384$ mm; 3) $c_0 = 12.792$ μg/g tissue

12.8 1) $D\frac{d^2c}{dz^2} = k''$

2) At $z = 0, c = c_0$; At $z = p, c = 0, \frac{dc}{dz} = 0$

3) $p = \sqrt{\frac{2Dc_0}{k''}}$

12.9 1.45

12.10 1) $c = -1860.28\ln r - 4725.45$ g/m^3

2) Flux $= 9.301 \times 10^{-3}$ g/m^2·s

12.11 1) $\frac{d^2c_A}{dx^2} - \frac{k''}{D}c_A = 0$;

2) $c_{A,0} = 300$ nmol/mL; $c_{A,\infty} = 0$;

3) $c_A = 300\exp\left(-2.89 \times 10^5 x\right)\left[\frac{\text{nmol}}{\text{mL}}\right]$; x in m; 4) 7.98 μm; 5) $1.73 \times 10^4\frac{\text{nmol}}{\text{m}^2\text{s}}$; 6) 5.64 μm; 7) Change $c_{A,0}$ to 779 nmol/mL. Change D to 4×10^{-10} m^2/s

12.13 1) $c = \frac{r_A}{2D}\left(L^2 - x^2\right) + c_s$

2) 15.67 g/m^3; 14.8 g/m^3

12.14 5.47 cm

12.15 0.8011 mg/s

12.17 blood side $= 10.69 \times 10^{-6}$ m/s; Overall $= 4.02 \times 10^{-6}$ m/s; 4580 tubes

12.18 150.94 mm Hg

12.19 3) $c = \left[\frac{c_0}{\left(1-\exp\left(\frac{uL}{D}\right)\right)}\right]\left(\exp\left(\frac{uz}{D}\right) - \exp\left(\frac{uL}{D}\right)\right)$

12.20 3) $c = \frac{k''}{2D}\left(\frac{r^2-R^2}{2} + S^2\ln\frac{R}{r}\right) + c_i$; 4) $r = 1.41$ mm; 5) 4.01×10^{-6} mg/s

12.21 **3)** $c = -\frac{RN}{2D}\left(L^2 - x^2\right) + c_s$; **5)** 5.78×10^{-4} million cell/ml; **6)** $c_s = 1.09 \times 10^{-5}$

Chap 13

13.2 $D = 1.3 \times 10^{-9} \mathrm{m^2/s}$

13.3 **1)** $c = 0.028$ kmol/m^3 at center; $c = 0.018$ kmol/m^3 at 5 mm from center
2) $c = 0.09$ kmol/m^3

13.4 $t = 5.58$ hours

13.5 **1)** $t = 5.85 \times 10^4$ s = 16.25 hours; **2)** $M_{equilibrium} = 0.03$ kg H$_2$O/kg dry matter

13.6 $D = 1.13 \times 10^{-9}$ m^2/s

13.7 0.33 s

13.8 **1)** $c = 0.148$; Additional time = 0.97 h

13.9 $t_2/t_1 = 1/4$; $t \propto R^2$ so when R is halved, t becomes a quarter

13.10 Total amount left in the domain $(t \to \infty) = A\left(c_1 + c_2\right) l \frac{4}{\pi^2}$

13.11 $x = 0.288$ m

13.12 **1)** $c_s = 0.2$ kmol/m^3; **2)** $c = 0.124$ kmol/m^3

13.13 **1)** Bi = 2.33×10^5 (assuming $K^* = 0.233$)
2) 9.0 kg H$_2$O/m^3

3) $\frac{\partial c}{\partial x} = -\frac{(c_s - c_i)}{2\sqrt{Dt}} \Rightarrow \eta_s = \sqrt{\frac{D}{t}}\frac{(c_s - c_i)}{2}$; **4)** -2.42 g/m$^2 \cdot$ s

13.14 $x = 11.8$ cm

13.15 7.61×10^6 s

13.16 $\frac{dM_t}{dt} = 2\sqrt{\frac{D}{L^2\pi t}}$

13.17 20%

13.18 1507%

13.19 **3)** $x = 0.387$ mm; **4)** $c = c_i$; **5)** $n_s = 4.04 \times 10^{-5}$ kg/m$^2 \cdot$ s

13.21 Twice as long.

13.22 Two solutions $\begin{cases} D_{AB} = 4.242 \times 10^{-12} \text{ m}^2/\text{s} \\ D_{AB} = 3.017 \times 10^{-11} \text{ m}^2/\text{s} \end{cases}$

13.23 352.4 days

13.24 **2)** 18.556 hr; **3)** 6.68 hr

13.25 **1)** 4.55×10^{-14}m^2/s; **2)** 1.89×10^{-13}m^2/s

13.26 **3)** 358/m^2; ≈ 1/m^2; ≈ 0/m^2; **4)** 0.745 m; **5)** 3.647 m

13.27 $c = c_s + \frac{k"}{6D}\left(r^2 - R^2\right)$; $\frac{R^2 k"}{c_s D} \geq 6$

13.28 no change

13.29 **1)** 3.73×10^{-10} m^2/s; **2)** 3.18×10^{-10} m^2/s; **3)** 2.2 g/m^3; **4)** 6.66×10^{-5} g/m$^2 \cdot$ s; **6)** The concept of K^*, as we studied, relates to concentration at the interface of two phases touching each other. Even though two patches have the same water content, their binding ability to water can be different. Thus, when the materials are in contact with skin, skin will be able to pull different amounts of water from the two materials, leaving the skin at two different moisture levels at the surface.

13.30 **2)** 0.33×10^{-8} m^2/s; **3)** $t > 1515$ s

13.31 **1)** 1166 years; **2)** Calculation would underestimate; **3)** $1.128 \times 10^{-10} c_s \sqrt{t}$ g/m^2; **4)** $\sqrt{2}$

13.31 **1)** 5.42×10^{-3} m/s; **2)** 11.3 min; **3)** 0.643 g/m^3; **4)** $n_s \propto \sqrt{D}$

Chap 14

14.2 **2)** $\frac{\partial c}{\partial t} + u\frac{\partial c}{\partial z} = D\frac{\partial^2 c}{\partial z^2} + r_A$; **3)** $c = c_i$ everywhere at $t = 0$; $c = c_s$ at $z = 0$; $c \to 0$ at $z \to \infty$; **4)** $\frac{c_s - c}{c_s - c_i} = \mathrm{erf}\left[\frac{z - ut}{2\sqrt{Dt}}\right]$

14.3 5.79×10^{-5} g/m$^2 \cdot$ s

14.4 **1)** 8.69×10^{-2} g/m$^2 \cdot$ s, much greater than 14.3

14.5 Moisture loss = 0.127 g/m^2·s

14.6 **1)** 0.0154 g/m^2·s

14.7 $\delta_{vel} = 14.1$ mm; $\delta_{thermal} = 15.4$ mm; $\delta_{mass} = 16.6$ mm

14.8   **1)** $\underbrace{\partial c/\partial t}_{=0} + u\partial c/\partial x = D\partial^2 c/\partial x^2 + \underbrace{r_A}_{=0}$

2) at $x = -\delta$, $c = c_b$; at $x = 0$,
$\left(-D\frac{\partial c}{\partial x} + uc\right)_{x=0} = 0$

3) $c = 2\exp\left(\frac{30\times 10^{-9}\,m/s((x+8)\times 10^{-9}m)}{2\times 10^{-10}m^2/s}\right)$; x in nm, c in mol/L

14.9   **1)** $\frac{h_m L}{D_{AB}} = 0.664\,Re_L^{1/2}Sc^{1/3}$; **2)** $h_m = 0.01$ m/s; **3)** 0.833 mg/h; **4)** 47%

14.10   8.61 g/m^2· s

14.11   **1)** $dm/dt = h_m(c_{sat} - c_0)4\pi r^2$; $dm/dt = 1.75\times 10^{-5}r$ kg/s; **2)** $t = 3.59\times 10^8(r_0^2 - r^2)$ s; $t = 3.59$ s

14.12 $\dot{m}_A = 4.33\times 10^{-5}$g Cl$_2$/s

14.13   **1)** $\frac{1}{r^2}\frac{\partial}{\partial r}\left(r^2\frac{\partial c}{\partial r}\right) = 0$; **2)** $c = 0$ at $r = R$; $c = c_0$ at $r \to \infty$; **3)** $c = c_0\left(1 - \frac{R}{r}\right)$; **4)** $c_0 = 0.265$ mol/m^3; **5)** Rate of O$_2$ consumption $= \frac{DC_0}{R} = 1.72\times 10^{-5}$ mol/m$^2\cdot$s

14.14   **1)** 0.752 mg/s; **2)** 0.716 mg/s

14.15   **1)** $\frac{\partial c}{\partial t} + u\frac{\partial c}{\partial z} = E\frac{\partial^2 c}{\partial z^2}$; $c(z \to \infty) = c_i$; $c(z = 0, t > 0) = 0$; $c(t = 0, 0 \le z \le z_0) = c_i$; **2)** Distance moved due to dispersion. **4)** The peak gets smaller because the same amount is being spread over larger distances (the dispersion effect). **5)** No, since dispersion is simply a measure of non-uniform flow; **6)** $\approx 0.154\,\mu$g/cm$^2\cdot$day

14.16   0.0856 m/s

14.17   **1)** $h(T - T_s)A$; **2)** $h_m(c_s - c)A$

14.18   5.34 mm

14.19   **1)** 8.79×10^{-5} m/s; **2)** 5.7×10^{-2} g/m^3·s; **3)** 3.44×10^{-4} m/s; **4)** Mass transfer coefficient will increase as D becomes smaller.

14.20   **2)** and **3)** $\frac{c}{c_s} = e^{-(k/u)x} = e^{-1.85\times 10^4 x}$

   4) and **5)** $\frac{c}{c_s} = e^{-\sqrt{\frac{k}{E}}x} = e^{-2.13\cdot 10^4 x}$

14.21   **1)** $h_m = 0.002$ m/s; **2)** Flux $= 7.28\times 10^{-6}$ mol/m$^2\cdot$s $= 1.24\times 10^{-4}$ g/m$^2\cdot$s; **3)** 2.236; **4)** 1.214

14.22   **1)** $h_{m1} = 6.59\times 10^{-3}$ cm/s; $h_{m2} = 9.32\times 10^{-3}$cm/s; **2)** $h_{m3} = 11.4\times 10^{-3}$ cm/s; **3)** No

14.23   **1)** $h_m = 6.7\times 10^{-4}$ m/s; **3)** $h_m = 5.33\times 10^{-5}$ m/s

INDEX